Lecture Notes in Computer Science 16364

Founding Editors

Gerhard Goos
Juris Hartmanis

Editorial Board Members

Elisa Bertino, *Purdue University, West Lafayette, IN, USA*
Wen Gao, *Peking University, Beijing, China*
Bernhard Steffen, *TU Dortmund University, Dortmund, Germany*
Moti Yung, *Columbia University, New York, NY, USA*

The series Lecture Notes in Computer Science (LNCS), including its subseries Lecture Notes in Artificial Intelligence (LNAI) and Lecture Notes in Bioinformatics (LNBI), has established itself as a medium for the publication of new developments in computer science and information technology research, teaching, and education.

LNCS enjoys close cooperation with the computer science R & D community, the series counts many renowned academics among its volume editors and paper authors, and collaborates with prestigious societies. Its mission is to serve this international community by providing an invaluable service, mainly focused on the publication of conference and workshop proceedings and postproceedings. LNCS commenced publication in 1973.

Enrico Formenti · Luca Manzoni

Editors

Unconventional Computation and Natural Computation

22nd International Conference, UCNC 2025
Nice, France, September 1–5, 2025
Proceedings

 Springer

Editors
Enrico Formenti
Université Côte d'Azur
Sophia Antipolis, France

Luca Manzoni
Università degli Studi di Trieste
Trieste, Italy

ISSN 0302-9743   ISSN 1611-3349 (electronic)
Lecture Notes in Computer Science
ISBN 978-3-032-15640-2   ISBN 978-3-032-15641-9 (eBook)
https://doi.org/10.1007/978-3-032-15641-9

© The Editor(s) (if applicable) and The Author(s), under exclusive license
to Springer Nature Switzerland AG 2026

This work is subject to copyright. All rights are solely and exclusively licensed by the Publisher, whether the whole or part of the material is concerned, specifically the rights of translation, reprinting, reuse of illustrations, recitation, broadcasting, reproduction on microfilms or in any other physical way, and transmission or information storage and retrieval, electronic adaptation, computer software, or by similar or dissimilar methodology now known or hereafter developed.
The use of general descriptive names, registered names, trademarks, service marks, etc. in this publication does not imply, even in the absence of a specific statement, that such names are exempt from the relevant protective laws and regulations and therefore free for general use.
The publisher, the authors and the editors are safe to assume that the advice and information in this book are believed to be true and accurate at the date of publication. Neither the publisher nor the authors or the editors give a warranty, expressed or implied, with respect to the material contained herein or for any errors or omissions that may have been made. The publisher remains neutral with regard to jurisdictional claims in published maps and institutional affiliations.

This Springer imprint is published by the registered company Springer Nature Switzerland AG
The registered company address is: Gewerbestrasse 11, 6330 Cham, Switzerland

If disposing of this product, please recycle the paper.

Preface

The 22nd International Conference on Unconventional Computation and Natural Computation (UCNC 2025) was held at the Université Côte d'Azur in Nice, France, from September 1 to 5, 2025.

The UCNC series of international conferences is a forum bringing together scientists from many different backgrounds who are united in their interest in novel forms of computation, human-designed computation inspired by nature, and computational aspects of natural processes. The 22nd conference of the series continued the tradition of focusing on current important theoretical and experimental results. Typical, but not exclusive, UCNC topics of interest include Molecular computing, Quantum computing, Optical computing, Chaos computing, Physarum computing, Collision-based computing, Self-assembling and self-organizing systems, Super-Turing computation, Cellular automata, Neural computation, Evolutionary computation, Swarm intelligence, Ant algorithms, Artificial immune systems, Artificial life, Membrane computing, Amorphous computing, Computational systems biology, Computational neuroscience, Synthetic biology, Cellular (in-vivo) computing, and other unconventional and natural models of computation with no limitations on their inspiration and origin.

The G. Rozenberg Natural Computing Award was established in 2023 to recognize outstanding achievements in the field of natural computing. The award is named after Professor Grzegorz Rozenberg to acknowledge his distinguished scientific achievements in many areas of science, including natural computing, as well as his crucial role in developing the UCNC conference series. Professor Rozenberg also invented the name and defined the scope of the natural computing area and continues to serve as Chair Emeritus of the UCNC Steering Committee.

This annual award is presented at the Unconventional Computation and Natural Computation conference. The recipient is invited to give the award lecture at UCNC. This year, the Award Committee selected as this year's recipient Eric Goles from the University Adolfo Ibañes, Santiago, Chile in recognition of his numerous and important contributions to Natural Computing and Unconventional Computing. During the conference he gave an inspiring lecture on his scientific journey lasting many decades.

The program committee of UCNC 2025 reviewed 43 regular and short paper submissions, of which 24 regular and 5 short papers were selected for presentation at the conference and are published in these proceedings. In addition, the conference program included 6 short presentations of late-breaking abstracts and a poster session. The conference was co-located with three thematic workshops on quantum computing and quantum infomation, reaction systems, and cellular automata and Boolean automata networks, which also included a doctoral school.

The conference included four keynote talks given by Marco Dorigo (Université Libre de Bruxelles, Belgium), Alessandra Carbone (CNRS - Sorbonne Université, France), Matthew Patitz (University of Arkansas, USA), and Alberto Dennunzio (University of

Milano-Bicocca, Italy). In addition, there were also two tutorials given by Nataša Jonoska (University of South Florida, USA) and Marc Antonini (Université Côte d'Azur, France).

Moreover, each of the co-located workshops had its own invited talks. The Quantum Computing and Quantum Information workshop had invited talks given by Mika Hirvensalo (Turku University, Finland) and Jason Haraldsen (University of North Florida, USA). The Reaction Systems workshop had invited talks given by Roberto Bruni (University of Pisa, Italy) and Wojciech Penczek (Polish Academy of Science, Poland).

We warmly thank the keynote, tutorial, and invited speakers, the workshop organizers, and all the authors of the contributed papers and posters.

We are also grateful to the members of the program committee and the external reviewers for their invaluable help in reviewing the submissions and selecting the papers to be presented at the conference. We thank the organizing committee of UCNC 2025 for their tireless help in taking care of all the numerous details in organizing the event. The conference would not have happened without the active support of the Laboratoire I3S, Université Côte d'Azur, and the national French founding agency (ANR). Finally, we thank the EasyChair conference system, and the LNCS team at Springer for helping in the process of making these proceedings.

We are grateful for the support of the natural and unconventional computing community, which made UCNC 2025 a success in terms of participation and scientific engagement.

September 2025   Enrico Formenti
                                                                   Luca Manzoni

Organization

Steering Committee

Thomas Bäck	University of Leiden, The Netherlands
Cristian S. Calude	University of Auckland, New Zealand
Enrico Formenti (Co-chair)	Université Côte d'Azur, France
Lov K. Grover	Bell Labs, USA
Mika Hirvensalo (Co-chair)	University of Turku, Finland
Natasha Jonoska	University of South Florida, USA
Jarkko Kari	University of Turku, Finland
Seth Lloyd	Massachusetts Institute of Technology, USA
Giancarlo Mauri	Università degli Studi di Milano-Bicocca, Italy
Gheorghe Paun	Institute of Mathematics of the Romanian Academy, Romania
Grzegorz Rozenberg (Emeritus Chair)	Leiden University, The Netherlands
Arto Salomaa	University of Turku, Finland
Shinnosuke Seki (Co-chair)	University of Electro-Communications, Japan
Tomasso Toffoli	Boston University, USA
Carme Torras	Institute of Robotics and Industrial Informatics, Spain
Jan van Leeuwen	Utrecht University, The Netherlands

Program Committee

Cristian S. Calude	University of Auckland, New Zealand
Ho-Lin Chen	National Taiwan University, Taiwan
Da-Jung Cho	Ajou University, South Korea
Moreno Falaschi (LBA & Posters Chair)	University of Siena, Italy
Paola Flocchini	University of Ottawa, Canada
Enrico Formenti (Co-chair)	Université Côte d'Azur, France
Giuditta Franco	University of Verona, Italy
Daniela Genova	University of North Florida, USA
Masami Hagiya	University of Tokyo, Japan
Mika Hirvensalo	University of Turku, Finland
Natasha Jonoska	University of South Florida, USA

Jarkko Kari	University of Turku, Finland
Jongmin Kim	Pohang University of Science and Technology, South Korea
Ian McQuillan	University of Saskatchewan, Canada
Luca Manzoni (Co-chair)	University of Trieste, Italy
Bruno Martin	Université Côte d'Azur, France
Ion Petre	University of Turku, Finland
Zornitza Prodanoff	University of North Florida, USA
Friedrich Simmel	Technical University of Munich, Germany
Susan Stepney	University of York, UK
Yukiko Yamauchi	Kyushu University, Japan

Organizing Committee

Kenza Benjelloun	University of Trieste, Italy
Filippo Dal Lago	Université Côte d'Azur, France
Enrico Formenti (Chair)	Université Côte d'Azur, France
Sandrine Julia	Université Côte d'Azur, France
Amélia Kunze	Université Côte d'Azur, France
Davide La Torre	SKEMA Business School, France
Bruno Martin	Université Côte d'Azur, France

Additional Reviewers

Artiom Alhazov	Enrique Rico Ortega
Julio Aracena	Hava Siegelmann
Elisabetta De Maria	Damien Vergnaud
Alessandra Di Pierro	Jean-Baptiste Yunès
Simone Furini	Katalin Anna Lazar
Silvère Gangloff	Da-Jung Cho
Léo Gayral	Giuseppe Di Molfetta
Caterina Graziani	Sandro Erba
Katsunobu Imai	Maximilien Gadouleau
Gary Kochenberger	Max Garzon
Amelia Kunze	Zsolt Gazdag
Brennan Lockinger	Shih-Han Hung
Nicolas Ollinger	Hendrik Jan Hoogeboom
Simon Perdrix	Iliya Kulbaka
Antonio E. Porreca	Pascal Lafourcade
Matthieu Rambaud	Cristopher Moore
Elisabeth Remy	Loïc Paulevé

Kévin Perrot
Thomas Prévost
Ivan Rapaport
Jiakai Ren

Heike Siebert
Guillaume Theyssier
Shizhao Wei
Dan Zhang

Contents

Invited Talk

Toward Non Abelian Scenarios for Additive Cellular Automata on Finite Groups

Alberto Dennunzio[(✉)] [iD]

Dipartimento di Informatica, Sistemistica e Comunicazione, Università degli Studi di Milano-Bicocca, Viale Sarca 336, 20126 Milano, Italy
`alberto.dennunzio@unimib.it`

Abstract. We consider Additive CA on a finite group, also called Group CA. First of all, we present the abelian situation in which easy-to-check algebraic characterizations of the main dynamical properties in terms of the CA local rule have been provided. Then, we move to the non abelian scenario. Precisely, we consider the dynamical behavior of Additive CA on a number of classes of specific finite groups and for each of those classes, we focus our attention to the non abelian scenarios, providing exact characterizations for some dynamical properties.

Keywords: Cellular Automata · Additive Cellular Automata · Discrete Dynamical Systems

1   Introduction

Cellular Automata (CA) are formal models for complex systems that have been widely studied and find application in a number of distinct disciplines with different applicative purposes. Although they are defined by a finite local rule on a finite set of states (alphabet), CA can display a rich and complex temporal global evolution (for an introduction to CA theory, see [12]). However, the main dynamical properties of general CA are undecidable ([8,10,13]). Fortunately the undecidability issue can be in a sense overcome by imposing some restrictions on the model. In the setting of our interest, the alphabet and the global updating map are a group and an additive function, respectively, giving rise to Additive CA on a finite group, also called Group CA [1].

In this paper we consider such CA. First of all, we present the abelian situation in which easy-to-check algebraic characterizations of the main dynamical properties in terms of the CA local rule have been provided. We start by the subclass of Linear CA, i.e., those Additive CA having the multidimensional set $(\mathbb{Z}/m\mathbb{Z})^n$ as alphabet and a local rule defined by $n \times n$ matrices over $\mathbb{Z}/m\mathbb{Z}$. We review how the dynamical properties of Linear CA are "hidden" inside the characteristic polynomial of the matrix defining a Linear CA and built on the

© The Author(s), under exclusive license to Springer Nature Switzerland AG 2026
E. Formenti and L. Manzoni (Eds.): UCNC 2025, LNCS 16364, pp. 3–13, 2026.
https://doi.org/10.1007/978-3-032-15641-9_1

basis of the above-mentioned $n \times n$ matrices. Then, we recall how the characterizations of the dynamical properties for Linear CA can be exploited to decide the dynamical behavior of the entire class of Additive CA on a finite abelian group.

Then, we move to the non abelian scenario. To do this, we consider the dynamical behavior of Additive CA on a number of classes of specific finite groups (such as, for instance, simple, symmetric, alternating, dihedral, quaternion and decomposable groups) and for each of those classes, we focus our attention to the non abelian scenarios, providing exact characterizations for some of the dynamical properties.

2   Basic Notions and Background

Let $\mathbb{K}$ be any commutative ring and let $A \in \mathbb{K}^{n \times n}$ be an $n \times n$-matrix over $\mathbb{K}$. We denote by χ_A the characteristic polynomial of A and by $\mathbb{K}[X, X^{-1}]$ the set of Laurent polynomials with coefficients in $\mathbb{K}$. When $\mathbb{K} = \mathbb{Z}/m\mathbb{Z}$ for some natural $m > 1$, we will write $\mathbb{L}_m$ instead of $\mathbb{Z}/m\mathbb{Z}[X, X^{-1}]$

Let $\mathbb{K} = \mathbb{Z}/m\mathbb{Z}$ for some natural $m > 1$ and let q be any natural such that $1 < q < m$. If P is any polynomial from $\mathbb{K}[t]$ (resp., a Laurent polynomial from $\mathbb{L}_m$) (resp., a matrix from $(\mathbb{L}_m)^{n \times n}$), $P \bmod q$ denotes the polynomial (resp., the Laurent polynomial) (resp., the matrix) obtained by P by taking all its coefficients modulo q.

Let Σ be a finite set (also called *alphabet*). A *CA configuration* (or, briefly, a *configuration*) is any function from $\mathbb{Z}$ to Σ. For any configuration $c \in \Sigma^{\mathbb{Z}}$ and any integer $i \in \mathbb{Z}$, the value of c in position i is denoted by c_i. The set $\Sigma^{\mathbb{Z}}$, called *configuration space*, is as usual equipped with the standard Tychonoff distance d. Whenever the term *linear* is involved the alphabet Σ is $\mathbb{K}^n$, where $\mathbb{K} = \mathbb{Z}/m\mathbb{Z}$ for some natural $m > 1$. Clearly, in that case both $\mathbb{K}^n$ and $(\mathbb{K}^n)^{\mathbb{Z}}$ become $\mathbb{K}$-modules in the entrywise way. On the other hand, whenever the term *additive* is involved the alphabet Σ is a finite group G and the configuration space turns $G^{\mathbb{Z}}$ turns out to be a group, too, where the group operation of $G^{\mathbb{Z}}$ is the componentwise extension of the group operation of G, both of them will be denoted by $+$.

We are going to deal with both an *abelian* and a *non abelian* group G, the abelian situation clearly including the linear one. Let G be any finite group. In the sequel, we will denote by 0 the identity (neutral) element of G. The set $\{0\}$ is called the trivial group and is a subgroup of any group. A set $N \subseteq G$ is a *normal subgroup* of G if for all $g \in G$ and for all $n \in N$ it holds that $g + n + (-g) \in N$, where $-g$ is the inverse element of g. Recall that the centralizer of a subset $S \subseteq G$ is the set $C_G(S) = \{x \in G : \forall g \in S, x + g = g + x\}$, while the center of G is $Z(G) = C_G(G) = \{z \in G : \forall g \in G, z + g = g + z\}$, i.e., the set of elements commuting with every element of G, and $Z(G)$ is an abelian and normal subgroup of G. Given two finite groups G and H, a *group homomorphism* from G to H is a function $h : G \rightarrow H$ such that for all $g_1, g_2 \in G$ it holds that $h(g_1 + g_2) = h(g_1) + h(g_2)$, where we used the same symbol $+$ of group operation for both G and H. If $G = H$ the homomorphism h is called group *endomorphism*. The image and the *kernel* of a homomorphism $h : G \rightarrow H$

are the sets $Img(h) = \{h(g) : g \in G\}$ and $Ker(h) = \{g \in G : h(g) = 0\}$, respectively. It is well-known that the kernel of any group homomorphism is a normal subgroup of G and any endomorphism h of a finite group G is injective if and only if it is surjective if and only if $Ker(h) = \{0\}$. An endomorphism h of a finite group G is said to be trivial if $Img(h) = \{0\}$. A *permutation group* is a group G with elements that are permutations of $\{1, \ldots, n\}$ and where the group operation is the composition of permutations in G. By Cayley's theorem, every group is isomorphic to some permutation group.

A *one-dimensional CA* (or, briefly, a *CA*) on Σ is a pair $(\Sigma^{\mathbb{Z}}, \mathcal{F})$, where $\mathcal{F} \colon \Sigma^{\mathbb{Z}} \to \Sigma^{\mathbb{Z}}$ is the uniformly continuous transformation (called *global rule*) defined as $\forall c \in \Sigma^{\mathbb{Z}}, \forall i \in \mathbb{Z}, \mathcal{F}(c)_i = f(c_{i-r}, \ldots, c_{i+r})$, for some fixed natural number $r \in \mathbb{N}$ (called *radius*) and some fixed function $f \colon \Sigma^{2r+1} \to \Sigma$ (called *local rule* of radius r). In the sequel, when no misunderstanding is possible, we will sometimes identify any CA with its global rule.

A CA $(\Sigma^{\mathbb{Z}}, \mathcal{F})$ is *topologically transitive*, or, simply *transitive*, if for any pair of nonempty open subsets $U, V \subseteq \Sigma^{\mathbb{Z}}$ there exists a natural $h > 0$ such that $\mathcal{F}^h(U) \cap V \neq \emptyset$, while it is said to be *topologically mixing*, or, simply *mixing*, if the latter intersection condition holds ultimately. A CA $(\Sigma^{\mathbb{Z}}, \mathcal{F})$ is *strongly transitive* if for any nonempty open subset $U \subseteq \Sigma^{\mathbb{Z}}$ it holds that $\bigcup_{h \in \mathbb{N}} \mathcal{F}^h(U) = \Sigma^{\mathbb{Z}}$. A CA $(\Sigma^{\mathbb{Z}}, \mathcal{F})$ has *dense periodic orbits* if the set of its periodic points is dense in $\Sigma^{\mathbb{Z}}$, where a periodic point for is any configuration $c \in \Sigma^{\mathbb{Z}}$ such that $F^h(c) = c$ for some natural $h > 0$.

A CA $(\Sigma^{\mathbb{Z}}, \mathcal{F})$ is *sensitive to the initial conditions* or, simply *sensitive*, if there exists $\epsilon > 0$ such that for any $\delta > 0$ and $c \in \Sigma^{\mathbb{Z}}$ there is a configuration $c' \in \Sigma^{\mathbb{Z}}$ with $0 < d(c', c) < \delta$ such that $d(\mathcal{F}^h(c'), \mathcal{F}^h(c')) \geq \epsilon$ for some natural h. A CA $(\Sigma^{\mathbb{Z}}, \mathcal{F})$ is said to be *equicontinuous* if for any $\epsilon > 0$ there exists $\delta > 0$ such that for all $c, c' \in \Sigma^{\mathbb{Z}}$, $d(c', c) < \delta$ implies that $\forall k \in \mathbb{N}, d(\mathcal{F}^k(c'), \mathcal{F}^k(c')) < \epsilon$. While equicontinuity is a strong form of stability, sensitivity is the basic component of chaos. As a matter of fact, a discrete time dynamical system is chaotic according to Devaney [7] if it is sensitive, transitive and has dense periodic orbits.

A CA $(\Sigma^{\mathbb{Z}}, \mathcal{F})$ is *positively expansive* if there exists a constant $\varepsilon > 0$ such that for any pair of configurations $c, c' \in \Sigma^{\mathbb{Z}}$ with $c \neq c'$ there exists a natural number ℓ satisfying $d(\mathcal{F}^{\ell}(c), \mathcal{F}^{\ell}(c')) \geq \varepsilon$. We stress that CA positive expansivity is a condition of very strong chaos. Indeed, on one hand, positive expansivity is a stronger condition than sensitivity. On the other hand, any positively expansive CA is also topologically mixing (and strongly transitive, too) and it has dense periodic orbits.

An *Additive CA on a finite group* G, also called *Group CA (GCA)* is a CA $(G^{\mathbb{Z}}, \mathcal{F})$ where the global rule $\mathcal{F} : G^{\mathbb{Z}} \to G^{\mathbb{Z}}$ is an endomorphism of $G^{\mathbb{Z}}$. We recall that the local rule $f : G^{2r+1} \to G$ of an Additive CA of radius r over any finite group G is always a group homomorphism [3].

2.1 Linear and Additive CA on a Finite Abelian Group

Let $\mathbb{K} = \mathbb{Z}/m\mathbb{Z}$ for some natural $m > 1$ and let $n \in \mathbb{N}$ with $n \geq 1$. Let G be a finite *abelian* group.

A local rule $f\colon (\mathbb{K}^n)^{2r+1} \to \mathbb{K}^n$ of radius r is said to be *linear* if it is defined by $2r + 1$ matrices $A_{-r}, \ldots, A_r \in \mathbb{K}^{n\times n}$ as follows: $\forall (x_{-r}, \ldots, x_r) \in (\mathbb{K}^n)^{2r+1}, f(x_{-r}, \ldots, x_r) = \sum_{i=-r}^{r} A_i \cdot x_i$. A one-dimensional *linear CA (LCA)* over $\mathbb{K}^n$ is a CA $\mathcal{F}$ based on a linear local rule. The Laurent polynomial (or matrix)

$$A = \sum_{i=-r}^{r} A_i X^{-i} \in \mathbb{K}^{n\times n}[X, X^{-1}] \cong (\mathbb{L}_m)^{n\times n}$$

is said to be the *the matrix associated with $\mathcal{F}$*.

Many years ago easy-to-check characterizations of the above mentioned dynamical properties for LCA over $\mathbb{Z}/m\mathbb{Z}$, i.e., LCA in the case $n = 1$, had been provided.

Theorem 1 ([4,9,14]). *Let $\mathcal{F}$ be any LCA on $\mathbb{Z}/m\mathbb{Z}$ and let $A_{-r}, \ldots, A_r \in \mathbb{Z}/m\mathbb{Z}$ define the 1×1 matrix $A \in \mathbb{L}_m$ associated with $\mathcal{F}$, where m is any natural with $m > 1$. An efficiently computable characterization in terms of $A_{-r}, \ldots, A_r$ and m exists for each of the following properties: injectivity, surjectivity, sensitivity, equicontinuity, transitive, positive expansivity.*

Indeed, each characterization allowing one to decide whether a LCA over $\mathbb{Z}/m\mathbb{Z}$ satisfies one of the properties from Theorem 1 only requires gcd computations and the corresponding decision algorithm has computational complexity which is polynomial in $\log m$ and in $2r + 1$ [14].

Among Additive CA on any finite group, we now focus on *Additive CA on a finite abelian group G*. Clearly, the class of such CA is wider than LCA. We stress that the local rule $f : G^{2r+1} \to G$ of an Additive CA of radius r on a finite abelian group G can be written as $\forall (g_{-r}, \ldots, g_r) \in G^{2r+1}, f(g_{-r}, \ldots, g_r) = \sum_{i=-r}^{r} h_i(g_i)$, where the functions h_i are endomorphisms of G (a precise characterization of the local rule of an Additive CA on any finite group is illustrated in Sect. 4. Moreover, as a consequence of the application of the fundamental theorem of finite abelian groups to Additive CA on a finite abelian group, as recalled in [5], *we can always assume* that

$$G = \mathbb{Z}/p^{k_1}\mathbb{Z} \times \ldots \times \mathbb{Z}/p^{k_n}\mathbb{Z}$$

for some naturals $k_1, \ldots, k_n$ with $k_1 \geq \ldots \geq k_n$.

Furthermore, if $(G^{\mathbb{Z}}, F)$ is any Additive CA on a finite group $G = \mathbb{Z}/p^{k_1}\mathbb{Z} \times \ldots \times \mathbb{Z}/p^{k_n}\mathbb{Z}$, where $k_1 \geq \ldots \geq k_n$, then a LCA $(\hat{G}^{\mathbb{Z}}, L)$ on $\hat{G} = (\mathbb{Z}/p^{k_1}\mathbb{Z})^n$ can be built on the basis of $(G^{\mathbb{Z}}, F)$ in such a way that $L \circ \Psi = \Psi \circ F$, where $\Psi : G^{\mathbb{Z}} \to \hat{G}^{\mathbb{Z}}$ is a suitable map which turns out to be injective and continuous. For this reason, $(\hat{G}^{\mathbb{Z}}, L)$ is said to be the *LCA associated with $(G^{\mathbb{Z}}, F)$ via the embedding Ψ*. In general, $(G^{\mathbb{Z}}, F)$ is not topologically conjugated (i.e., homeomorphic) to $(\hat{G}^{\mathbb{Z}}, L)$ but $(G^{\mathbb{Z}}, F)$ is a subsystem of $(\hat{G}^{\mathbb{Z}}, L)$ and the latter condition alone is not enough in general to immediately lift dynamical properties from a one system to the other one.

Finally, transitivity and mixing are equivalent properties for Additive CA on a finite abelian group. The same also holds for surjectivity and dense periodic orbits. Since all transitive CA are both surjective and sensitive, as a consequence, chaos turns out to be equivalent to transitivity for Additive CA on a finite abelian group.

3   Dynamics of Additive CA on a Finite Abelian Group

For a comprehensive bibliography concerning all the results recalled in this section see [5]. Let us start reviewing those results about the set theoretic and dynamical properties of LCA on $(\mathbb{Z}/m\mathbb{Z})^n$ in terms of algebraic characterizations for them.

In order to deal with the characterization of positive expansivity, we need to introduce the following notions. The *positive* (resp., *negative*) degree of any given polynomial $\alpha \in \mathbb{L}_m$ with $\alpha \neq 0$, denoted by $\deg^+(\alpha)$ (resp., $\deg^-(\alpha)$), is the maximum (resp, minimum) value among the degrees of the monomials of α. Furthermore, the previous notions are extended to $\alpha = 0$ as follows: $\deg^+(0) = -\infty$ and $\deg^-(0) = +\infty$.

Let $\pi(t) = \alpha_0 + \alpha_1 t + \cdots + \alpha_{n-1} t^{n-1} + t^n$ be any polynomial from $\mathbb{L}_m[t]$. We say that $\pi(t)$ is *expansive* if both the following two conditions are satisfied:

(i) $\deg^+(\alpha_0) > 0$ and $\deg^+(\alpha_0) > \deg^+(\alpha_i)$ for every $i \in \{1, \ldots, n-1\}$;
(i) $\deg^-(\alpha_0) < 0$ and $\deg^-(\alpha_0) < \deg^-(\alpha_i)$ for every $i \in \{1, \ldots, n-1\}$;

The following summarizes the characterization of the set theoretic and dynamical properties of LCA on $(\mathbb{Z}/m\mathbb{Z})^n$ (see [2,11] for the set theoretic properties).

Theorem 2. *Let $\mathcal{F}$ be any LCA on $(\mathbb{Z}/m\mathbb{Z})^n$ and let $A \in \mathbb{L}_m^{n \times n}$ be the matrix associated with $\mathcal{F}$, where m and n are any two naturals with $m > 1$ and $n > 1$. Let $\chi_A(t) = \alpha_0 + \alpha_1 t + \cdots + \alpha_{t-1} t^{n-1} + t^n$ be the characteristic polynomial of A, where each $\alpha_i \in \mathbb{L}_m$ and as such is the 1×1 matrix associated with a LCA on $\mathbb{Z}/m\mathbb{Z}$. Let $p_1, \ldots, p_l$ be all the primes appearing inside the prime factor decomposition of m (i.e., $m = p_1^{k_1} \cdots p_l^{k_l}$). The following characterizations hold:*

- *$\mathcal{F}$ is injective iff α_0 is the matrix associated with an injective LCA on $\mathbb{Z}/m\mathbb{Z}$*
- *$\mathcal{F}$ is surjective iff α_0 is the matrix associated with a surjective LCA on $\mathbb{Z}/m\mathbb{Z}$*
- *$\mathcal{F}$ is sensitive (resp., equicontinuous) iff at least one (resp., each) α_i is the matrix associated with a sensitive (resp., equicontinuous) LCA on $\mathbb{Z}/m\mathbb{Z}$.*
- *$\mathcal{F}$ is transitive iff for every $j \in \{1, \ldots, l\}$ the LCA on $(\mathbb{Z}/p_j^{k_j}\mathbb{Z})^n$ having $A \bmod p_j^{k_j}$ as associated matrix is transitive iff both the following conditions hold*

$$\det(A \bmod p_j) \neq 0$$

$$\gcd(\chi_{A \bmod p_j}(t), t^{p^h - 1} - 1) = 1 \quad \text{for all } h \in \{1, \ldots, n\}$$

- *$\mathcal{F}$ is positively expansive iff $\chi_{A \bmod p_j}(t)$ is expansive for every $j \in \{1, \ldots, l\}$*

We stress that all the characterizations from Theorem 2 are efficiently computable except those for transitivity and positive expansivity since these require the prime factor decomposition of m. Furthermore, for any term p^k (where p is a prime) appearing in such a decomposition, the algorithm deciding transitivity for a LCA on $(\mathbb{Z}/p^k\mathbb{Z})^n$ is not efficient at all. However, efficient algorithms for deciding positive expansivity and transitivity for LCA on $(\mathbb{Z}/m\mathbb{Z})^n$ were provided very recently. In both of them the prime factor decomposition of m is bypassed.

3.1   Dynamics of Additive CA on a Finite Abelian Group

We now recall that the characterization results regarding all the mentioned properties for LCA on $(\mathbb{Z}/m\mathbb{Z})^n$ can be lifted to the whole class Additive CA on any finite abelian group. We stress that the characterization results are stated for Additive CA on $G = \mathbb{Z}/p^{k_1}\mathbb{Z} \times \ldots \times \mathbb{Z}/p^{k_n}\mathbb{Z}$, but this is not a restriction, as already mentioned in Sect. 2.

Theorem 3. *Let $\mathcal{F} : G^{\mathbb{Z}} \to G^{\mathbb{Z}}$ be any Additive CA on a finite abelian group G, where $G = \mathbb{Z}/p^{k_1}\mathbb{Z} \times \ldots \times \mathbb{Z}/p^{k_n}\mathbb{Z}$ for some prime p and some non zero naturals $k_1, \ldots, k_n$ with $k_1 \geq k_2 \geq \ldots \geq k_n$. Let $\mathcal{L}$ be the LCA over $\hat{G}$ associated with $\mathcal{F}$ via the embedding Ψ, where $\hat{G} = (\mathbb{Z}/p^{k_1}\mathbb{Z})^n$. It holds that*

(i) $\mathcal{F}$ is injective (resp., surjective) if and only if $\mathcal{L}$ is injective (resp., surjective);

(ii) $\mathcal{F}$ is sensitive to the initial conditions (resp., equicontinuous) if and only if $\mathcal{L}$ is sensitive to the initial conditions (resp., equicontinuous);

(iii) $\mathcal{F}$ is topologically transitive if and only if $\mathcal{L}$ is topologically transitive;

(iv) $\mathcal{F}$ is positively expansive if and only if $\mathcal{L}$ is positively expansive.

4   Additive CA over a Non Abelian Finite Group

We now deal with Additive CA on any finite group G, with a special focus on non abelian groups. All the results from this section come from [6], where a characterization of the local rule Additive CA on any finite group G was also provided. According to that, $f : G^{2r+1} \to G$ is the local rule of a r-radius Additive CA on a finite group G if and only if there exist $2r+1$ endomorphisms $h_{-r}, \ldots, h_r$ of G such that both the following properties hold:

$$\forall (g_{-r}, \ldots, g_r) \in G^{2r+1}, \ f(g_{-r}, \ldots, g_r) = h_{-r}(g_{-r}) + \cdots + h_r(g_r) \ , \quad (1)$$

and

$$\forall i, j \in \{1, \ldots, k\} \text{ with } i \neq j, \ Img(h_i) \subseteq C_G(Img(h_j)) \quad (2)$$

From now on, any local rule $f : G^{2r+1} \to G$ a r-radius Additive CA on a finite group G will be identified as a $(2r+1)$-tuple $f = (h_{-r}, \ldots, h_r)$ of endomorphisms of G satisfying properties (1) and (2).

We now recall two classes of Additive CA on a finite group that play an important role for this class of CA, namely, the shift-like and identity-like ones. An r-radius Additive CA on a finite group G with local rule $f = (h_{-r}, \ldots, h_r)$ is a *shift-like* if and only if $Img(h_k) \neq \{0\}$ for some h_k with $k \neq 0$ and $Img(h_i) = \{0\}$ for all $i \neq k$, while it is an *identity-like GCA* if and only if $r = 0$ and $Img(h_0) \neq \{0\}$. Note that any shift-like or identity-like Additive CA on a finite group is surjective if and only if its local rule is surjective if and only if its local rule is injective if and only if it is injective. Furthermore, any surjective shift-like Additive CA on a finite group is topologically transitive.

The following is an important result showing that the non abelianness of the group imposes strong constraints on the dynamical behavior of Additive CA on that group.

Theorem 4. *Let $\mathcal{F}$ be any r-radius Additive CA on a finite non-abelian group G and let $f = (h_{-r}, \ldots, h_r)$ be its local rule. If $Img(h_i) = G$ for some $i \in \{-r, \ldots, r\}$ then the following facts hold:*

(i) $\mathcal{F}$ is surjective (resp., injective) iff $\mathcal{F}_Z$ is surjective (resp., injective), where $\mathcal{F}_Z$ is the CA $\mathcal{F}$ restricted to $Z(G)^{\mathbb{Z}}$;

(ii) if at least one of the two following conditions holds then $\mathcal{F}$ is either a bijective shift-like or a bijective identity-like:
- every endomorphism of G is either surjective or trivial;
- $Z(G) = \{0\}$.

(iii) $\mathcal{F}$ is neither strongly transitive nor positively expansive.

We stress that $\mathcal{F}$ restricted to $Z(G)^{\mathbb{Z}}$ is an Additive CA on a finite abelian group and as such it can be easily investigated.

We are now going to see what happens as far as some specific groups are concerned with a focus on the non abelian scenario.

4.1 Simple Groups

Simple groups are among the most studied groups and the classification of finite simple groups is an important result in group theory. A simple group is a group G that has no proper non-trivial normal subgroups. Hence, every endomorphism of G is either surjective or trivial. As a consequence, Additive CA on a finite non-abelian simple group behave as stated in the following.

Theorem 5. *Let $\mathcal{F}$ be a GCA on a finite non-abelian simple group. The following facts hold*

(i) $\mathcal{F}$ is either a bijective shift-like or a bijective identity-like;
(ii) $\mathcal{F}$ is topologically transitive iff it is a bijective shift-like;
(iii) $\mathcal{F}$ is neither strongly transitive nor positively expansive.

4.2 Decomposable groups

We now deal with Additive CA on a finite group G which is the internal direct product of two normal subgroups G_1 and G_2, or, equivalently, $G \cong G_1 \times G_2$. Let $\mathcal{F}$ be Additive CA on G such that G is the internal direct product of two normal subgroups G_1 and G_2. Hence, $\mathcal{F} \cong \mathcal{F}_1 \times \mathcal{F}_2$, where $\mathcal{F}_1$ and $\mathcal{F}_2$ are the two Additive CA of $\mathcal{F}$ on G_1 and G_2, induced by the projections of G on G_1 and G_2, respectively. In this way, one can investigate the dynamical behavior of $\mathcal{F}$ by separately studying the behaviors of $\mathcal{F}_1$ and $\mathcal{F}_2$. Indeed, except for sensitivity, $\mathcal{F}$ has a certain property if and only if both $\mathcal{F}_1$ and $\mathcal{F}_2$ have the same property, too (while $\mathcal{F}$ is sensitive if and only if at least one between $\mathcal{F}_1$ and $\mathcal{F}_2$ is sensitive).

4.3 Alternating Groups A_n

An alternating group is the group of the even permutations of a finite set. When it has n elements the alternating group is denoted by A_n. We known that A_n is abelian if and only if $n \leq 3$ and A_n is simple if and only if $n = 3$ or $n \geq 5$. Hence, the normal subgroups of A_n are $\{0\}$ and A_n for $n = 3$ or $n \geq 5$, while A_4 admits a further normal subgroup besides $\{0\}$ and A_4. Regarding the center of the alternating groups, $Z(A_n) = A_n$ if $n \leq 3$, and $Z(A_n) = \{0\}$, otherwise.

Theorem 6. *Additive CA on any non abelian group A_n with $n \geq 5$ behave exactly as Additive CA on a non abelian simple group, according to Theorem 6.*

Theorem 7. *Let $\mathcal{F}$ be an Additive CA on the non abelian group A_4. The following facts hold:*

   (i) $\mathcal{F}$ is surjective iff F is injective iff F is either a bijective shift-like or a bijective identity-like GCA;
  (ii) $\mathcal{F}$ is topologically transitive iff F is a bijective shift-like GCA;
 (iii) $\mathcal{F}$ is neither strongly transitive nor positively expansive.

We stress that there exist Additive CA on A_4 that are neither shift-like nor identity-like and then, in view of Theorem 7, thery are not surjective.

4.4 Symmetric Groups S_n

A symmetric group, denoted by S_n, is any group consisting of all bijective functions from a set of n elements to itself (i.e., all permutations of a set with n elements). It is known that a symmetric group S_n is non-abelian if and only if $n \geq 3$ and the center is $Z(S_n) = S_n$ if $n = 2$, and $Z(S_n) = \{0\}$, otherwise. The normal subgroups of S_n are $\{0\}$, A_n and S_n for all natural $n \geq 3$ with $n \neq 4$, while S_4 admits a further normal subgroup besides $\{0\}$, A_4, and S_4.

Theorem 8. *Additive CA on any non abelian group S_n with $n \geq 3$ behave exactly as Additive CA on the non abelian group A_4, according to Theorem 7.*

There also exist non transitive but sensitive Additive CA on S_4. Clearly, by Theorem 8 and since identity-like CA are not sensitive, they are not surjective.

4.5 Dihedral Groups D_n

A dihedral group is the group of symmetries of a regular polygon, which includes rotations and reflections. In the sequel, D_n refers to the symmetries of the n-gon, a group of order $2n$. For $n \geq 6$, D_n is decomposable if and only if $n = 2k$ where k is odd and $D_n \cong C_2 \times D_{\frac{n}{2}}$ or, equivalently, $D_{2k} \cong C_2 \times D_k$ (where C_2 is the cyclic group of order 2). In addition, the center is $Z(D_n) = \{0\}$, if n is odd, $Z(D_n) \cong C_2$, otherwise.

Theorem 9. *Additive CA on any non abelian group D_n where n is an odd integer with $n \geq 3$ behave exactly as Additive CA on the non abelian group A_4, according to Theorem 7.*

Theorem 4 does not hold for Additive CA on a non abelian finite group D_n, where n is even. However, something can be said. Indeed, consider any Additive CA $\mathcal{F}$ on a non abelian finite group D_{2k} where k is an odd natural with $k \geq 3$. Since $D_{2k} \cong C_2 \times D_k$, $\mathcal{F}$ can be decomposed as the product of two Additive CA $\mathcal{F}_1$ on C_2 and $\mathcal{F}_2$ on D_k. Therefore, one can study set theoretic and dynamical properties of $\mathcal{F}$ by the investigation of $\mathcal{F}_1$ and $\mathcal{F}_2$ (see Sect. 4.2).

4.6 Quaternion Group Q_8

The quaternion group noted by Q_8 is Hamiltonian, i.e., Q_8 is not abelian, but every subgroup of Q_8 is normal. By Cayley's theorem, we see Q_8 as subgroup of all the permutations of $\{1, \ldots, 8\}$. Let π be the permutation $(1, 5)(2, 6)(3, 7)(4, 8)$. It holds that $Z(Q_8) = \{0, \pi\}$. There are 28 endomorphisms of Q_8, 4 of them are non-surjective. If h is any non-surjective endomorphism of Q_8 then $h(\pi) = 0$ and $Img(h)$ is either $\{0\}$ or $\{0, \pi\}$. Regarding any surjective endomorphisms h of Q_8 h it holds that $h(\pi) = \pi$. We want to stress that any $2r + 1$ endomorphisms of Q_8 satisfy property (2) and so the $(2r + 1)$-tuple of such endomorphisms gives rise to a local rule of an Additive CA on Q_8.

Theorem 10. *Let $\mathcal{F}$ be an r-radius Additive CA on Q_8 and let $f = (h_{-r}, \ldots, h_r)$ be its local rule. The following facts hold:*

(i) *$\mathcal{F}$ is surjective iff $\mathcal{F}$ is injective iff at least one h_i is surjective iff $\mathcal{F}_Z$ is either a CA shift or the CA identity;*

(ii) *$\mathcal{F}$ is topologically transitive iff $\mathcal{F}$ is bijective and there exists at least one surjective h_i with $i \neq 0$;*

(iii) *$\mathcal{F}$ is neither strongly transitive nor positively expansive.*

5 Conclusions

We reviewed the mostly recent easy to check characterizations of the main set-theoretic and dynamical properties for Linear CA over $(\mathbb{Z}/m\mathbb{Z})^n$ and recalled how such characterizations can be exploited to decide the same properties for the whole class of Additive CA over a finite abelian group. Then, we moved on the non abelian scenario of Additive CA on certain classes of specific finite groups and the corresponding recent results. The latter put in evidence that the non abelianness of the group imposes strong constraints on the dynamical behaviors of Additive CA. A first step for further research is identifying other classes of finite group for which the set-theoretic and dynamical properties can be characterized. The main goal is indeed finding out easy to check characterizations of those properties for Additive CA on any finite group.

Acknowledgments. This work was supported by the HORIZON-MSCA-2022-SE-01 project 101131549 "Application-driven Challenges for Automata Networks and Complex Systems (ACANCOS)" and by the PRIN 2022 PNRR project "Cellular Automata Synthesis for Cryptography Applications (CASCA)" (P2022MPFRT) funded by the European Union – Next Generation EU.

Disclosure of Interests. The author has no competing interests to declare that are relevant to the content of this article.

References

1. Béaur, P., Kari, J.: Effective projections on group shifts to decide properties of group cellular automata. Int. J. Found. Comput. Sci. **35**(1&2), 77–100 (2024)
2. Bruyn, L.L., den Bergh, M.V.: Algebraic properties of linear cellular automata. Linear Algebra Appl. **157**, 217–234 (1991)
3. Castillo-Ramirez, A., Mata-Gutiérrez, O., Zaldivar-Corichi, A.: Cellular automata over algebraic structures. Glasg. Math. J. **64**(2), 306–319 (2022)
4. Cattaneo, G., Formenti, E., Manzini, G., Margara, L.: Ergodicity, transitivity, and regularity for linear cellular automata over $\mathbb{Z}_m$. Theoret. Comput. Sci. **233**(1–2), 147–164 (2000)
5. Dennunzio, A.: Easy to check algebraic characterizations of dynamical properties for linear CA and additive CA over a finite abelian group. In: Gadouleau, M., Castillo-Ramirez, A. (eds.) Cellular Automata and Discrete Complex Systems - 30th IFIP WG 1.5 International Workshop, AUTOMATA 2024, Durham, UK, July 22-24, 2024, Proceedings. Lecture Notes in Computer Science, vol. 14782, pp. 23–34. Springer (2024)
6. Dennunzio, A., Formenti, E., Margara, L.: On the dynamical behavior of cellular automata on finite groups. IEEE Access **12**, 122061–122077 (2024)
7. Devaney, R.L.: An Introduction to Chaotic Dynamical Systems. Addison-Wesley, Addison-Wesley advanced book program (1989)
8. Durand, B., Formenti, E., Varouchas, G.: On undecidability of equicontinuity classification for cellular automata. In: Morvan, M., Rémila, E. (eds.) Discrete Models for Complex Systems, DMCS'03, Lyon, France, June 16-19, 2003. Discrete Mathematics and Theoretical Computer Science Proceedings, vol. AB, pp. 117–128. DMTCS (2003)
9. Ito, M., Osato, N., Nasu, M.: Linear cellular automata over $\mathbb{Z}_m$. J. Comput. Syst. Sci. **27**, 125–140 (1983)
10. Kari, J.: Rice's theorem for the limit sets of cellular automata. Theoret. Comput. Sci. **127**(2), 229–254 (1994)

11. Kari, J.: Linear Cellular Automata with Multiple State Variables. In: Reichel, H., Tison, S. (eds.) STACS 2000. LNCS, vol. 1770, pp. 110–121. Springer, Heidelberg (2000). https://doi.org/10.1007/3-540-46541-3_9
12. Kari, J.: Theory of cellular automata: a survey. Theor. Comput. Sci. **334**(1–3), 3–33 (2005)
13. Lukkarila, V.: Sensitivity and topological mixing are undecidable for reversible one-dimensional cellular automata. J. Cell. Autom. **5**(3), 241–272 (2010)
14. Manzini, G., Margara, L.: A complete and efficiently computable topological classification of d-dimensional linear cellular automata over $\mathbb{Z}_m$. Theoret. Comput. Sci. **221**(1–2), 157–177 (1999)

Regular Papers

Reversibility, Balance and Expansivity of Non-uniform Cellular Automata

Katariina Paturi$^{(\boxtimes)}$

University of Turku, Vesilinnantie 5, 20500 Turku, Finland
`katariina.paturi@utu.fi`

Abstract. Non-uniform cellular automata (NUCA) are an extension of cellular automata (CA), which transform cells according to multiple different local rules. A NUCA is defined by a configuration of local rules called a local rule distribution. We examine what properties of uniform CA can be recovered by restricting the rule distribution to be (uniformly) recurrent, focusing on only 1D NUCA.

We show that a bijective NUCA with a uniformly recurrent rule distribution is reversible. We also show that if a NUCA is surjective and has a recurrent rule distribution, or if it is bijective, then it is balanced. We present an example of a NUCA which has a non-empty and non-residual set of equicontinuity points, and one which is not sensitive but has no equicontinuity points. Finally, we show that (positively) expansive NUCA are sensitive.

Keywords: Non-uniform cellular automata · Symbolic dynamics · Reversibility

1 Introduction

Non-uniform cellular automata (NUCA) are an extension of cellular automata (CA), which are machines that operate on configurations - grids of cells coloured by an alphabet. In both, each cell simultaneously observes its neighbouring cells and changes its state depending on the state of the neighbourhood. A uniform CA transforms all cells with the same local rule, while a NUCA uses different local rules in different cells. The local rules of a NUCA are given by a fixed local rule distribution.

Cellular automata are a powerful tool in simulating systems based on local transformations, and NUCA provide a generalization capable of simulating systems with variation in their local behaviour. The dynamical properties of uniform CA have been studied extensively, but many of them rely on the translation invariance of CA, and no-longer apply in the broader setting of NUCA. For example, the well-known Garden of Eden theorem for uniform CA does not hold for NUCA. However, some properties, such as the Garden of Eden theorem, can be

---

Supported by the Magnus Ehrnrooth foundation and the Research Council of Finland.

© The Author(s), under exclusive license to Springer Nature Switzerland AG 2026
E. Formenti and L. Manzoni (Eds.): UCNC 2025, LNCS 16364, pp. 17–32, 2026.
https://doi.org/10.1007/978-3-032-15641-9_2

recovered in NUCA when the distribution of local rules is recurrent or uniformly recurrent [3].

The properties we examine are reversibility, balance, sensitivity and expansivity. There are implications known to hold for CA which don't necessarily hold for NUCA in general: all bijective CA are reversible, all surjective CA are balanced, a CA is either sensitive or has a dense set of equicontinuity points, and a (positively) expansive CA is sensitive. We find (positively) expansive NUCA are still sensitive, but the other statements all have counter-examples. We also show that the reversibility of bijective NUCA can be recovered when the rule distribution is uniformly recurrent, and the balance of surjective NUCA when the distribution is recurrent. Additionally we show that bijective NUCA are balanced.

2   Definitions

In this section, we define a number of concepts used throughout the work. Non-uniform cellular automata may be defined over any group, but in this work we only consider the 1-dimensional case of NUCA over $\mathbb{Z}$. One-dimensional NUCA are dynamical systems that operate in a space of configurations, bi-infinite tapes of symbols.

Definition 1. *Let Σ be a finite set called a* state set. *A* configuration (on $\mathbb{Z}$) *is an element $c \in \Sigma^{\mathbb{Z}}$. A* cell *is an element $x \in \mathbb{Z}$. A* domain *is any subset $D \subseteq \mathbb{Z}$. If D is finite, we call it a* finite domain.

The space $\Sigma^{\mathbb{Z}}$ equipped with Cantor's topology and is known to be compact [1]. In contrast with uniform CA, which are defined by a single local rule, a NUCA is defined by a local rule distribution. A rule distribution is a configuration over a set of rules rather than a state set, which determines the local rule used in each cell.

Definition 2. *Let Σ be a state set. A* radius r local rule *is a function $f : \Sigma^{2r+1} \to \Sigma$. A finite set of* local rules $\mathcal{R}$ *is called a* rule set.

Note that defining local rules with radial neighbourhoods is equivalent to defining them with arbitrarily shaped neighbourhoods. Now we are ready to define NUCA.

Definition 3. *A* non-uniform cellular automaton *(or NUCA) is a tuple $A = (\Sigma, \mathcal{R}, \theta)$, where Σ is a state set, $\mathcal{R}$ is a rule set, and $\theta \in \mathcal{R}^{\mathbb{Z}}$ is a* rule distribution. *The* global transition function *of A is the function $H_\theta : \Sigma^{\mathbb{Z}} \to \Sigma^{\mathbb{Z}}$ such that for $c \in \Sigma^{\mathbb{Z}}$, $x \in \mathbb{Z}$,*

$$H_\theta(c)(x) = \theta(x)(c(x - r), \ldots, c(x + r)),$$

where $r \in \mathbb{Z}_+$ is the radius of the local rule $\theta(x)$. If $|\mathcal{R}| = 1$ then the NUCA is a uniform cellular automaton *(or CA).*

We generally equate a NUCA with its transition function. Let us define a few more concepts used throughout the work. Since NUCA are machines based on local transformations, we often consider the neighbourhood of a cell.

Definition 4. *Let $\theta \in \mathcal{R}^{\mathbb{Z}}$ be a rule distribution, and suppose $\theta(x)$ is a radius r rule. We call the domain $N_\theta(x) = [x-r, x+r]$ the* neighbourhood of x. *If $D \subseteq \mathbb{Z}$ is a domain, we call the domain $N_\theta(D) = \bigcup_{x \in D} N_\theta(x)$ the* neighbourhood of D.

A shift is a function that moves a configuration (or rule distribution left or right. A major feature of uniform CA is that they are translation invariant: if G is a CA and σ^k is a shift, then $\sigma^{-k} \circ G \circ \sigma^k = G$. Formally, shifts are defined as follows.

Definition 5. *Let S be a state set or rule set. We denote the* left-shift *as a function $\sigma : S^{\mathbb{Z}} \to S^{\mathbb{Z}}$ which maps $c \in S^{\mathbb{Z}}$ such that $\sigma(c)(x) = c(x+1)$ for all $x \in \mathbb{Z}$. Similarly, we call the function σ^{-1} the* right shift. *Note that $\sigma \circ H_\theta = H_{\sigma'(\theta)} \circ \sigma$, where σ' is the left-shift over the set of rule distributions.*

Of course, NUCA are not translation invariant, and hence lack many useful properties of CA. Sometimes translation invariance isn't strictly necessary, and it's simply enough for a NUCA's rule distribution to be self-similar. The principal forms of self-similarity we investigate are recurrence and uniform recurrence. To formally define them, we must consider patterns and translated copies. A pattern is a (usually finite) snippet of a configuration or rule distribution. Unlike words, patterns are tied to specific domains.

Definition 6. *Let S be a state set or rule set and $c \in S^{\mathbb{Z}}$ a configuration or rule distribution. A* pattern *is any $p \in S^D$ for some domain D. If D is finite, we call p a* finite pattern. *We use $c_{|D} \in S^D$ to denote the pattern such that for any $x \in D$, $c_{|D}(x) = c(x)$.*

Definition 7. *Let S be a state set or rule set. If $p \in S^D$ and $q \in S^C$ are patterns and there is $k \in \mathbb{Z}$ such that $C = D + k$ and $p(x) = q(x+k)$ for any $x \in D$, then we call q a* translated copy *(or simply* copy*) of p. If $D \cap (D+k) = \emptyset$, the copies are* disjoint.

We call a rule distribution recurrent when every pattern it contains has a copy, and we call it uniformly recurrent if there is a lower bound on the frequency of such copies. The definitions of recurrence and uniform recurrence are equivalent with the usual topological definitions for subshifts.

Definition 8. *Let $\theta \in \mathcal{R}^{\mathbb{Z}}$ be a rule distribution. The distribution θ is (spatially)* recurrent *if for any finite $D \subseteq \mathbb{Z}$ there is $C \subseteq \mathbb{Z}$ such that $D \neq C$ and $\theta_{|C}$ is a copy of $\theta_{|D}$. The rule distribution θ is (spatially)* uniformly recurrent *if for any finite $D \subseteq \mathbb{Z}$ there is $n \in \mathbb{Z}_+$ such that for any $x \in \mathbb{Z}$, there is $C \subseteq [x, x+n]$ such that $\theta_{|C}$ is a copy of $\theta_{|D}$.*

3   Reversibility

Informally, we call a NUCA H reversible if there exists an inverse NUCA H^{-1}, such that $H \circ H^{-1} = H^{-1} \circ H$ is the identity function. In uniform CA reversibility is equivalent with bijectivity, which in turn is equivalent with injectivity.

Definition 9. *Let Σ be a state set. By* id $: \Sigma^{\mathbb{Z}} \to \Sigma^{\mathbb{Z}}$ *we denote the identity function.*

Definition 10. *Let $\mathcal{R}_1$ be a rule set and $\theta \in \mathcal{R}_1^{\mathbb{Z}}$. If there exists finite rule set $\mathcal{R}_2$ and $\phi \in \mathcal{R}_2^{\mathbb{Z}}$ such that $H_\phi = H_\theta^{-1}$, then the automaton H_θ is* reversible.

Obviously, a reversible NUCA is bijective. For uniform CA, the converse requires a non-trivial proof based on the compactness of the configuration space $\Sigma^{\mathbb{Z}}$. If we allow NUCA to have infinitely many rules, the converse is also simple. This follows from the fact that NUCA (allowing infinitely many rules) are exactly the continuous functions over the Cantor topology [2], and the fact that the inverse of a bijective continuous function is also continuous. Then the inverse of a bijective NUCA is a continuous function over the Cantor topology, thus also being a NUCA.

Since the usual proof requires translation invariance, it is natural to ask if this is true restricting ourselves to NUCA with finitely many rules only. The following is an example of a bijective NUCA which is not reversible.

Definition 11. *Let $\theta \in \mathcal{R}^{\mathbb{Z}}$ be a rule distribution of rules over state set Σ. Configurations $c, e \in \Sigma^{\mathbb{Z}}$ are* asymptotic, *if there are finitely many cells $x \in \mathbb{Z}$ such that $c(x) \neq c(e)$. The NUCA H_θ is* pre-injective, *if for any asymptotic $c, e \in \Sigma^{\mathbb{Z}}$, $c \neq e$, it holds that $H_\theta(c) \neq H_\theta(e)$.*

Lemma 1 (Garden of Eden theorem). *Let $\theta \in \mathcal{R}^{\mathbb{Z}}$ be a recurrent rule distribution. Then H_θ is surjective if and only if it is pre-injective [3].*

Example 1. We denote $a \oplus b = a + b \mod 2$. Let $\Sigma = \mathbb{Z}_2$ and $\mathcal{R} = \{f_L, f_R, g\}$ be a set of radius 1 rules over Σ, which map

$$f_L(a, b, c) = a \oplus b,$$
$$f_R(a, b, c) = b \oplus c,$$
$$g(a, b, c) = b,$$

where $a, b, c \in \Sigma$. For all $n \in \mathbb{Z}_+$, let $u_n = (f_R)^n g(f_L)^n$ and let $w_n = u_1 \ldots u_n$. Then, let $\theta = \ldots w_3 w_2 w_1 w_2 w_3 \ldots$ be the rule distribution.

The distribution θ is clearly recurrent. First, let's show that H_θ is injective. Let $c, e \in \Sigma^{\mathbb{Z}}$ be such that $c(x) \neq e(x)$ for some $x \in \mathbb{Z}$. If $\theta(x) = g$, then $H_\theta(c)(x) = c(x) \neq e(x) = H_\theta(e)(x)$. Suppose then $\theta(x) = f_R$. Then there is $k \in \mathbb{Z}_+$ such that $\theta(x + i) = f_R$ when $0 \leq i < k$ and $\theta(x + k) = g$.

Now suppose $H_\theta(c)(x+i) = H_\theta(e)(x+i)$ when $0 \le i < k$. If $c(x+j) \ne e(x+j)$ where $0 \le j < k$, then $c(x + j + 1) \ne e(x + j + 1)$ also, because otherwise it would hold that

$$H_\theta(c)(x + j) = c(x + j) \oplus c(x + j + 1)$$
$$= c(x + j) \oplus e(x + j + 1)$$
$$\ne e(x + j) \oplus e(x + j + 1) = H_\theta(e)(x + j)$$

Now inductively, $c(x + i) \ne e(x + i)$ when $0 \le i \le k$. Then because $c(x + k) \ne e(x + k)$ and $\theta(x + k) = g$, we have $H_\theta(c)(x + k) \ne H_\theta(e)(x + k)$. In any case, $H_\theta(c) \ne H_\theta(e)$.

The case that $\theta(x) = f_L$ is symmetrical. Therefore H_θ is injective. Because θ is recurrent, by the Lemma 1, H_θ is then also bijective. Next we show that H_θ is not reversible with a finite rule set.

Suppose there is a finite rule set $\mathcal{R}'$ and $\phi \in \mathcal{R}'^{\mathbb{Z}}$ such that $H_\phi = H_\theta^{-1}$. Then there is some $r \in \mathbb{Z}_+$ such that any rule in $\mathcal{R}'$ is at most radius r. Somewhere in θ, there is a length $2r + 1$ segment containing only the rule f_R; let $x \in \mathbb{Z}$ be the centermost cell of such a segment. Let then $c, e \in \Sigma^{\mathbb{Z}}$ be such that $c(y) = 0$ and $e(y) = 1$ for all $y \in \mathbb{Z}$. Then for all $z \in [x - r, x + r]$,

$$H_\theta(c)(z) = 0 \oplus 0 = 1 \oplus 1 = H_\theta(e)(z).$$

Suppose $\phi(x)(0, \ldots, 0) = a$ for some $a \in \Sigma$. Then

$$c(x) = (H_\phi \circ H_\theta)(c)(x) = a = (H_\phi \circ H_\theta)(e)(x) = e(x),$$

which is a contradiction. Therefore H_θ is not finitely reversible.

Note that the rule distribution in the previous counter-example is recurrent. A simpler, yet non-recurrent counter-example in the notation of the previous example would be distribution $(f_R)^\infty g(f_L)^\infty$. Next we show that if a bijective NUCA is defined by a uniformly recurrent rule distribution, then it is reversible. To do so, we use the notion of stable injectivity.

Definition 12. *Let $\theta \in \mathcal{R}^{\mathbb{Z}}$ be a rule distribution. The* orbit *of θ is the set $\mathcal{O}(\theta) = \{\sigma^k(\theta) \mid k \in \mathbb{Z}\}$. The* orbit closure $\overline{\mathcal{O}(\theta)}$ *is the topological closure of $\mathcal{O}(\theta)$. We call H_θ* stably injective *if for any $\phi \in \overline{\mathcal{O}(\theta)}$, the NUCA H_ϕ is injective [8].*

Lemma 2. *Let $\theta \in \mathcal{R}^{\mathbb{Z}}$ be a uniformly recurrent rule distribution such that H_θ is injective. Then H_θ is stably injective.*

Proof. Let $\phi \in \overline{\mathcal{O}(\theta)}$. Suppose H_ϕ is not injective. Then there are two cases: suppose first that H_ϕ is not pre-injective. Let then $c'', e'' \in \Sigma^{\mathbb{Z}}$ be asymptotic configurations such that $c'' \ne e''$ and $H_\phi(c'') = H_\phi(e'')$, and $x, y \in \mathbb{Z}$ such that for any cell $z \notin [x, y]$, $c''(z) = e''(z)$. Because $\phi \in \overline{\mathcal{O}(\theta)}$, any pattern in ϕ has a copy in θ. Denote $D = [x - r, y + r]$ and let $m \in \mathbb{Z}$ be such that $\theta_{|(D+m)}$ is a copy of $\phi_{|D}$. Then because for any cell $z \in (D + m)$, $\theta(z) = \phi(z - m)$, so

$$H_\theta(\sigma^{-m}(c''))(z) = H_\phi(c'')(z - m) = H_\phi(e'')(z - m) = H_\theta(\sigma^{-m}(e''))(z),$$

and for any cell $z \notin (D + m)$, $\sigma^{-m}(c'')_{|[z-r,z+r]} = \sigma^{-m}(e'')_{|[z-r,z+r]}$, we have $H_\theta(\sigma^{-m}(c'')) = H_\theta(\sigma^{-m}(e''))$. Then H_θ is not pre-injective, which is a contradiction.

Suppose then that H_ϕ is pre-injective. Then either every length $2r+1$ segment to the right of the origin 0 contains a cell where c' and e' differ, or the same holds for every $2r + 1$ segment to the left of the origin. Assume without loss of generality the first case, that is for every $k > 0$ there is $i \in [-r, r]$ such that $c'(k + i) \neq e'(k + i)$.

Because θ is uniformly recurrent, so is ϕ and $\theta \in \overline{\mathcal{O}(\phi)}$. Hence every finite segment of θ appears in ϕ in every segment of some length $n > 0$. Then there exists $k_1, k_2, k_3, \ldots > 0$ such that $\sigma^{k_1}(\phi), \sigma^{k_2}(\phi), \ldots$ converges to θ. Select then a subsequence $k_{i_1}, k_{i_2}, \ldots$ such that $c'^{k_{i_1}}, c'^{k_{i_2}}, \ldots$ and $e'^{k_{i_1}}, e'^{k_{i_2}}, \ldots$ converge to some c'' and e'' respectively. Then by the previous we have $c''_{|[-r,r]} \neq e''_{|[-r,r]}$, and by continuity of NUCA, $H_\theta(c'') = H_\theta(e'')$. Then H_θ is not injective, which is a contradiction. $\qquad\qquad\square$

Note that Lemma 2 doesn't hold in general, even in $\mathbb{Z}^2$. Since bijective, stably injective NUCA are reversible [8], we can prove the following theorem.

Theorem 1. *Let $\mathcal{R}$ be a finite rule set and $\theta \in \mathcal{R}^{\mathbb{Z}}$ a uniformly recurrent rule distribution such that H_θ is injective. Then H_θ is reversible.*

Proof. By Lemma 2, H_θ is stably injective. Because θ is recurrent and H_θ is injective, then H_θ is bijective [3]. Then there is a finite rule set $\mathcal{R}_2$ and $\phi \in \mathcal{R}_2^{\mathbb{Z}}$, such that $H_\phi \circ H_\theta = \mathrm{id}$ [8]. Now because H_θ is surjective, we have $H_\phi = H_\theta^{-1}$. $\qquad\square$

We can also show that the inverse automaton of a reversible NUCA with a uniformly recurrent rule distribution will also have a uniformly recurrent rule distribution, up to renaming rules.

Definition 13. *Let $f : \Sigma^{2r+1} \to \Sigma$ and $g : \Sigma^{2R+1} \to \Sigma$ be radius r and R local rules respectively for some $r, R \in \Sigma^{\mathbb{Z}}$. Assume $r \leq R$. The rules f and g are considered identical, if for any $a_{-R}, \ldots, a_R \in \Sigma$ it holds that*

$$g(a_{-R}, \ldots, a_R) = f(a_{-r}, \ldots, a_r).$$

Theorem 2. *Let $\mathcal{R}_1$ be a finite rule set and $\theta \in \mathcal{R}_1^{\mathbb{Z}}$ uniformly recurrent. Let $\mathcal{R}_2$ be a rule set containing no pair of identical rules, and $\phi \in \mathcal{R}_2^{\mathbb{Z}}$ a distribution such that $H_\phi = H_\theta^{-1}$. Then ϕ is uniformly recurrent.*

Proof. Let $r \in \mathbb{Z}_+$ be at least as large as the radius of any rule in $\mathcal{R}_1$ or $\mathcal{R}_2$. We show that for any $x, y \in \mathbb{Z}$, if $\theta_{|[x-r,x+r]} = \sigma^k(\theta_{|[y-r,y+r]})$, where $k = y - x$, then $\phi(x) = \phi(y)$. For all $c \in \Sigma^{\mathbb{Z}}$ we have $H_\theta(c)_{[x-r,x+r]} = H_{\sigma^k(\theta)}(c)_{[x-r,x+r]}$. Then applying the local rule $\phi(x)$, we have

$$H_\phi(H_\theta(c))(x) = H_\phi(H_{\sigma^k(\theta)}(c))(x).$$

Since $H_\phi \circ H_\theta = \mathrm{id} = H_{\sigma^k(\phi)} \circ H_{\sigma^k(\theta)}$, we then have

$$H_{\sigma^k(\phi)}(H_{\sigma^k(\theta)}(c))(x) = H_\phi(H_{\sigma^k(\theta)}(c))(x).$$

Because $H_{\sigma^k(\theta)}$ is surjective, for any $e \in \Sigma^{\mathbb{Z}}$ there is $c \in \Sigma^{\mathbb{Z}}$ such that $H_{\sigma^k(\theta)}(c) = e$. Therefore for any $e \in \Sigma^{\mathbb{Z}}$,

$$H_{\sigma^k(\phi)}(e)(x) = H_\phi(e)(x).$$

Since there are no identical rules in $\mathcal{R}_2$, $\sigma^k(\phi)(x) = \phi(x)$. Now because θ is uniformly recurrent, for any $m \in \mathbb{Z}_+$, and any segment $D \subseteq \mathbb{Z}$ of length $|D| = m + 2r$, there is $n \in \mathbb{Z}_+$ such that a copy of $\theta_{|D}$ appears in every length n segment of θ. Then by the previous, for every segment D' of length $|D'| = m$ there is $n \in \mathbb{Z}_+$ such that a copy of $\phi_{|D'}$ appears in every segment of length n. Hence ϕ is uniformly recurrent. $\qquad\square$

4   Balance

Balance is a property found in surjective CA, where every finite pattern in a given domain has the same number of pre-images. Balance is equivalent to invariance of the uniform measure. This makes it a particularly useful property, as it gives access to the Poincaré recurrence theorem. Due to the lack of translation invariance, surjective NUCA are not necessarily balanced, as shown by the following example.

Definition 14. *Let $\theta \in \mathcal{R}^{\mathbb{Z}^d}$ be a rule distribution with state set Σ. Let $D \subseteq \mathbb{Z}^d$ be a finite domain. The function $H_{\theta|D} : \Sigma^{N_\theta(D)} \to \Sigma^D$ is a* non-uniform CA over finite domain D*, which maps $c_{|N_\theta(D)}$, where $c \in \Sigma^{\mathbb{Z}}$, to*

$$H_{\theta|D}(c_{|N_\theta(D)}) = H_\theta(c)_{|D}.$$

Definition 15. *Let H_θ be a* NUCA*. We say H_θ is* balanced*, when for any finite domain $D \subset \mathbb{Z}$ and pattern $p \in \Sigma^D$, if $E = N_\theta(D)$ and $\mathcal{A} = \{q \in \Sigma^F \mid H_{\theta|D}(q) = p\}$, then $|\mathcal{A}| = |\Sigma|^{|E| - |D|}$.*

Example 2. We denote $a \oplus b = a + b \mod 2$. Let $\Sigma = \mathbb{Z}_2$ and $\mathcal{R} = \{f, g\}$ be a set of radius 1 rules, where

$$f(a, b, c) = a \oplus b \oplus c, \quad g(a, b, c) = \max(b, c),$$

Let $\theta \in \mathcal{R}^{\mathbb{Z}}$ be the rule distribution such that

$$\theta(x) = \begin{cases} g, & \text{if } x = 0, \\ f, & \text{if otherwise.} \end{cases}$$

First, let's show H_θ is surjective. Let $c \in \Sigma^{\mathbb{Z}}$ be any configuration. Then let $e \in \Sigma^{\mathbb{Z}}$ be a configuration such that

$$e(x) = \begin{cases} 0, & \text{if } x = -1 \text{ or } x = 1 \\ c(x), & \text{if } x = 0, \\ c(x+1) \oplus e(x+1) \oplus e(x+2), & \text{if } x < -1, \\ c(x-1) \oplus e(x-1) \oplus e(x-2), & \text{if } x > 1, \end{cases}$$

where $x \in \mathbb{Z}$. Now if $x \neq 0$ clearly $H_\theta(e)(x) = c(x)$, and if $x = 0$, $H_\theta(e)(x) = \max(c(x), 0) = c(x)$. Therefore $H_\theta(e) = c$, meaning H_θ is surjective.

On the other hand, let $D = \{0\}$ and $p' \in \Sigma^D$ be such that $p'(0) = 1$. Now the patterns $p \in \Sigma^{N_\theta(D)}$ such that $H_{\theta|D}(p) = p'$ are $001, 010, 011, 101, 110$ and 111, for a total of $t = 6$ patterns. Then clearly $t = 6 \neq 4 = 2^2 = |\Sigma|^{|N(D)|-|D|}$.

Note that the NUCA in the previous example is also pre-injective. Unlike in uniform CA, pre-injectivity and surjectivity aren't necessarily equivalent in NUCA [3].

It can be shown that a surjective NUCA of any dimension with a uniformly recurrent rule distribution has the balance property, with a proof very similar to the proof on uniform CA. In the one-dimensional case, it is enough for the rule distribution to be recurrent. Note that surjectivity and pre-injectivity are equivalent in NUCA with recurrent rule distributions [3].

Lemma 3. *Let $s, n, r \in \mathbb{Z}_+$. For all sufficiently large $m \in \mathbb{Z}_+$, it holds that*

$$(s^n - 1)^m s^{k-nm} < s^{k-2r},$$

for all $k \in \mathbb{Z}_+$.

Theorem 3. *Let $\theta \in \mathcal{R}^{\mathbb{Z}}$ be a recurrent rule distribution such that H_θ is surjective. Then H_θ is balanced.*

Proof. Let $D \subseteq \mathbb{Z}$ be a finite domain and $E = N_\theta(D)$. Without loss of generality, we can assume all rules in $\mathcal{R}$ have radius r. Suppose there is $p \in \Sigma^D$ such that the number of $H_{\theta|D}$ pre-images is $t \neq s^{|E|-|D|}$. We can select p such that $t < s^{|E|-|D|}$, because if every pattern in domain D were to have at least $s^{|E|-|D|}$ patterns, since there are only s^E patterns in domain E, every pattern would have exactly $s^{|E|-|D|}$ pre-images, contradicting our assumption.

Because θ is recurrent, it contains infinitely many copies of $\theta_{|D}$. Let $n \in \mathbb{Z}_+$ be such that $|D| = n - 2r$. Let then $m \in \mathbb{Z}_+$ be large enough that $(s^n - 1)^m s^{k-nm} < s^{k-2r}$ for any $k \in \mathbb{Z}_+$. By Lemma 3 such a number exists.

Let D' be a $k - 2r$ wide segment where $k \in \mathbb{Z}_+$ is large enough that a $\theta_{|D'}$ contains m disjoint copies of $\theta_{|D}$ that are at least $2r$ apart, and let $E' = N_\theta(D')$. Of course now $|E'| = k$. Let $\mathcal{B} \subseteq \Sigma^{D'}$ be the set of patterns such that there is a copy of p on each disjoint copy of $\theta_{|D}$. Because the rest of the cells may be freely chosen, $|\mathcal{B}| = s^{|D'|-m|D|} = s^{k-2r-m(n-2r)}$.

Let then $\mathcal{A} = \{q \in \Sigma^{E'} \mid H_{\theta|D'}(q) \in \mathcal{B}\}$. Then $|\mathcal{A}| = t^m \cdot s^{|E'|-m|E|} = t^m \cdot s^{k-mn}$, because on each n wide segment centered the copies of $\theta_{|D}$ must be a copy of one of the t pre-images of p, and the other cells may be chosen freely. Now it remains to show that $|\mathcal{A}| < |\mathcal{B}|$. Now because $t \leq s^{2r} - 1$, using the previous, we have

$$|\mathcal{A}| = t^m \cdot s^{k-mn} \leq (s^{2r} - 1)^m \cdot s^{k-mn} \leq (s^{2r} - s^{-(n-2r)})^m \cdot s^{k-mn}$$
$$= (s^n - 1)^m \cdot s^{k-mn} \cdot s^{-m(n-2r)} < s^{k-2r} \cdot s^{-m(n-2r)} = |\mathcal{B}|.$$

Hence H_θ is not surjective. Then if H_θ is surjective, the theorem statement holds. $\qquad\square$

Though surjectivity is not enough to guarantee balance, bijectivity is. In fact, it is enough to assume that every configuration has the same number of pre-images. Through a slightly modified proof, this can also be showed to hold in any dimension.

Theorem 4. *Let $\theta \in \mathcal{R}^{\mathbb{Z}}$ be a rule distribution such that H_θ is bijective. Then H_θ is balanced.*

Proof. Let $D \subseteq \mathbb{Z}$ be a finite segment and $p \in \Sigma^D$ a pattern. Let $E = N_\theta(D)$ and $\mathcal{A}_p = \{q \in \Sigma^E \mid H_{\theta|D}(q) = p\}$. Without loss of generality, we can assume D is a contiguous segment. Let $D' = \mathbb{Z} \setminus N_\theta(E)$ and $E' = \mathbb{Z} \setminus E$. Let $C = \mathbb{Z} \setminus (D \cup D')$. This is illustrated in Fig. 1.

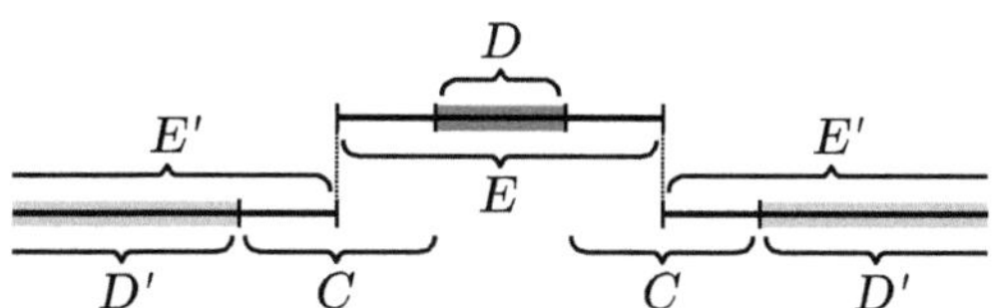

Fig. 1. Naming scheme of the domains used in the proof of Theorem 4.

Let $p' \in \Sigma^{D'}$ be a fixed (infinite) pattern and let $\mathcal{B}_{p'} = \{q' \in \Sigma^{E'} \mid H_{\theta|D'}(q') = p'\}$. Let $p \in \Sigma^D$ be a finite pattern. Then consider $\mathcal{C}_{p,p'} = \{c \in \Sigma^{\mathbb{Z}} \mid c_{|D} = p, c_{|D'} = p'\}$ and $\mathcal{C}'_{p,p'} = \{c' \in \Sigma^{\mathbb{Z}} \mid H_\theta(c') \in \mathcal{C}_{p,p'}\}$. Obviously $|\mathcal{C}_{p,p'}| = |\Sigma|^{|C|}$. On the other hand, we have $|\mathcal{C}'_{p,p'}| = |\mathcal{A}_p| \cdot |\mathcal{B}_{p'}|$. Finally, because H_θ is bijective, we have $|\mathcal{C}_{p,p'}| = |\mathcal{C}'_{p,p'}|$. Therefore

$$|\mathcal{A}_p| \cdot |\mathcal{B}_{p'}| = |\mathcal{C}'_{p,p'}| = |\mathcal{C}_{p,p'}| = |\Sigma|^{|C|},$$

which gives $|\mathcal{A}_p| = \frac{|\Sigma|^{|C|}}{|\mathcal{B}_{p'}|}$. Choice of p' and C was independent of the choice of p, and therefore this is true for any $p \in \Sigma^D$, and therefore every pattern in the domain D has the same number of pre-images. Hence we have $|\mathcal{A}_p| = |\Sigma|^{|E|-|D|}$. $\qquad\qquad\qquad\square$

5   Sensitivity and Equicontinuity

Sensitivity and equicontinuity are properties of general dynamical systems. Sensitivity in particular is a principal component of chaotic systems. To define them, some topological concepts are needed.

Definition 16. *Let Σ be a state set, $c \in \Sigma^{\mathbb{Z}}$ a configuration and $D \subseteq \Sigma$ a finite domain. We denote a* cylinder *as*

$$\mathrm{Cyl}(c, D) = \{e \in \Sigma^{\mathbb{Z}} \mid c_{|D} = e_{|D}\},$$

when Σ is clear from context.

It's well known that cylinders form a topological basis for the Cantor space [4], that is, every open set is a union of cylinders. Topologically, a set $A \subseteq \Sigma^{\mathbb{Z}}$ is dense if every cylinder contains an element of A, and residual if it is a countable intersection of dense open sets. Because the Cantor space is a Baire space, all residual sets are also dense.

The following definitions are equivalent with the topological definitions of equicontinuity and sensitivity. They can naturally be extended to any dimension, but here we focus on only 1-dimensional NUCA.

Definition 17. *Let H_θ be a NUCA with the state set Σ. The configuration $c \in \Sigma^{\mathbb{Z}}$ is an* equicontinuity point *if for every finite $E \subseteq \mathbb{Z}$ there is a finite $D \subseteq \mathbb{Z}^d$ such that*

$$\forall e \in \mathrm{Cyl}(c, D), n \in \mathbb{Z}_+ : H_\theta^n(e) \in \mathrm{Cyl}(H_\theta^n(c), E).$$

We denote the set of equicontinuity points as $\mathcal{E}$. We say H_θ is equicontinuous, *if $\mathcal{E} = \Sigma^{\mathbb{Z}}$. If $\mathcal{E}$ is residual, we say H_θ is* almost equicontinuous.

In uniform CA we find that only eventually temporally periodic CA are equicontinuous. However, there are (reversible) equicontinuous NUCA which have no periodic configurations [5]. Next we define sensitivity.

Definition 18. *Let H_θ be a NUCA with the state set Σ. The NUCA is called* sensitive *if there exists a finite $E \subseteq \mathbb{Z}$ such that for every $c \in \Sigma^{\mathbb{Z}}$ and every finite $D \subseteq \mathbb{Z}$,*

$$\exists e \in \mathrm{Cyl}(c, D), n \in \mathbb{Z}_+ : H_\theta^n(e) \notin \mathrm{Cyl}(H_\theta^n(c), E).$$

In 1-dimensional uniform CA, we find that either $\mathcal{E} = \emptyset$, which is equivalent with sensitivity, or $\mathcal{E}$ is residual. However, there are 2-dimensional uniform CA whose set of equicontinuity points is neither empty nor residual, as well as ones which have no equicontinuity points but are not sensitive [6].

In the 1-dimensional non-uniform case, the dichotomy found in uniform CA does not hold. In fact there exist both 1D NUCA whose set of equicontinuity points is neither empty nor residual, and 1D NUCA which have no equicontinuity points but are not sensitive.

Theorem 5. *There is a rule set $\mathcal{R}$ and rule distribution $\theta \in \mathcal{R}^{\mathbb{Z}}$ such that if $\mathcal{E}$ is the set of equicontinuity points of H_θ, then $\mathcal{E} \neq \emptyset$ and $\mathcal{E}$ is not residual.*

Proof. Let $\Sigma = \{0, 1\}$ and $\mathcal{R} = \{\text{id}, \tau\}$ where id is the identity rule and and τ is the leftward traffic rule, a radius 1 which maps

$$\tau(a, b, c) = \begin{cases} 0, & \text{if } a = 0, b = 1, \\ 1, & \text{if } b = 0, c = 1, \\ b, & \text{otherwise}, \end{cases}$$

for any $a, b, c \in \Sigma$. Let $\theta \in \mathcal{R}^{\mathbb{Z}}$ be such that $\theta(x) = \tau$ if $x > 0$ and $\theta(x) = \text{id}$ if $x \leq 0$. Let $\mathcal{E}$ be the set of equicontinuity points of H_θ. The operation of H_θ is illustrated in Fig. 2.

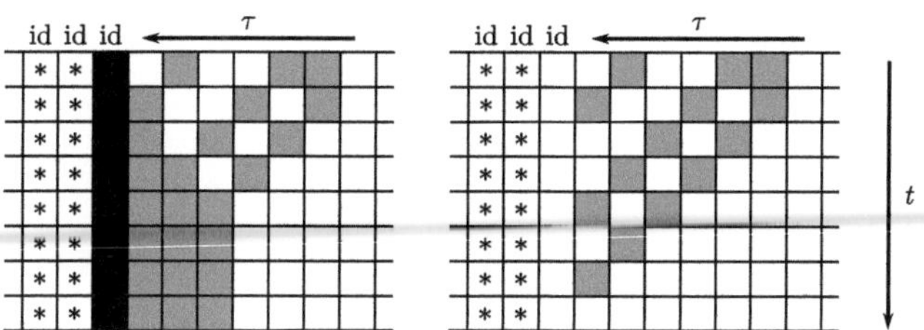

Fig. 2. Space-time diagram of the NUCA defined in the proof of Theorem 5. On the left, cell 0 is set to "1" and doesn't let traffic pass, whereas on the right it's set to "0" and annihilates traffic.

First to show that $\mathcal{E} \neq \emptyset$. Let $c_1 \in \Sigma^{\mathbb{Z}}$ be a configuration of all 1's. Because every cell in c_1 sees only 1's, clearly $H_\theta(c_1) = c_1$. Let $E \subseteq \mathbb{Z}$ be a finite segment and let $k \in \mathbb{Z}$ be the rightmost cell of E and let $D = E \cup [0, k]$. Let then $e \in \text{Cyl}(c_1, D)$. For all $x \leq 0$ obviously $H_\theta^n(e)(x) = e(x)$ for any $n \geq 1$. Then if $k \leq 0$, we have $H_\theta^n(e)(x) = e(x) = c(x) = H_\theta^n(c)(x)$ for all $x \in E$.

Suppose then $k > 0$. For any cell $x > 0$, if $e(x) = 1$, then $H_\theta(e)(x) = 0$ only if $e(x - 1) = 0$. However because for all $x \in [0, k]$, $e(x) = 1$, and by the previous $H_\theta(e)(0) = 1$, then also $H_\theta(e)(x) = 1$. Now we have $H_\theta(e) \in \text{Cyl}(c_1, D)$ and because e was arbitrary, then $H_\theta(\text{Cyl}(c_1, D)) \subseteq \text{Cyl}(c_1, D)$ and hence

$H_\theta^n(e) \in \mathrm{Cyl}(c_1, D) \subseteq \mathrm{Cyl}(c_1, E)$ for all $n > 0$. Therefore $c_1 \in \mathcal{E}$.

Now to show that $\mathcal{E}$ is not residual. Because residual sets in a Baire space are dense, it is enough to show that $\mathcal{E}$ is not dense. Let $c \in \Sigma^{\mathbb{Z}}$ be any configuration such that $c(0) = 0$. We show that $c \notin \mathcal{E}$. Let $E = \{1\}$ and $D \subseteq \mathbb{Z}$ be any finite domain. Without loss of generality we can assume $1 \in D$. Informally, the idea is that the leftward traffic is annihilated at cell 0 and traffic from arbitrarily far to the left can then arrive at E.

Let $e_0, e_1 \in \mathrm{Cyl}(c, D)$ be such that for all $x \in \mathbb{Z} \setminus D$, $e_0(x) = 0$ and $e_1(x) = 1$. Firstly, there are finitely many $x > 0$ such that $e_0(x) = 1$. Also the number of such cells can only decrease, since under the rule τ, when a "0" cell turns into a "1" cell, another "1" cell turns into a "0" cell. Then for any $t \in \mathbb{Z}_+$ such that $H_\theta^t(e_0)(1) = 1$, the number of cells $x > 0$ such that $H_\theta^{t+1}(e_0)(x) = 1$ is greater than the number of such cells with $H_\theta^t(e_0)(x) = 1$. On the other hand if for some $k > 0$ and $H_\theta^t(e_0)(i) = 0$ for all $i \in [1, k]$ and $H_\theta^t(e_0)(i) = 1$ and $H_\theta^t(e_0)(k+1) = 1$, then $H_\theta^{t+k}(e_0)(1) = 1$. Hence there is some $n \geq 0$ such that for all $t \geq n$, $H_\theta^t(e_0)(1) = 0$, because the number of positive "1" cells decreases to 0.

On the other hand, there are infinitely many $t \geq 0$ such that $H_\theta^t(e_1)(1) = 1$, because there are infinitely many "1" cells to the right of 1. Now there must be some $t \geq 0$ such that $H_\theta^t(e_0)(1) \neq H_\theta^t(e_1)(1)$. Therefore either $H_\theta^t(e_0) \notin \mathrm{Cyl}(c, E)$ or $H_\theta^t(e_1) \notin \mathrm{Cyl}(c, E)$. In either case, because D was arbitrary, this means c is not an equicontinuity point. Now since no configuration with "0" at cell 0 is an equicontinuity point, $\mathcal{E}$ is not dense, and therefore not residual.

□

Theorem 6. *There is a rule set $\mathcal{R}$ and rule distribution $\theta \in \mathbb{Z}$ such that H_θ is not sensitive, and if $\mathcal{E}$ is the set of equicontinuity points of H_θ, then $\mathcal{E} = \emptyset$.*

Proof. Let $\Sigma = \{0, 1, 2, 3\}$ and $\mathcal{R} = \{\mathrm{id}, f\}$ a rule set, where id is the identity rule, and $f : \Sigma^3 \to \Sigma$ is a radius-1 rule that maps

$$(a, 0, 3) \mapsto 3, (1, 1, 1) \mapsto 1, (1, 1, 2) \mapsto 1, (a, 1, 3) \mapsto 3,$$
$$(1, 2, c) \mapsto 2, (a, 3, 3) \mapsto 3, \text{ and } (a, b, c) \mapsto 0 \text{ otherwise,}$$

where $a, b, c \in \Sigma$. Let $\theta \in \mathcal{R}^{\mathbb{Z}}$ be such that $\theta(x) = \mathrm{id}$ if $x \leq 0$ and $\theta(x) = f$ if $x > 0$. The operation of this NUCA is illustrated in Fig. 3.

First to show that H_θ is not sensitive: let $E \subseteq [i, j] \subseteq \mathbb{Z}$ be an arbitrary finite domain. Let $c \in \Sigma^{\mathbb{Z}}$ be such that $c(j) = 2$ and $c(x) = 1$ if $x \neq j$, and let then $D = [\min(i, 0), j]$. Let $e \in \mathrm{Cyl}(c, D)$. If $x \in [i, 0]$, then obviously $H_\theta(e)(x) = c(x)$, because $\theta(x) = \mathrm{id}$. For every $x \in [1, j-1]$, $H_\theta(e)(x) = f(1, 1, a) = 1 = c(x)$, where $a \in \{1, 2\}$ and $H_\theta(e)(j) = f(1, 2, b) = 2 = c(j)$, where $b \in \Sigma$. Then $H_\theta(e) \in \mathrm{Cyl}(c, D)$, and because e was an arbitrary element of $\mathrm{Cyl}(c, D)$, then $H_\theta^n(e) \in \mathrm{Cyl}(c, D)$ for all $n > 0$. Since also $c \in \mathrm{Cyl}(c, D)$, and $E \subseteq D$, we have $H_\theta^n(e) \in \mathrm{Cyl}(H_\theta^n(c), D) \subseteq \mathrm{Cyl}(H_\theta^n(c), E)$ for all $n > 0$. Therefore H_θ is not sensitive.

id	id	id	f	f	f	f	f	f	f	f
*	*	1	1	1	2	0	3	0	0	3
*	*	1	1	1	2	3	0	0	3	0
*	*	1	1	1	2	0	0	3	0	0
*	*	1	1	1	2	0	3	0	0	0
*	*	1	1	1	2	3	0	0	0	0
*	*	1	1	1	2	0	0	0	0	0
*	*	1	1	1	2	0	0	0	0	0

f	f	f	f	f	f	f	f	f	f	f
1	2	1	1	1	2	0	3	0	0	3
1	2	0	1	1	2	3	0	0	3	0
1	2	0	0	1	2	0	0	3	0	0
1	2	0	0	0	2	0	3	0	0	0
1	2	0	0	0	0	3	0	0	0	0
1	2	0	0	0	3	0	0	0	0	0
1	2	0	0	3	0	0	0	0	0	0

(right diagram annotated with t pointing downward)

Fig. 3. Space-time diagrams of the NUCA defined in the proof of Theorem 6. The left diagram illustrates how a segment arbitrarily far from the origin can be "protected" from 3-cells with a blocking pattern, and the right diagram illustrates why said blocking pattern can't be copied to protect every cell from 3-cells.

Then to show that $\mathcal{E} = \emptyset$. First we show that if a "2"-cell (on the right side) has any non-"1"-cell between it and the origin, it will eventually become a "0"-cell. Let $c \in \Sigma^{\mathbb{Z}}$ be a configuration with $x_1, x_2 \in \mathbb{Z}$, $0 \leq x_1 < x_2$ such that $c(x_1) \neq 1$, $c(x_2) = 2$ and $c(x) = 1$ when $x_1 < x < x_2$. Suppose for some $n \in [0, x_2 - x_1 - 2]$ we have

$$H_\theta^n(c)(x_1 + n) \neq 1, \; H_\theta^n(c)(x_2) = 2, \; H_\theta^n(c)(x) = 1,$$

if $x_1 + n < x < x_2$. Then under the local rule f,

$$H_\theta^{n+1}(c)(x_1 + n + 1) = 0, \; H_\theta^{x+1}(c)(x_2) = 2, \; H_\theta^{n+1}(c)(x) = 1,$$

if $x_1 + n + 1 < x < x_2$. Then by induction, $H_\theta^{x_2 - x_1 - 1}(c)(x_2 - 1) = 0$ and $H_\theta^{x_2 - x_1 - 1}(c)(x_2) = 2$, and hence $H_\theta^{x_2 - x_1}(c)(x_2) = 0$.

Let now $c \in \Sigma^{\mathbb{Z}}$. We show that $c \notin \mathcal{E}$. Let $x \geq 1$ be the smallest integer such that $c(x - 1) = 2$. If no such cell exists, we let $x = 0$. Let $E = \{x\}$, and let $D = [y, z]$ with any $y, z \in \mathbb{Z}$, $y \leq z$. Without loss of generality, we can assume $x \in D$. Let $e_1, e_2 \in \mathrm{Cyl}(c, D)$ be such that if $k \notin D$, then $e_1(k) = 0$ and $e_2(k) = 3$.

By the previous, because no non-"2"-cell can become a "2"-cell, there is $n \geq 0$ such that for all $k \in [x, z]$, $H_\theta^n(e_1)(k) \neq 2$ and $H_\theta^n(e_2)(k) \neq 2$. Then there are only finitely many "3"-cells to the right of x in $H_\theta^n(e_1)$, so for some $n_1 \geq n$, for all $t \geq n_1$ we have $H_\theta^t(e_1)(x) \neq 3$. On the other hand, because there are infinitely many "3"-cells to the right of x in $H_\theta^n(e_2)$, for some $n_2 \geq n$, for infinitely many $t > n_2$, $H_\theta^t(e_2)(x) = 3$. Then there is some $t > 0$ such that $H_\theta^t(e_1)(x) \neq 3 = H_\theta^t(e_2)(x)$, and therefore either $H_\theta^t(e_1) \notin \mathrm{Cyl}(H_\theta^t(c), E)$ or $H_\theta^t(e_2) \notin \mathrm{Cyl}(H_\theta^t(c), E)$. Since this is true for any finite segment D, $c \notin \mathcal{E}$, and therefore $\mathcal{E} = \emptyset$. $\square$

It is still open whether there exist counter-examples that have uniformly recurrent rule distributions.

6   Expansivity

In CA, expansivity is a much stronger notion of sensitivity. A (NU)CA is expansive if there's a fixed window where any difference between configurations propagates.

Definition 19. *Let $\theta \in \mathcal{R}^{\mathbb{Z}}$ be a rule distribution. The* NUCA H_θ *is* positively expansive *if there's some finite domain $E \subseteq \mathbb{Z}$ such that for any $c, e \in \Sigma^{\mathbb{Z}}$, $c \neq e$, there is $n \in \mathbb{Z}_+$ such that $H_\theta^n(e) \notin \mathrm{Cyl}(H_\theta^n(c), E)$.*

The NUCA H_θ *is* expansive *if it's bijective and there's a finite domain $E \subseteq \mathbb{Z}$ such that for any $c, e \in \Sigma^{\mathbb{Z}}$, $c \neq e$, there is $n \in \mathbb{Z}$ such that $H_\theta^n(e) \notin \mathrm{Cyl}(H_\theta^n(c), E)$.*

In uniform CA, the usual proof that expansivite CA are sensitive utilizes the denseness of periodic points. In bijective NUCA, periodic points are not necessarily dense [5], so we use the Poincaré recurrence theorem instead. Let's first recall some definitions from ergodic theory.

Definition 20. *Let Σ be a state set. A Borel measure $\mu : \mathcal{B} \to \mathbb{R}$, where $\mathcal{B}$ is a Borel algebra over $\Sigma^{\mathbb{Z}}$, is* uniform *if it maps*

$$\mathrm{Cyl}(c, D) \mapsto \left(\frac{1}{|\Sigma|} \right)^{|D|} ,$$

where $c \in \Sigma^{\mathbb{Z}}$ and $D \subseteq \mathbb{Z}$ is a finite domain.

Let $H_\theta : \Sigma^{\mathbb{Z}} \to \Sigma^{\mathbb{Z}}$ be a NUCA. *We denote $H_\theta(\mu) : \mathcal{C} \to \mathbb{R}$ as the measure that gives $H_\theta(\mu)(A) = \mu(H_\theta^{-1}(A))$ for any $A \in \mathcal{B}$. We say that measure μ is* invariant *under H_θ if $H_\theta(\mu) = \mu$.*

It's well known that the value of a measure on cylinders uniquely determines it, that is, we need to only define a measure on cylinders [7]. The following proof is essentially identical to the proof on uniform automata.

Lemma 4. *Let $H_\theta : \Sigma^{\mathbb{Z}} \to \Sigma^{\mathbb{Z}}$ be a bijective* NUCA. *The uniform probability measure μ is invariant under H_θ.*

Proof. Let $C = \mathrm{Cyl}(c, D)$ for some finite domain $D \subseteq \mathbb{Z}$ and $c \in \Sigma^{\mathbb{Z}}$. By Theorem 4, $H_\theta^{-1}(C) = C_1 \cup C_2 \cup \ldots \cup C_n$, where C_i are disjoint cylinders with some finite domain $E \subseteq \mathbb{Z}$, and $n = |\Sigma|^{|E|-|D|}$. Then due to the additivity of measures [7], we have

$$\mu(H_\theta^{-1}(C)) = \mu(C_1) + \mu(C_2) + \ldots + \mu(C_n) = n \left(\frac{1}{|\Sigma|} \right)^{|E|} = \left(\frac{1}{|\Sigma|} \right)^{|D|} = \mu(C).$$

Therefore $H_\theta(\mu) = \mu$. $\qquad\qquad\square$

It is well know that if a dynamical system has an invariant measure of full support, temporally recurrent points in it are dense. Therefore we state the following lemma without proof [7].

Definition 21. *Let $H_\theta : \Sigma^{\mathbb{Z}} \to \Sigma^{\mathbb{Z}}$ be a* NUCA. *A configuration $c \in \Sigma^{\mathbb{Z}}$ is* temporally recurrent *if for any $D \subseteq \mathbb{Z}$ there exists $n \in \mathbb{Z}_+$ such that $H_\theta^n(c) \in \mathrm{Cyl}(c, D)$.*

Lemma 5 (Poincaré). *Let $H_\theta : \Sigma^{\mathbb{Z}} \to \Sigma^{\mathbb{Z}}$ be a bijective NUCA. Then temporally recurrent configurations of H_θ are dense, that is, for any cylinder C there is a configuration $c \in C$ that is temporally recurrent under H_θ.*

Now we are ready to show that expansive NUCA are indeed sensitive.

Theorem 7. *Let $\theta \in \mathcal{R}^{\mathbb{Z}}$ be a rule distribution such that H_θ is expansive or positively expansive. Then H_θ is sensitive.*

Proof. If H_θ is positively expansive, it is clearly sensitive by definition. Suppose then that (bijective) H_θ is expansive. Let $E \subseteq \mathbb{Z}$ be the observation window confirming expansivity, that is, a finite domain such that for any $c_1, c_2 \in \Sigma^{\mathbb{Z}}$, $c_1 \neq c_2$, there is $n \in \mathbb{Z}$ such that $H_\theta^n(c_2) \notin \mathrm{Cyl}(H_\theta^n(c_1), E)$. Let $D \subseteq \mathbb{Z}$ be a finite domain and $c_1 \in \Sigma^{\mathbb{Z}}$ a configuration. It is enough to show that for some $c_2 \in \mathrm{Cyl}(c_1, D)$, $m \in \mathbb{Z}_+$ and $x \in E$, we have $H_\theta^m(c_1)(x) \neq H_\theta^m(c_2)(x)$.

By expansivity there is $n \in \mathbb{Z}$ such that $H_\theta^n(c_1)(x) \neq H_\theta^n(c_2)(x)$ for some $x \in E$. If $n \geq 0$ the proof is complete, so assume $n < 0$. Consider the space $(\Sigma \times \Sigma)^{\mathbb{Z}}$ and NUCA $H_\phi : (\Sigma \times \Sigma)^{\mathbb{Z}} \to (\Sigma \times \Sigma)^{\mathbb{Z}}$ such that for $e \in (\Sigma \times \Sigma)^{\mathbb{Z}}$ and $e_1, e_2 \in \Sigma^{\mathbb{Z}}$ where $e(y) = (e_1(y), e_2(y))$ for all $y \in \mathbb{Z}$, H_ϕ maps e such that

$$H_\phi(e)(y) = (H_\theta(e_1)(y), H_\theta(e_2)(y)),$$

for all $y \in \mathbb{Z}$. Obviously such e_1, e_2 exist for any e, and H_ϕ is a NUCA.

Because H_θ is bijective, clearly H_ϕ is too. Then by Lemma 5, temporally recurrent configurations of H_ϕ are dense. Then there's a temporally recurrent $e \in (\Sigma \times \Sigma)^{\mathbb{Z}}$ with $e_1, e_2 \in \mathrm{Cyl}(c, D)$ such that $e(y) = (e_1(y), e_2(y))$ for all $y \in \mathbb{Z}$ and $H_\theta^n(e_1)(x) = H_\theta^n(c_1)(x)$, $H_\theta^n(e_2)(x) = H_\theta^n(c_2)(x)$ by the continuity of H_θ. Because e is temporally recurrent, there is some $m > 0$ such that $H_\theta^m(e_1)(x) = H_\theta^n(e_1)(x)$ and $H_\theta^m(e_2)(x) = H_\theta^n(e_2)(x)$. Then because

$$H_\theta^m(e_1)(x) = H_\theta^n(c_1)(x) \neq H_\theta^n(c_2)(x) = H_\theta^m(e_2)(x),$$

either $H_\theta^m(e_1)(x) \neq H_\theta^m(c_1)(x)$ or $H_\theta^m(e_2)(x) \neq H_\theta^m(c_1)(x)$. Therefore H_θ is sensitive. $\square$

7   Conclusions

In summary, we've shown that a bijective NUCA with a uniformly recurrent rule distribution is reversible and that this is not true of bijective NUCA in general, even with recurrent rule distributions. We've shown that a bijective NUCA is balanced, and that a surjective NUCA with a recurrent rule distribution is balanced, once again not true of all surjective NUCA.

We've also presented counter-examples for the usual dichotomy of sensitivity and equicontinuity in 1D CA, that is, a 1D NUCA which is not sensitive and has no equicontinuity points, and one which has a non-residual and non-empty set of equicontinuity points. The question of whether there are such counter-examples with (uniformly) recurrent rule distributions remains open. Finally, we've shown that (positively) expansive NUCA are sensitive.

Acknowledgements. We would like to acknowledge funding from the Magnus Ehrnrooth foundation and the Research Council of Finland, grant 354965. This work was partially supported by the HORIZON-MSCA-2022-SE-01 project 101131549 "Application-driven Challenges for Automata Networks and Complex Systems (ACAN-COS)".

References

1. Cecchirini-Silberstein, T., Coornaert, M.: Cellular Automata and Groups. Springer, Berlin, Heidelberg (2010)
2. Dennunzio, A., Formenti, E., Provillard, J.: Non-uniform cellular automata: classes, dynamics, and decidability. Inf. Comput. **215**, 32–46 (2012)
3. Paturi, P.: On the Surjunctivity and the Garden of Eden theorem for non-uniform cellular automata. In: Manzoni, L., Mariot, L., Roy Chowdhury, D. (eds) Cellular Automata and Discrete Complex Systems. AUTOMATA 2023. Lecture Notes in Computer Science, vol. 14152. Springer, Cham (2023). https://doi.org/10.1007/978-3-031-42250-8_2
4. Kari, K.: Theory of cellular automata: a survey. Theoret. Comput. Sci. **334**(1–3), 3–33 (2005)
5. Kamilya, S., Kari, J.: Nilpotency and periodic points in non-uniform cellular automata. Acta Informatica **58**(4), 319–333 (2020). https://doi.org/10.1007/s00236-020-00390-7
6. Sablik, M., Theyssier, G.: Topological dynamics of cellular automata: dimension matters. Theory Comput. Syst. **48**, 693–714 (2011). https://doi.org/10.1007/s00224-010-9255-x
7. Choe, G.: Computational Ergodic Theory, 1st edn. Springer, Berlin, Heidelberg (2005)
8. Phung, X.: On invertible and stably reversible non-uniform cellular automata. Theoretical Comput. Sci. **940**, Part B, 43–59 (2023)

Simulating Virtual Players for UNO Without Computers

Suthee Ruangwises[1]([✉]) [iD] and Kazumasa Shinagawa[2,3] [iD]

[1] Department of Computer Engineering, Faculty of Engineering, Chulalongkorn University, Bangkok, Thailand
suthee@cp.eng.chula.ac.th
[2] University of Tsukuba, Tsukuba, Japan
shinagawa@cs.tsukuba.ac.jp
[3] National Institute of Advanced Industrial Science and Technology, Tokyo, Japan

Abstract. UNO is a popular multiplayer card game. In each turn, a player has to play a card in their hand having the same number or color as the most recently played card. When having few people, adding virtual players to play the game can easily be done in UNO video games. However, this is a challenging task for physical UNO without computers. In this paper, we propose an unconventional protocol that can simulate virtual players using nothing but physical UNO cards. In particular, our protocol can uniformly select a valid card to play from each virtual player's hand at random, or report that none exists, without revealing the rest of its hand. The protocol can also be applied to simulate virtual players in other turn-based card or tile games where each player has to select a valid card or tile to play in each turn.

Keywords: card-based cryptography · UNO · card game · simulation · randomization

1 Introduction

UNO is a multiplayer card game invented by Merle Robbins in 1972, and is now commercially produced by a game company Mattel. It is one of the world's most popular card games.

UNO consists of a deck of 108 cards. Except for a few special cards, an UNO card has one of the four colors: red, yellow, green, and blue, and has a single-digit number between 0 and 9 written on the front side. All players have several cards in their hands and take turn to *play* a card, discarding it from the hand face-up.

In each turn, a player plays a card with the same number or color as the most recently played card (with the exception of some special cards). If the player does not have a valid card or does not wish to play, they have to draw a card from the deck and may play that card if it is valid. The first player who plays all cards in their hand wins the game.

© The Author(s), under exclusive license to Springer Nature Switzerland AG 2026
E. Formenti and L. Manzoni (Eds.): UCNC 2025, LNCS 16364, pp. 33–46, 2026.
https://doi.org/10.1007/978-3-032-15641-9_3

UNO can be played by as few as two players, but it is typically played by a large group of players. When having few people, a possible way to make the game more fun is to add virtual players to the game. Most video game versions of UNO provide an option to add bots to the game. However, simulating virtual players is a challenging task for physical UNO without computers.

We consider the most basic model of a virtual player: a randomized one. In each virtual player's turn, it should always play a card if possible. It plays a random card uniformly chosen from all valid cards in its hand, without revealing the rest of the hand. If there is no valid card to play, it should correctly report that without revealing any card in the hand. The main difficulty is how to operate virtual players so that everyone can verify that their moves are correct, while always keeping their hands secret to all players.

1.1   Related Work

Card-based cryptography. is a research area studying unconventional cryptographic protocols that use a deck of physical cards. There are two main lines of research in this area.

Secure Multi-Party Computation: This area involves protocols that can securely compute functions with private input from multiple parties, without revealing the inputs to other parties. Card-based protocols to compute various Boolean functions, including a logical AND function [5,7], a logical XOR function [7], a *majority function* [9,19], and an *equality function* [13], have been developed. Nishida et al. [8] proved that any n-variable Boolean function can be computed using $2n + 6$ cards. Shinagawa and Nuida [16] proved that any Boolean function can be computed using one shuffle.

Zero-Knowledge Proof: A zero-knowledge proof is an interactive protocol between a prover and a verifier, which allows the prover to show that they know a solution of a specific problem without revealing the solution itself [1]. Card-based zero-knowledge proof protocols for a wide range of problems have been developed, including computational problems such as graph isomorphism [4] and pancake sorting [3], pencil puzzles such as Sudoku [10,14,18] and Nonogram [11], and mobile games such as Ball Sort Puzzle [12].

Very recently, Shinagawa et al. [15] developed a card-based player simulation protocol for a card game Old Maid, which can simulate virtual players to play the game with real human players. In particular, their protocol can remove pairs of cards having the same number from each virtual player's hand without revealing the rest of its hand. This result created a new possible line of research: simulating virtual players in card games.

1.2   Our Contribution

In this paper, we propose a card-based protocol that can simulate virtual players to play UNO. In particular, our protocol can uniformly select a valid card (according to UNO's rules) to play from each virtual player's hand at random, or report that none exists, without revealing the rest of its hand. Our protocol is the second card-based player simulation protocol after the Old Maid protocol of Shinagawa et al. [15].

While Old Maid is purely based on luck, UNO is a strategic game. Knowing hands of virtual players will greaty affect the other players' strategies. Therefore, keeping the hands of virtual players secret is much more crucial in UNO than in Old Maid, highlighting the importance of security of our protocol.

Our protocol is a generic one. We show that it can also be applied to simulate virtual players in other turn-based card or tile games such as *Sevens*, *Hearts*, and *Dominoes*, where each player has to select a valid card or tile to play in each turn.

2　Preliminaries

2.1　Cards

There are 54 different types of UNO card. 52 of them have color red, yellow, green, or blue on the front side and are called *color cards*. Two of them are black on the front side and are called *black cards*. All cards have indistinguishable back sides.

A color card may have a single-digit number between 0 and 9, a skip sign, a reverse sign, or a +2 sign written on the front side. All 13 distinct characters can be combined with all four colors, resulting in the total of 52 different types. A black card may have nothing written on the front side (called a *wild card*) or a +4 sign written on the front side (called a *+4 card*). See Fig. 1 for examples of UNO cards.

We use $\boxed{a^b}$ to denote a color card, where a is either a single-digit number, an 'S' for a skip sign, an 'R' for a reverse sign, or a '+' for a +2 sign, and b is an initial of a color (R for red, Y for yellow, G for green, and B for blue). We also use $\boxed{W}$ and $\boxed{D}$ to denote a wild card and a +4 card, respectively.

Fig. 1. From left to right: $\boxed{7^R}$, $\boxed{S^Y}$, $\boxed{R^G}$, $\boxed{+^B}$, $\boxed{W}$, and $\boxed{D}$.(Color figure online)

An UNO deck consists of two copies of each type of color card (except for cards with a number 0 with only one copy), and four copies of each type of black card, thus consisting of 108 cards.

Note that our protocol uses additional cards besides the ones in the currently played deck, so several UNO decks have to be prepared beforehand.

2.2 Pile-Scramble Shuffle

Given mk cards divided into m piles, each with k cards, a *pile-scramble shuffle for m piles* [2] rearranges all piles by a uniformly random permutation unknown to all parties. It can be implemented in real world by putting all cards in each pile into an envelope, and scrambling all envelopes together completely randomly on a table.

The pile-scramble shuffle is denoted by $[p_1|p_2|...|p_m]$, where $p_1, p_2, ..., p_m$ are the m piles. For example, a pile-scramble shuffle for two piles:

$$\left[\begin{array}{ccc|ccc} 1 & 2 & 3 & 4 & 5 & 6 \\ ? & ? & ? & ? & ? & ? \end{array} \right]$$

results in either $\begin{array}{cccccc} 1 & 2 & 3 & 4 & 5 & 6 \\ ? & ? & ? & ? & ? & ? \end{array}$ or $\begin{array}{cccccc} 4 & 5 & 6 & 1 & 2 & 3 \\ ? & ? & ? & ? & ? & ? \end{array}$, each with probability 1/2.

2.3 Encoding a Boolean Bit

Let $\boxed{\alpha}$ and $\boxed{\beta}$ denote any two different pre-designated types of UNO card (e.g. one may designate a $\boxed{1^R}$ as $\boxed{\alpha}$ and a $\boxed{2^G}$ as $\boxed{\beta}$). A Boolean bit is encoded by a sequence of two cards:

$$\boxed{\alpha}\,\boxed{\beta} = 0, \qquad \boxed{\beta}\,\boxed{\alpha} = 1.$$

2.4 Six-Card AND Protocol

Given two pairs of face-down $\boxed{\alpha}$ and $\boxed{\beta}$ encoding $x, y \in \{0, 1\}$ and two additional cards $\boxed{\alpha}\,\boxed{\beta}$, a *six-card AND protocol* outputs two pairs of face-down cards encoding $x \wedge y$ and $\bar{x} \wedge y$.

$$\underbrace{?\,?}_{x}\;\underbrace{?\,?}_{y}\;\alpha\,\beta \quad\xrightarrow{\text{protocol}}\quad \underbrace{?\,?}_{x\wedge y}\;\underbrace{?\,?}_{\bar{x}\wedge y}\;\alpha\,\beta$$

This protocol was developed by Mizuki and Sone [7]. The procedures are as follows.

1. Place the two pairs of cards encoding x and y, and two additional cards $\boxed{\alpha}\,\boxed{\beta}$ encoding 0 as follows.

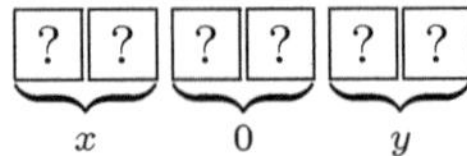

$$\underbrace{?\,?}_{x}\;\underbrace{?\,?}_{0}\;\underbrace{?\,?}_{y}$$

2. Rearrange the sequence as follows.

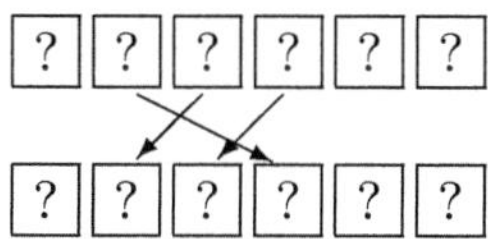

3. Apply the pile-scramble shuffle for two piles.

$$\left[\,\boxed{?}\,\boxed{?}\,\boxed{?}\,\Big\|\,\boxed{?}\,\boxed{?}\,\boxed{?}\,\right] \rightarrow \boxed{?}\,\boxed{?}\,\boxed{?}\,\boxed{?}\,\boxed{?}\,\boxed{?}$$

4. Rearrange the sequence as follows.

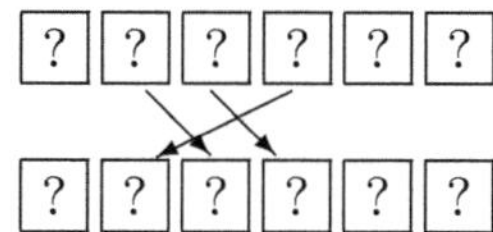

5. Turn over the two leftmost cards. If they are $\boxed{\alpha}\,\boxed{\beta}$, the middle and right pairs encode $x \wedge y$ and $\bar{x} \wedge y$, respectively. If they are $\boxed{\beta}\,\boxed{\alpha}$, the middle and right pairs encode $\bar{x} \wedge y$ and $x \wedge y$, respectively.

(i) $\boxed{\alpha}\,\boxed{\beta}\,\underbrace{\boxed{?}\,\boxed{?}}_{x \wedge y}\,\underbrace{\boxed{?}\,\boxed{?}}_{\bar{x} \wedge y}$ (ii) $\boxed{\beta}\,\boxed{\alpha}\,\underbrace{\boxed{?}\,\boxed{?}}_{\bar{x} \wedge y}\,\underbrace{\boxed{?}\,\boxed{?}}_{x \wedge y}$

2.5 Modified Covert Lottery Protocol

We slightly modify the *covert lottery protocol* developed by Shinoda et al. [17].

In the modified protocol, m face-down arbitrary cards $c_1, c_2, ..., c_m$ and m pairs of face-down $\boxed{\alpha}$ and $\boxed{\beta}$ encoding $x_1, x_2, ..., x_m \in \{0, 1\}$ are given, where each x_i denotes whether c_i is a valid card that can be selected. If there is at least one valid card, this protocol uniformly selects one of them at random (i.e. if there are k valid cards, each one will have probability $1/k$ to be selected) without revealing any card or the number of valid cards. If there is no valid card, i.e. $c_i = 0$ for every i, this protocol reports that none is valid. The protocol uses four additional cards $\boxed{\alpha}\,\boxed{\alpha}\,\boxed{\beta}\,\boxed{\beta}$.

The only difference from the original protocol is that in the original one, if there is no valid card, the protocol nevertheless uniformly selects one of all m cards at random.

The procedure of this protocol are as follows.

1. Place each pair of cards encoding x_i below the card c_i, making m piles of three cards.

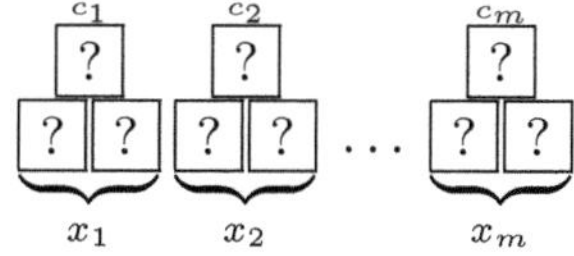

2. Apply the pile-scramble shuffle for m piles.

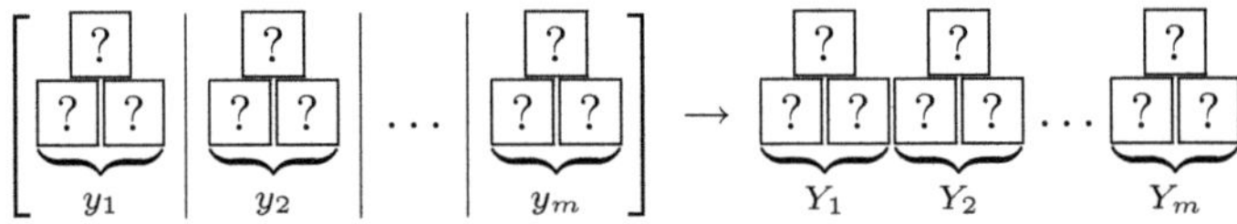

Let $(X_1, X_2, ..., X_m)$ denote the permutation of $(x_1, x_2, ..., x_m)$ after the shuffle.

3. Place two additional cards $\boxed{\beta}\,\boxed{\alpha}$ to encode a bit $t = 1$. This pair of cards is called a *token*.

4. For each $i = 1, 2, ..., m$, perform the following steps.

   (a) Apply the six-card AND protocol in Sect. 2.4 with inputs X_i and t (using two additional cards $\boxed{\alpha}\,\boxed{\beta}$) to create $X_i \wedge t$ and $\bar{X}_i \wedge t$.

   (b) Place a pair of cards encoding $y_i = X_i \wedge t$ in the pile previously containing X_i. Let the other pair of cards encoding $\bar{X}_i \wedge t$ be the new token t to use in the next iteration.

5. After m iteration, we now have m piles of three cards, with the bottom pair of each pile encoding $y_1, y_2, ..., y_m$, where $y_i = x_i \wedge x_{i-1}^- \wedge x_{i-2}^- \wedge ... \wedge \bar{x}_1$ for every i.

6. Apply the pile-scramble shuffle for m piles.

Let $(Y_1, Y_2, ..., Y_m)$ denote the permutation of $(y_1, y_2, ..., y_m)$ after the shuffle.

7. Turn over all cards encoding $Y_1, Y_2, ..., Y_m$. There can be at most one pair of cards encoding 1. If exactly one pair encodes 1, select the card above that pair; if every pair encodes 0, report that no card is valid.

This protocol uses $m + 2$ shuffles, one in the AND protocol per each iteration plus two in the main protocol.

Note that in the original protocol of Shinoda et al. [17], Step 4 is performed for only $m - 1$ iterations. Then, y_m is set to be t after the $(m - 1)$-th iteration to ensure that exactly one of $y_1, y_2, ..., y_m$ is 1.

3 Overview

3.1 Rules of UNO

Due to the popularity of the game, there are dozens of variants of UNO, each one with slightly different rules. The following set of rules is one of the simplest and one of the most played variants.

All players are located around a circle. Each player begins with seven cards in hand randomly drawn from the deck. The rest of the deck, called the *unplayed deck*, are placed face-down at the center.

First, turn over the topmost card of the unplayed deck and put it in the *discard pile*, designating it as the most recently played card. Then, randomly pick a player to start the game. Play initially proceeds in the clockwise order around the circle.

In each turn, a player plays a card in their hand, discarding it face-up to the discard pile. The played card has to be a color card with the same number or color as the most recently played card, or can be any black card ($\boxed{W}$ or $\boxed{D}$).

- If a card with a number has been played, the turn ends with no additional action.
- If a card with a skip sign is played, the next player misses a turn.
- If a card with a reverse sign is played, the play order changes direction (from clockwise to counterclockwise, and vice versa).
- If a card with a +2 sign is played, the next player has to draw two cards from the unplayed deck and misses a turn.
- If a $\boxed{W}$ is played, the current player can designate one of the four colors to it; the next player has to play a card with that color (or a black card).
- If a $\boxed{D}$ is played, the next player has to draw four cards from the unplayed deck and misses a turn. Also, the current player can designate one of the four colors to it; the next player has to play a card with that color (or a black card).

If the player does not have a valid card in hand to play, or does not wish to play a card, they have to draw a card from the unplayed deck and may optionally play that card if it is valid.

The first player who plays all cards in their hand wins the game.

3.2 Simulating Virtual Players

First, we will show the overview of how to simulate virtual players. Full details of the *card selection protocol*, our most important protocol, will be later explained in Sect. 4.

For each virtual player P, all people jointly operate P as follows.

- At the start of the game, draw seven cards from the deck and put them face-down in P's hand without looking at them.

- When P's previous player plays a $+2$ (resp. $+4$) card, draw two (resp. four) cards from the unplayed deck and put them face-down in P's hand without looking at them. Note that P's turn is skipped.
- In P's turn, apply the card selection protocol in Sect. 4 to select a random valid card from P's hand to play, or report that none exists.
  - If a color card has been played, no further action is necessary.
  - If a black card ($\boxed{W}$ or $\boxed{D}$) has been played, designate a color of that card by uniformly selecting one of the four colors: red, yellow, green, and blue, at random (which can easily be simulated by shuffling a pile of any four different cards and randomly picking one of them).
  - If no card has been played, draw one card from the unplayed deck and put it face-down in P's hand without looking at it. Then, apply the card selection protocol in Sect. 4 again to play the newly drawn card if it is valid (this time if no card is played, no further action is necessary).

4   Card Selection Protocol

Suppose there are $n-1$ players. Let P_1 be a virtual player in consideration, and $P_2, P_3, ..., P_{n-1}$ be other (human or virtual) players. We consider the unplayed deck as the hand of a hypothetical n-th player; the cards in this pile must not be revealed to anyone.

Let $\boxed{\theta_1}, \boxed{\theta_2}, ..., \boxed{\theta_n}$ denote any n different pre-designated types of UNO card, which are also different from $\boxed{\alpha}$ and $\boxed{\beta}$.

All people jointly operate P_1 as follows.

1. Arrange all players' hands and the unplayed deck face-down horizontally, where the unplayed deck is regarded as P_n's hand.

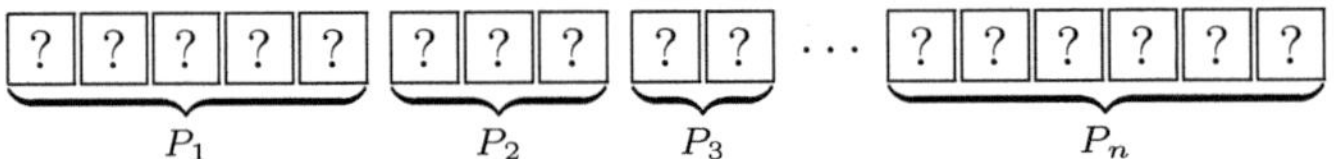

2. Place a $\boxed{\theta_i}$ below each of P_i's cards $(1 \le i \le n)$.

3. Turn over all face-up cards.

4. Apply the pile-scramble shuffle to the matrix, with each column as each pile.

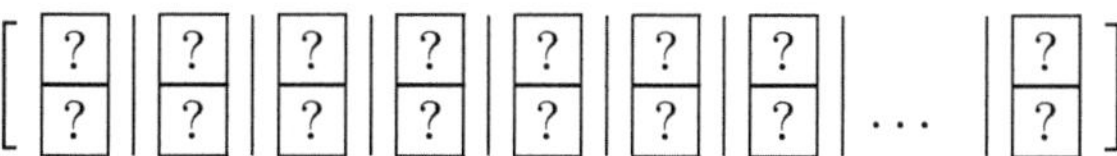

5. Turn over all cards in the first row.

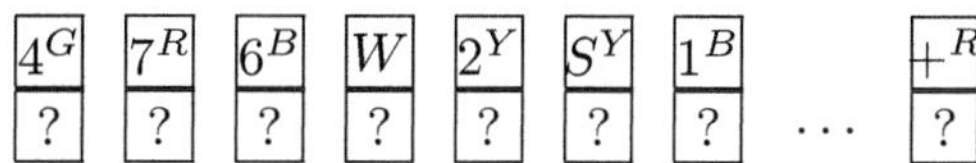

6. For each column, if the card in the first row is valid, place $\boxed{\beta}\,\boxed{\alpha}$ below that column; otherwise, place $\boxed{\alpha}\,\boxed{\beta}$ below that column. (In the following example, the most recently played card is $\boxed{2^R}$.)

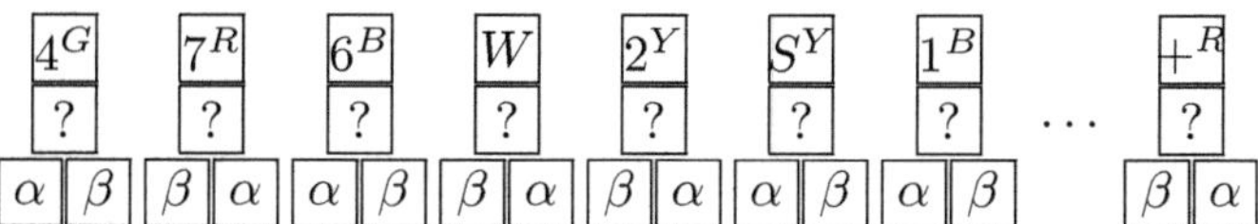

7. Turn over all face-up cards.

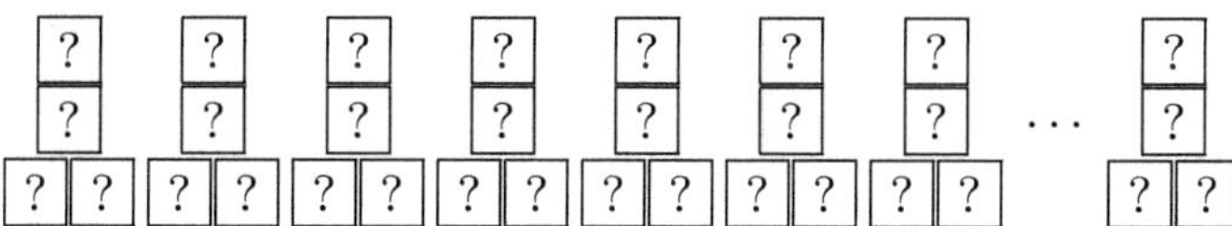

8. Apply the pile-scramble shuffle to the matrix, with each column as each pile.

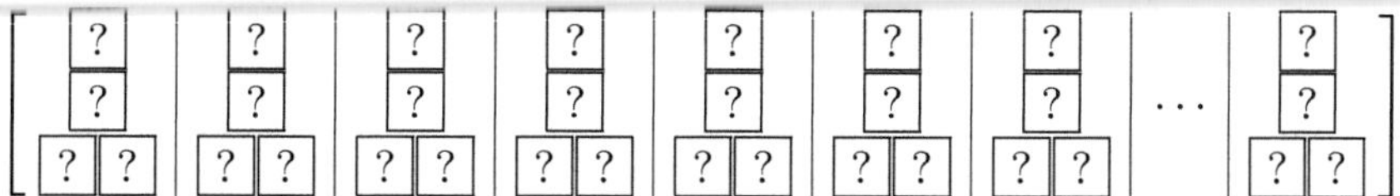

9. Turn over all cards in the second row.

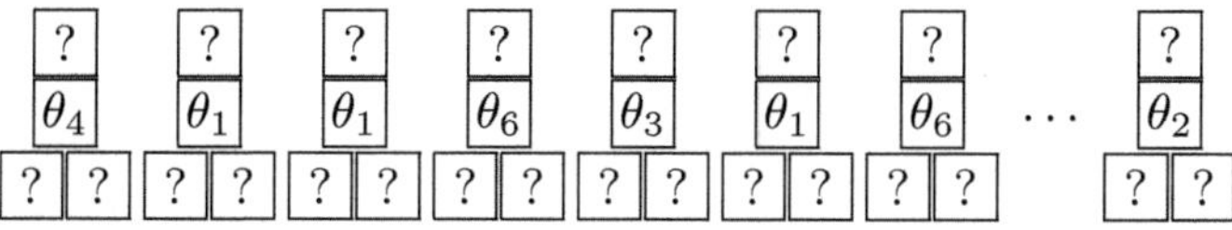

10. Consider only columns containing a $\boxed{\theta_1}$. In such a column, the card in the first row is P_1's card, and the pair of cards in the third row indicates whether the card in the first row is valid. Remove these columns from the matrix and use the cards in the first and third rows of the these columns as inputs for the modified covert lottery protocol in Sect. 2.5 to randomly select a valid card, or report that none exists.

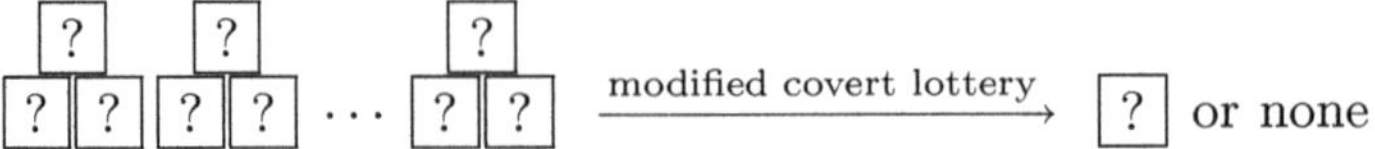

11. Put the unselected cards from the modified covert lottery protocol back into P_1's hand.
12. For the rest of the matrix, put each card above a $\boxed{\theta_i}$ back into P_i's hand.

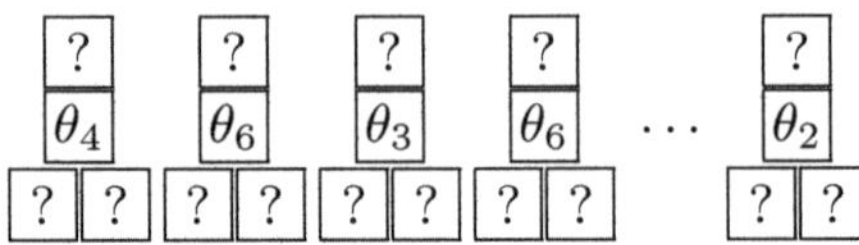

Let k and k_1 be the number of remaining cards in the game and the number of cards in P_1's hand, respectively. This protocol uses $3k + 4$ additional cards and uses $k_1 + 4$ shuffles ($k_1 + 2$ in the modified covert lottery protocol and two in the main protocol).

Note that Step 5 of this protocol reveals the remaining cards in the game, as the sequence in Step 1 excludes the discard pile. Since the discarded cards are publicly known, the remaining cards are also public, so this does not cause a security problem. However, typical human players may not remember which cards remain unless they possess perfect memory. To enhance the enjoyment of the game, it may be preferable to consider the discard pile as P_{n+1}'s hand in Step 1 and put a θ_{n+1} card below each discarded card in Step 2, thereby concealing the remaining cards from all players as well.

4.1 Proof of Correctness and Security

First, consider the covert lottery protocol. Correctness and security of the original protocol has been proved in [17]. Our modified protocol is almost the same as the original one, except that it reports that no card is valid when that is the case. In such cases, the virtual player has to draw a card from the unplayed deck, so the information that no valid card exists is public. Therefore, the modified covert lottery protocol is correct and secure.

Next, we will prove the correctness of the card selection protocol. Due to the way we place cards in Steps 2 and 6, for each card c in the first row, the card $\boxed{\theta_i}$ in the second row indicates that c is P_i's card, and the pair of cards in the third row indicates whether c is valid ($\boxed{\alpha}\,\boxed{\beta}$ for yes and $\boxed{\beta}\,\boxed{\alpha}$ for no). Therefore, in Step 10 we correctly select all cards in P_1's hand and the cards indicating whether they are valid as inputs for the modified covert lottery protocol. Moreover, in Step 12 each card above a $\boxed{\theta_i}$ is P_i's card, so we correctly return it to P_i. Hence, our protocol is correct.

The card selection protocol follows the computational model of card-based cryptography [6], where the security is proved based on information theory. To prove the security of this protocol, it is sufficient to prove that the cards that are turned face-up during the protocol does not reveal any information beyond the public information.

- In Step 5, all cards in the first row, i.e. all remaining cards in the game, are revealed. Due to the pile-scramble shuffle, the order of the k columns are uniformly distributed among all $k!$ permuation. Therefore, the only information revealed is the set of these cards, which is public information because it is equal to a full UNO deck minus the public discard pile.
- In Step 9, all cards in the second row are revealed. Due to the pile-scramble shuffle, the order of the k columns are uniformly distributed among all $k!$ permuation. Therefore, the only information revealed is the number of $\boxed{\theta_i}$ s for each $i = 1, 2, ..., n$, which is public information.

Hence, the protocol is correct and secure.

5   Application to Other Games

The card selection protocol in Sect. 4 is a generic protocol. It can be used to simulate virtual players in other turn-based card or tile games where each player has to select a valid card to play in each turn. We hereby show three examples of Sevens, Hearts, and Dominoes.

5.1   Sevens

Sevens is a card game played with a standard 52-card deck. The whole deck is divided (roughly) equally to all players. A player who has a $\boxed{7\diamond}$ starts the game by playing the $\boxed{7\diamond}$. Then, the next player has to play a $\boxed{6\diamond}$, a $\boxed{8\diamond}$, or a $\boxed{7}$ of any other suit. In general, in each turn a player may either

1. play a $\boxed{7}$ of any unplayed suit, or
2. play a card of a played suit, maintaining that the played cards in that suit still form a set of consecutive ranks. For example, if $\boxed{5\diamond}\boxed{6\diamond}\boxed{7\diamond}\boxed{8\diamond}$ have already been played, a player may only play a $\boxed{4\diamond}$ or a $\boxed{9\diamond}$.

If a player does not have a valid card to play, they have to pass their turn. The first player who plays all cards in their hand wins the game.

Our card selection protocol works in the same way for Sevens, except that there is no unplayed deck. In each virtual player's turn, all people jointly apply the card selection protocol to select a random valid card in its hand to play, or pass the turn if none exists.

5.2   Hearts

Hearts is a four-player card game also played with a standard 52-card deck. Every player begins each round with 13 cards in hand.

A round of Hearts consists of 13 *tricks*. In each trick, every player plays a card in the clockwise order. The first player can play any card. The subsequent players have to play a card having the same suit as the first card if possible; if they do not have such a card, they can play any card. A player who plays the highest-ranked card with the same suit as the first card *wins* the trick, and receives points equal to the points of all four played cards combined. That player also starts the next trick.

Each heart card has one point, and a $Q\spadesuit$ has 13 points. After the end of each round, the points are tallied and the next round begins. The first player who reaches 100 points loses the game.

Our card selection protocol works in the same way for Hearts (without the unplayed deck), except that the original covert lottery protocol [17] is used instead of the modified one. In each virtual player's turn, all people jointly apply the card selection protocol to select a random valid card in its hand, or randomly select any card if no card is valid.

5.3   Dominoes

Dominoes is a deck of rectangular tiles, each with two numbers between 0 and 6 on its face. The backs of the dominoes are indistinguishable, just like playing cards. Tiles with the same number (such as 6 and 6) are called doublets. Although there are several games of dominoes, *Muggins* (also known as *All Fives*) is one of the most commonly played games of dominoes.

At the beginning of the game, the deck of dominoes is completely shuffled face down and each player starts with five tiles drawn from the deck. The player with the highest doublet plays first, and turns proceed in a clockwise direction. The first player plays any domino, and each player plays a matching tile on one of the endpoints. For example, if the first player plays a tile of $(5,6)$, the second player can play the tile of $(5,x)$ or $(6,y)$. A player who cannot play must repeatedly draw a tile domino from the deck until they can play. A player scores points when playing a tile whose sum of all open endpoints is a multiple of five. The game ends when a player has no tiles left or when all players have passed. The player with the highest score wins the game.

Our card selection protocol works the same way for the game. On each virtual player's turn, all people jointly apply the protocol to select a random valid tile in its hand to play. If there is no valid tile, a tile is drawn from the deck and the protocol is run again.

Note that the selection of the first player can also be solved by our card selection protocol as follows. For simplicity, assume that the number of virtual players is one. First, each real player announces the largest doublet in their hand. For example, suppose that the largest doublet among the real players is $(4,4)$. Then, by specifying that the valid tiles are doublets of $(5,5)$ or $(6,6)$, our card selection protocol tells whether the virtual player has the largest doublet or not. Even if there are two or more virtual players, we can determine the player who has the largest doublet by applying our card selection protocol multiple times.

6   Future Work

We developed a card-based protocol that can simulate virtual players to play UNO, as well as other turn-based card games. An interesting future work is to design a protocol to simulate virtual players that can randomly select a card following a non-uniform distribution depending on the available cards, making them more fun and more difficult to play against. Other possible future work includes developing a simuation protocol for card games where a player can simultaneously play multiple cards in each turn, such as *Dominion*.

References

1. Goldwasser, S., Micali, S., Rackoff, C.: The knowledge complexity of interactive proof systems. SIAM J. Comput. **18**(1), 186–208 (1989)
2. Ishikawa, R., Chida, E., Mizuki, T.: Efficient Card-Based Protocols for Generating a Hidden Random Permutation Without Fixed Points. In: Calude, C.S., Dinneen, M.J. (eds.) UCNC 2015. LNCS, vol. 9252, pp. 215–226. Springer, Cham (2015). https://doi.org/10.1007/978-3-319-21819-9_16
3. Y. Komano and T. Mizuki. Card-Based Zero-Knowledge Proof Protocol for Pancake Sorting. In *Proceedings of the 15th International Conference on Security for Information Technology and Communications (SecITC)*, pp. 222–239 (2023)
4. Miyahara, D., Haneda, H., Mizuki, T.: Card-Based Zero-Knowledge Proof Protocols for Graph Problems and Their Computational Model. In: Huang, Q., Yu, Yu. (eds.) ProvSec 2021. LNCS, vol. 13059, pp. 136–152. Springer, Cham (2021). https://doi.org/10.1007/978-3-030-90402-9_8
5. Mizuki, T.: Card-based protocols for securely computing the conjunction of multiple variables. Theoret. Comput. Sci. **622**, 34–44 (2016)
6. Mizuki, T., Shizuya, H.: A formalization of card-based cryptographic protocols via abstract machine. Int. J. Inf. Secur. **13**(1), 15–23 (2014)
7. Mizuki, T., Sone, H.: Six-Card Secure AND and Four-Card Secure XOR. In: Deng, X., Hopcroft, J.E., Xue, J. (eds.) FAW 2009. LNCS, vol. 5598, pp. 358–369. Springer, Heidelberg (2009). https://doi.org/10.1007/978-3-642-02270-8_36
8. Nishida, T., Hayashi, Y., Mizuki, T., Sone, H.: Card-Based Protocols for Any Boolean Function. In: Jain, R., Jain, S., Stephan, F. (eds.) TAMC 2015. LNCS, vol. 9076, pp. 110–121. Springer, Cham (2015). https://doi.org/10.1007/978-3-319-17142-5_11
9. Nishida, T., Mizuki, T., Sone, H.: Securely Computing the Three-Input Majority Function with Eight Cards. In: Dediu, A.-H., Martín-Vide, C., Truthe, B., Vega-Rodríguez, M.A. (eds.) TPNC 2013. LNCS, vol. 8273, pp. 193–204. Springer, Heidelberg (2013). https://doi.org/10.1007/978-3-642-45008-2_16
10. T. Ono, S. Ruangwises, Y. Abe, K. Hatsugai and M. Iwamoto. Single-Shuffle Physical Zero-Knowledge Proof for Sudoku Using Interactive Inputs. In *Proceedings of the 12th ACM ASIA Public-Key Cryptography Workshop (APKC)*, pp. 1–8 (2025)

11. Ruangwises, S.: An Improved Physical ZKP for Nonogram and Nonogram Color. J. Comb. Optim. **45**(5), 122 (2023)
12. S. Ruangwises. Physical Zero-Knowledge Proof for Ball Sort Puzzle. In *Proceedings of the 19th Conference on Computability in Europe (CiE)*, pp. 246–257 (2023)
13. Ruangwises, S., Itoh, T.: Securely Computing the n-Variable Equality Function with $2n$ Cards. Theoret. Comput. Sci. **887**, 99–110 (2021)
14. Sasaki, T., Miyahara, D., Mizuki, T., Sone, H.: Efficient card-based zero-knowledge proof for Sudoku. Theoret. Comput. Sci. **839**, 135–142 (2020)
15. K. Shinagawa, D. Miyahara and T. Mizuki. How to Play Old Maid with Virtual Players. In *Proceedings of the 18th International Conference on Frontiers of Algorithmic Wisdom (FAW)*, pp. 53–65 (2024)
16. Shinagawa, K., Nuida, K.: A single shuffle is enough for secure card-based computation of any Boolean circuit. Discret. Appl. Math. **289**, 248–261 (2021)
17. Shinoda, Y., Miyahara, D., Shinagawa, K., Mizuki, T., Sone, H.: Card-Based Covert Lottery. In: Maimut, D., Oprina, A.-G., Sauveron, D. (eds.) SecITC 2020. LNCS, vol. 12596, pp. 257–270. Springer, Cham (2021). https://doi.org/10.1007/978-3-030-69255-1_17
18. K. Tanaka, S. Sasaki, K. Shinagawa and T. Mizuki. Only Two Shuffles Perform Card-Based Zero-Knowledge Proof for Sudoku of Any Size. In *Proceedings of the 8th SIAM Symposium on Simplicity in Algorithms (SOSA)*, pp. 94–107 (2025)
19. K. Toyoda, D. Miyahara and T. Mizuki. Another Use of the Five-Card Trick: Card-Minimal Secure Three-Input Majority Function Evaluation. In *Proceedings of the 22nd International Conference on Cryptology in India (INDOCRYPT)*, pp. 536–555 (2021)

Entropy Guarantees for Quantum Boolean Functions

Zornitza Prodanoff$^{(\boxtimes)}$ and David Wisnosky

School of Computing, University of North Florida, Jacksonville, FL 32224, USA
{zprodano,n01153911}@unf.edu

Abstract. This study demonstrates how to provide specific guarantees on the entropy produced by quantum circuits implementing Boolean functions. More specifically, we study guarantees on von Neumann entropy for subsystems in such circuits. Our findings indicate that input state initialization by rotating the target qubit around the X-axis of the Bloch sphere, combined with arbitrary but fixed individual qubit rotations of the control qubits around any axis or a combination of thereof, preserves the von Neumann entropy of reduced density matrices across subsystem traces. In contrast, similar rotations around the Y-axis cause entropy variation with rotation angle. Moreover, ceteris paribus, arbitrary qubit rotations on all qubits, including the target, result similarly in variability of subsystem entropy. We apply these findings to secure the classical communication channel in a novel multi-qubit quantum teleportation protocol. We depict a full IBM Qiskit implementation and its corresponding graphical-empirical example for the execution of a three-qubit quantum teleportation protocol. Overall, our results reveal insights into system entanglement and demonstrate that Boolean functions provide a structured approach for the purposes of controlling the dynamics of quantum entropy. This has implications beyond the secure obfuscation of classical communication in quantum teleportation and warrants additional investigation of quantum Boolean functions in the broad context of quantum information scrambling and beyond.

Keywords: Boolean function · quantum computing · Disjunctive Normal Form (DNF) · Algebraic Normal Form (ANF) · reversible computing · Zhegalkin polynomial · von Neumann entropy · teleportation

1 Introduction

David Deutsch was the first to suggest the use of Boolean functions in quantum computing [3]. Numerous works have since extended the idea by presenting specific quantum circuit constructions using linear algebra and physical principles to analyze those [1,4–6]. A particular type of Boolean function representation, widely referred to as *algebraic normal form* (ANF), described in Sect. 2, plays a key role in the study of such circuits. ANF has historically been attributed to

This study was supported through the IBM Quantum Researchers program.

© The Author(s), under exclusive license to Springer Nature Switzerland AG 2026
E. Formenti and L. Manzoni (Eds.): UCNC 2025, LNCS 16364, pp. 47–62, 2026.
https://doi.org/10.1007/978-3-032-15641-9_4

Zhegalkin [12], however, its relevance to quantum computing was first proposed by Hirvensalo in [6]—a study that provides a comprehensive treatment of the algebraic structure of such functions in the context of their representation degree (algebraic degree) and function complexity. Since ANF is a less commonly used representation in classical computation studies, methods to convert a Boolean function of other representations into ANF (e.g., from the widely popular in classical computation disjunctive normal form) and vice versa have been proposed already, starting with the brute-force algorithm introduced in [8].

In this work, we focus on the concept of quantum entropy in quantum circuits that are specifically designed to implement Boolean functions. We explore some ideas on "controlling" the entropy of their final state, i.e. providing fixed-level guarantees, under imposing specific restrictions on the input states. Additionally, we suggest a potential application for the purposes of protecting the classical communication channel of quantum teleportation protocols.

Quantum teleportation was introduced by Bennett et al. [2]. Building on this original scheme, several studies extended the idea to multi-qubit teleportation and addressed its security. For example, Sisodia et al. [9] proposed the teleportation of an arbitrary n-qubit states using a single 3-qubit GHZ state and a controller who authorizes recovery. Recent efforts introduced Boolean function teleportation through quantum channels, rather than simply utilizing them throug local gates. Kadry et al. proposed a protocol to teleport an oracle encoding a multivariate Boolean function, leveraging the Extended Deutsch-Jozsa algorithm to classify the function post-teleportation [7]. This approach not only demonstrates the viability of transmitting functional information rather than quantum states but also suggests applications in distributed quantum computing and key generation. Separately, Yoshida and Yao developed a teleportation-based scrambling diagnostic that distinguishes genuine quantum information delocalization from decoherence, highlighting teleportation as a probe of entropy and mutual information across subsystems [11]. These results collectively illustrate the role of quantum teleportation as more than a communication protocol. It can be viewed more generally as a framework for understanding the dynamics of entanglement, function transmission, and entropy control. This realization sparked our interest in applying the findings presented in this study to multi-qubit quantum teleportation.

The paper is organized as follows. Section 2 provides necessary preliminaries. The main results are presented in Sects. 3 and 4. An application to multi-qubit quantum teleportation is proposed in Sect. 5 which presents a novel protocol and its execution circuit as well as proof of protocol correctness. The security of the protocol's classical channel is discussed in Sect. 6. A brief summary is included last, in Sect. 7.

2   Preliminaries

A Boolean function, that is, $f : \{0,1\}^n \to \{0,1\}$, is specified through a Boolean *formula*—an expression, using n variables $x_1, x_2, \ldots, x_n$ and logical connectives.

Quantum Boolean function circuits are designed most intuitively by directly using Boolean formulas in Algebraic Normal Form (ANF) to arrange reversible computation gates. ANF is a representation of Boolean functions over a finite (Galois) field of two elements $GF(2) = \{0, 1\}$ and the logical *exclusive or* and *and* operations.

Every Boolean function $f : \{0,1\}^n \to \{0,1\}$ has a unique ANF representation:

$$f(x_1, \ldots, x_n) = \bigoplus_{S \subseteq [n]} c_S \cdot \prod_{i \in S} x_i, \quad \text{with } c_S \in \mathrm{GF}(2),$$

where $\bigoplus$ denotes addition modulo 2 (XOR) and each *monomial* $\prod_{i \in S} x_i$ represents a conjunction (logical and) over the variables in subset S. The coefficients $c_S \in \{0, 1\}$ indicate whether the monomial is present (1) or not (0). That is, ANF is a sum-of-product expansion form, structured as follows:

$$f(x_1, ..., x_n) = a_1 \mathbf{1} \oplus a_2 x_1 \oplus a_3 x_2 \oplus a_{n+1} x_n \oplus a_{n+2} x_1 x_2 \oplus ...$$
$$\oplus a_k x_{n-1} x_n \oplus a_{k+1} x_1 x_2 x_3 \oplus ... \oplus a_{2^n} x_1 x_2 x_3 ... x_n$$

where $\oplus$ denotes addition modulo 2, logical conjunction is presented as juxtaposition, Boolean coefficients $a_1, ..., a_{2^n} \in \{0, 1\}$, and $x_1, x_2, ..., x_n$ represent the Boolean input variables, while $\mathbf{1} = \begin{bmatrix} 1 \\ 1 \end{bmatrix}$ is the true term. Each monomial corresponds to a subset of the variables (e.g., $x_1 x_3$ corresponds to the subset $\{x_1, x_3\}$). Hence, the set of all possible monomials corresponds to the power set of the variable set $\{x_1, \ldots, x_n\}$. One example of a Boolean function in ANF is $f = \mathbf{1} \oplus x_3 \oplus x_2 x_3 x_4$, that is, f has two monomials x_3 and $x_2 x_3 x_4$ as well as a true term $\mathbf{1}$.

If U is a unitary operator on $\mathcal{H}$, then the transformation $\rho' = U \rho U^\dagger$ preserves both trace and purity: $\mathrm{Tr}(\rho') = \mathrm{Tr}(\rho)$, $\mathrm{Tr}(\rho'^2) = \mathrm{Tr}(\rho^2)$, where the quantity $\mathrm{Tr}(\rho^2) \in (0, 1]$ is called the *purity* of the quantum state, and it is a real number that measures how "mixed" the state is. A state is pure if and only if its purity is exactly 1; otherwise, it is mixed. A density matrix ρ on a Hilbert space $\mathcal{H}$ represents a *pure* state if it can be written as

$$\rho = |\psi\rangle\langle\psi|$$

for some normalized state vector $|\psi\rangle \in \mathcal{H}$. Equivalently, a pure state satisfies $\rho^2 = \rho$ and $\mathrm{Tr}(\rho^2) = 1$.

When the eigenvalues of ρ' and ρ are equal, ρ' is said to be *spectrally invariant*, denoted as $\mathrm{Spec}(\rho') = \mathrm{Spec}(\rho)$, where $\mathrm{Spec}(\cdot)$ denotes the multiset of eigenvalues. If $\rho = \rho^\dagger$, then $\rho'^\dagger = (U\rho U^\dagger)^\dagger = U\rho^\dagger U^\dagger = U\rho U^\dagger = \rho'$, so ρ' is also Hermitian.

If set $A \subseteq \{0, 1, \ldots, k-1\}$ is used to index a subset of all qubits and $\bar{A} = \{0, 1, \ldots, k-1\} \setminus A$ is the complement of A with a physical meaning of degrees of freedom outside the observer's subsystem A, then, the *reduced density matrix* on subsystem A is defined by $\rho_A = \mathrm{Tr}_{\bar{A}}(\rho)$, where $\mathrm{Tr}_{\bar{A}}$ denotes the partial trace over the complement subsystem $\bar{A}$.

The execution of a *quantum circuit* on n qubits can be summarized as a unitary transformation $U : \mathcal{H}_{2^n} \to \mathcal{H}_{2^n}$, where $\mathcal{H}_{2^n}$ denotes the 2^n-dimensional Hilbert space associated with n-qubit quantum states. At the lowest level of abstarction, such circuits can be deconstructed as finite sequences of elementary quantum gates from a universal set that are applied to the quantum system in separate time steps. In this study, we only consider the highest level of logical representation, using reversible logical gates and ignoring other operations such as qubit measurement due to the assumption of noiseless conditions for the purposes of our analysis.

Quantum Boolean Function Circuits (QBFCs) are quantum circuits that implement classical, reversible Boolean functions $f : \{0,1\}^n \to \{0,1\}^n$ as unitary operators U_f acting on n-qubit quantum states. Each such function f induces a permutation on the computational basis of the Hilbert space $\mathcal{H} = (\mathbb{C}^2)^{\otimes n}$, and the corresponding unitary U_f is defined as a permutation matrix of size $2^n \times 2^n$.

The matrix elements of U_f in the computational basis are given by $(U_f)_{j,k} = \delta_{j,f(k)}$, where the permutation induced by f maps the index k of basis state $|k\rangle$, interpreted as the binary string in $\{0,1\}^n$ to the index j corresponding to $f(k)$.

Given an input density matrix $\rho \in \mathbb{C}^{2^n \times 2^n}$, the QBFC applies the transformation $\rho' = U_f \rho U_f^\dagger$. The output state ρ' may be entangled, even if the input ρ is pure and separable. The transformation of individual matrix elements is given by $(\rho')_{j,k} = \rho_{f^{-1}(j),f^{-1}(k)}$, where j and k index computational basis states, interpreted as binary strings under f^{-1}. This corresponds to a simultaneous permutation of rows and columns of ρ according to the inverse of the Boolean function f, while preserving amplitude magnitudes..

Because U_f is a permutation unitary, it preserves the eigenvalue spectrum of ρ and hence its *von Neumann entropy*, definied as $S(\rho) = -\operatorname{Tr}(\rho \log \rho)$. However, entanglement and subsystem purity may change, depending on how f redistributes computational basis amplitudes across qubit partitions.

In QBFCs, the global unitary U_f is constructed from a sequence of these basic gates. Each gate acts non-trivially on a small subset of the qubits and trivially (as the identity) on the remaining ones. The full operator is constructed by embedding each gate into the larger Hilbert space using tensor products. For instance, a single-qubit gate G acting on qubit i of a n-qubit system in time step j is represented as:

$$U_t = I^{\otimes(i-1)} \otimes G \otimes I^{\otimes(n-i)},$$

where I is the 2×2 identity matrix and $t \in [1,T]$. A sequence of such gates is then applied one after another in discrete time steps, where each gate transforms the system state at its respective step. These gate applications are composed via matrix multiplication, i.e., $|\psi_{\text{out}}\rangle = U_T \cdots U_2 U_1 |\psi_{\text{in}}\rangle$. Hence, the overall $2^n \times 2^n$ unitary matrix implementing the desired Boolean function can be written as $U_f = U_T \cdots U_2 U_1$.

Each U_t transformation can be constructed logically using a combination of the Pauli-X (NOT) gate $X = \begin{bmatrix} 0 & 1 \\ 1 & 0 \end{bmatrix}$, the controlled-NOT (CNOT) gate

$$\text{CNOT} = \begin{bmatrix} 1 & 0 & 0 & 0 \\ 0 & 1 & 0 & 0 \\ 0 & 0 & 0 & 1 \\ 0 & 0 & 1 & 0 \end{bmatrix}, \text{ the } \textit{Toffoli (CCNOT) gate, introduced in [10], similar to}$$

the CNOT gate, but acting on a taget qubit with a quantum probability of both control qubits being set to $|1\rangle$ simultaneously, and multi-control Toffoli gates, an extension with more than two controls.

QBFC circuits preserve the computational basis structure, meaning that they map computational basis states to other computational basis states without introducing superposition (i.e., map any $|\psi\rangle \in [|0\rangle, |1\rangle]$ to $|\phi\rangle \in [|0\rangle, |1\rangle]$, where those two vectors are represented as linear combinations $\alpha|0\rangle + \beta|1\rangle$, where $\alpha, \beta \in [0, 1]$. Important to note here is that the inputs may not be restricted to basis state $|0\rangle$ and $|1\rangle$ (as in the quantum simulation of a classical Boolean function execution). Instead, they can be a tensor product of superpositions, achieved through individual qubit rotations, or alternatively and more generically, arbitrary entangled n-qubit states, including mixed states.

Throughout the rest of this study, we assume the former, that is, pure and separable input states. Note that this restriction does not limit the impact of our findings but is rather necessary in order to ensure the entropy guarantees in QBFCs that we describe next.

3 Subsystem Entropy Behavior Under Input Restrictions in QBFCs

In this section we outline some observations about subsystem entropy in QBFC. Note that throughout this entire study, we assume noiseless circuits and no quantum decoherence.

Theorem 1 (Conditional Entropy Invariance in QBFC).

Let $\mathcal{H}_k = \bigotimes_{j=0}^{k-1} \mathcal{H}_j$, where each $\mathcal{H}_j \cong \mathbb{C}^2$, be the Hilbert space of k qubits. Consider a pure separable state of the form:

$$|\Psi\rangle = \left(\cos\left(\frac{\alpha}{2}\right)|0\rangle + \gamma\sin\left(\frac{\alpha}{2}\right)|1\rangle\right) \otimes \bigotimes_{j=1}^{k-1}\left(\cos\left(\frac{\theta_j}{2}\right)|0\rangle + e^{i\phi_j}\sin\left(\frac{\theta_j}{2}\right)|1\rangle\right),$$

where $\theta_j \in [0, \pi]$, $\phi_j \in [0, 2\pi)$, and $\gamma = -i$ if the target qubit is prepared via $R_x(\alpha)$, and $\gamma = +1$ if prepared via $R_y(\alpha)$.

Let U_f be a unitary operator implementing a quantum Boolean function derived from a classical reversible Boolean function $f : \{0, 1\}^{k-1} \rightarrow \{0, 1\}$, such that:

$$U_f|q_0, q_1, \ldots, q_{k-1}\rangle = |q_0 \oplus f(q_1, \ldots, q_{k-1}), q_1, \ldots, q_{k-1}\rangle.$$

Then, defining $\rho' = U_f|\Psi\rangle\langle\Psi|U_f^\dagger$, the von Neumann entropy of any reduced subsystem $A \subseteq \{1, \ldots, k-1\}$, given by:

$$S(\rho'_A) = -\operatorname{Tr}(\rho'_A \log \rho'_A), \quad \text{where } \rho'_A = \operatorname{Tr}_0(\rho'),$$

is invariant with respect to α when $\gamma = -i$ (i.e., the target is initialized via $R_x(\alpha)$), and varies with α when $\gamma = +1$ (i.e., for $R_y(\alpha)$).

Proof (Direct) We expand the state:

$$|\Psi\rangle = \sum_{x \in \{0,1\}^{k-1}} c_x \left[\cos\left(\frac{\alpha}{2}\right) |0,x\rangle + \gamma \sin\left(\frac{\alpha}{2}\right) |1,x\rangle \right],$$

leading to a density matrix:

$$\rho = \sum_{x,x'} c_x c_{x'}^* \left[\cos^2\left(\frac{\alpha}{2}\right) |0,x\rangle\langle 0,x'| + \gamma \cos\left(\frac{\alpha}{2}\right) \sin\left(\frac{\alpha}{2}\right) |0,x\rangle\langle 1,x'| + \right.$$

$$\left. + \gamma^* \cos\left(\frac{\alpha}{2}\right) \sin\left(\frac{\alpha}{2}\right) |1,x\rangle\langle 0,x'| + |\gamma|^2 \sin^2\left(\frac{\alpha}{2}\right) |1,x\rangle\langle 1,x'| \right].$$

Applying U_f, which permutes the computational basis without altering coefficients, we obtain $\rho' = U_f \rho U_f^\dagger$. When we trace out qubit 0 to obtain ρ'_A, cross-terms between $|0,x\rangle\langle 1,x'|$ and $|1,x\rangle\langle 0,x'|$ contribute only if they are Hermitian conjugates.

The matrix elements of ρ'_A take the form:

$$(\rho'_A)_{x,x'} = c_x c_{x'}^* \left[\cos^2\left(\frac{\alpha}{2}\right) + |\gamma|^2 \sin^2\left(\frac{\alpha}{2}\right) + (\gamma + \gamma^*) \cos\left(\frac{\alpha}{2}\right) \sin\left(\frac{\alpha}{2}\right) \right].$$

We observe that for $\gamma = -i$, we have $\gamma + \gamma^* = -i + i = 0$, and $|\gamma|^2 = 1$, so that $\cos^2\left(\frac{\alpha}{2}\right) + \sin^2\left(\frac{\alpha}{2}\right) = 1$, leaving: $(\rho'_A)_{x,x'} = c_x c_{x'}^*$, which is independent of α, so the spectrum of ρ'_A and hence its von Neumann entropy remain constant. For $\gamma = +1$, we have $\gamma + \gamma^* = 2$, so the expression becomes $(\rho'_A)_{x,x'} = c_x c_{x'}^* \left[1 + 2\cos\left(\frac{\alpha}{2}\right)\sin\left(\frac{\alpha}{2}\right)\right]$, which does depend on α, causing the spectrum of ρ'_A and thus its entropy to vary with the target rotation.

An analogous argument applies to any reduced subsystem $A \subseteq \{1,\ldots,k-1\}$; we omit the details for brevity.

Hence, the von Neumann entropy $S(\rho'_A)$ is constant in α when using $R_x(\alpha)$ (i.e., $\gamma = -i$), and varies in the case of $R_y(\alpha)$ (i.e., $\gamma = +1$).

$\square$

Remark 1 Note that rotations of the target qubit about the Z-axis introduce only a global phase and are therefore undetectable by Bob. Specifically, recall that the Pauli-Z operator is defined as $Z = \begin{bmatrix} 1 & 0 \\ 0 & -1 \end{bmatrix}$, and that the computational basis state $|0\rangle$ is an eigenvector of Z with eigenvalue $+1$. A rotation about the Z-axis by angle θ_t is implemented by the unitary operator $R_z(\theta_t) = e^{-i\frac{\theta_t}{2}Z}$, where the exponential denotes the matrix exponential, defined via the convergent power series $e^A = \sum_{n=0}^\infty \frac{A^n}{n!}$ for any square matrix A. Because $Z^2 = I$, we can simplify this exponential using the identity:

$$e^{-i\frac{\theta_t}{2}Z} = \cos\left(\frac{\theta_t}{2}\right) I - i \sin\left(\frac{\theta_t}{2}\right) Z.$$

Acting on $|0\rangle$, this gives:

$$R_z(\theta_t)|0\rangle = \left[\cos\left(\frac{\theta_t}{2}\right) I - i\sin\left(\frac{\theta_t}{2}\right) Z\right]|0\rangle = e^{-i\frac{\theta_t}{2}}|0\rangle,$$

since $Z|0\rangle = |0\rangle$. Hence, the action of $R_z(\theta_t)$ on $|0\rangle$ could be viewed as a multiplication by a global phase $e^{-i\theta_t/2}$. Such a phase has no observable consequence: it cancels out in all measurement probabilities, inner products, and density matrices. Consequently, the subsystem entropies, which depend only on the eigenvalues of the reduced density matrices, are also unchanged. Therefore, a rotation about the Z-axis leaves $|0\rangle$ physically unchanged from Bob's perspective and cannot be detected through his measurement.

Corollary 1. *Let the target qubit be initialized by a combined application of $R_x(\alpha)$ and $R_z(\beta)$ in the form:*

$$|\psi_0\rangle = R_z(\beta)R_x(\alpha)|0\rangle \quad or \quad |\psi_0\rangle = R_x(\beta)R_z(\alpha)|0\rangle.$$

Then, the von Neumann entropy of any reduced subsystem $A \subseteq \{0,1,\ldots,k-1\}$, after applying a Boolean function unitary U_f, generally varies with α and β.

Proof. The rotation $R_x(\alpha)$ produces a superposition state with imaginary relative phase:

$$R_x(\alpha)|0\rangle = \cos\left(\frac{\alpha}{2}\right)|0\rangle - i\sin\left(\frac{\alpha}{2}\right)|1\rangle,$$

which leads to off-diagonal terms in the density matrix that are complex conjugates with opposite signs. When tracing over the target qubit, these terms cancel due to the identity $\gamma + \gamma^* = 0$, resulting in a reduced density matrix ρ_A' whose entropy is independent of α.

However, applying $R_z(\beta)$ afterwards introduces an additional phase difference between $|0\rangle$ and $|1\rangle$, yielding:

$$R_z(\beta)R_x(\alpha)|0\rangle = e^{-i\beta/2}\cos\left(\frac{\alpha}{2}\right)|0\rangle - ie^{i\beta/2}\sin\left(\frac{\alpha}{2}\right)|1\rangle.$$

The resulting off-diagonal terms in the density matrix are:

$$ab^* = i\cos\left(\frac{\alpha}{2}\right)\sin\left(\frac{\alpha}{2}\right)e^{-i\beta}, \quad ba^* = -i\cos\left(\frac{\alpha}{2}\right)\sin\left(\frac{\alpha}{2}\right)e^{i\beta},$$

which are not complex conjugates unless $\beta = 0$. Thus, their cancellation under partial trace no longer occurs. The reduced density matrix ρ_A' gains α- and β-dependent off-diagonal terms, resulting in eigenvalue shifts and entropy variation.

$\square$

Figure 1 shows empirical results from a noiseless simulation of QBFC for $f(x_2, x_1, x_0) = 1 \oplus x_2 \oplus x_1 x_2$ under various inputs. Each graph, in a specific color, depicts a specific rotation angle combination under fixed rotations of individual control qubits in the range $[0 - 2\pi]$, while 100 target qubit rotations are plotted on the X-axis, with a step of approximately:

$$\Delta\theta = \frac{2\pi - 0}{99} \approx 0.0635 \text{ radians.}$$

Fig. 2 depicts the same empirical set up but with rotations along the Y-axis.

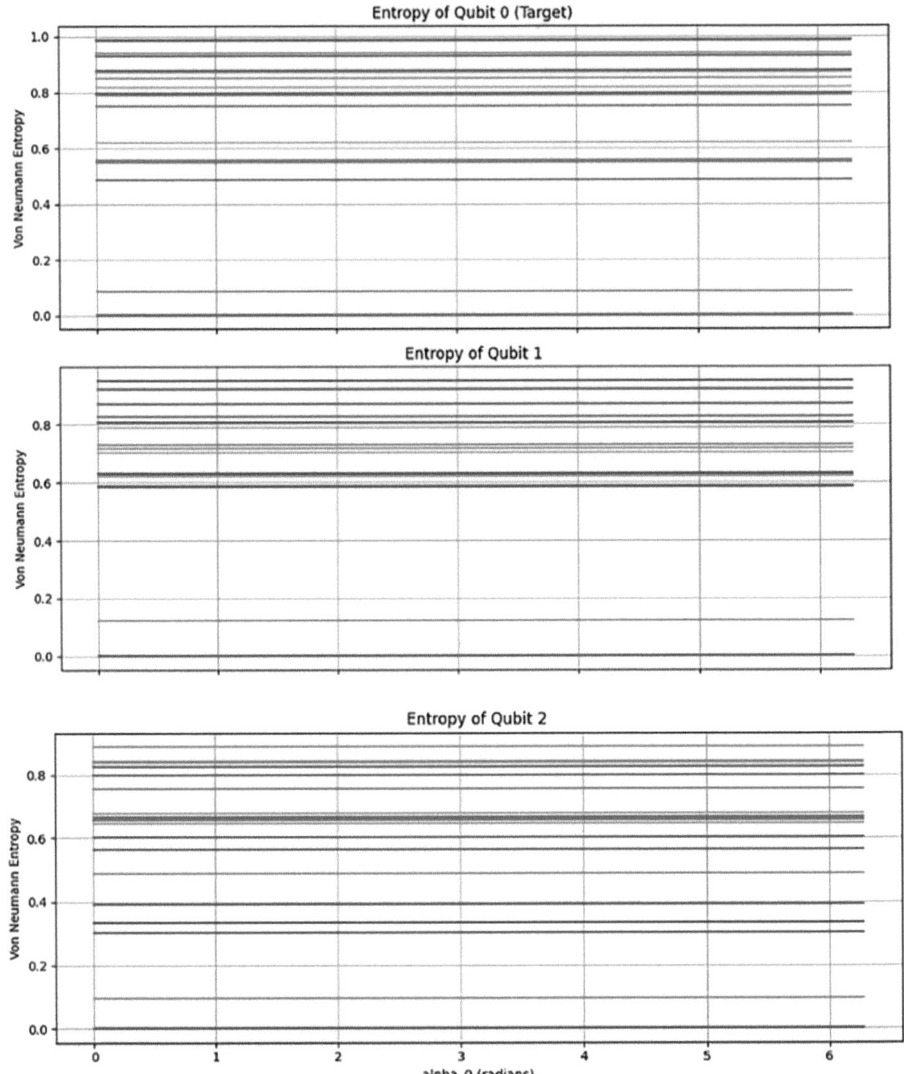

Fig. 1. Subsystem entanglement for 3-qubit quantum teleportation with input state restrictions, incl. R_x on target qubit.

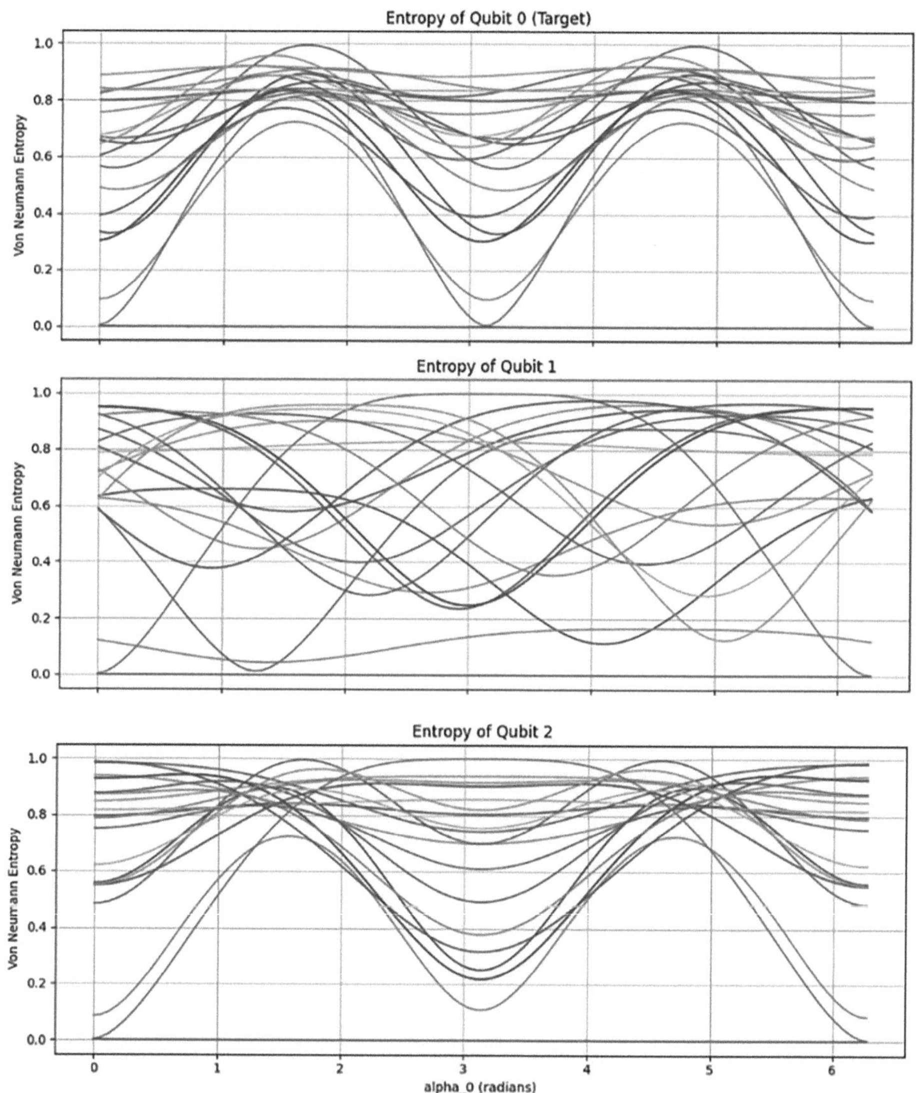

Fig. 2. Subsystem entanglement for 3-qubit quantum teleportation with input state restrictions, incl. R_y on target qubit.

4 Removing and Creating Entropy in QBFC Under Input Constraints

Theorem 2. (Conditional Preservation of Separability in QBFC). *Let* $\mathcal{H}_k = \bigotimes_{j=0}^{k-1} \mathcal{H}_j$, *with* $\mathcal{H}_j \cong \mathbb{C}^2$, *be the Hilbert space of* k *qubits. Let the target qubit* q_0 *be initialized as* $|\psi_0\rangle = R_y(\alpha)|0\rangle = \cos\left(\frac{\alpha}{2}\right)|0\rangle + \sin\left(\frac{\alpha}{2}\right)|1\rangle$, *and let the* *remaining qubits $q_1, \ldots, q_{k-1}$ be initialized in an arbitrary product state:*

$$|\psi_C\rangle = \bigotimes_{j=1}^{k-1} \left(\cos\left(\frac{\theta_j}{2}\right)|0\rangle + e^{i\phi_j}\sin\left(\frac{\theta_j}{2}\right)|1\rangle \right)$$

with $\theta_j \in [0, \pi]$, $\phi_j \in [0, 2\pi)$.

For an arbitrary Boolean function f, let U_f act as:

$$U_f|q_0, x\rangle = |q_0 \oplus f(x)\rangle \otimes |x\rangle, \quad x \in \{0,1\}^{k-1}.$$

Suppose α satisfies $\alpha = \frac{\pi}{2} + m\pi$ for some integer m, so that $|\psi_0\rangle = R_y(\alpha)|0\rangle = \frac{1}{\sqrt{2}}(|0\rangle \pm |1\rangle)$, i.e., $|\psi_0\rangle \in \{|+\rangle, |-\rangle\}$.

Then:

1. *The composite state $|\Psi'\rangle = U_f(|\psi_0\rangle \otimes |\psi_C\rangle)$ remains a separable product state.*
2. *All reduced density matrices ρ'_S, for any subset $S \subseteq \{0, \ldots, k-1\}$, are pure, i.e., $S(\rho'_S) = 0$.*

Proof. We examine the only two valid initialization cases for qubit 0, that is, $|\psi_0\rangle = |+\rangle$ and $|\psi_0\rangle = |-\rangle$. Let first review the case of $|\psi_0\rangle = |+\rangle = \frac{1}{\sqrt{2}}(|0\rangle + |1\rangle)$. The initial composite state is then

$$|\Psi\rangle = |+\rangle_0 \otimes \sum_{x \in \{0,1\}^{k-1}} \beta_x |x\rangle_{1,\ldots,k-1},$$

where the coefficients β_x encode the amplitudes of the arbitrary product state $|\psi_C\rangle$ over the standard basis.

Here U_f acts as: $U_f|b\rangle_0|x\rangle = |b \oplus f(x)\rangle_0|x\rangle$, $\quad b \in \{0,1\}$, $x \in \{0,1\}^{k-1}$. Thus, the action of U_f on the state $|\Psi\rangle$ is:

$$U_f|\Psi\rangle = \sum_x \beta_x \cdot \frac{1}{\sqrt{2}}\left(|0 \oplus f(x)\rangle + |1 \oplus f(x)\rangle\right)_0 \otimes |x\rangle$$

$$= \frac{1}{\sqrt{2}} \sum_x \beta_x \left(|f(x)\rangle + |1 \oplus f(x)\rangle\right)_0 \otimes |x\rangle.$$

Now observe that for any Boolean value $f(x) \in \{0,1\}$, that is, irrespective of the Boolean function f, we have: $|f(x)\rangle + |1 \oplus f(x)\rangle = |0\rangle + |1\rangle = \sqrt{2}|+\rangle$. This identity holds uniformly for all x, so:

$$U_f|\Psi\rangle = \sum_x \beta_x |+\rangle_0 \otimes |x\rangle = |+\rangle_0 \otimes \sum_x \beta_x |x\rangle = |+\rangle_0 \otimes |\psi_C\rangle.$$

Hence, the output state $|\Psi'\rangle$ is a product (separable) state across the partition $\{0\} \cup \{1, \ldots, k-1\}$. Therefore, any reduced density matrix ρ'_S, for any subset $S \subseteq \{0, \ldots, k-1\}$, is pure, with $S(\rho'_S) = 0$.

The case $|\psi_0\rangle = |-\rangle = \frac{1}{\sqrt{2}}(|0\rangle - |1\rangle)$ yields the same conclusion, as the unitary action maps $|-\rangle_0 \otimes |x\rangle \mapsto |-\rangle_0 \otimes |x\rangle$ for all x, since:

$$|f(x)\rangle - |1 \oplus f(x)\rangle = |0\rangle - |1\rangle = \sqrt{2}|-\rangle.$$

Thus, separability and purity are preserved in both cases. $\qquad\qquad\square$

Theorem 3. (Entanglement Induction via Bloch Sphere Equatorial Initialization in QBFC). *Let the system of $k \geq 2$ qubits be initialized as a product state:* $|\Psi_{\text{in}}\rangle = |\psi_0\rangle \otimes |\psi_C\rangle$, *where the target qubit (qubit 0) is initialized as:*

$$|\psi_0\rangle = R_x(\alpha)|0\rangle = \cos\left(\frac{\alpha}{2}\right)|0\rangle - i\sin\left(\frac{\alpha}{2}\right)|1\rangle,$$

for any $\alpha \notin \{0, \pi\}$, *and the remaining* $k - 1$ *control qubits are initialized in a separable state:*

$$|\psi_C\rangle = \bigotimes_{j=1}^{k-1}\left(\cos\left(\frac{\theta_j}{2}\right)|0\rangle + e^{i\phi_j}\sin\left(\frac{\theta_j}{2}\right)|1\rangle\right),$$

with $\theta_j \in [0, \pi]$, $\phi_j \in [0, 2\pi)$.

Let U_f *be the unitary implementation of a reversible quantum Boolean function induced by a non-constant classical function* $f : \{0,1\}^{k-1} \to \{0,1\}$. *Then the output state* $|\Psi_{\text{out}}\rangle = U_f|\Psi_{\text{in}}\rangle$ *is generally entangled across the target/control bipartition, and there exists at least one strict subsystem* $S \subsetneq \{0, \ldots, k-1\}$ *for which the reduced density matrix* ρ_S *is mixed, i.e.,* $S(\rho_S) > 0$.

Proof. Let the initial state be $|\Psi_{\text{in}}\rangle = \left(\cos\left(\frac{\alpha}{2}\right)|0\rangle - i\sin\left(\frac{\alpha}{2}\right)|1\rangle\right) \otimes \sum_x \beta_x|x\rangle$, where $\beta_x \in \mathbb{C}$, $\sum_x |\beta_x|^2 = 1$, and at least two β_x are nonzero.

Applying U_f, we obtain the output state:

$$|\Psi_{\text{out}}\rangle = \sum_x \beta_x\left(\cos\left(\frac{\alpha}{2}\right)|f(x)\rangle - i\sin\left(\frac{\alpha}{2}\right)|1 \oplus f(x)\rangle\right) \otimes |x\rangle.$$

Let further define:

$$|\phi_x\rangle = \cos\left(\frac{\alpha}{2}\right)|f(x)\rangle - i\sin\left(\frac{\alpha}{2}\right)|1 \oplus f(x)\rangle,$$

so that $|\Psi_{\text{out}}\rangle = \sum_x \beta_x|\phi_x\rangle \otimes |x\rangle$. The latter is an entangled state across the bipartition $\{0\} \cup \{1, \ldots, k-1\}$, unless all $|\phi_x\rangle$ are equal up to global phase. Since f is non-constant, there exist x, x' such that $f(x) \neq f(x')$, implying $|\phi_x\rangle \neq |\phi_{x'}\rangle$ and hence distinguishability among the $|\phi_x\rangle$.

The reduced density matrix on subsystem $\{1, \ldots, k-1\}$ is $\rho_S = \text{Tr}_0(\rho) = \sum_{x,x'} \beta_x\beta_{x'}^*\langle\phi_{x'}|\phi_x\rangle|x\rangle\langle x'|$. Since at least one inner product $\langle\phi_{x'}|\phi_x\rangle \neq 1$, ρ_S is not a rank-one projector and is therefore mixed.

By the Schmidt decomposition, any entangled pure state $|\Psi\rangle \in \mathcal{H}_A \otimes \mathcal{H}_B$ can be written as $|\Psi\rangle = \sum_{j=1}^r \sqrt{\lambda_j}|e_j\rangle_A \otimes |f_j\rangle_B$, where $\lambda_j > 0$, $\sum_j \lambda_j = 1$, and $r \geq 2$ is the Schmidt rank. The reduced state on subsystem A is then $\rho_A = \sum_j \lambda_j|e_j\rangle\langle e_j|$, which is mixed whenever $r > 1$. Applying this to the output state $|\Psi_{\text{out}}\rangle$, which is entangled across the target/control bipartition, it follows that the reduced state on the control subsystem $\{1, \ldots, k-1\}$ must be mixed. Since the reduced state $\rho_{\{1,\ldots,k-1\}}$ is mixed, it has a spectral decomposition $\rho_{\{1,\ldots,k-1\}} = \sum_j \lambda_j|\psi_j\rangle\langle\psi_j|$, and by linearity of the partial trace, the reduced state on any proper subsystem $S \subsetneq \{0, \ldots, k-1\}$ satisfies $\rho_S = \sum_j \lambda_j\rho_S^{(j)}$, where $\rho_S^{(j)} = \text{Tr}_{\bar{S}}(|\psi_j\rangle\langle\psi_j|)$.

For contradiction, let's suppose that ρ_S is pure, i.e., $\rho_S = |\chi\rangle\langle\chi|$ for some normalized $|\chi\rangle \in \mathcal{H}_S$. Then, being a convex combination of states $\rho_S^{(j)}$, the only possibility is that each $\rho_S^{(j)} = |\chi\rangle\langle\chi|$ as well. This implies that all eigenstates $|\psi_j\rangle$ of $\rho_{\{1,\ldots,k-1\}}$ have the same reduced density matrix on S, i.e., $\mathrm{Tr}_{\bar{S}}(|\psi_j\rangle\langle\psi_j|) = |\chi\rangle\langle\chi|$ for all j. In other words, the restriction of each $|\psi_j\rangle$ to subsystem S is indistinguishable, which contradicts the assumption that the global state $|\Psi_{\mathrm{out}}\rangle$ is entangled across the target/control bipartition. Entanglement implies that the reduced states on subsystems corresponding to different components of the Schmidt decomposition should not all be identical. If they were, it would indicate that the subsystem carries no information distinguishing among the global state's components, contradicting the presence of entanglement. Therefore, our assumption must be false, and we conclude that ρ_S is mixed. $\qquad\square$

5   Multi-Qubit Teleportation Protocol

We next propose an extension of the single-qubit protocol ([2]) to the multi-qubit case and show its correctness under the assumption of ideal (noiseless) decoherence conditions.

Let $|\psi\rangle_I \in \mathcal{H}_{0,1,\ldots,k-1}$ be an arbitrary k-qubit pure state, expressed as $|\psi\rangle_I = \sum_{x \in \{0,1\}^k} c_x |x\rangle_{0,1,\ldots,k-1}$. Alice first applies a reversible k-qubit Boolean function represented by a unitary operator U_f, yielding the encoded state $|\phi\rangle_I = U_f |\psi\rangle_I$. Simultaneously, Alice and Bob share k Bell pairs between qubits $(k, \ldots, 2k - 1)$ (Alice) and $(2k, \ldots, 3k - 1)$ (Bob), where each Bell pair is given by:

$$|\beta_{00}\rangle_{k+j,2k+j} = \frac{1}{\sqrt{2}} \left(|0\rangle_{k+j}|0\rangle_{2k+j} + |1\rangle_{k+j}|1\rangle_{2k+j} \right), \quad j = 0, \ldots, k-1.$$

The total initial state is $|\Psi_0\rangle = |\phi\rangle_{0,\ldots,k-1} \otimes \bigotimes_{j=0}^{k-1} |\beta_{00}\rangle_{k+j,2k+j}$.

Alice then applies a CNOT gate from each information qubit j to its corresponding Bell qubit $k + j$, followed by a Hadamard gate on each qubit j. She subsequently measures all of her qubits 0 through $2k - 1$ in the computational basis, obtaining $2k$ classical bits.

After measurement, Bob's qubits $(2k, \ldots, 3k-1)$ are left in the product state $\bigotimes_{j=0}^{k-1} X^{n_j} Z^{m_j} |\phi_j\rangle$, where $|\phi_j\rangle$ denotes the j-th qubit of $|\phi\rangle_I$.

Upon receiving the classical bits from Alice, Bob applies $Z^{m_j} X^{n_j}$ to each qubit $2k + j$ to recover the encoded state $|\phi\rangle_B = |\phi\rangle_{2k,\ldots,3k-1}$. Bob then applies the adjoint Boolean function operator $U_f^\dagger$, which reverses the U_f scrambling applied by Alice:

$$|\psi\rangle_B = U_f^\dagger |\phi\rangle_B = U_f^\dagger U_f |\psi\rangle_I = |\psi\rangle_I.$$

The initial state is now reflected in the measurement results after multiple protocol executions, each following a nondeterministic measurement operations performed by Alice and transmitted over an insecure classical channel. Figure 3 shows a Qiskit implementation of the protocol.

Theorem 4. (Correctness of k-Qubit Teleportation with Boolean Function Obfuscation).

Let $|\psi\rangle_I \in \mathcal{H}_{0,1,\ldots,k-1}$ be an arbitrary pure k-qubit state, and let U_f be a unitary corresponding to a reversible $k-1$-qubit Boolean function. Then, the teleportation protocol effectively transmits $|\psi\rangle_I$ to Bob, provided that Bob applies the appropriate local Pauli corrections determined by Alice's measurement outcomes, followed by the application of $U_f^\dagger$.

Proof. (Direct) We proceed step-by-step through the protocol and verify that Bob reconstructs ρ_I, the density matrix corresponding to the initial state.

Step 1 (Information Encoding): Alice begins with an arbitrary state. (Note that for the purposes of this proof we are making an exception to our default assumption for separable product states, as the protocol is correct for any arbitrary state preparation.)

$$\rho_I = \sum_{x \in \{0,1\}^k} \alpha_x |x\rangle_{0,1,\ldots,k-1}\langle x|_{0,1,\ldots,k-1}.$$

She applies U_f to obtain the new state $\rho_I' = U_f \rho_I U_f^\dagger$.

Step 2 (Entanglement): Alice and Bob share k Bell pairs, prepared as:

$$\bigotimes_{j=0}^{k-1} \frac{1}{\sqrt{2}} \left(|0\rangle_j |0\rangle_{j+k} + |1\rangle_j |1\rangle_{j+k} \right).$$

Thus, the full initial state of the system (Alice's $2k$ and Bob's k qubits) is:

$$\rho_{\text{init}} = \rho_I' \otimes \left(\bigotimes_{j=0}^{k-1} \frac{1}{\sqrt{2}} \left(|0\rangle_{k+j} |0\rangle_{2k+j} + |1\rangle_{k+j} |1\rangle_{2k+j} \right) \right).$$

Step 3 (Bell Measurement): For each $j = 0, \ldots, k-1$, Alice applies a CNOT gate with control qubit j and target qubit $k+j$, followed by a Hadamard gate on qubit j. Then, she measures her $2k$ qubits $(0, \ldots, 2k-1)$ in the computational basis. Letting the measurement outcomes be (m_j, n_j) for each j, where m_j is the measurement result on qubit j (Hadamard qubit) and n_j is the measurement result on qubit $k+j$ (target qubit), after measurement, the state of Bob's qubits $(2k, 2k+1, \ldots, 3k-1)$ collapses to:

$$\rho_B' = \left(\bigotimes_{j=0}^{k-1} X^{n_j} Z^{m_j} \right) \rho_B \left(\bigotimes_{j=0}^{k-1} X^{n_j} Z^{m_j} \right)^\dagger, \tag{1}$$

where ρ_B is the encoded state ρ_I' now existing on Bob's qubits $(2k, \ldots, 3k-1)$.

Step 4 (Pauli Corrections): Upon receiving the classical bits (m_j, n_j) from Alice, Bob applies $Z^{m_j-2k} X^{n_j-2k}$ for each qubit $j = 2k, \ldots, 3k-1$. This has the effect of choosing one of the possible states based on Alice's measurements.

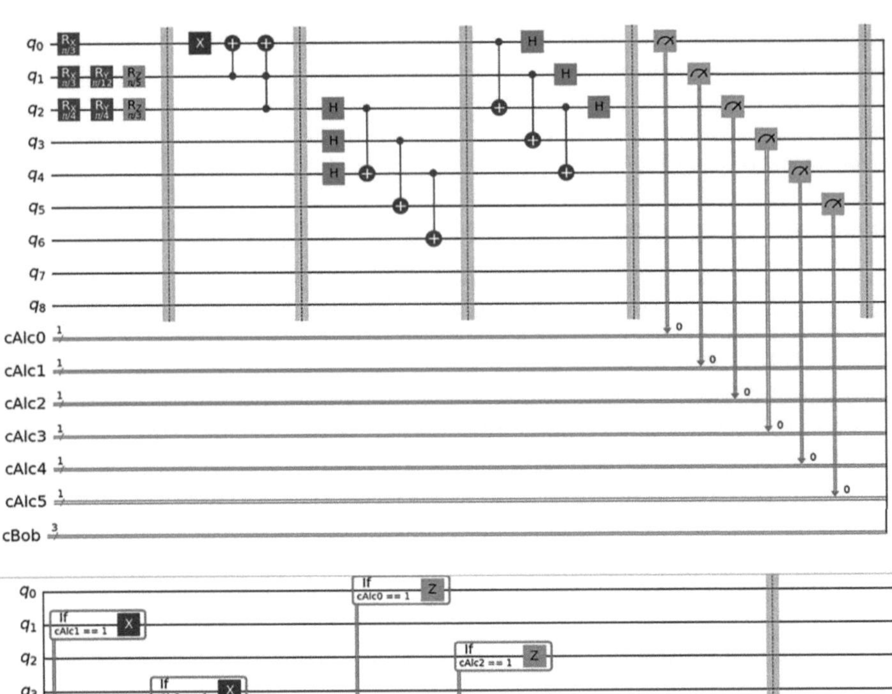

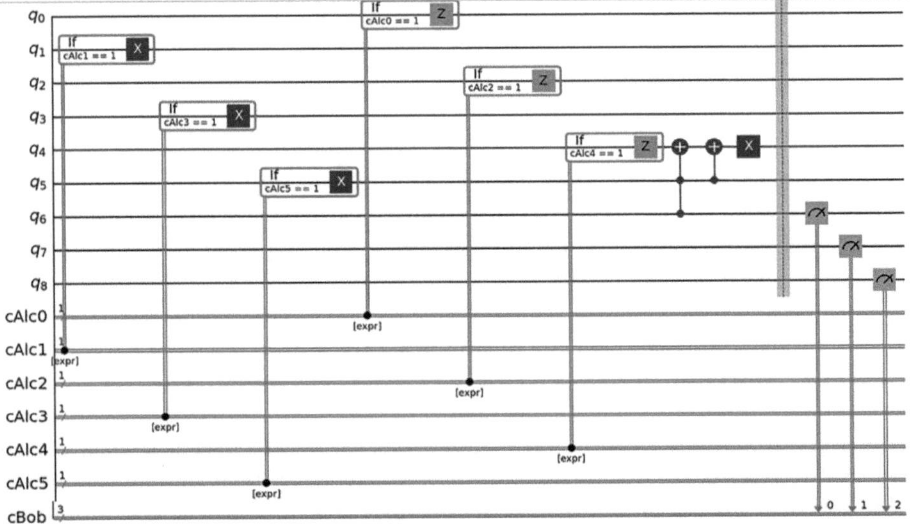

Fig. 3. 3-qubit teleportation circuit (q0, q1, and q2) are initialized via individual qubit rotations.

Thus, after correction, Bob holds: $\rho_B = U_f \rho'_B U_f^\dagger$, where his qubits are relabeled as B.

Step 5 (Boolean Function Inversion): Since the Boolean function operator is unitary, Bob finally applies $U_f^{-1} = U_f^\dagger$ to recover ρ_B. This ensures that applying $U_f^\dagger$ precisely reverses the transformation applied earlier, thereby recovering the original information state. More specifically, Bob must apply the inverse of each operation in the reverse order since for most Boolean functions $U_f^2 \neq I$, but $X^2 = I$, $\text{CNOT}^2 = I$, and $\text{Toffoli}^2 = I$. The inequality arises because quantum Boolean function circuits (QBFCs) involve entanglement and phase changes through controlled operations, such as the Toffoli gates, which depend on the states of control qubits.

Thus, the protocol correctly "transmits" the initial k-qubit state to Bob. That is, the teleported state ρ_I is reconstructed at Bob's location as reflected in the distribution of Bob's measurement outcomes. $\square$

6 Securing the Protocol's Classical Communication

Next, we impose constraints by design in order to ensure that the protocol provides *constant* levels of quantum entropy for the purposes of obfuscating the classical communication sent by Alice.

Based on Theorem 1, it is important to note that for the purposes of securing quantum teleportation, the target qubit can be restricted to initializations with rotations around the X-axis on the target qubit and only for any given specific fixed individual qubit rotations on the control qubits. We note that target qubit rotations along the Y-axis break the invariance condition on the entropy of reduced density matrices. Additionally, rotations of the target qubit about the Z-axis introduce only a global phase and are therefore undetectable by Bob as elaborated on previously in Remark 1. Consequently, the subsystem entropies, which depend only on the eigenvalues of the reduced density matrices, are unchanged. Therefore, the preparation of input states via Z-axis rotations on the target qubit does not affect classical channel scrambling.

7 Concluding Remarks and Future Work

We analyzed how input qubit rotations affect entropy behavior in quantum Boolean function circuits (QBFCs). We showed that target qubit rotation along the X-axis of the Bloch sphere at fixed control input preserves subsystem entropy under Boolean function unitaries, while rotation along Y-axis does not. We used those finding to design a multi-qubit teleportation protocol with Boolean function obfuscation and proved its correctness. We showed that secure classical communication is achievable using an insecure classical channel and a Boolean function-based unitary transformation under specific input constraints. Future work includes the investigation of entropy guarantees for QBFC with mixed and entangled input states.

References

1. Barenco, A., et al.: Smolin, and harald weinfurter. Elementary gates for quantum computation. Phys. Rev. A, 52:3457–3467 (1995). https://doi.org/10.1103/PhysRevA.52.3457
2. Bennett, C. H., et al.: Teleporting an unknown quantum state via dual classical and einstein-podolsky-rosen channels. Phys. Rev. Lett., 70:1895–1899 (1993). https://doi.org/10.1103/PhysRevLett.70.1895
3. Deutsch, D.: Quantum theory, the church-turing principle and the universal quantum computer. Proc. Royal Soc. London A **400**(1818), 97–117 (1985). https://doi.org/10.1098/rspa.1985.0070
4. Fastovets, D.V., Bogdanov, Y. I., Bogdanova, N.A., Lukichev, V.F.: Representation of boolean functions in terms of quantum computation. In: Vladimir, F., Lukichev, Konstantin, V., Rudenko., (eds) International Conference on Micro- and Nano-Electronics 2018. SPIE, (2019). https://doi.org/10.1117/12.2522053
5. Stuart Hadfield. On the representation of boolean and real functions as hamiltonians for quantum computing. ACM Trans. Quantum Comput. 2(4) (2021). https://doi.org/10.1145/3478519
6. Hirvensalo, M.: Studies on Boolean Functions Related to Quantum Computing. PhD thesis, Turku Centre for Computer Science, University of Turku (2003)
7. Kadry, H., Eldin, M. G., Zidane, M. G.: Decoding a teleported boolean function based on the extended deutsch-jozsa algorithm. Azerbaijan J. High Perf. Comput. 4(1):48–52 (2021). https://doi.org/10.32010/26166127.2021.4.1.48-52
8. Manev, K., Bakoev, V.: Algorithms for performing the zhegalkin transformation. In: Proceedings of the XXVII Spring Conference of the Union of Bulgarian Mathematicians, pp. 229–233, Pleven, Bulgaria, 1998. Union of Bulgarian Mathematicians. https://www.uni-vt.bg/userinfo/32/pub/22626/zheg.pdf
9. Sisodia, M.: A theoretical study of controlled quantum teleportation scheme for n-qubit quantum state. Int. J. Theor. Phys. **61**(12), 270 (2022). https://doi.org/10.1007/s10773-022-05260-1
10. Toffoli, T.: Reversible computing. In: Bakker, J. d., Leeuwen, J. v., (eds) Automata, Languages and Programming, pp. 632–644, Berlin, Heidelberg, 1980. Springer Berlin Heidelberg. https://doi.org/10.1007/3-540-10003-2_104
11. Yoshida, B., Yao, N.Y.: Disentangling scrambling and decoherence via quantum teleportation. Physical Review X, 9(1):011006 (2019). https://doi.org/10.1103/PhysRevX.9.011006
12. Žegalkin, I. I.: Sur le calcul des propositions dans la logique symbolique. Rec. Math. Moscou, 34:9–28 (1927). http://mi.mathnet.ru/eng/msb7433

Balance-Based Cryptography: Physically Computing Any Boolean Function

Suthee Ruangwises[(✉)][iD]

Department of Computer Engineering, Faculty of Engineering, Chulalongkorn
University, Bangkok, Thailand
`suthee@cp.eng.chula.ac.th`

Abstract. Secure multi-party computation is an area in cryptography
which studies how multiple parties can compare their private information without revealing it. Besides digital protocols, many unconventional
protocols for secure multi-party computation using physical objects have
also been developed. The vast majority of them use playing cards as the
main tools. In 2024, Kaneko et al. introduced the use of a balance scale
and coins in zero-knowledge proof protocols for pencil puzzles. In this
paper, we extend the use of these tools to secure multi-party computation. In particular, we develop four protocols that can securely compute
any n-variable Boolean function using a balance scale and coins.

Keywords: secure multi-party computation · physical cryptography ·
card-based cryptography · Boolean function · balance scale

1 Introduction

Secure multi-party computation (MPC) is one of the most actively studied areas
in cryptography. It investigates how multiple parties can compare their private
information without revealing it.

One of the best-known classical examples of MPC involves the following problem. Alice and Bob want to know whether they both like each other. However,
no one wants to confess first due to fear of embarrassment of getting rejected.
They need a protocol that only distinguishes the two cases where they both like
each other and otherwise without leaking any other information. Theoretically,
this setting is equivalent to computing a logical AND function of two input bits,
one from each player.

Besides the AND function, other widely studied Boolean functions include a
logical XOR function, a *majority function* (deciding whether there are more 1 s
than 0 s in the inputs), and an *equality function* (deciding whether all inputs are
equal).

Instead of digital protocols, many researchers have developed unconventional
protocols for MPC using physical objects found in everyday life such as cards,
coins, and envelopes. These protocols have the benefit that they do not require
computers and also allow external observers to verify that all parties truthfully

© The Author(s), under exclusive license to Springer Nature Switzerland AG 2026
E. Formenti and L. Manzoni (Eds.): UCNC 2025, LNCS 16364, pp. 63–72, 2026.
https://doi.org/10.1007/978-3-032-15641-9_5

execute them (which is often a challenging task for digital protocols). In addition, they are easier to understand and to verify the correctness and security, even for non-experts, and thus can be used for didactic purposes.

The vast majority of existing physical protocols use playing cards as the main tools, thus this area of research is often called *card-based cryptography*. During the past decade, card-based cryptography has rapidly gained interest and has been extensively studied, with dozens of papers published [7,8,19].

1.1   Related Work

Research in card-based cryptography began in 1989 when den Boer [4] introduced the *five-card trick* protocol to solve the classical Alice-and-Bob problem. This protocol can compute the AND function of two input bits from two players using five cards: three identical black cards and two identical red cards.

Mizuki and Sone [14] later improved the AND protocol to accommodate n inputs. Apart from the AND function, card-based protocols to compute other Boolean functions have also been proposed, including the XOR function [14], the majority function [18,22], and the equality function [20]. Nishida et al. [17] proved that any n-variable Boolean function can be computed using $2n+6$ cards, and any such function that is symmetric can be computed using $2n+2$ cards.

Besides cards, other physical objects such as coins [9,10], envelopes [15], combination locks [12], balls and bags [11], printed transparencies [3], and PEZ candy dispensers [1,2,16] have also been used in cryptographic protocols.

In 2024, Kaneko et al. [5] were the first ones to use a balance scale and coins in a cryptographic protocol.[1] Their protocols, however, were not designed to compute functions in the MPC setting, but are zero-knowledge proof protocols to verify solutions of pencil puzzles like Sudoku. Very recently, Kaneko et al. [6] also proposed a voting protocol using the same tools, which can compute a specific function in the MPC setting.

1.2   Our Contribution

In this paper, we extend the use of a balance scale and coins as physical tools to securely compute arbitrary Boolean functions, providing a new physical model for a generic MPC setting. Note that coins can be replaced by any small objects with significant weights like balls or marbles, but we use coins to preserve the setting proposed by Kaneko et al. [5].

In particular, we develop four protocols that securely compute n-variable Boolean functions using a balance scale and coins. The first protocol can compute the AND function. The second one can compute any *threshold function* (including the AND function and the majority function). The third protocol can compute any symmetric function, while the fourth one can compute any Boolean

---

[1] While some earlier protocols [9,10] also used coins, only faces (head or tail) were used to encode inputs without considering the weight of the coins.

function. See Table 1 for the number of coins, bags, and comparisons (times using the balance scale to weigh objects) required for each protocol.

Our protocols are also practical. Unlike the two existing balance-based protocols [5,6] which use same-size coins with n different weights, our protocols only use coins with two different weights, removing a practical challenge to implement them in real world.

Table 1. Number of coins, bags, and comparisons required for each protocol to compute an n-variable Boolean function

Protocol	Function	#Coins	#Bags	#Comparisons	Notes
Protocol 1 (Sect. 3)	AND	$3n$	0	1	
Protocol 2 (Sect. 4)	threshold	$2n$	0	1	Uses a custom weight
Protocol 3 (Sect. 5)	symmetric	at most $n\lceil\frac{n}{2}+2\rceil$	at most $\lceil\frac{n}{2}+1\rceil$	at most $\lceil\frac{n}{2}\rceil$	Uses a pen
Protocol 4 (Sect. 6)	any	at most $n(2^n+1)$	at most 2^{n-1}	at most 2^{n-1}	

2 Preliminaries

2.1 Balance Scale

Our protocols use a double-pan balance scale (see Fig. 1). When placing objects on both pans, the scale will tilt towards the heavier side at a constant angle (regardless of weight difference), or will stay put if the objects on both pans have equal weight.

Note that in reality, each balance scale may have a tolerance ε, where the scale will stay put (due to friction) if the difference of weights on both pans is at most ε.

2.2 Coins

Two types of coin, called *heavy coins* and *light coins*, are used in our protocols. All coins have the same size and color (see Fig. 1). Each heavy coin weighs w, while each light coin weighs $w - \delta$. We assume that δ is much larger than ε, so the balance scale can detect the difference between heavy and light coins.

2.3 Custom Weights

A custom weight is only used in the threshold function protocol in Sect. 4. The exact weight depends on the function being computed, but is always in the form $pw + q(w - \delta) + \frac{\delta}{2}$, where $p, q \in \mathbb{Z}^+ \cup \{0\}$.

When computing multiple different threshold functions, instead of preparing a lot of different custom weights, it is sufficient to prepare only one custom weight of $\frac{\delta}{2}$, as all weights in such form can be obtained by combining this custom weight with p heavy coins and q light coins.

Fig. 1. Examples of a balance scale and coins (left) and bags (right) used in our protocols.

2.4   Bags and Shuffle Operation

Bags used in our protocols are small opaque bags with the same size, color, and weight, and can be closed at the top (see Fig. 1). Each bag can contain multiple coins and can be placed on a balance scale.

Given k bags, each with the same number of coins inside (but with slightly different weights), the players may jointly *shuffle* all bags into a uniformly random permutation unknown to all players. This operation has previously been used in [11], where there are balls inside each bag.

The shuffle operation can be done in real world by all players scrambling the bags together on a table. We assume that no player can determine the exact weight of any bag just by touching it.

3   Computing AND Function

First, we propose the following protocol for one of the most fundamental functions: the AND function. In a setting with n players, where the i-th player has a bit x_i, this protocol securely computes $x_1 \wedge x_2 \wedge ... \wedge x_n$.

Each player is given a heavy coin (with weight w) and a light coin (with weight $w - \delta$). Each player then chooses one of the two coins to place on the same side of a balance scale. If the player's bit is 1, the player chooses the heavy coin; if the player's bit is 0, the player chooses the light coin.

We now have n coins placed on one side of the scale. Then, the players jointly place n heavy coins on the other side of the scale. If the weights of the two sides are equal, return 1; otherwise, return 0.

This protocol is correct because the only case where the weights are equal is when all players' coins are heavy coins (and the players' coins are lighter in all other cases), which occurs when all players' bits are 1 s. Moreover, each player learns no information other than whether the n coins combined have weight nw

or less than nw. Therefore, the protocol is correct and secure. This protocol uses $3n$ coins (two given to each player plus n additional heavy coins) and uses one comparison.

4   Computing Threshold Function

For a positive integer $k \leq n$ and bits $x_1, x_2, ..., x_n \in \{0, 1\}$, a Boolean *threshold function* T_k returns 1 if and only if at least k of the n input bits are 1 s. Formally,

$$T_k(x_1, x_2, ..., x_n) := \begin{cases} 1, & \text{if } \sum_{i=1}^{n} x_i \geq k; \\ 0, & \text{otherwise.} \end{cases}$$

Note that T_k is the AND function when $k = n$, and is the majority function when $k = \lfloor \frac{n}{2} \rfloor + 1$.

We can slightly modify the AND protocol in Sect. 3 to compute any threshold function. Similarly to the AND protocol, each player is given a heavy coin and a light coin, and then chooses the heavy (resp. light) coin if the player's bit is 1 (resp. 0) to place on the same side of a balance scale.

The players then place a custom weight $W = (k-1)w + (n-k+1)(w-\delta) + \frac{\delta}{2}$ on the other side of the scale.[2] If the n coins combined are heavier than W, return 1; otherwise, return 0.

Observe that W is heavier than $k - 1$ heavy coins plus $n - k + 1$ light coins, but is lighter than k heavy coins plus $n - k$ light coins. As a result, the n coins are heavier than W if and only if at least k of them are heavy coins (and the n coins are lighter in all other cases). This occurs when at least k of the players' bits are 1 s, i.e. $\sum_{i=1}^{n} x_i \geq k$, which is the exact condition for $T_k(x_1, x_2, ..., x_n) = 1$. Moreover, each player learns no information other than whether the n coins are heavier or lighter than W. Therefore, the protocol is correct and secure.

This protocol uses $2n$ coins and a custom weight W (or $3n$ coins and a custom weight of $\frac{\delta}{2}$), and uses one comparison.

5   Computing Symmetric Function

A Boolean function $f : \{0, 1\}^n \to \{0, 1\}$ is called symmetric if

$$f(x_1, x_2, ..., x_n) = f(x_{\sigma(1)}, x_{\sigma(2)}, ..., x_{\sigma(n)})$$

for any $x_1, x_2, ..., x_n \in \{0, 1\}$ and any permutation $\sigma : \{1, 2, ..., n\} \to \{1, 2, ..., n\}$. Note that for any such function f, the value of $f(x_1, x_2, ..., x_n)$ only depends on the number of inputs being 1 s, i.e. the integer sum $\sum_{i=1}^{n} x_i$.

We denote an n-variable symmetric Boolean function by S_X^n for some $X \subseteq \{0, 1, ..., n\}$. A function S_X^n is defined by

---

[2] Alternatively, the weight W can be obtained by placing $k - 1$ heavy coins, $n - k + 1$ light coins, and a custom weight of $\frac{\delta}{2}$.

$$S^n_X(x_1, x_2, ..., x_n) := \begin{cases} 1, & \text{if } \sum_{i=1}^n x_i \in X; \\ 0, & \text{otherwise.} \end{cases}$$

For example, the AND function $x_1 \wedge x_2 \wedge ... \wedge x_n$ is denoted by $S^n_{\{n\}}$, and the XOR function $x_1 \oplus x_2 \oplus ... \oplus x_n$ is denoted by $S^n_{\{1,3,5,...,2\lceil \frac{n}{2} \rceil - 1\}}$.

To compute S^n_X, we prepare $|X|$ bags, each one corresponding to each element of X. A bag corresponding to $k \in X$ contains k heavy coins and $n-k$ light coins.

We also prepare an additional bag, called a *special bag*, for players to put coins in. Each player is given a heavy coin and a light coin, and then chooses the heavy (resp. light) coin if the player's bit is 1 (resp. 0) to put in this bag.

The players use a pen to make a small mark inside of the special bag. This mark must not visible from the outside without opening the bag. After that, all players together shuffle all $|X|$ non-special bags randomly. Then, the players jointly perform the following steps.

1. Pick an arbitrary non-special bag. Shuffle this bag with the special bag randomly.
2. Place each of these two bags on each side of the scale. If the weights of the two sides are equal, return 1 and end the protocol.
3. Take both bags from the scale. Shuffle them randomly again.
4. Open both bags and find a pen mark to locate the special bag. Then, close both bags and disregard the non-special bag.

Repeat the above steps for a new non-special bag. After $|X|$ iterations, if no non-special bag has weight equal to the special bag, return 0.

5.1   Proofs of Correctness and Security

The protocol returns 1 when there is a non-special bag with weight equal to the special bag (and there can be at most one such bag). Suppose this non-special bag corresponds to an element $k \in X$ (containing k heavy coins and $n-k$ light coins). This case occurs if and only if exactly k of the n players' coins are heavy coins, which is equivalent to $\sum_{i=1}^n x_i = k$. Therefore, the protocol returns 1 if and only if $\sum_{i=1}^n x_i \in X$, which is the exact condition for $S^n_X(x_1, x_2, ..., x_n) = 1$.

Consider the process of comparing a special bag and a non-special bag in each iteration. Due to the shuffles both before and after weighing, no player can distinguish which side of the scale is a special bag. Therefore, all information each player learns is whether or not both bags have equal weight.

Also, due to the shuffle of all $|X|$ non-special bags into a uniformly random permutation, no player can learn about which non-special bag corresponds to which element of X. Therefore, all information each player learns is whether or not there is a non-special bag with weight equal to the special bag. Therefore, the protocol is correct and secure.

5.2   Optimization

In the protocol, we need $|X| + 1$ bags to compute S_X^n. However, if $|X| > \frac{n+1}{2}$, we can instead compute the function $S_{\{0,1,\dots,n\}-X}^n$ and then negate the output. In such cases, we have $|\{0, 1, \dots, n\} - X| = n + 1 - |X| < \frac{n+1}{2}$. Therefore, the number of required bags can be reduced to at most $\lfloor \frac{n+1}{2} \rfloor + 1 = \lceil \frac{n}{2} \rceil + 1$.

The number of comparisons is equal to the number of non-special bags $|X|$, which can be reduced to at most $\lceil \frac{n}{2} \rceil$. The number of required coins is $n(|X|+2)$ (two given to each player plus n in each non-special bag), which can be reduced to at most $n(\lceil \frac{n}{2} \rceil + 2)$.

6   Computing Any Boolean Function

Finally, we propose the following protocol to compute an arbitrary n-variable Boolean function among n players, where the i-th player has a bit x_i.

For any Boolean function $f : \{0,1\}^n \to \{0,1\}$, we can write f as Boole's expansion (or Shannon expansion) [21]:

$$
\begin{aligned}
f(x_1, x_2, \dots, x_n) &= \bar{x}_1 \bar{x}_2 \dots \bar{x}_n f(0, 0, \dots, 0) \vee x_1 \bar{x}_2 \dots \bar{x}_n f(1, 0, \dots, 0) \\
&\quad \vee \bar{x}_1 x_2 \dots \bar{x}_n f(0, 1, \dots, 0) \vee x_1 x_2 \dots \bar{x}_n f(1, 1, \dots, 0) \\
&\quad \vee \dots \vee x_1 x_2 \dots x_n f(1, 1, \dots, 1) \\
&= \bigvee_{(b_1, b_2, \dots, b_n) \in \{0,1\}^n} (x_1 \leftrightarrow b_1)(x_2 \leftrightarrow b_2)\dots(x_n \leftrightarrow b_n) f(b_1, b_2, \dots, b_n),
\end{aligned}
$$

where $\vee$ and $\leftrightarrow$ are Boolean OR and XNOR (i.e. if and only if) operations, respectively. Observe that we only need to consider the terms where $f(b_1, b_2, \dots, b_n) = 1$, since the rest of the terms are all zeroes. The output $f(x_1, x_2, \dots, x_n)$ is 1 if and only if at least one such term is 1.

Let $B_f = \{(b_1, b_2, \dots, b_n) \in \{0,1\}^n | f(b_1, b_2, \dots, b_n) = 1\}$. We prepare $|B_f|$ bags, each one corresponding to each element of B_f. Also, each player is given 2^{n-1} heavy coins and 2^{n-1} light coins.

For each bag corresponding to $(b_1, b_2, \dots, b_n)$, the i-th player puts into that bag a heavy coin if $x_i = b_i$, and a light coin if $x_i \neq b_i$.[3]

After that, all players together shuffle all $|B_f|$ bags randomly. Then, for each bag, they pour the coins inside it on one side of a balance scale, and put n heavy coins on the other side. Repeat this for every bag. If there is a bag where the weights of the two sides are equal, return 1; otherwise, return 0.

Note that for a special case where $|B_f| = 1$, a bag is not necessary; the players can directly place coins on the scale similarly to the AND protocol.

---

[3] We assume a *semi-honest model* [13], where each player puts a coin into every bag consistently according to the rule and does not try to sabotage or hack the protocol.

6.1   Example: Three-Input XOR Function

To clearly illustrate the protocol, we show an example to compute the three-variable XOR function $f(x_1, x_2, x_3) = x_1 \oplus x_2 \oplus x_3$.

First, prepare four bags corresponding to (1,0,0), (0,1,0), (0,0,1), and (1,1,1), respectively. Each player is given two heavy coins and two light coins.

Each player then puts a coin in each bag according to Table 2. For instance, if $x_1 = 1$, then Player 1 puts a heavy coin in Bag (1,0,0) and Bag (1,1,1), and a light coin in Bag (0,1,0) and Bag (0,0,1).

Table 2. The coin each player putting in each bag

Player		Bag			
		(1,0,0)	(0,1,0)	(0,0,1)	(1,1,1)
Player 1	$x_1 = 1$	heavy	light	light	heavy
	$x_1 = 0$	light	heavy	heavy	light
Player 2	$x_2 = 1$	light	heavy	light	heavy
	$x_2 = 0$	heavy	light	heavy	light
Player 3	$x_3 = 1$	light	light	heavy	heavy
	$x_3 = 0$	heavy	heavy	light	light

After that, all players together shuffle all four bags randomly. Then, for each bag, they pour the coins inside it on one side of a balance scale, and put three heavy coins on the other side. Repeat this for every bag. If there is a bag where the weights of the two sides are equal, return 1; otherwise, return 0.

6.2   Proofs of Correctness and Security

The protocol returns 1 when there is a bag corresponding to $(b_1, b_2, ..., b_n) \in B_f$ containing n heavy coins, which occurs when $x_i = b_i$ for every $i = 1, 2, ..., n$ (and there can be at most one such bag). Therefore, the protocol returns 1 if and only if $(x_1, x_2, ..., x_n) \in B_f$, which is the exact condition for $f(x_1, x_2, ..., x_n) = 1$.

Consider the process of comparing n heavy coins with the n coins inside each bag. Each player learns no information other than whether the n coins inside the bag have weight nw or less than nw, i.e. whether or not they are all heavy coins. Also, due to the shuffle of all $|B_f|$ bags into a uniformly random permutation, no player can learn about which bag correspond to which element of B_f. Therefore, all information each player learns is whether or not there is a bag containing all heavy coins. Therefore, the protocol is correct and secure.

6.3   Optimization

In the protocol, we need $|B_f|$ bags to compute f. However, if $|B_f| > 2^{n-1}$, we can instead compute the function $\bar{f}(x_1, x_2, ..., x_n) := 1 - f(x_1, x_2, ..., x_n)$ and then negate the output. In such cases, we have $|B_{\bar{f}}| = 2^n - |B_f| < 2^{n-1}$. Therefore, the number of required bags can be reduced to at most 2^{n-1}. As the number of comparisons is equal to the number of bags, it can also be reduced to at most 2^{n-1}.

For $1 \leq i \leq n$, let p_i (resp. q_i) be the numbers of $(b_1, b_2, ..., b_n) \in B_f$ such that $b_i = 0$ (resp. $b_i = 1$). Observe that the i-th player has to use at most $\max(p_i, q_i)$ heavy coins and at most $\max(p_i, q_i)$ light coins (we have $\max(p_i, q_i) \leq |B_f| \leq 2^{n-1}$). As a result, we can give only $\max(p_i, q_i)$ heavy coins and $\max(p_i, q_i)$ light coins to the i-th player (instead of the maximum 2^{n-1} heavy coins and 2^{n-1} light coins). For instance, in the example in Sect. 6.1, each player receives only two heavy coins and two light coins (instead of four heavy coins and four light coins). The total number of required coins is at most $n(2^n + 1)$ (at most 2^n given to each player plus n additional heavy coins).

7   Conclusion

We developed four balance-based protocols that can compute arbitrary n-variable Boolean functions, and proved the correctness and security of these protocols. By doing so, we extended the use of a balance scale and coins to a generic MPC setting, implying that they have computational power equal to cards, PEZ candy dispensers, and other existing physical tools.

References

1. Abe, Y., Iwamoto, M., Ohta, K.: Efficient private PEZ protocols for symmetric functions. In: Hofheinz, D., Rosen, A. (eds.) TCC 2019. LNCS, vol. 11891, pp. 372–392. Springer, Cham (2019). https://doi.org/10.1007/978-3-030-36030-6_15
2. Balogh, J., Csirik, J.A., Ishai, Y., Kushilevitz, E.: Private computation using a PEZ dispenser. Theor. Comput. Sci. **306**(1–3), 69–84 (2003)
3. D'Arco, P., De Prisco, R.: Secure computation without computers. Theor. Comput. Sci. **651**, 11–36 (2016)
4. Boer, B.: More efficient match-making and satisfiability *The Five Card Trick*. In: Quisquater, J.-J., Vandewalle, J. (eds.) EUROCRYPT 1989. LNCS, vol. 434, pp. 208–217. Springer, Heidelberg (1990). https://doi.org/10.1007/3-540-46885-4_23
5. Kaneko, S., Lafourcade, P., Mallordy, L.-B., Miyahara, D., Puys, M., Sakiyama, K.: Balance-based ZKP protocols for pencil-and-paper puzzles. In: Proceedings of the 27th Information Security Conference (ISC), pp. 211–231 (2024)
6. Kaneko, S., Lafourcade, P., Mallordy, L.-B., Miyahara, D., Puys, M., Sakiyama, K.: Secure voting protocol using balance scale. In: Proceedings of the 17th International Symposium on Foundations and Practice of Security (FPS), pp. 365–376 (2025)
7. Koch, A.: The landscape of optimal card-based protocols. Math. Cryptol. **1**(2), 115–131 (2021)
8. Koch, A.: The landscape of security from physical assumptions. In: *Proceedings of the 2021 IEEE Information Theory Workshop (ITW)*, pp. 1–6 (2021)
9. Komano, Y., Mizuki, T.: Coin-based secure computations. Int. J. Inf. Secur. **21**(4), 833–846 (2022)

10. Minamikawa, Y., Shinagawa, K.: Coin-based cryptographic protocols without hand operations. IEICE Trans. Fund. **107**(8), 1178–1185 (2024)
11. Miyahara, D., Komano, Y., Mizuki, T., Sone, H.: Cooking cryptographers: secure multiparty computation based on balls and bags. In: Proceedings of the 34th IEEE Computer Security Foundations Symposium (CSF), pp. 1–16 (2021)
12. Mizuki, T., Kugimoto, Y., Sone, H.: Secure multiparty computations using a dial lock. In: Cai, J.-Y., Cooper, S.B., Zhu, H. (eds.) TAMC 2007. LNCS, vol. 4484, pp. 499–510. Springer, Heidelberg (2007). https://doi.org/10.1007/978-3-540-72504-6_45
13. Mizuki, T., Shizuya, H.: Practical card-based cryptography. In: Proceedings of the 7th International Conference on Fun with Algorithms (FUN), pp. 313–324 (2014)
14. Mizuki, T., Sone, H.: Six-card secure AND and four-card secure XOR. In: Deng, X., Hopcroft, J.E., Xue, J. (eds.) FAW 2009. LNCS, vol. 5598, pp. 358–369. Springer, Heidelberg (2009). https://doi.org/10.1007/978-3-642-02270-8_36
15. Moran, T., Naor, M.: Polling with physical envelopes: a rigorous analysis of a human-centric protocol. In: Vaudenay, S. (ed.) EUROCRYPT 2006. LNCS, vol. 4004, pp. 88–108. Springer, Heidelberg (2006). https://doi.org/10.1007/11761679_7
16. Murata, S., Miyahara, D., Mizuki, T., Sone, H.: Public-PEZ cryptography. In: Susilo, W., Deng, R.H., Guo, F., Li, Y., Intan, R. (eds.) ISC 2020. LNCS, vol. 12472, pp. 59–74. Springer, Cham (2020). https://doi.org/10.1007/978-3-030-62974-8_4
17. Nishida, T., Hayashi, Y., Mizuki, T., Sone, H.: Card-based protocols for any Boolean function. In: Jain, R., Jain, S., Stephan, F. (eds.) TAMC 2015. LNCS, vol. 9076, pp. 110–121. Springer, Cham (2015). https://doi.org/10.1007/978-3-319-17142-5_11
18. Nishida, T., Mizuki, T., Sone, H.: Securely computing the three-input majority function with eight cards. In: Dediu, A.-H., Martín-Vide, C., Truthe, B., Vega-Rodríguez, M.A. (eds.) TPNC 2013. LNCS, vol. 8273, pp. 193–204. Springer, Heidelberg (2013). https://doi.org/10.1007/978-3-642-45008-2_16
19. Ruangwises, S.: The landscape of computing symmetric n-variable functions with $2n$ cards. In: Proceedings of the 20th International Colloquium on Theoretical Aspects of Computing (ICTAC), pp. 74–82 (2023)
20. Ruangwises, S., Itoh, T.: Securely computing the n-variable equality function with 2n cards. Theor. Comput. Sci. **887**, 99–110 (2021)
21. Sasao, T.: Switching Theory for Logic Synthesis, 1st edn. Kluwer Academic Publishers, Norwell (1999)
22. Toyoda, K., Miyahara, D., Mizuki, T.: Another use of the five-card trick: card-minimal secure three-input majority function evaluation. In: Proceedings of the 22nd International Conference on Cryptology in India (INDOCRYPT), pp. 536–555 (2021)

Benchmarking ORCA PT-1 Boson Sampler in Simulation

Jessica Park[1](✉) [iD], Susan Stepney[1] [iD], and Irene D'Amico[2] [iD]

[1] Department of Computer Science, University of York, York, UK
jlp567@york.ac.uk
[2] Department of Physics, University of York, York, UK

Abstract. Boson Sampling, a non-universal computing paradigm, has resulted in impressive claims of quantum supremacy. ORCA Computing have developed a time-bin interferometer (TBI) that claims to use the principles of boson sampling to solve a number of computational problems including optimisation and generative adversarial networks [5]. We solve a dominating set problem with a surveillance use case on the ORCA TBI simulator to benchmark the use of these devices against classical algorithms. Simulation has been used to consider the optimal performance of the computing paradigm without having to factor in noise, errors and scaling limitations. We show that the ORCA TBI is capable of solving moderately sized ($n < 250$) dominating set problems with comparable success to linear programming and greedy methods. Wall clock timing shows that the simulator has worse scaling than the classical methods, but this is unlikely to carry over to the physical device where the outputs are measured rather than calculated.

Keywords: Boson Sampling · Quantum Computing · Benchmarking

1 Introduction

Boson sampling is a proposed non-universal form of quantum computing. It has often been considered a toy problem: Potentially proving quantum supremacy[1] over its circuit model counterparts, but effectively without practical applications [2]. However, there is an increasing school of thought that boson sampling does have the potential to bring quantum advantage in a number of real world, computationally challenging problems [6].

Aside from quantum advantage and quantum supremacy, there are other measures of utility that may motivate the development of photonic quantum computers. There is growing concern over the environmental impact of supercomputers and large data centres and quantum computers may be more energy

---

[1] The exact definition of quantum supremacy and its relation to the phrases 'quantum advantage' or 'quantum utility' are not the topic of this section and is assumed to be uncontroversial here.

© The Author(s), under exclusive license to Springer Nature Switzerland AG 2026
E. Formenti and L. Manzoni (Eds.): UCNC 2025, LNCS 16364, pp. 73–87, 2026.
https://doi.org/10.1007/978-3-032-15641-9_6

efficient [3,20]. This is especially likely to be true of boson samplers as photons are stable at room temperature and do not require cryogenic cooling like superconducting qubits.

Here we use a boson sampling simulator to solve a dominating set problem with a surveillance use case on the ORCA TBI simulator, as a benchmark for the use of these devices against classical algorithms. The use of simulation allows us to consider the optimal performance of the computing paradigm without having to factor in noise, errors and scaling limitations. Future work should extend this study to the real hardware to consider both optimisation success and the timing compared to both the simulator and the classical algorithms.

2   Background

2.1   Boson Sampling Theory

Boson sampling involves sampling from a distribution of identical bosons (typically photons) that have been through a linear interferometer.

A linear interferometer contains a number of beamsplitters, optical components used to split a beam of light into transmitted and reflected components. When a photon reaches a beamsplitter, it can either be reflected or transmitted, traversing a different path through the interferometer, with a probability depending on the *angle* of the beamsplitter. Photon detectors are placed at the end of each possible paths; the number of possible end locations is the number of *modes* in the system. The count from an end location is the occupation of that particular mode. Two key parameters in practical boson sampling setups are the number of input photons, N, and the number of modes in the system, M. The total occupation over all the modes should be equal to the number of input photons, however these are often thresholded to given a binary readout of occupied and unoccupied modes.

When a photon hits a beamsplitter, the probability of reflection or transmission is controlled by a parameter called the *angle* of the beamsplitter. With multiple indistinguishable photons and multiple beamsplitters, there are a number of ways that each possible output state could be obtained; calculating the expected distribution in advance is a computationally expensive problem, equivalent to calculating the permanent[2] of the matrix that characterises the interferometer [11]. Figure 1 shows an example small interferometer.

The Hong-Ou-Mandel effect describes the non-classical outcome of two indistinguishable photons coincident on a beam splitter. It states that the two photons always exit in the same output mode, and never take different paths [15]. The outcomes where the photons would end in opposite modes destructively interfere, and therefore have a probability of 0. This occurs only if the two photons are identical and coincident, which stresses the importance of controlled photon

---

[2] The permanent of a matrix is similar to the determinant in that it is a polynomial of the matrix coefficients [16], but unlike the determinant, there is no known efficient classical algorithm to compute it.

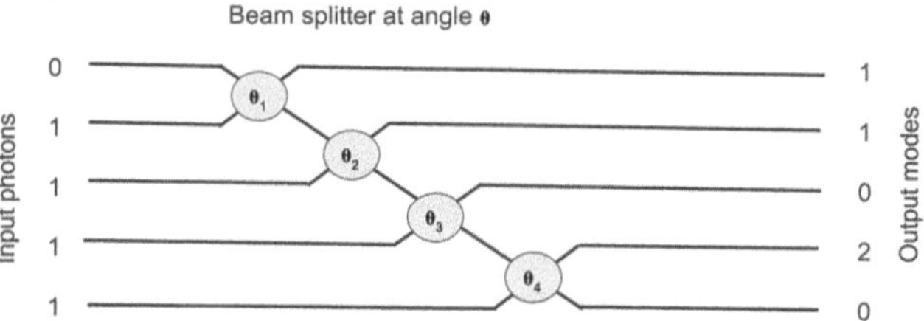

Fig. 1. A simple boson sampling set up with 4 input photons, 5 output modes and 4 variable beam splitters each with angle θ_i.

generation and tightly controlled distances between components. This has also been generalised to multi-photon interference [23].

The Hong-Ou-Mandel effect is the key element in making boson-sampling a quantum process because it relies on the quantised photonic nature of light. The resulting path of the photonic pair is unknown until detection; we can think of the photons taking a superposition of the possible paths. Once detected, the superposition collapses and the path taken by the photon is known through the mode it was detected in.

Calculating the amplitude of all terms in the output state is #P-complete, and there are an exponential number of terms in the output [2]. Even approximating these amplitudes up to a multiplicative error is #P hard. It is challenging to classically simulate boson sampling for more than 10 s of photons.

Photonics is a promising medium for quantum computation: photons are stable at room temperature, so do not require expensive cryogenics. They do not react strongly with the environment. They move at light speed, which presents both benefits and challenges when working with photonic qubits. The high speed can make operations difficult to control, and it requires high precision to get indistinguishable photons in terms of timings.

A large source of error in photonic systems are 'lossy components'. Theoretical beamsplitters as lossless; there is no chance of photon absorption In reality, beamsplitters can absorb photons, resulting in fewer output than input photons [4]. Another source of error is 'dark counts', where a photon detector registers a detection even though no photon is present. This is likely to be a random error, and its likelihood of occurring should be determined by the manufacturer of the device. It is also possible for the performance of detectors to deteriorate over time. Therefore, detectors should be calibrated frequently, to ensure the error rate of dark counts is known and of an acceptable level.

In simulation and in theory, beam splitter angles can be set to arbitrary precision, but in experimental set up, there is a limit to the achievable precision. There may also be systematic errors in the setting of beam splitter angles, such as a constant offset. This can be tested and calibrated in small setups.

The photons needs to be indistinguishable, both temporally and in frequency. Any temporal mismatch will cause errors. Small temporal displacement of the wavepacket will manifest as dephasing, but when this becomes larger, it may

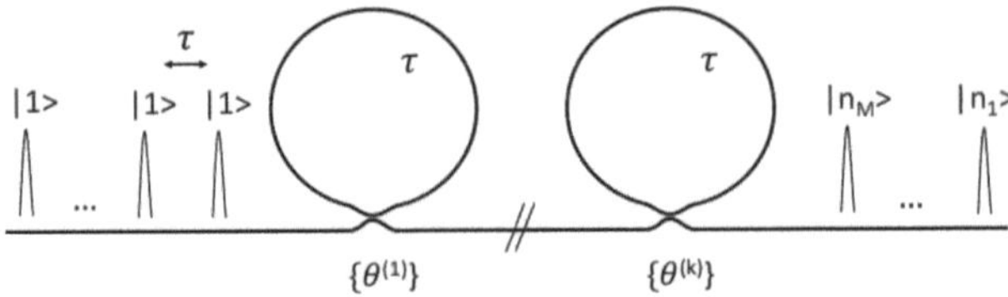

Fig. 2. Diagram showing time-bin interferometry. Taken from [5].

become ambiguous which time-bin the photon should be in. Temporal displacement can be caused by errors in the path lengths, implemented by fibre loops, or by jitter in the photon source. Errors caused by fibre loop path lengths could potentially be replaced by a quantum memory or free space delay line.

2.2   ORCA PT Series

ORCA computing have designed and built a boson sampler that uses a time-bin interferometer [5] (unlike the spatial binning of Fig. 1), see Fig. 2. Time-binning uses only a single photon source and a single detector. Photons are inserted into the interferometer at regular time intervals; beam splitters control the access to delay loops. The photon modes are determined by the time-bins over which the detector may or may not receive a photon [14,17].

To translate the spatial boson sampler into a time-bin interferometer, consider the input photons entering a single path at regular time intervals, called the characteristic time-step of the interferometer. Where there is a photon on a spatial input path, there is a photon entering the time-bin interferometer at that time-step. When a photon reaches a beamsplitter, there is a probability (dependent on beamsplitter angle) that the photon is transmitted and continues on its path, and a corresponding probability the photon is reflected and enters a loop of fibre, returning to the same beamsplitter at a multiple of the characteristic time-step later. Photons can interact at a beamsplitter when an early photon in the sequence has been suitably delayed within a loop such that its arrival is coincident with a later photon. The lengths of the loops are precisely defined such that the photons can be coincident on the beamsplitter. The output modes of the time-bin interferometer are determined by a single photon detector at the output recording the number of photons that arrive at each time step. Photons arrive at the detectors at time steps corresponding to the characteristic length of the input sequence. As with the spatial form, the photons need to be indistinguishable: when they interact at the beam splitters it is not possible to trace which output mode corresponds to the input mode (at least in cases with multiple loops).

Using time-bin sampling rather than spatial bins allows the interferometer to be compact. Optical fibres and other telecommunication components can be purchased off-the-shelf. These factors have allowed ORCA Computing to build their first prototype, the PT-1, in a standard server rack. The PT-1 is a simple test of the hardware components and software processes, and is formed of a single loop and beam splitter. As soon as a photon is detected at the end of the interferometer, the user knows which path the photon has taken, which collapses the quantum state. For a single loop, then, there are no 'quantum' effects, since the presence (or lack) of a detection collapses the state at every time step. This limits the capability of this test-bed device in providing quantum utility, but it is useful for experimenting with the kinds of problems that these devices may be able to solve, and learning more about the parameters involved. Future iterations of the technology include multiple loops, which is where the quantum effects are expected, taking the capability beyond that which is classically simulatable. Multiple loops and multiple beam splitters is also likely to increase the errors in the system.

ORCA Computing have developed and released a Python-based Software Development Kit (SDK), which leverages the widely-used PyTorch package for machine learning applications [5]. The SDK includes a simulation capability and functions for solving problems on either the inbuilt simulator or on a real device. The SDK simulator calculates output distributions given an interferometer configuration (number of loops and beam splitter angles) and an input distribution of photons. The simulator assumes perfect photon generation, detection and interactions. It would also be possible to build in realistic loss and noise models on top of the base simulator.

3   Methodology

3.1   Benchmarking Design

Here we use the SDK to benchmark a particular use case. The design phase is guided by the methodology laid out by the authors [18] and other work on the principles and key steps of quantum benchmarking [7,25]. It first clarifies the purpose of the benchmark, then defines success and test selection based on the key principles of generalisability, robustness and expressivity [18]. The key factors influencing the choice of benchmark problem are:

1. As our first exploration in boson sampling, we aim to to understand how it works and what it was designed to do.
2. Quantifying to what extent the device does what it has been designed to do, in this case, solve optimisation problems.
3. Binary variables are native for the solver
4. Quadratic Unconstrained Binary Optimisation (QUBO) problems are common benchmark problems and have been cited as a use case for boson solvers
5. A problem with a generalisable structure with potential applications
6. A problem that is easily scalable and can be tested with a number of examples of different sizes
7. A problem with known bounds for classical and/or alternative quantum algorithms.

A problem that fits the criteria listed and chosen for the benchmark assessment is the NP-hard *Surveillance Coverage Problem*. In its simplest form, this reduces to the problem of determining a minimum dominating set of a graph: given a graph $G = (V, E)$, find a subset V' of V such that every node in V is adjacent, via an edge in E, to at least one node in V' [13].

The nodes V of the graph may model points of interest (POI) over a large region, where an edge represents line of sight between POIs, which could be affected by distance and obstacles. The minimal dominating set represents the minimal number and optimal positions of surveillance equipment such that every POI is under surveillance.

This fits our requirements, as it can be extended in size and complexity by adding more nodes or assigning some targets more value than others. It is an optimisation function with a well defined cost function that fits in with ORCA PT series designed use cases. As a well-studied problem in graph theory, we have bounds for the various classical algorithms that we can compare against.

3.2   Solving the Dominating Set Problem

Consider a set V' defined by a bit string $\boldsymbol{x} = \{x_0, x_1, ..., x_{n-1}\}$, where n is the number of vertices in the graph G, and $x_i = 1$ if the vertex is in the set and 0 otherwise. (Hence the set V of all vertices in G is the all-ones bitstring.) The neighbouring vertices connected to vertex i by some edge in the graph are denoted by the set $N(i)$. The cost function F to be minimised is:

$$F(\boldsymbol{x}) = \sum_{i=0}^{n-1} (x_i + A\, P_i) \tag{1}$$

where the x_i term gives the size of the set, A is a scaling factor (typically $A = 2$), and P_i is a penalty term:

$$P_i = \begin{cases} 0, & \text{if } (x_i + \sum_{j \in N(i)} x_j) - 1 \geq 0 \\ 1, & \text{otherwise} \end{cases} \tag{2}$$

The penalty P_i term checks whether the vertex i is either in the set $\boldsymbol{x}$, or is a neighbour of some vertex in the set. For every vertex in the graph that passes this check, there is no penalty; for every vertex that fails, a penalty of 1 is added. If no penalty terms are added, the set in question is a dominating set and the result of the cost function is the size of the set. We refer to the result of the cost function as the *state energy* of that bit string.[3]

---

[3] Here we are using the definition in which a node can dominate itself. This models a surveillance coverage problem in which a sensor at a particular POI monitors its own location as well as other POIs within its line of sight. For problems where this is not the case and a node cannot dominate itself, we have a strongly-dominating set or a total dominating set. For such a case the cost function would not have the x_i term in Eq. 2. This then requires that each vertex be a neighbour of a member of the set without considering whether it is part of the set itself.

Algorithm 1. Variational Quantum Algorithm

Inputs: cost function $F(\boldsymbol{x})$, input state $|10\rangle \otimes \frac{n}{2}$
Outputs: Optimum state x_{min}, Minimum energy E_{min}

 1: Initialise interferometer parameters to random values $\boldsymbol{\theta} = \{\theta_i\}$
 2: Initialise flipping probabilities to random values $\boldsymbol{p} = \{p_i\}$
 3: **repeat**
 4:    **repeat**
 5:        Pass input state through the interferometer and measure output photons in each mode $|out\rangle$
 6:        Convert $|out\rangle$ to binary bit string via Threshold Mapping, $\boldsymbol{x}$
 7:        Flip or hold bits in $\boldsymbol{x}$ depending on $\boldsymbol{p}$
 8:        Calculate state energy, $E = F(\boldsymbol{x})$
 9:    **until** $maxSamp$ is reached
10:    Use classical optimisation algorithm (SPSA) to update $\boldsymbol{\theta}$ and $\boldsymbol{p}$
11: **until** $maxIter$ has been reached
12: **Return** b_{min}, E_{min}

The minimisation is completed via a variational quantum algorithm shown in Algorithm 1, which involves a learning process to find the problem-specific angles θ_i. The input cost function, $F(x)$, is derived from the graph and is calculated via Eqs. 1 and 2. The input state used in this algorithm is a state with number of input modes equal to the number of vertices in the graph, n. The occupancy of the input modes alternates between 1 and 0 photons. For example when $n = 6$, the input state is $|101010\rangle$. This is the default input state provided in the ORCA SDK and advice from the ORCA Computing team[4] suggests that they believe it to provide the most non-deterministic behaviour and therefore more efficiently explore the search space. As far as we aware, there are no published results on the effect that changing the input state has on the results. There may be input states that are more suited to this (or any particular problem) but the state that maximises output entropy is likely to be a generically good choice. This is an area of active research at ORCA Computing and an open question for further research.

Algorithm 1, line 6, the conversion from output state to a binary bit-string, is performed using the Threshold Mapping, in which any mode that has at least one photon detected is given a value 1 and modes with no photon detected are given the value 0. Line 7 describes the second part of the threshold mapping process, which involves probabilistically flipping the bits of the bit-string x to generate a new candidate set. The probabilities of flipping each bit are updated in each iteration of the training process.

The classical optimisation, line 10, is Simultaneous Perturbation Stochastic Approximation (SPSA), which is often used to optimise multiple unknown parameters [21], and is the standard optimisation algorithm in the ORCA SDK. It follows a similar optimisation rule to standard gradient descent, calculating the gradient for all of the parameters together using two function estimations

---

[4] Personal communication.

per iteration. SPSA has been shown to perform better when optimising noisy or error-prone functions [9]: every parameter is shifted stochastically in each step even in the perfect case, making the whole process more robust to noise in the cost function [22].

We run this for a set number of iterations, $maxIter$, with a fixed learning rate (representing how much a parameter is changed per iteration) and number of samples, $maxSamp$. At the end of the training, the bit-string with the lowest energy is returned. This is classically checked for being a dominating set.

Test graphs are generated using the NetworkX python package and the function $fast\_gnp\_random\_graph$. This function takes in three parameters: n is the number of nodes in the graph (here called the problem size); p is the probability for edge creation (graph density); s is the random seed. For each combination of (n, p), different seeds are used to produce alternate graphs for repeat tests. Aside from setting the number of input photons, the specific problem being solved does not affect the initial configuration of interferometer and there is no associated problem mapping cost. The beam splitter angles are changed iteratively based on the result of the interferometer outputs fed into the problem specific cost function.

The following values are recorded for each run:

1. Is the solution a dominating set? (Y/N)
2. Size of the found set (# of nodes)
3. Total time to train (for one run, no restarts, over fixed number of iterations)
4. Time per update (total time divided by number of iterations)
5. Iterations required to converge
6. Time require to converge (the product of the two previous measures)

It is computationally expensive to verify whether a dominating set is minimal, so we do not use this as a stand-alone measure. We use the size of the dominating sets found to compare between different implementations run on the same graph.

The ORCA SDK does not provide any timing metrics natively, so these are built on top of the training algorithm. This inefficiency might add some time to the measure, but we assume that it is negligible compared to the time to train. This assumption is monitored throughout the experiment.

We define training to have converged when the minimum energy has not decreased over the previous 50 iterations. This method for testing convergence was deemed appropriate after some preliminary tests on example problems.

It is possible to formulate the dominating set problem as a QUBO. However, that formulation would require additional bits to encode the constraints involved, increasing the dimensionality of the problem, and limiting the problems possible to solve on existing intermediate scale hardware [12]. The VQA algorithms allows us to work with the inequalities directly in the cost function, circumventing the need for increased dimensionality.

For the classical comparison, the measures also include success rate (does the algorithm find a dominating set), the size of the found set and the time taken to execute the algorithm. As for many NP-hard problems, there are approximate algorithms that are more efficient than exact algorithms, but have a non-zero probability of failure.

The first classical algorithm is based on integer linear programming methods [10]. We use the python package PULP with the default solver.

The second classical algorithm is a basic greedy algorithm. It loops over vertices not yet in the set, adding the one with highest number of neighbours to the set [8]. All of the neighbours are then marked as 'visited'. The loop ends when all the vertices in the network have been visited. This algorithm never fails to find a dominating set, but it may be much larger than the minimal set.

The third classical algorithm is the approximation algorithm in the NetworkX python package, which runs in $\mathcal{O}(\#E)$ time. It follows the k-centre algorithm of [24], which uses the relationship between this problem and maximally independent sets.

These three classical methods are compared to the simulated ORCA PT-1 in terms of the size of the dominating set found and in the wall clock time to run. The simulated PT-1 works on a perfect hardware, noiseless model, including perfectly repeatable and controllable photon generation, allowing us to consider the ideal performance of boson samplers and therefore the theoretical limit of their utility for optimisation problems. All the results in this paper have been produced with an Intel i7 processor with 64GB of RAM.

4 Results and Discussion

4.1 ORCA TBI Simulator

This section covers the results from the Time-Bin Interferometer Simulator provided in the ORCA SDK. In this case, the TBI Simulator is used to simulate a PT-1 type machine with a single loop and beam splitter. Figure 3 shows a summary of the individual tests carried out. Each test has a unique graph associated with it that is characterised by the problem size n, number of vertices, and graph density p, the probability of an edge existing between any two vertices. Higher n and lower p values characterise harder problems. In Fig. 3 the markers represent that all of the tests performed successfully found a dominating set. The size of the markers indicate the size of the dominating set found by the TBI simulator. The measure of success here is whether alternative, classical, methods fail to find dominating sets smaller in size than the quantum simulator.

The TBI simulator outputs a log that documents the training process. Examples of these logs are shown in Fig. 4. Figure 3 shows that the size of the dominating set tends to decrease as graph density p increases; this can also be seen in Fig. 4a. This is expected because a higher number of edges (within graphs of the same size) should mean that fewer vertices are needed to dominate the graph. In the surveillance coverage problem, a higher density of connections might imply a flatter landscape where a POI has line of sight to many more POIs than in a rugged landscape.

The time taken to run training also scales with both problem size and graph density. Problem size defines the length of the bit-string, and graph density

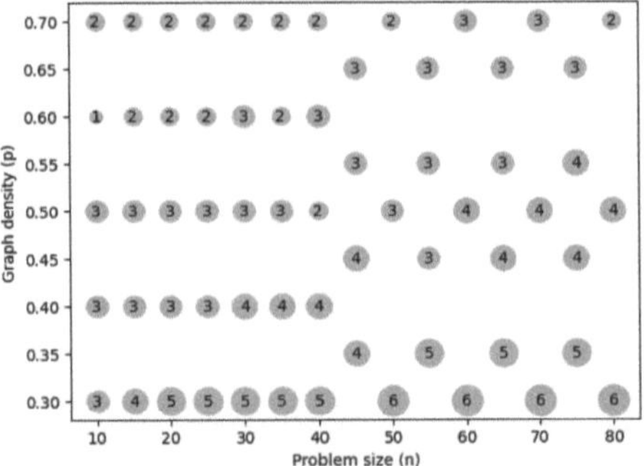

Fig. 3. Summary of test runs on the TBI Simulator. Markers are sized and labelled with the size of the dominating set found.

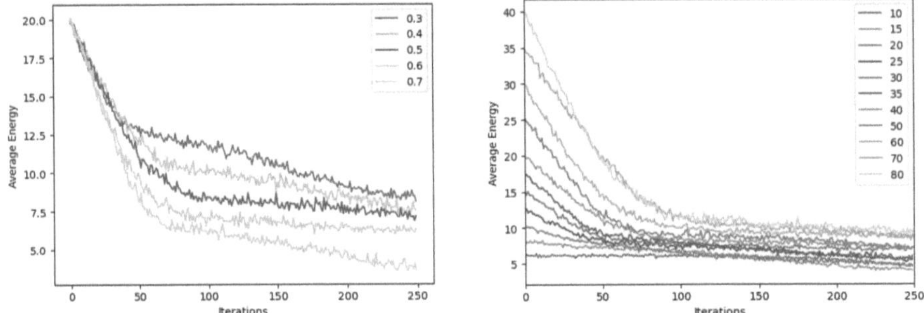

(a) Problem size fixed ($n = 30$) with vary- (b) Fixed $p = 0.5$ for a range of n values.
ing graph density (p).

Fig. 4. The average energy of the samples, as given by Eq. 1, at each step in the training process.

defines the probability that each possible edge will be included in the graph ($p = 1$ produces a complete graph). Larger values extend the run time of the cost function. Figure 5 follows the same format as Fig. 3, except now the markers are sized and labelled according to the number of iterations required for convergence. As stated in the methodology, convergence has occurred when the minimum energy has not decreased for 50 consecutive iterations. All of the tests here converge before the 250th iteration.

4.2 Classical Method Comparison

Figure 6 shows a comparison of the three classical methods used in this test against the TBI Simulator. The left graph is a comparison in terms of the size of dominating sets found; the right graph compares the methods based on their wall clock runtime. The TBI method is the only one that runs for a set number of iterations, so to ensure a fair comparison, the time taken for convergence

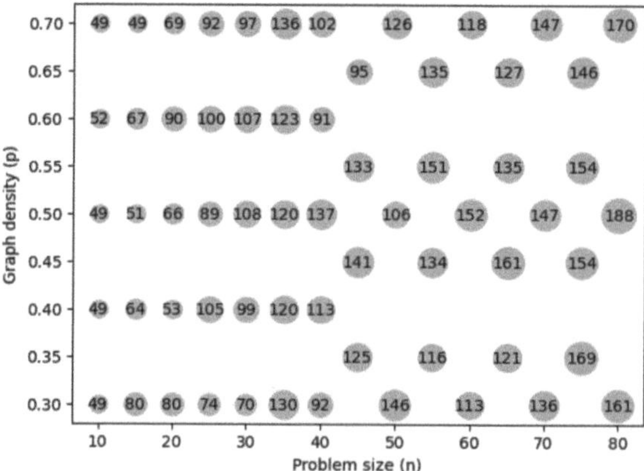

Fig. 5. Summary of test runs on the TBI Simulators. Markers are sized and labelled with the number of iterations required for convergence according to the convergence test laid out in the methodology section.

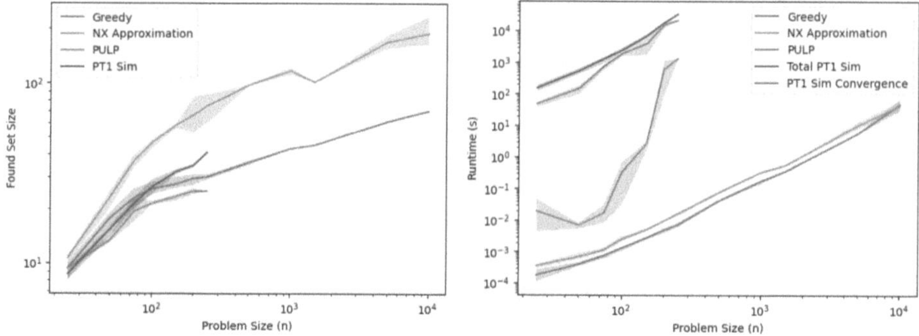

Fig. 6. Comparisons between the three classical methods for solving the dominating set problem and the TBI Simulator. Note the log scaling. The lines represent the average value over three different graphs (of the same n and p values) and the shaded areas shows the spread of results. The left-hand graph shows the minimum dominating set size found; the right-hand graph shows the wall clock runtime. "PT1 Sim Convergence" is calculated to be the time at which the minimum energy has not decreased from the previous 50 iterations.

according to the methodology detailed in the previous section has also been plotted ("PT1 Sim Convergence"). The runtime is measured just for performing the algorithm and does not include the time to save out the results. For all of the graphs in these tests, the graph density parameter is set to $p = 0.05$. As seen in Fig. 5, the time taken scales with both n and p. Fixing p to be small in this case ensures feasible runtimes with an increasingly large range of n.

The Greedy and NetworkX algorithms both run in the approximate millisecond regime for the problem sizes tested here, and also show good scaling that would allow the solving of much larger problems in times less than the PULP solver, even at modest problem sizes. However, the NX Approximation algorithm consistently finds larger dominant sets than the other method. The Greedy algorithm finds dominant sets of an approximately equal size to that of the TBI Simulator. The PULP method typically finds the smallest dominant set out of all the methods tested here, however the time taken is approximately an order of magnitude larger than the other two classical methods for $n < 100$ and scales increasingly badly beyond this. Neither the TBI Simulator or the PULP methods could produce results beyond $n > 250$, due to prohibitive runtimes.

Both the runtime and the minimum set size of the TBI simulation results are affected by the parameters chosen in the algorithm, such as the number of samples and the learning rate. The results shown here are not claimed to be an optimal setting, but may be indicative of likely performance results that could be achieved. The optimal parameters for the simulator will likely depend on the problem and its formulation, as well as the priorities of the user. In the case where a near-minimal dominating set is sufficient, the run time could be reduced by doing fewer samples or iterations.

5   PT-2 Preliminary Results

ORCA Computing have recently released the PT-2 series of Time-Bin Interferometers [1] with two loops (two beam splitters) and upgraded photon generation. A single loop cannot exploit superposition, as either the photon is detected upon leaving the beam splitter or it is not detected and we are certain it remains in the loop. With two loops, there is ambiguity in the location of the photons and it is likely that the effects of superposition can be exploited to solve problems.

In preparation for this new hardware, we perform some preliminary investigations using the TBI Simulator to emulate a PT-2 device, with the aim to compare these both to PT-1 simulation results described above, and to experimental results in the future.

The TBI simulator can be used to simulate time-bin interferometers with any number of loops of differing sizes. However, there are limitations on the size of systems that can be simulated. The power of the computer the simulator is run on affects both the time the simulator takes to run and the maximum problem size that can be run on the simulator. Due to this, only problem sizes $n \leq 40$ are able to be tested here. To compare these to the PT-1 results, Fig. 7 shows a zoom in of Fig. 6 with additional lines for the PT-2 simulation. As in Fig. 6, the left-hand plot also shows the time to convergence for both the PT-1 and PT-2 simulations (the purple and pink lines). This is the time at which the minimum energy had not decreased over the previous 50 iterations.

The two-loop simulation shows comparable quality results to the single loop simulation, but the limitation on solvable problem sizes has allowed only a small sample size of results. The runtime is approximately a factor of 10 higher (for the problem size ~ 40) than the single-loop scenario, but shows a similar scaling. In terms of the found set size, the two-loop scenario appears to slightly outperform

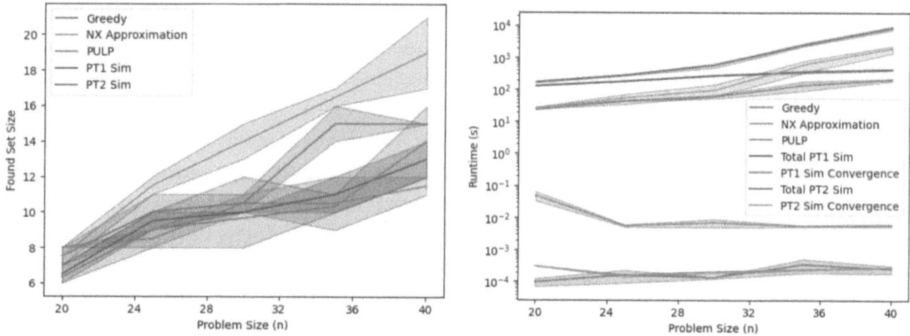

Fig. 7. Comparisons between the three classical methods for solving the dominating set problem and the TBI Simulator in both PT- 1 and PT-2 configurations. Note the log scaling on the y-axes of the runtime graph only. The lines represent the average value over three different graphs (of the same n and p values) and the shaded areas shows the spread of results.

the other methods, but does show considerable variability, making it difficult to conclude whether there is additional benefit in using the second loop. It could be the case that the additional benefit from using the second loop does not appear in the simulation case: if the quantum effects of superposition and entanglement that could allow a better exploration of the search space are only noticeable in larger problem sizes, they would be out of range of the simulation capability. We aim to test this hypothesis in future work completed on the real ORCA PT-2 hardware.

6   Experimental Conclusions and Future Work

The results from the benchmarking experiment show that, at least as modest problem sizes, boson sampling is capable of solving a minimum dominating set problem. This problem can be interpreted in a problem specific context as a surveillance coverage problem, showing how boson sampling might be operationalised to provide benefit in such scenarios. In a less specific, more general sense, these results show that boson sampling has potential uses for a range of graph and optimisation problems.

Although the results show some potential utility in the boson sampling methodology in simulation, the timing results show that these use cases might be niche and focused on areas in which minimising the size of the dominating set is really important and justifies the extra time that would be required to optimise the parameters and run the algorithms compared to the faster classical algorithms. Although we chose the problem to be representative of graph problems in general, we do not know whether the results seen here generalise to other problems. We might expect the real device to have much faster runtimes that would expand the potential use cases, but this will be tested in future work.

Limitations of this study are that it considers only relatively simple time-bin interferometer set ups with one or two loops. ORCA have a roadmap for further devices that use multiple loops of various lengths; the current emulator is capable of simulating these set ups. A more complex interferometer, by providing more parameterised components, should be able to solve more complex problems in fewer iterations [5]. Understanding how the solution quality and runtime scale with the depth and complexity of the interferometer circuit will provide more information on the potential for boson sampling to perform comparatively with classical algorithms [19]. This is currently an open question and a valuable option for further research.

It would also be an interesting topic of further work to consider this dominating set problem on other quantum computing paradigms such as quantum annealing (whereby a QUBO formulation would be required) and gate-model processors (potentially also in a variational type algorithm).

Acknowledgments. The authors acknowledge Defence Science Technical Laboratory (Dstl) who are funding this research. The contents include material subject to © Crown copyright (2025), Dstl. This information is licensed under the Open Government Licence v3.0. To view this licence, visit https://www.nationalarchives.gov.uk/doc/open-government-licence/. Where we have identified any third party copyright information you will need to obtain permission from the copyright holders concerned. Any enquiries regarding this publication should be sent to: centralenquiries@dstl.gov.uk.

Disclosure of Interests. The authors have no competing interests.

References

1. ORCA computing unveils the PT-2: Delivering quantum-enhanced generative AI capabilities (2024). https://orcacomputing.com/orca-computing-unveils-the-pt-2-delivering-quantum-enhanced-generative-ai-capabilities/. Accessed 3 Apr 2025
2. Aaronson, S., Arkhipov, A.: The computational complexity of linear optics. In: STOC 2011, pp. 333–342. ACM (2011)
3. Arora, N., Kumar, P.: Sustainable quantum computing: Opportunities and challenges of benchmarking carbon in the quantum computing lifecycle. arxiv:2408.05679 [quant ph] (2024)
4. Barnett, S.M., Jeffers, J., Gatti, A., Loudon, R.: Quantum optics of lossy beam splitters. Phys. Rev. A **57**(3), 2134 (1998)
5. Bradler, K., Wallner, H.: Certain properties and applications of shallow bosonic circuits. arXiv:2112.09766 [quant-ph] (2021)
6. Bromley, T.R., et al.: Applications of near-term photonic quantum computers: software and algorithms. Quant. Sci. Technol. **5**(3), 034010 (2020)
7. Eisert, J., et al.: Quantum certification and benchmarking. Nat. Rev. Phys. **2**(7), 382–390 (2020)
8. Esfahanian, A.H.: Connectivity algorithms. In: Topics in Structural Graph Theory, pp. 268–281. Cambridge University Press (2013)
9. Finck, S., Beyer, H.G., Melkozerov, A.: Noisy optimization: a theoretical strategy comparison of ES, EGS, SPSA & IF on the noisy sphere. In: GECCO 2011, pp. 813–820. ACM (2011)

10. Forrest, J., Lougee-Heimer, R.: CBC user guide. In: Emerging Theory, Methods, and Applications, pp. 257–277. INFORMS (2005)
11. Gard, B.T., et al.: An introduction to boson-sampling. In: From Atomic to Mesoscale, pp. 167–192. World Scientific (2015)
12. Glover, F., Kochenberger, G., Du, Y.: Quantum bridge analytics I: a tutorial on formulating and using QUBO models. Ann. Oper. Res. **314**, 141–183 (2022)
13. Grandoni, F.: A note on the complexity of minimum dominating set. J. Disc. Algor. **4**(2), 209–214 (2006)
14. He, Y., et al.: Time-bin-encoded boson sampling with a single-photon device. Phys. Rev. Lett. **118**(19), 190501 (2017)
15. Hong, C.K., Ou, Z.Y., Mandel, L.: Measurement of subpicosecond time intervals between two photons by interference. Phys. Rev. Lett. **59**(18), 2044–2046 (1987)
16. Marcus, M., Minc, H.: Permanents. Am. Math. Mon. **72**(6), 577–591 (1965)
17. Motes, K.R., et al.: Scalable boson sampling with time-bin encoding using a loop-based architecture. Phys. Rev. Lett. **113**(12), 120501 (2014)
18. Park, J., Stepney, S., D'Amico, I.: A methodology for comparing and benchmarking quantum devices. In: UCNC 2024. LNCS, vol. 14776, pp. 28–42. Springer, Heidelberg (2024). https://doi.org/10.1007/978-3-031-63742-1_3
19. Slysz, M., Kurowski, K., Waligóra, G.: Solving combinatorial optimization problems on a photonic quantum computer. arXiv:2409.13781 [quant-ph] (2024)
20. Sood, V., Chauhan, R.P.: Quantum computing: impact on energy efficiency and sustainability. Expert Syst. Appl. **255**(124401), 124401 (2024)
21. Spall, J.C.: Multivariate stochastic approximation using a simultaneous perturbation gradient approximation. IEEE Trans. Autom. Control **37**(3), 332–341 (1992)
22. Szava, A., Wierichs, D.: Optimization using SPSA (2023). https://pennylane.ai/qml/demos/tutorial_spsa. Accessed 28 Nov 2024
23. Tillmann, M., et al.: Generalized multiphoton quantum interference. Phys. Rev. X **5**(4), 041015 (2015)
24. Vazirani, V.V.: Approximation Algorithms. Springer, Heidelberg (2001)
25. Wang, J., Guo, G., Shan, Z.: SoK: benchmarking the performance of a quantum computer. Entropy **24**(10), 1467 (2022)

Short and Useful Quantum Proofs
for Sublogarithmic-Space Verifiers

Ahmet Celal Cem Say[(✉)] [iD]

Department of Computer Engineering, Boğaziçi University,
Bebek, 34342 İstanbul, Türkiye
`say@bogazici.edu.tr`

Abstract. Quantum Merlin-Arthur proof systems are believed to be
stronger than both their classical counterparts and "stand-alone" quantum computers when Arthur is assumed to operate in $\Omega(\log n)$ space.
No hint of such an advantage over classical computation had emerged
from research on smaller space bounds, which had so far concentrated on
constant-space verifiers. We initiate the study of quantum Merlin-Arthur
systems with space bounds in $\omega(1) \cap o(\log n)$, and exhibit a problem family $\mathcal{F}$, whose yes-instances have proofs that are verifiable by polynomial-
time quantum Turing machines operating in this regime. We show that no
problem in $\mathcal{F}$ has proofs that can be verified classically or is solvable by a
stand-alone quantum machine in polynomial time if standard complexity
assumptions hold. Unlike previous examples of small-space verifiers, our
protocols require only subpolynomial-length quantum proofs.

Keywords: Quantum proofs · Quantum Turing machines ·
Merlin-Arthur games · Space-bounded quantum computation ·
Sublogarithmic space complexity

1 Introduction

The study of quantum advantage aims to distinguish setups where a model of
quantum computation outperforms its classical counterpart from those where
no such superiority is exhibited. Shor's celebrated factorization algorithm [12]
provides one hint in this regard, although an unconditional conclusion of quantum advantage would also require a proof that no polynomial-time classical
algorithm exists for that problem. Another conjectured difference between the
power of classical and quantum machines, formulated as $\mathsf{MA} \subsetneq \mathsf{QMA}$, pertains
to their use as verifiers for purported proofs associated with yes-instances in
promise problems.

Both MA and QMA are generalizations of the complexity class NP, which
is the set of promise problems that have polynomial-time deterministic Turing
machines that take an input string w and an additional *certificate* string c and
accept if and only if c constitutes a proof that w is a yes-instance of the problem under consideration. When the verifier in the NP scenario is substituted

© The Author(s), under exclusive license to Springer Nature Switzerland AG 2026

E. Formenti and L. Manzoni (Eds.): UCNC 2025, LNCS 16364, pp. 88–99, 2026.
https://doi.org/10.1007/978-3-032-15641-9_7

by a probabilistic Turing machine, and is allowed to output the wrong verdict with a small probability, one obtains the class MA, whose name is inspired by the mighty but unreliable Merlin, who prepares the purported proof, and the resource-bounded Arthur, who attempts to verify it. Upgrading the verifier to a quantum machine and allowing the certificate to be a quantum state yields the class QMA. [14] As mentioned above, it is believed that QMA is a proper superset of MA.

The imposition of stricter space bounds on the polynomial-time verifiers in the setups described above has also been considered. Condon [1] proved that the version of MA where Arthur is further restricted to use logarithmic space equals NP. Recently, it was also proven [5] that the logspace restriction on the (quantum) Arthur leaves the class QMA unchanged. Note that, when modeling small-space verifiers, one assumes that the "proof", which can be polynomially long,[1] and therefore cannot be stored in its entirety in the verifier's memory, is streamed bit by bit (or qubit by qubit) into the verifier from its original storage location.

Our focus is on quantum verifiers with sublogarithmic space bounds. Unlike the case with $s(n) \in \Omega(\log n)$, one is not able to use $s(n)$-space uniform families of quantum circuits and $s(n)$-space quantum Turing machines as equivalent models of computation in this even stricter regime, and has to adopt a machine-based model. To date, studies on quantum Merlin-Arthur proof systems in this regard have only considered verifiers using $O(1)$ space. [15,18] No hint of any advantage of two-way quantum finite automata over their classical counterparts has emerged from that work, and all example protocols presented by Villagra and Yamakami ([15], see also Lemma 5.2 in [9]) involve languages which also have Merlin-Arthur systems with classical constant-space verifiers.[2] [10]

In this paper, we initiate the study of quantum Merlin-Arthur systems with space bounds in $\omega(1) \cap o(\log n)$. In Sect. 2, we provide the relevant background information for our results and define a quantum Turing machine model suitable for representing verifiers under these extreme space bounds. Section 3 describes our main result, namely, a problem family $\mathcal{F}$, whose members have yes-instances with proofs that are verifiable by polynomial-time quantum Turing machines operating in $\omega(\log \log n) \cap o(\log n)$ space. We obtain the members of $\mathcal{F}$ by padding QMA-complete problems in a fashion that is decodable in small space. We show that no problem in $\mathcal{F}$ has proofs that can be verified classically or is solvable by a stand-alone quantum machine in polynomial time if a standard complexity assumption about the difficulty of QMA-complete problems holds. Unlike previous examples [10,15] of small-space verifiers, our protocols require only subpolynomial-length quantum proofs. Section 4 is a conclusion.

---

[1] In both the classical and quantum [3,4,7] cases, logarithmic-length proofs are useless for logspace verifiers, i.e., such a verifier can be replaced by a stand-alone algorithm that runs within $O(\log n)$ space and solves the same problem.

[2] When one considers *interactive* proof systems, where Merlin and Arthur can exchange several messages, constant-space quantum verifiers allowed to run for exponential time outperform their classical counterparts. [2,11,17].

2 Definitions and Background

Early work on quantum Merlin-Arthur systems [6,7] did not consider the imposition of subpolynomial space bounds on the polynomial-time verifiers. The study of space-bounded quantum computation involves additional difficulties which do not manifest themselves when one merely focuses on time complexity. A case in point is the principle of deferred measurement. [8] While it has been known for decades that making intermediate measurements during the execution of a quantum algorithm provides no time complexity advantage, this equivalence was only recently proved [4] for $(\Omega(\log n))$ space complexity.

One complication that arises when sublogarithmic space bounds are considered is the breakdown of the equivalence between the uniform circuit family (UCF) and Turing machine (TM) models. Many works on quantum computation with $\Omega(\log n)$ space are presented in terms of the quantum circuit model, which is favored by quantum complexity theorists and physicists. We shall see shortly that this framework is not suitable for modeling computation with smaller space bounds. Before proceeding with the definition of the TM model to be used in this study, we will give an overview of the UCF-based logarithmic-space verification model of Gharibian and Rudolph [5]. The fact that TM's are able to simulate UCF's with no space overhead will enable us to use a scaled-down version of a protocol from [5] in our main result.

Gharibian and Rudolph's definition [5] of a proof system with a verifier using $s(n)$ space for inputs of length n stipulates the existence of an $s(n)$-space uniform family of quantum circuits realizing the verification. When specialized to the case $s(n) \in \Theta(\log n)$, this requires a deterministic logspace Turing machine that prints circuit descriptions with $n^{O(1)}$ gates (from a fixed set that is universal for quantum computation) for any input of length n. These circuits operate on three registers named the *input, ancilla*, and *proof* registers, containing n, $s(n)$, and $n^{O(1)}$ qubits, respectively. Every such circuit Q_n is structured so that it does not alter the contents of the input register, and the proof is "read" in a streaming fashion in the following manner: For a proof of m qubits, the computation modeled by the circuit is sliced into $m+1$ stages, named V_0 through V_m. No gate in any V_i subcircuit accesses the proof register. For each $i \in \{0, \dots, m-1\}$, Q_n contains a gate that swaps the first ancilla qubit with the $(i+1)$st proof qubit between the subcircuits V_i and V_{i+1}. Since the input register is never changed and each proof qubit is accessed only once, the only "real" memory used by the verifier is the collection of logarithmically many qubits (all initialized to $|0\rangle$) in the ancilla register. It is within this framework that Gharibian and Rudolph have proven that all problems in **QMA** have logspace quantum verifiers inspecting streamed quantum proofs in polynomial time.

It is easy to see that the above-mentioned definition of an $s(n)$-space verifier is unsuitable when $s(n)$ is a sublogarithmic function of n. Although a deterministic Turing machine using memory in $\Omega(\log n)$ can output circuit descriptions with polynomially many gates, where all the polynomially many qubits in the circuit are named as needed, a machine with $o(\log n)$ space, which cannot fit even the index of an individual qubit in its worktape, seems too weak for this purpose.

For this reason, we formulate a quantum Turing machine model (based on [16]) that can be used to study any space bound in order to represent Arthur directly in our protocols.

In addition to a constant-sized central memory represented by a finite set of classical states, every verifier V to be modeled as a quantum Turing machine will have a finite *quantum register* consisting of a fixed number of qubits, a classical *measurement register* of fixed size, and four one-way infinite tapes:

- The *input tape* contains the input string w sandwiched between the special endmarker symbols $\triangleright$ and $\triangleleft$. The read-only, two-way input head is located on the left endmarker $\triangleright$ at the beginning of the computation.
- The *classical work tape* is accessed by a dedicated two-way read-write head which is located on its leftmost cell at the beginning of the computation. All cells contain the blank symbol initially.
- The *quantum work tape* is accessed by a dedicated two-way head which is located on its leftmost cell at the beginning of the computation. Each cell of this tape holds a qubit, and all these qubits are in their zero states initially.
- The *proof tape* is accessed by a dedicated one-way head (which cannot move to the left) that is located on its leftmost cell at the beginning of the computation. We assume that the m qubits whose state is a purported proof that the input string w is a yes-instance of the problem associated with V are situated in the first m cells of this tape at the start, and the head cannot move beyond the area occupied by the proof qubits.

Definition 1. *A* quantum verifier *is an 11-tuple*

$$\left(S, \Sigma, K, T, k, \delta_1, \delta_2, s_0, \tau_0, s_{\mathrm{acc}}, s_{\mathrm{rej}}\right),$$

where

- *S is the finite set of classical states,*
- *Σ is the finite input alphabet, not containing $\triangleright$ and $\triangleleft$,*
- *K is the finite classical work alphabet, including the blank symbol $\sqcup$,*
- *T is the finite set of possible values of the measurement register,*
- *k, a positive integer, is the number of qubits in the quantum register,*
- *$\delta_1 : (S \setminus \{s_{\mathrm{acc}}, s_{\mathrm{rej}}\}) \times (\Sigma \cup \{\triangleright, \triangleleft\}) \times K \to \Delta$, where Δ is the set of all selective quantum operations [16] that have operator-sum representations consisting of matrices of algebraic numbers,[3] is the first-substep transition function,*
- *$\delta_2 : (S \setminus \{s_{\mathrm{acc}}, s_{\mathrm{rej}}\}) \times (\Sigma \cup \{\triangleright, \triangleleft\}) \times K \times T \to S \times K \times \{-1, 0, 1\}^3 \times \{0, 1\}$ is the second-substep transition function, which determines the updates to be performed on the classical components of the machine, including the positions of all heads, and whether a new proof qubit will be swapped in or not, as described below,*

---

[3] It is well known [8] that using only the members of a fixed, small set of algebraic numbers as the transition amplitudes (i.e., the entries of these matrices) is sufficient for universal quantum computation.

- $s_0 \in S$ *is the initial state,*
- $\tau_0 \in T$ *is the initial value of the measurement register, and*
- $s_{acc}, s_{rej} \in S$, *such that* $s_{acc} \neq s_{rej}$, *are the accept and reject states, respectively.*

At the start of computation, a quantum verifier V is in state s_0, the measurement register contains τ_0, all qubits of the finite quantum register are in their zero states, and the heads are initialized as described above. Every computational step of V consists of a *first substep*, which operates on the quantum register and the quantum work tape and concludes by updating the measurement register, and a *second substep*, which can change the classical state, classical work tape content and the head positions, and read a new proof qubit, based on the presently scanned input and classical work tape symbols and the value in the measurement register: If V is in some non-halting state s, with the input head scanning the symbol σ and the classical work tape head scanning the symbol κ, the selective quantum operation $\delta_1(s, \sigma, \kappa)$ acts on the quantum register and the qubit presently scanned by the quantum work tape head. This action may be a unitary transformation or a measurement, resulting in those $k + 1$ qubits evolving, and some output τ being stored in the measurement register with an associated probability. The ensuing second-substep transition $\delta_2(s, \sigma, \kappa, \tau) = (s', \kappa', d_I, d_C, d_Q, d_P)$ updates the classical state to s', changes the presently scanned classical work tape symbol to κ', and moves the input, classical work, quantum work and proof heads in the directions indicated by the increments d_I, d_C, d_Q, and d_P, respectively. If $d_P = 1$, the second-substep transition concludes by swapping the qubit at the new position of the proof tape head with the first qubit of the finite quantum register. We assume that δ_2 is restricted so that it never attempts to move the input head off the tape area containing $\triangleright w \triangleleft$, or to move any work head to the left of the leftmost cell of the corresponding tape. Computation halts with acceptance (resp. rejection) when V enters the state s_{acc} (resp. s_{rej}).

A quantum verifier is said to *run within space* $s(n)$ if, on any input of length n, at most $s(n)$ cells are scanned on both the classical and quantum work tapes during the computation.

Definition 2. *A promise problem* $A = (A_{yes}, A_{no})$ *is in* QMASPACE($s(n)$) *if there exists a polynomial-time quantum verifier* V *which runs within space* $O(s(n))$, *such that, for every input string* $w \in \Sigma^*$,

- *if* $w \in A_{yes}$, *there exists an m-qubit quantum proof* $|\psi\rangle$ *such that* V *accepts with probability at least* $\frac{2}{3}$ *when started with* $\triangleright w \triangleleft$ *on the input tape and* m *qubits whose state is* $|\psi\rangle$ *on the proof tape, and*
- *if* $w \in A_{no}$, V *rejects with probability at least* $\frac{2}{3}$ *when started with* $\triangleright w \triangleleft$ *on the input tape, regardless of the content of the proof tape.*

Note that Definition 2 requires the verifier to respect the time and space bounds even when the input string is in $\Sigma^* \setminus (A_{yes} \cup A_{no})$.

For any $s(n) \in \Omega(\log n)$, machines conforming to Definition 1 can simulate $s(n)$-space uniform circuit families within the same bound: The deterministic algorithm preparing the circuit for the given input size uses only $s(n)$ cells on the classical work tape, and the quantum register and the quantum work tape accommodate the $O(s(n))$ qubits of the simulated ancilla register. Although the overall computation of the quantum verifier as described is not necessarily unitary (because the classical part can evolve irreversibly), we note that the quantum part evolves unitarily if the simulated circuit contains no intermediate measurements. Recalling Fefferman and Remscrim's demonstration [4] that intermediate measurements can be eliminated without increasing space or time complexity, we can restate the relationship of Gharibian and Rudolph's result to the original definition of QMA in the terminology of Definition 2:

Theorem 1. *[5, 6]* QMA = QMASPACE$(n^{O(1)})$ = QMASPACE$(O(\log n))$.

Analogously to the well-known definition of NP-completeness, we say that a promise problem A is QMA-*complete* if A is both in QMA and QMA-hard under polynomial-time classical (Karp) reductions. Several interesting problems have been shown to be QMA-complete. [14]

The known relationships between QMA and other complexity classes can be summarized as follows. Since quantum machines can simulate their classical counterparts, we have

$$\text{MA} \subseteq \text{QMA}.$$

Since Arthur can ignore Merlin's message while solving the problem,

$$\text{BQP} \subseteq \text{QMA}.$$

We also know [14] the following inclusions:

$$\text{QMA} \subseteq \text{PP} \subseteq \text{PSPACE} \subseteq \text{EXPTIME} \tag{1}$$

Unfortunately, as is often the case in computational complexity theory, it is not known whether any of the inclusions above is strict. We will therefore base our claim about the "difficulty" of the problem family to be introduced in Sect. 3 on the following hypothesis, which we believe reflects the prevailing opinion among experts:

The Exponential Time Hypothesis for QMA-Complete Problems. Informally, the exponential time hypothesis for QMA-complete problems (ETH-QMA for short) states that neither a classical Merlin-Arthur proof system nor a standalone quantum algorithm can handle any QMA-complete problem in subexponential time. Let MATIME$(t(n))$ and BQTIME$(t(n))$ denote respectively the generalizations of the complexity classes MA and BQP, where the runtime budget of the machines is the (not necessarily polynomial) function $t(n)$. ETH-QMA is then the claim that the set

$$\text{MATIME}(t(n)) \cup \text{BQTIME}(t(n))$$

contains no QMA-complete problems if $t(n)$ is a subexponential function, i.e. if $t(n) \in 2^{n^{o(1)}}$.[4]

We conclude this section by recalling a classical result by Stearns et al. [13] about sublogarithmic-space computation that will be useful in Sect. 3: For any integer $l \geq 0$, let $\mathrm{bin}(l)$ denote the shortest string representing l in the binary notation. The language

$$L_S = \{\mathrm{bin}(0)\$\mathrm{bin}(1)\$\mathrm{bin}(2)\$\cdots\$\mathrm{bin}(k) \mid k > 0\}$$

is in the complexity class $\mathsf{SPACE}(\log\log n)$. The proof uses the fact that the rightmost "segment" (i.e. the postfix following the rightmost \$) of any n-symbol member of L_S is of length $m \in \Theta(\log n)$. A polynomial-time Turing machine compares consecutive segments bit by bit, and halts when it detects an inconsistency or when all comparisons are successful. Since each comparison involves a counter whose value can be at most m, the space complexity is $\Theta(\log\log n)$. Yes-instances of the problem family to be defined in the next section will be based on the pattern of L_S.

3   A Family of Problems

This section presents the description of a family $\mathcal{F}$ of promise problems and an examination of $\mathcal{F}$'s complexity-theoretic properties. We give a quick overview: Every member of $\mathcal{F}$ is obtained by padding a different QMA-complete problem. Since the padding is superpolynomially long, yes-instances of problems in $\mathcal{F}$ need correspondingly shorter quantum proofs compared to yes-instances of the unpadded versions. Since the padding is not *too* long, if the ETH-QMA holds, those seemingly simpler problems are still too difficult for stand-alone quantum computers or classical verifiers to handle in polynomial time. All of this would be straightforward if Arthur did not have to worry about space bounds, but he has to, and the discussion below concentrates on the construction of verifiers that use only sublogarithmic amounts of space to carry out their protocols.

3.1   Padded Problems

Consider the language

$$L_u = \{\mathrm{bin}(0)\$\mathrm{bin}(1)\$\mathrm{bin}(2)\$\cdots\$\mathrm{bin}(k) \mid k > 0, \mathrm{bin}(k) \in 1^+\}.$$

Note that $L_u \subsetneq L_S$ (Sect. 2), such that L_u contains only the strings whose rightmost segments represent some integer of the form $2^m - 1$, where m is a

---

[4] By Eq. 1, we know that QMA is included in both $\mathsf{MATIME}(t(n))$ and $\mathsf{BQTIME}(t(n))$ for $t(n) \in 2^{n^{O(1)}}$. Every QMA-complete problem A solvable in time 2^{n^k} for some constant $k > 0$ can be "padded" to obtain another QMA-complete problem A' which can then be solved by a machine whose runtime is proportional to $2^{n^{k/c}}$ for any positive constant c. The task of "unpacking" the padded strings is easy for machines employing at least logarithmic space.

positive integer. In the following, let u_m denote the member of L_u which ends with the postfix $\$1^m$.

We are now ready to describe the problem family $\mathcal{F}$. $\mathcal{F}$'s alphabet Σ is the set $\{0, 1, \$\} \times \{0, 1, \#\}$. Strings written using Σ are to be interpreted as containing two tracks, with the symbol (σ_u, σ_l) indicating that the upper (resp. lower) track contains the symbol σ_u (resp. σ_l) at that position.

A string on Σ whose upper track equals u_m and whose lower track is a member of the finite set

$$\left\{ w\#^{|u_m|-j} \;\middle|\; w = \{0,1\}^j,\; j \leq 2^{\frac{m}{\lceil \text{bin}(m) \rceil}} \right\} \tag{2}$$

is said to be a *padding of* w. For any language L, $pad(L)$ is the set of all strings which are paddings of the members of L.

Definition 3. $\mathcal{F}$ *is the set of all promise problems of the form* $A' = (A'_{yes}, A'_{no})$, *such that*

$$A'_{yes} = pad(A_{yes}), \quad \text{and}$$

$$A'_{no} = pad(A_{no}) \cup (\Sigma^* \setminus pad(\{0,1\}^*))$$

for some QMA-*complete promise problem* $A = (A_{yes}, A_{no})$ *on the alphabet* $\{0,1\}$.

Essentially, A' is a heavily padded version of A, with strings which are not syntactically correct paddings also considered as no-instances. Although it will be seen that the precise amount of padding can be selected from a wide variety of superpolynomial (but subexponential) functions, we have fixed that parameter to a specific function to simplify the exposition below.

3.2   The Verifier

Consider a member A' of $\mathcal{F}$. Mirroring Definition 3, we describe a polynomial-time quantum verifier V for A' that decides whether its input of length n is a well-formed padding of some binary string w and is able to check a quantum proof which claims w is a yes-instance of A.

V starts by running the $O(\log \log n)$-space deterministic algorithm of Stearns et al. [13] that was described in Sect. 2, followed by an easy check to see if the final segment is composed entirely of 1's, to determine whether the upper track of its input is some member u_m of L_u. If this check fails, V halts by declaring that the input is a no-instance. Otherwise, V proceeds to check whether the lower track is of the form in Expression (2). Note that V already has a $O(\log \log n)$-bit representation of the length of the last segment of the upper track in its classical work tape at this point. The computation and storage of the integer $i = \lceil \frac{m}{\lfloor \log m \rfloor + 1} \rceil$ is also performed in $O(\log \log n)$ space.[5] V then writes the number 2^i in binary notation in its classical work tape, using $\lceil \frac{m}{\lfloor \log m \rfloor + 1} \rceil + 1$, i.e. $O(\frac{\log n}{\log \log n})$ bits. It then uses this variable as a counter to check whether

---

[5] Note that $|\text{bin}(m)| = \lfloor \log m \rfloor + 1$. (Logarithms are to the base 2.).

the lower track is of the form $\{0,1\}^j \#^*$, where $j \leq 2^i$, and rejects if this check reveals a malformed padding.

If the input is verified to be a legitimate padding of some binary string w, V runs the streaming quantum Merlin-Arthur protocol (which must exist by Theorem 1) to verify the quantum proof's claim that w is a yes-instance of A. As noted in Sect. 2, this protocol uses only logarithmically many qubits in terms of the length of w, which corresponds to just $O(\frac{\log n}{\log \log n})$ space in the quantum work tape of V. Since the runtime of the protocol is bounded by a polynomial in terms of the (already very small) length of w, V's runtime as a verifier for proofs about A' is also polynomially bounded.

We note that the quantum proofs read by V are significantly shorter than those considered previously in the study of constant-space Merlin-Arthur systems. Those protocols [10,15] have proofs whose lengths are at least linear in terms of the input. V requires its proofs to be only polynomially long [5] in terms of the length of w. Since

$$2^{\frac{\log n}{\log \log n}} = n^{\frac{1}{\log \log n}}, \tag{3}$$

any power of $|w|$ is necessarily in $o(n)$. We argue that V handles significantly more difficult problems than those of [10,15] in the next subsection.

3.3   The Superiority of Quantum Verifiers for $\mathcal{F}$

In this subsection, we will provide evidence that quantum verifiers using sublogarithmic space can handle problems that are impossible for both stand-alone (polynomial-space) quantum algorithms and classical verifiers under comparable time bounds.

Theorem 2. *If the ETH-QMA holds, neither* BQP *nor* MA *contains any member of $\mathcal{F}$.*

Proof. Assume that the ETH-QMA holds, and that some member A' of $\mathcal{F}$ is solved in polynomial time by a stand-alone quantum algorithm M'. (The proof works verbatim for a verifier in a classical Merlin-Arthur system playing the role of M' as well.) We show how to build a quantum algorithm M that solves the QMA-complete problem A, from which A' was derived, in subexponential time, thereby violating the ETH-QMA.

M starts by running the deterministic algorithm in Fig. 1 to convert its input w to a padding of w in the format described in Sect. 3.1. In that algorithm, Stage 1 takes polynomial time. To see that it is easy to find an m satisfying the equation depicted in Stage 2, consider searching for a number d (the "length" of m) that satisfies $2^{d-1} \leq i \times d \leq 2^d - 1$, that is, $\frac{2^{d-1}}{d} \leq i \leq \frac{2^d-1}{d}$. The smallest integer d which satisfies $i \leq \frac{2^d-1}{d}$ is guaranteed to also satisfy $\frac{2^{d-1}}{d} \leq i$, since otherwise we would have $i \leq \frac{2^{d-1}-1}{d-1}$, contradicting the minimality of d.[6] Since d is much smaller than i, which is itself logarithmic in terms of the length of the input, Stage 2 also takes polynomial time.

---

6 $\frac{2^{d-1}}{d} \leq \frac{2^{d-1}-1}{d-1}$ for all $d > 1$. Very short inputs are handled specially in Stage 1.

On input w:

1. Store the integer i such that $i = \lceil r \rceil$, where $|w| = 2^r$. (If $|w| < 3$, let $i = 2$.)
2. Find the smallest integer m satisfying the equation $m = i \times |\operatorname{bin}(m)|$.
3. Print the padding of w as follows:
    3.1. The upper track will contain u_m, i.e. the member of L_u that ends with the segment 1^m.
    3.2. The lower track will contain the member of $w\#^*$ whose length matches the string in the upper track.

Fig. 1. The reduction of A to A'.

Since the computation required to determine the next symbol to print during the execution of Stage 3 is trivial, the runtime of this stage is linear in terms of the length of the string it prints. Noting that the algorithm of Fig. 1 clearly prints a padding of w, we examine the relationship between the length n of the final printed string and the length j of w to bound this runtime.

Stage 1 ensures that $j > 2^{i-1}$ for all sufficiently long w. Since $m \in \Theta(\log n)$, we have

$$j \in \Theta(n^{\frac{1}{\log \log n}})$$

by the observation (3) at the end of Sect. 3.2. We "invert" that relationship as follows: Taking logarithms and rearranging, one gets

$$\log n = \log j \times \log \log n. \tag{4}$$

Taking logarithms again, we have

$$\log \log n = \log \log j + \log \log \log n. \tag{5}$$

Clearly, $\log \log j < \log \log n$. Since $\frac{1}{2} \log \log n > \log \log \log n$ for all sufficiently large n, Eq. 5 yields

$$\log \log j < \log \log n < 2 \log \log j \tag{6}$$

for all such n. Using Eqs. 4 and 6, we get

$$\log n = \Theta(\log j \times \log \log j),$$

and conclude that

$$n = j^{\Theta(\log \log j)}. \tag{7}$$

It is easy to see from Eq. 7 that n is a subexponential function of j. The algorithm of Fig. 1, which maps members of the three sets A_{yes}, A_{no}, and $\{0,1\}^* \setminus (A_{yes} \cup A_{no})$ to their respective counterparts (Definition 3), thereby reduces A to A' in subexponential time. Algorithm M then submits the output of this reduction as the input of the hypothetical polynomial-time algorithm M', and reports the verdict of M' as its own decision. Since any polynomial in n remains subexponential in j (Eq. 7), M handles problem A in subexponential time, contradicting the ETH-QMA.

When comparing our verifiers with stand-alone quantum algorithms, Theorem 2 indicates that the quantum proofs read by our machines are "useful". If the comparison is with classical verifiers, Theorem 2 is a potential quantum advantage result.

4   Concluding Remarks

For the first time, we have demonstrated the existence of a task at which sublogarithmic-space quantum Merlin-Arthur systems seem to outperform both their classical counterparts and stand-alone quantum computers, even when those competitors are allowed to use polynomial amounts of space in their computations. Treatment of such small space bounds required a verifier definition based directly on the quantum Turing machine model. The protocol presented in Sect. 3.2 is also novel in the sense that no sublinear-length proofs were considered for such small-space verifiers until now.

As mentioned during the presentation of $\mathcal{F}$ in Sect. 3.1, it is possible to obtain similar families of problems by varying the amount of padding in the definition. We note that the padding should not be selected to be too long: If Expression (2) is modified so that w is too short, e.g., if $|w| \leq \log \log n$ where n is the length of the padding, even a stand-alone "brute-force" deterministic algorithm can solve the associated problem in time polynomial in n. (Equation 1) It is therefore still an open question whether there exists a potential quantum advantage as the one we demonstrate here for verifiers respecting even smaller space bounds like $O(1)$.

Acknowledgments. The author thanks Ryan O'Donnell, Abuzer Yakaryılmaz and Utkan Gezer for all the useful discussions, and the anonymous reviewers for their helpful remarks. This work is supported by TUBITAK under program 2224-A.

Disclosure of Interests. The author has no competing interests to declare that are relevant to the content of this article.

References

1. Condon, A.: The complexity of the max word problem and the power of one-way interactive proof systems. Comput. Complex. **3**(3), 292–305 (1993)
2. Dwork, C., Stockmeyer, L.: Finite state verifiers I: the power of interaction. J. ACM **39**(4), 800–828 (1992)
3. Fefferman, B., Kobayashi, H., Lin, C.Y.Y., Morimae, T., Nishimura, H.: Space-efficient error reduction for unitary quantum computations. In: Proceedings of the 43rd International Colloquium on Automata, Languages, and Programming, pp. 14:1–14:14 (2016)
4. Fefferman, B., Remscrim, Z.: Eliminating intermediate measurements in space-bounded quantum computation. In: Proceedings of the 53rd Annual ACM SIGACT Symposium on Theory of Computing, pp. 1343–1356 (2021)

5. Gharibian, S., Rudolph, D.: Quantum space, ground space traversal, and how to embed multi-prover interactive proofs into unentanglement. In: 14th Innovations in Theoretical Computer Science Conference (ITCS 2023), pp. 53:1–53:23. Leibniz International Proceedings in Informatics (LIPIcs) (2023). https://arxiv.org/abs/2206.05243

6. Kitaev, A., Shen, A., Vyalyi, M.: Classical and Quantum Computation. American Mathematical Society (2002)

7. Marriott, C., Watrous, J.: Quantum Arthur-Merlin games. Comput. Complex. **14**, 122–152 (2005)

8. Nielsen, M.A., Chuang, I.L.: Quantum Computation and Quantum Information. Cambridge University Press, Cambridge (2002)

9. Nishimura, H., Yamakami, T.: An application of quantum finite automata to interactive proof systems. J. Comput. Syst. Sci. **75**(4), 255–269 (2009)

10. Say, A.C.C., Yakaryılmaz, A.: Finite state verifiers with constant randomness. Logical Methods Comput. Sci. **10**(3) (2014)

11. Say, A.C.C., Yakaryılmaz, A.: Magic coins are useful for small-space quantum machines. Quant. Inf. Comput. **17**(11–12), 1027–1043 (2017)

12. Shor, P.W.: Polynomial-time algorithms for prime factorization and discrete logarithms on a quantum computer. SIAM J. Comput. **26**(5), 1484–1509 (1997)

13. Stearns, R.E., Hartmanis, J., Lewis, P.M.: Hierarchies of memory limited computations. In: 6th Annual Symposium on Switching Circuit Theory and Logical Design (SWCT 1965), pp. 179–190 (1965). https://doi.org/10.1109/FOCS.1965.11

14. Vidick, T., Watrous, J.: Quantum proofs. Found. Trends Theor. Comput. Sci. **11**(1–2), 1–215 (2015)

15. Villagra, M., Yamakami, T.: Quantum and reversible verification of proofs using constant memory space. In: Dediu, A.H., Lozano, M., Martín-Vide, C. (eds.) Theory and Practice of Natural Computing, pp. 144–156. Springer, Cham (2014)

16. Watrous, J.: On the complexity of simulating space-bounded quantum computations. Comput. Complex. **12**, 48–84 (2003)

17. Yakaryılmaz, A.: Public qubits versus private coins. In: Workshop on Quantum and Classical Complexity, pp. 45–60. University of Latvia Press, Riga (2013). ECCC:TR12-130

18. Yakaryılmaz, A.: Classical and quantum Merlin-Arthur automata (2022). https://arxiv.org/abs/2212.13801

Probabilistic Spiking Neural Networks: Formal Verification and Simulation

Zhen Yao[1]([✉])[iD], Elisabetta De Maria[1][iD], and Robert De Simone[2][iD]

[1] Université Côte d'Azur, Nice, France
zhen.yao@etu.univ-cotedazur.fr , elisabetta.de-maria@univ-cotedazur.fr
[2] Centre Inria d'Université Côte d'Azur, Nice, France
robert.de_simone@inria.fr

Abstract. Spiking Neural Networks (SNNs) offer a biologically inspired approach to modeling neuronal computation, emphasizing timed latency and probabilistic activation over the numerical computations typical of traditional deep-learning models. This paper introduces a modeling framework for SNNs that incorporates the RP-LI&F (Refractory-evolve Probabilistic Leaky Integrate-and-Fire) neuron model to capture realistic neuronal dynamics. We present a prototype, SuNNy, to translate elementary neural bundles and their synaptic connections into formal models compatible with PRISM, enabling model checking of stochastic timing properties using Probabilistic Computation Tree Logic (PCTL). Additionally, our framework integrates with Nengo for simulation, allowing for experimentation on large-scale SNNs. A key challenge addressed is the study of how compound SNN models can meet global reaction requirements based on the stochastic behaviors of individual neural bundles. While the framework supports parametric variations to represent different neuronal states, this work focuses on the core modeling, verification, and simulation techniques. Specific applications, such as simulating impaired neurons, are not explored in this paper but are left for future research. Our approach provides a foundation for both formal verification and simulation of SNNs, bridging the gap between theoretical models and practical tools for analyzing neuronal computations.

Keywords: Spiking Neural Networks · Bio-inspired Computing · Probabilistic Model Checking · Neuronal Circuits · Formal Methods

1 Introduction

Spiking Neural Networks constitute the third generation of neural network models, marrying computational efficiency with biological plausibility. Unlike traditional rate-based Artificial Neural Networks (ANNs), SNNs employ event-driven spikes—discrete voltage pulses—to encode and transmit information, focusing on the timed latency and probabilistic nature of neuronal activation. This mechanism greatly reduces energy consumption by activating neurons only when spikes occur, and naturally captures temporal dynamics, making SNNs well-suited for

© The Author(s), under exclusive license to Springer Nature Switzerland AG 2026
E. Formenti and L. Manzoni (Eds.): UCNC 2025, LNCS 16364, pp. 100–114, 2026.
https://doi.org/10.1007/978-3-032-15641-9_8

tasks involving time-varying inputs such as signal processing, motor control, and event-based vision [16].

Although various formalisms such as Timed Automata, DTMCs, and hybrid Petri nets have been used to model SNNs for verification [3,10,18], most existing approaches are tailored to specific domains or simplified architectures. Recent works, including those using Nengo and UPPAAL [19], demonstrate feasibility but lack generality. In contrast, our work focuses on a generic DTMC-based modeling of spiking archetypes and verifies their dynamics using PCTL, which, to our knowledge, has not been systematically explored. This highlights the novelty and general applicability of our approach.

Early demonstrations have shown that SNNs can match ANNs on both classical benchmarks (e.g., MNIST, N-MNIST, N-Caltech101) [11,15] and more challenging datasets like ImageNet and CIFAR-10 [20], while preserving their energy-saving advantages. Beyond performance, the bio-inspired nature of SNNs renders them invaluable for neuroscientific inquiry, offering a computational window into the dynamics of real neural circuits. Central to SNN performance is the choice of neuron model, which must balance biological realism against computational tractability. Leaky Integrate-and-Fire (LI&F) neurons afford efficiency but neglect some dynamics; biophysically detailed Hodgkin-Huxley models capture a wealth of ionic processes at the cost of prohibitive computational load [9]. Recent advances—in particular, Generalized Leaky Integrate-and-Fire (GLIF) [21] and Noisy LI&F [17] models—have introduced stochastic thresholds and membrane-potential noise to better mimic cortical variability. However, existing variants still fall short of emulating absolute and relative refractory behaviors observed in biological neurons.

In this paper, we introduce the **Refractory-evolve Probabilistic Leaky Integrate-and-Fire (RP-LI&F)** neuron model based on previous Boolean Probabilistic Spiking Neural Network [4]. RP-LI&F embeds discrete-time refractory dynamics and probabilistic spike generation into the LI&F framework, yielding a compact yet biologically grounded unit. We support RP-LI&F with a domain-specific format, Spiking Neural Networks Representation File (SNN-RF), that automatically produces discrete-time Markov chain models for formal verification in PRISM and high-performance simulations in Nengo [1].

SNN models require parametric constructs for elementary neural bundles and synaptic connections, enabling the assembly of complex data-flow network patterns. These constructs must support the instantiation of latency and probability parameters, which remain constant during runtime but can vary to simulate conditions such as neuronal fatigue or pharmacological effects. A key challenge is ensuring that compound SNN models meet global reaction requirements under stochastic timing constraints, given similar provisions for individual neural bundles. This necessitates a temporal logic language to express assume/guarantee contracts, facilitating formal verification for medium-sized models and testing through simulation for larger ones.

This preliminary work aims to develop a brain-like neural network by archetypes [2], which are specific neuronal micro-circuits, that can be combined in the future to form complicated, brain-inspired systems, distinct from second-generation networks focused solely on data processing efficiency. The remainder of this paper is structured as follows. Section 2 introduces our RP-LI&F neuron model. Section 3 describes the PRISM model checker with the logic PCTL and the Nengo simulator. Section 4 presents how we automatically generate a network model in both Nengo and PRISM via SNN-RF. Section 5 details case studies demonstrating RP-LI&F in contralateral inhibition and convergent excitatory circuit, two fundamental archetypes [2]. Finally, Sect. 6 concludes and outlines directions for future work. The code can be found at https://github.com/zane2077/RP-LIF.git.

2   Refractory-Evolve Probabilistic LI&F Model

Real neurons maintain a resting voltage that slowly drifts back after synaptic inputs, fire when they reach a threshold, exhibit random channel-driven fluctuations, and pause briefly after each spike before regaining full excitability. In our RP-LI&F abstraction, the leak factor $r = \exp(-\Delta t/\tau)$ models the passive return to rest, the threshold τ marks the voltage for spike initiation, the membrane potential $p(t)$ sums incoming spikes with exponential decay, discrete probability levels $\{p_i\}$ capture channel noise that makes firing stochastic, the absolute-refractory countdown ARP enforces a short window of zero spiking, and the relative-refractory countdown RRP with scaling factor α represents the gradual restoration of firing probability after an action potential.

We define the **Refractory-evolve Probabilistic Leaky Integrate and Fire** (RP-LI&F) neuron as a tuple $v = (\tau, r, \alpha, p, y, s, \mathtt{aref}, \mathtt{rref})$, where:

- $\tau \in \mathbb{N}$: reference firing threshold
- $r \in [0,1] \cap \mathbb{Q}$: leak factor, governing exponential decay of potential.
- $\alpha \in [0,1]$: scaling factor modulating neuronal firing probability during the relative refractory period.
- $p\colon \mathbb{N} \to \mathbb{Q}_0^+$: membrane potential at discrete time t, with

$$
p(t) = \begin{cases} \displaystyle\sum_{i=1}^{m} w_i\, x_i(t), & \text{if } y(t-1) = 1, \\ \displaystyle\sum_{i=1}^{m} w_i\, x_i(t) \; + \; r\, p(t-1), & \text{otherwise,} \end{cases} \tag{1}
$$

  where $p(0) = 0$, m is the number of inputs, w_i the synaptic weight, and $x_i(t) \in \{0,1\}$ the incoming spike.
- $y\colon \mathbb{N} \to \{0,1\}$: output spike function, $y(t) = 1$ iff a spike is emitted at t.
- $s \in \{0,1,2\}$: state flag, taking values 0 for Normal period, 1 for Absolute refractory period, and 2 for Relative refractory period.
- $\mathtt{aref} \in \{0,\ldots,\mathtt{ARP}\}$: **A**bsolute **R**efractory **P**eriod countdown,
- $\mathtt{rref} \in \{0,\ldots,\mathtt{RRP}\}$: **R**elative **R**efractory **P**eriod countdown.

2.1 Spike-Probability Law

Let $\Delta(t) = p(t) - \tau$. Let us discretize Δ into $2k+1$ intervals $\{-l_k, \ldots, 0, \ldots, l_k\}$. We define a base firing probability as follows:

$$P\big(y(t) = 1\big) = \begin{cases} 0, & s = 1, \\ \alpha P_{\text{base}}(y(t) = 1), & s = 2, \\ P_{\text{base}}(y(t) = 1), & s = 0. \end{cases}$$

$$P_{\text{base}}(y(t) = 1) = \begin{cases} 1, & \Delta(t) \geq l_k, \\ p_{2k}, & l_{k-1} \leq \Delta(t) < l_k, \\ \vdots & \vdots \\ p_{k+1}, & 0 \leq \Delta(t) < l_1, \\ p_k, & -l_1 \leq \Delta(t) < 0, \\ \vdots & \vdots \\ p_1, & -l_k \leq \Delta(t) < -l_{k-1}, \\ 0, & \Delta(t) < -l_k. \end{cases}$$

Here $P\big(y(t) = 1\big)$ is the actual spike probability based on state, and then $\{p_k\}$ are predefined probabilities over Δ/τ.

2.2 Refractory Dynamics

Absolute Refractory Period (s = 1) For this step after a spike, $y(t) = 0$ always, $p(t) \equiv 0$, and $\texttt{aref} \to \texttt{aref} - 1$ each step. When $\texttt{aref} = 0$, transition to $s = 2$.
Relative Refractory Period (s = 2) For this step, the neuron may fire at a reduced rate $\alpha\, P_{\text{base}}$. $\texttt{rref}$ decrements until zero; if a spike occurs, reset $\texttt{aref}$ and re-enter $s = 1$, else when $\texttt{rref} = 0$, return to $s = 0$.

Figure 1 illustrates the three-state finite-state machine governing transitions among Normal, Absolute, and Relative refractory periods based on spike events and state events.

3 SuNNy

This section introduces the prototype we implemented, SuNNy. SuNNy incorporates formal verification and runtime verification via PRISM and Nengo. First, SuNNy translates SNN-RFs to corresponding models and verifiers via a template engine. PRISM enables the modeling of systems with probabilistic aspects and the verification of dynamic properties. Nengo leverages the Neural Engineering Framework (NEF) to build and simulate mechanistic neuronal circuits. Combining them, SuNNy offers a comprehensive approach to designing, validating, and analyzing SNNs (Fig. 2).

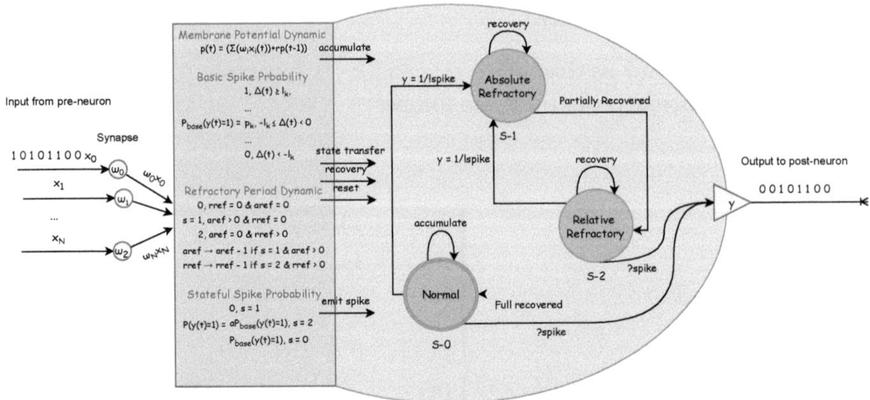

Fig. 1. RP-LI&F Neuron.

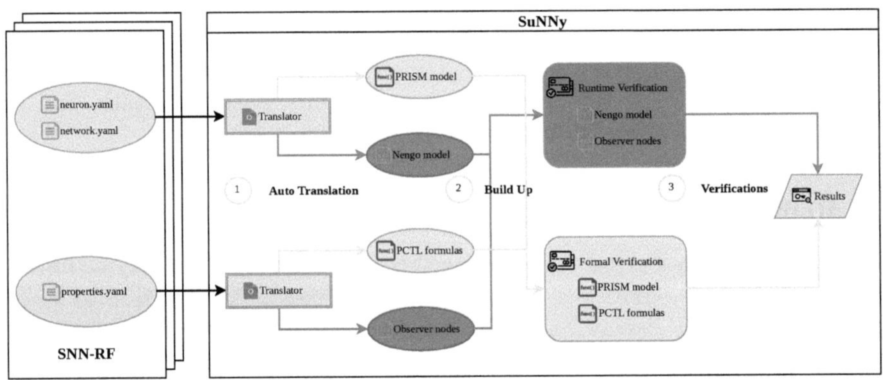

Fig. 2. Software Architecture.

3.1 PRISM

PRISM [14] is a powerful framework designed for probabilistic modeling and verification, enabling the analysis of diverse probabilistic systems such as discrete and continuous-time Markov chains (DTMCs and CTMCs), Markov Decision Processes (MDPs). In addition to modeling, PRISM supports the formal specification of system behavior using probabilistic temporal logics, including PCTL [8] (Probabilistic Computation Tree Logic).

Syntax and Modeling. We model our SNN as a DTMC by representing each neuron or population as a PRISM module with local variables capturing membrane potential and spike events. Guarded commands of the form:

$$[action]\ guard \rightarrow prob_1 : update_1 + \cdots + prob_n : update_n;$$

encode probabilistic spiking and rest behaviors. Synchronization labels *action* ensure coordinated updates across neurons at each time step. Reward structures are added to track metrics such as total spikes or energy consumption, enabling queries of the form: $\mathbf{R}\{"y"\} =? [\mathbf{C}^{\leq 100}]$ to compute expected spike counts within a time bound.

PCTL and Model Checking. PRISM employs PCTL to define the DTMCs with probabilistic and temporal constraints. PCTL extends CTL with a probabilistic path quantifier (e.g. $P_{\geq p}[F\,\phi]$ and time bounds).

All quantifiers except $\mathbf{X}$ support time-bounded variants (e.g., $\mathbf{F}^{\leq t}\,\phi$). The $\mathbf{P} =? [prop]$ form computes the exact probability of a property.

In our workflow, we automatically generate the PRISM model from a high-level description of the network topology: connectivity matrices, synaptic weights, and firing thresholds are parsed to instantiate modules and commands. We then specify PCTL queries for key properties and invoke PRISM's model checker to compute exact probabilities.

3.2   Nengo

Nengo is a neural modeling and simulation platform based on the Neural Engineering Framework (NEF), which describes how spiking neural populations can implement computations [5]. Models are constructed through a Python API by defining Ensembles (neuron groups), Connections (information flow), and Probes (monitoring tools).

We construct each SNN circuit in Nengo to align with the PRISM model: ensembles correspond to PRISM modules, and synaptic filters model transition probabilities. Simulation run on CPU or Liohi hardware generates spike trains and state trajectories over time. Probe data is exported to show how neurons in the neural network work. Both the DTMC abstraction and the NEF implementation are important to ensure our neural network works as we suppose.

4   Automated SNNs Generation and Modeling

In this section, we present a structured approach to modeling SNNs, encompassed in a dedicated representation file format (SNN-RF), itself compliant for translation to both PRISM and Nengo analysis frameworks. By defining both individual neuron archetypes and combining network properties in SNN-RF, we enable the automatic generation of models for both runtime verification and formal verification. This method simplifies the configuration of complex neural architectures, such as those based on our custom RP-LI&F neurons, while ensuring consistency across different modeling tools.

We begin by outlining the structure and purpose of these representation files, followed by detailing how they are used to generate PRISM and Nengo models. Finally, we provide insights into the implementation details within each tool, highlighting the custom neuron models and their behaviors.

4.1 SNN-RF

The SNN-RF format essentially provides a parametric description of the elementary neuron archetypes (in the present case RP-LI&F), and the synaptic connections to combine and associate them in larger structures. Distinct neurons can be instantiated by fixing local values for the parameters (timing and probability thresholds of reactive spikes). SNN-RF is designed to allow simple and correct translation into PRISM and Nengo, and also to promote easily readable configuration of parameters and network topologies through lists and dictionaries.

Importantly, the single source ensures that the resulting PRISM and Nengo descriptions have identical behaviors and therefore allows cross-validation between simulation and model-checking results.

Currently, correctness requirements are expressed only directly using PCTL probabilistic temporal logic in PRISM, but in the future, our SNN-RF format should also provide simple parametric requirement patterns to be translated both in PCTL for PRISM and as runtime observers for Nengo.

Listing 1.1 illustrates the structure of the SNN-RF. Key sections include inputs for defining input parameters, n_neurons for specifying neuron properties, and edges for describing network topologies and connection weights.

```
 1  network:
 2      name: "STRING"  # type of neural network
 3      simulate:
 4          steps: INTEGER
 5      inputs:
 6          - id: INTEGER
 7            value: INTEGER
 8          ......
 9      n_neurons:
10          - id: INTEGER
11            # More properties for neurons in the network
12          ......
13      edges:
14          - from:
15                type: STRING
16                id: INTEGER
17            to:
18                type: STRING
19                id: INTEGER
20            weight: INTEGER
21          ......
```

Listing 1.1. Spiking Neuron Networks Representation File

4.2 Neuron for PRISM

Our SNNs comprise multiple modules that collectively form a complete DTMC model. Typically, the model includes an input layer, a synapse layer, and an output layer. Each neuron receives a single input signal, and all inputs are grouped into the Input module. Each neuron receives a set of binary inputs $x_i \in \{0, 1\}$ where each input x_i is associated with a corresponding synaptic weight ω_i. Neuron modules are named by combining their type with a unique identifier (e.g., Neuron1), enabling the expansion of our model into three-layer architectures.

Neurons across different layers are interconnected through synapses, which are implemented as transfer modules. The Input, neuron, and transfer modules serve as the foundational building blocks of our network, enabling the generation of various archetypes. The core neuron behavior is encapsulated in Listing 1.2.

```
1  module Neuron1
2    aref: [0..ARP] init 0; //absolute refractory
3    rref: [0..RRP] init 0; //relative refractory
4
5    s: [0..2] init 0;  //state
6    y: [0..1] init 0;  //spike
7    p: [-500..500] init P_rest;  // membrane potential
8
9    [to] s = 0 & y = 0 & p < threshold1 ->
10        1.0: (y' = 0) & (p' = newPotential);
11   [to] ...
12   [to] s = 0 & y = 0 &
13        p >= threshold5 & p < threshold6 ->
14        0.5: (y' = 0) & (p' = newPotential) +
15        0.5: (y' = 1);
16   [to] ...
17   [to] s = 0 & y = 0 & p > threshold10 ->
18        1.0: (y' = 1);
19 endmodule
```

Listing 1.2. Formal specification of RP-LIF neuron

The neuron spikes ($y = 1$) when its potential exceeds a threshold, resetting to a basic value and entering an absolute refractory period ($s = 1$). We implement the update in Eq. 1, mentioned in Sect. 2, by first taking the weighted sum plus the leak term, rounding it down to an integer, and then clipping the result to lie between MAX and MIN, where MAX and MIN denote the upper and lower bounds of the discretized membrane potential: newPotential $= \max\Big(\min\big(\lfloor\sum_{i=1}^{m} w_i x_i + r \cdot p\rfloor, \text{MAX}\big), \text{MIN}\Big)$, which ensures our discrete membrane potential remains within the specified bounds. Moreover, a rewards structure counts spikes:

```
rewards "spike1_count"
   y1 = 1 : 1;
endrewards
```

enabling precise analysis of network activity.

4.3   Neuron for Nengo

Nengo's standard neurons lack probabilistic spiking and refractory periods, so we introduce ProbNeuron, based on our RP-LI&F model, to address these limitations. Using Nengo's API, we build networks with single-neuron ensembles (encoder=1) and a step delay to align with PRISM models. We probe spike activity, membrane potential, and state, synchronizing the time step dt with

PRISM's Markov chain for consistent, high-precision verification and simulation.

We implemented RP-LI&F neuron for Nengo as **ProbNeuron** class, which inherits from `nengo.neurons.NeuronType`. Based on this class, we construct SNNs using standard Nengo components such as `Ensemble`, `Connection`, and `Node`. As shown in Listing 1.3.

Listing 1.3. Nengo network with ProbNeuron

```python
with nengo.Network() as model:
    # input node
    input = [1]
    input_node = nengo.Node(input)
    # neuron group with ProbNeuron
    ens = nengo.Ensemble(
        n_neurons=1,
        dimensions=1,
        neuron_type=ProbNeuron(
        P_rth=40, ARP=2, RRP=4, r=0.9, dt=0.001),
        encoders=np.array([[1]])
    )
    # weights of input
    weights = np.array([[2]])
    nengo.Connection(input_node, ens, transform=weights
        )
    # add probes
    ...
with nengo.Simulator(model, dt=0.001) as sim:
    sim.run(simulation_time)
```

5   Case Study

We validate our framework with one RP-LI&F neuron, two archetypes: contralateral inhibition and Convergent circuit. These demonstrate our approach for probabilistic verification and simulation, using SNN-RF to generate consistent DTMCs and Nengo Ensembles with shared dt time steps, ensuring precise cross-domain validation (Figs. 4, 6). All experiments were performed on a machine with an `Intel(R) Core(TM) i5-8300H CPU @ 2.30 GHz`, 16 GB RAM, running ARCH LINUX (kernel version `6.14.6.arch1-1`).

5.1   RP-LI&F Neuron

Before we go into those two cases, first, we present the verification and simulation of only one RP-LI&F neuron to better understand how our neuron works. Here, we checked the correct implementation of the RP-LI&F neuron. The input of this neuron is a constant one (see Fig. 3a).

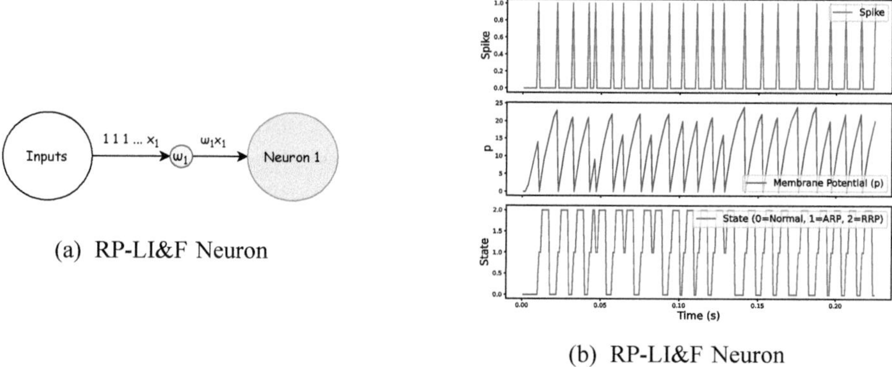

(a) RP-LI&F Neuron

(b) RP-LI&F Neuron

Fig. 3. RP-LI&F Neuron and Simulation (the X-axis represents time, and the Y-axis, from top to bottom, represents spike, membrane potential, and states).

PCTL Properties and Results. We validated the fundamental characteristics of the RP-LI&F neuron, specifically its transition rules and spiking probabilities, which are essential for all subsequent developments.

P_1 : $P_{>=1}[G\,((y_1 = 1) \longrightarrow (X\,(s = 1)))]$ verified that whenever a spike occurs, the very next state is the absolute refractory period, in which the neuron does not emit.

P_2 : $P_{>=1}[G\,((s = 1 \land \mathtt{aref{=}0}) \longrightarrow (X\,(s = 2 \land \mathtt{rref{=}RRP})))]$ asserts that when **aref** reaches 0 in state 1, the next state is 2 with **rref** reset to its maximum.

P_3 : $P_{=?}[((p > threshold5) \land (p <= threshold6)) \longrightarrow (X\,(y = 1))]$ checks the correctness of one of our probabilistic transitions. All the others can also be proved to be 1 in probability.

Table 1 shows the outcomes and execution time of Properties P_1 to P_3.

Table 1. Verification Results of RP-LI&F Neuron

Properties	Specifications	Outcomes	Time
P_1	$P_{>=1}[G\,((y_1 = 1) \longrightarrow (X\,(s = 1)))]$	True	45 ms
P_2	$P_{>=1}[G\,((s = 1 \land aref = 0) \longrightarrow (X\,(s = 2 \land rref = RRS)))]$	True	19 ms
P_3	$P_{=?}[((p > threshold5) \land (p <= threshold6)) \longrightarrow (X\,(y = 1))]$	0.5	9 ms

Nengo Simulation. The Nengo simulation results (see Fig. 3b) illustrate a single neuron's spiking behavior and state transitions.

5.2   Contralateral Inhibition

Contralateral Inhibition models neural populations that mutually suppress activity, a mechanism critical to sensory processing. They provide an elementary computation that can be used for decision-making (not only simple forms of behavioral choices but also more complex choices) [13]. As another example, active and passive fear responses are mediated by distinct and mutually inhibitory central amygdala neurons [6]. Here we present a network in which a constant input of 1 is fed to two RP-LI&F neurons. Although both neurons receive the same excitatory input, neuron 1 inhibits neuron 2 more strongly than vice versa. As a result, a winner-takes-all dynamic is expected to emerge, with neuron 1 more likely to dominate the activity. The connectivity is illustrated in Fig. 4.

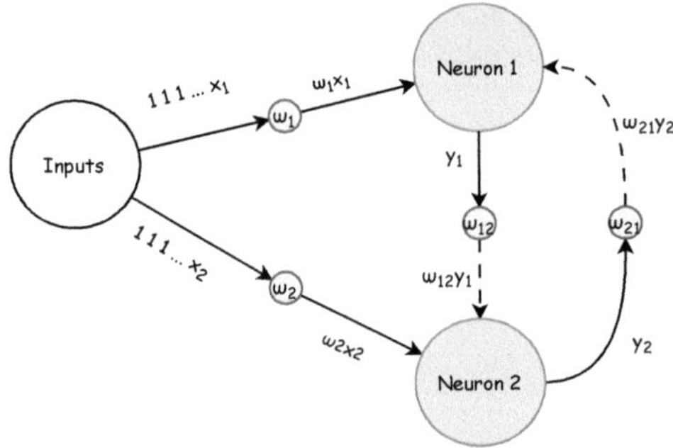

Fig. 4. Contralateral Inhibition (solid lines denote excitatory connections and dashed lines indicate inhibitory ones).

PCTL Properties and Results. Then we checked two of the most important properties for contralateral inhibition to show that it is well implemented.

P_4 : $P_{=?}[F\,G\,(y_2 = 0 \wedge (F\,(y_1 = 1)))]$ This specification ensures that in all future times, Neuron 2 stops spiking forever (becomes permanently inactive), and from that point onward, Neuron 1 continues to spike infinitely often, which means Neuron 1 is the winner and takes all priorities.

P_5 : $P_{>=1}[G > T\,(y_2 = 0)]$ This specification ensures that in all possible evolutions of the system, after an initial period of T time steps where the system may oscillate, Neuron 2 ceases to spike entirely and remains inactive for all subsequent time.

Table 2 shows the outcomes and execution time of P_4 and P_5. Those two properties demonstrate the correctness of our contralateral inhibition model and the fidelity of its biological simulation. These formal guarantees align with empirical observations of contralateral inhibitory circuits in sensory systems, where one neuronal population reliably dominates after mutual competition, thereby validating both the structural integrity and functional realism of our RP-L&IF-based network model.

Table 2. Verification Results of Contralateral Inhibition

Properties	Specifications	Outcomes	Time
P_4	$P_{=?}[F\,G\,(y_2 = 0 \land (F\,(y_1 = 1)))]$	1.0	372 ms
P_5	$P_{>=1}[G > T\,(y_2 = 0)]$	True (T=100)	165 ms

Nengo Simulation. Thanks to SNN-RF, we can automatically generate the Nengo RP-L&IF neural network model from identical parameters and topology. For contralateral inhibition, our simulation results illustrate the evolution of each neuron's internal state variables alongside the PRISM model-checking outcomes rendered visually (see Fig. 5).

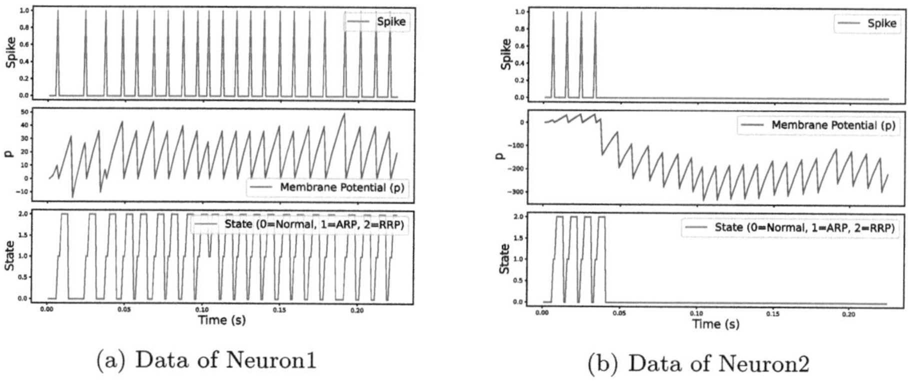

(a) Data of Neuron1   (b) Data of Neuron2

Fig. 5. Simulation of Contralateral Inhibition.

As shown in Fig. 5b, Neuron2 only emits spikes at the very start of the simulation and remains silent thereafter. In contrast, in Fig. 5a, Neuron1 persists in spiking throughout the entire simulation, passing through its normal period, and two refractory periods, with spikes occurring at irregular intervals.

5.3   Convergent Circuit

Convergent circuits-characterized by multiple presynaptic inputs converging onto common postsynaptic targets-amplify downstream spiking through spatial summation. This mechanism is crucial for precise temporal coordination, which is fundamental for generating rhythmic motor patterns like locomotion [7,12]. We model such a network with one constant input and two RP-LI&F neurons exciting a postsynaptic target (Fig. 6), to analyze temporal coherence using our framework.

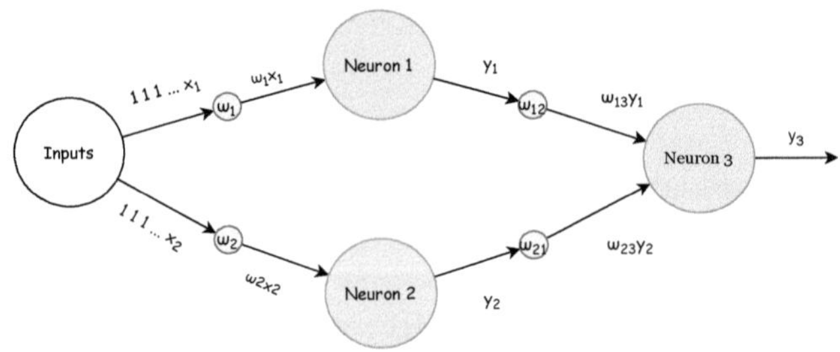

Fig. 6. Convergent circuit (solid lines indicate positive weights).

PCTL Properties and Results. Here we verify two gating and activation properties under different synaptic configurations. Those two properties illustrate different types of convergent excitatory motifs:

$P_6 : P_{\geq 1}\big[(\,G(y_2 = 0) \wedge F(y_1 = 1)\,) \longrightarrow F(y_3 = 1)\big]$ asserts that with probability one, if Neuron 2 remains silent throughout and Neuron 1 spikes at least once, then Neuron 3 will eventually spike. It guarantees the conditional activation of Neuron 3 by Neuron 1 in the absence of activating input from Neuron 2.

$P_7 : P_{\geq 1}\big[G((y_1 = 0 \vee y_2 = 0)) \longrightarrow \neg F(y_3 = 1)\big]$ ensures that with probability one, whenever either Neuron 1 or Neuron 2 never spikes, Neuron 3 will never spike. It enforces that a single missing presynaptic activation is sufficient to block downstream firing.

Table 3 shows the results and execution time of P_6 and P_7.

Table 3. Verification Results of Convergent circuit

Properties	Specifications	Outcomes	Time
P_6	$P_{\geq 1}\big[(\,G(y_2 = 0) \wedge F(y_1 = 1)\,) \longrightarrow F(y_3 = 1)\big]$	True	9 ms
P_7	$P_{\geq 1}\big[G((y_1 = 0 \vee y_2 = 0)) \longrightarrow \neg F(y_3 = 1)\big]$	True	1 ms

Nengo Simulation. In the Nengo part, we simulate two synaptic configurations that satisfy the desired properties, respectively. To better demonstrate our simulation results, here we display the comparisons of spikes of each neuron (see Fig. 7).

Property 6 ensures Neuron 3 activates with just one presynaptic spike, mimicking the brain's reflex-like response to stimuli (Fig. 7a). Property 7 requires spikes from both Neuron 1 and Neuron 2 for Neuron 3 to spike, akin to decision-making needing dual sensory inputs (Fig. 7).

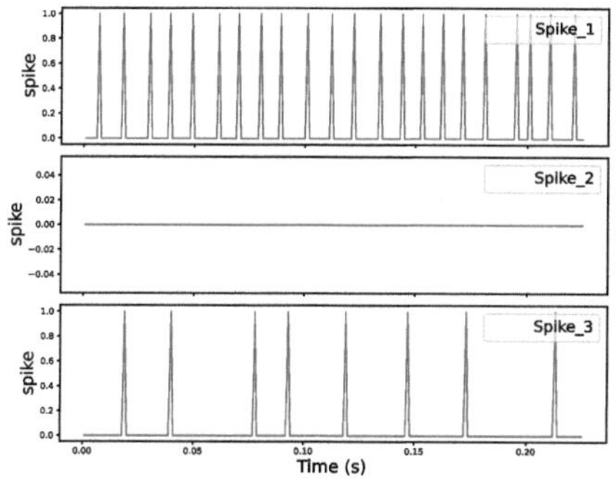

(a) Simulation of neuron 1, 2, and 3 with parameters respecting P_6

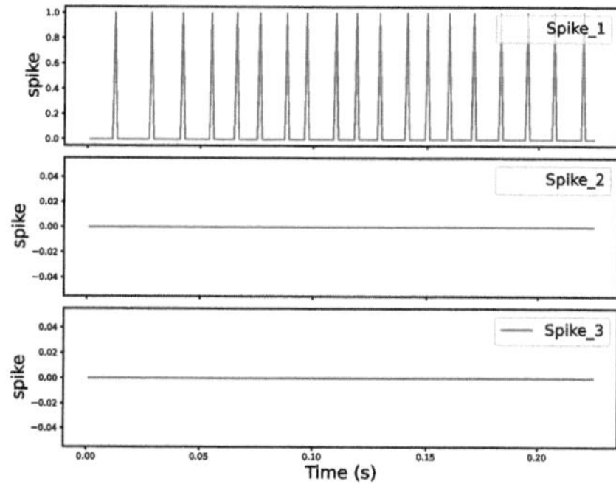

(b) Simulation of neuron 1, 2, and 3 with parameters respecting P_7

Fig. 7. Simulation of Convergent circuit (the X-axis represents time, and the Y-axis represents spike.

6 Conclusion and Future Work

This work establishes a co-verification framework for SNNs that bridges formal verification and large-scale simulation. We demonstrate three key advances:

First, the **RP-LI&F neuron model** integrates discrete refractory periods with probabilistic firing dynamics. Second, the **SNN-RF specification** enables automated consistency between PRISM (for PCTL verification) and Nengo (for parallel simulation), eliminating toolchain discrepancies. Third, we provide **SuNNy** for incorporating formal verification and simulation, which auto translates SNN-RF to models and properties leveraging a template engine.

Building on the established link between elementary neuron properties and global stochastic requirements, we will pursue two key directions:

1. Scalable Verification Frameworks: Extend our approach to large-scale SNNs using PRISM and Nengo, developing:
   - Assume/Guarantee contracts for reactive constraints
   - Observer nodes for runtime Observation in Nengo
2. Parameter Synthesis: Design algorithms to automatically derive SNN parameters satisfying probabilistic dynamic properties [3].

These steps will enable rigorous verification of brain-like neuromorphic systems solving complex tasks.

References

1. Bekolay, T., et al.: Nengo: a python tool for building large-scale functional brain models. Front. Neuroinf. **7**, 48 (2014)
2. De Maria, E., et al.: On the use of formal methods to model and verify neuronal archetypes. Front. Comput. Sci. **16**(3), 1–22 (2022). https://doi.org/10.1007/s11704-020-0029-6

3. De Maria, E., Di Giusto, C., Laversa, L.: Spiking neural networks modelled as timed automata: with parameter learning. Nat. Comput. **19**, 135–155 (2020)

4. De Maria, E., Gaffé, D., Ressouche, A., Riboulleau, C.G.: A model-checking approach to reduce spiking neural networks. In: BIOINFORMATICS 2018-9th International Conference on Bioinformatics Models, Methods and Algorithms, pp. 1–8 (2018). https://doi.org/10.5220/0006572000890096

5. DeWolf, T., Jaworski, P., Eliasmith, C.: Nengo and low-power AI hardware for robust, embedded neurorobotics. Front. Neurorobot. **14**, 568359 (2020)

6. Fadok, J.P., et al.: A competitive inhibitory circuit for selection of active and passive fear responses. Nature **542**(7639), 96–100 (2017)

7. Grillner, S.: Biological pattern generation: the cellular and computational logic of networks in motion. Neuron **52**(5), 751–766 (2006)

8. Hansson, H., Jonsson, B.: A logic for reasoning about time and reliability. Formal Aspects Comput. **6**, 512–535 (1994)

9. Hodgkin, A.L., Huxley, A.F.: A quantitative description of membrane current and its application to conduction and excitation in nerve. J. Physiol. **117**(4), 500 (1952)

10. Hofestädt, R., Thelen, S.: Quantitative modeling of biochemical networks. In Silico Biol. **1**(1), 39–53 (1998)

11. Kabilan, R., Muthukumaran, N.: A neuromorphic model for image recognition using snn. In: 2021 6th International Conference on Inventive Computation Technologies (ICICT), pp. 720–725. IEEE (2021)

12. Kiehn, O.: Decoding the organization of spinal circuits that control locomotion. Nat. Rev. Neurosci. **17**(4), 224–238 (2016)

13. Koyama, M., Pujala, A.: Mutual inhibition of lateral inhibition: a network motif for an elementary computation in the brain. Curr. Opin. Neurobiol. **49**, 69–74 (2018)

14. Kwiatkowska, M., Norman, G., Parker, D.: PRISM 4.0: verification of probabilistic real-time systems. In: Gopalakrishnan, G., Qadeer, S. (eds.) CAV 2011. LNCS, vol. 6806, pp. 585–591. Springer, Heidelberg (2011). https://doi.org/10.1007/978-3-642-22110-1_47

15. LeCun, Y., Bottou, L., Bengio, Y., Haffner, P.: Gradient-based learning applied to document recognition. Proc. IEEE **86**(11), 2278–2324 (1998)

16. Maass, W.: Networks of spiking neurons: the third generation of neural network models. Neural Netw. **10**(9), 1659–1671 (1997). https://doi.org/10.1016/s0893-6080(97)00011-7

17. Maio, V., Lánský, P., Rodriguez, R.: Different types of noise in leaky integrate-and-fire model of neuronal dynamics with discrete periodical input (2004)

18. de Maria, E., Lapijover, B., L'Yvonnet, T., Moisan, S., Rigault, J.P.: A formal probabilistic model of the inhibitory control circuit in the brain. In: BIOINFORMATICS 2023 - 14th International Conference on Bioinformatics Models, Methods and Algorithms, Lisbonne, Portugal (2023). https://inria.hal.science/hal-03999574

19. Pradhan, A., King, J., Pinisetty, S., Roop, P.S.: Model based verification of spiking neural networks in cyber physical systems. IEEE Trans. Comput. **72**(9), 2426–2439 (2023)

20. Sengupta, A., Ye, Y., Wang, R., Liu, C., Roy, K.: Going deeper in spiking neural networks: VGG and residual architectures. Front. Neurosci. **13**, 95 (2019)

21. Teeter, C., et al.: Generalized leaky integrate-and-fire models classify multiple neuron types. Nat. Commun. **9**(1), 709 (2018)

The Domino Problem Is Decidable
for Robust Tilesets

Nathalie Aubrun[1], Manon Blanc[1,2](✉), and Olivier Bournez[2]

^{1} Université Paris-Saclay, LISN, Gif-sur-Yvette, France
^{2} École Polytechnique, LIX, Palaiseau, France
`blanc@lix.polytechnique.fr`

Abstract. One of the most fundamental problems in tiling theory is the domino problem: given a set of tiles and tiling rules, decide if there exists a way to tile the plane. The problem is known to be undecidable in general. In this paper, we focus on Wang tilesets. We prove that the domino problem is decidable for robust tilesets, *i.e.* tilesets that either cannot tile the plane or can by provably satisfying some particular invariant. We establish that several famous tilesets considered in the literature are robust. We give arguments this is true for all tilesets unless they are produced from non-robust Turing machines: a Turing machine is said to be non-robust if it does not halt and furthermore does so non-provably. As a side effect of our work, we provide a sound, relatively complete method for proving that a tileset can tile the plane. Our analysis also provides explanations for the similarities between proofs in the literature for various tilesets, as well as of phenomena that have been observed experimentally in the systematic study of tilesets using computer methods.

1 Introduction

Wang tiles were introduced in the early '60 s to study the undecidability of the $\forall\exists\forall$ fragment of first-order logic [16]. They consist of unit square tiles with colours on their edges. A tiling of the infinite Euclidean plane is obtained using arbitrarily many copies of the tiles in the given finite tileset. Tiles are placed on the integer lattice points of the plane so that they cover the whole plane with no overlap. The tiling is *valid* if everywhere the contiguous edges have the same colour. The *domino problem* is the decision problem which takes as input τ a finite set of Wang tiles, and outputs `Yes` if there exists a valid tiling by τ and `No` otherwise. A famous result by Berger states that the Domino problem is undecidable [3]. Berger's original proof relies on the construction of an aperiodic tileset, that is used to perform Turing machine computations inside tilings. This proof has subsequently been simplified [15] and alternative proofs were exhibited [1,6,7,12]. All known proofs rely on the existence of an aperiodic tileset, even if the techniques can be different. For most of the known aperiodic tilesets, once the tileset has been explicitly built, proving aperiodicity is not that complicated, but proving a tiling exists may reveal more difficulty. Yet, no general

© The Author(s), under exclusive license to Springer Nature Switzerland AG 2026
E. Formenti and L. Manzoni (Eds.): UCNC 2025, LNCS 16364, pp. 115–131, 2026.
https://doi.org/10.1007/978-3-032-15641-9_9

technique to prove an aperiodic tileset admits a valid tiling is known. One objective of this paper is to set a formalism unifying proofs of the existence of tilings for aperiodic tilesets. We also go one step further, studying several tilesets in this new light and proving they exhibit a particular form of recurrence invariant.

Using the formalism of transducers, we introduce two notions of *robust* tilesets (Definitions 6 and 7), according to whether some semantic or provably invariant holds. These concepts are motivated by our exploration of Robinson's tileset with transducers: our result on this particular tileset that inspires the rest of the paper is Theorem 1. This theorem expresses in an extremely simple way the fact that a valid tiling exists: the recurrence relation (1), which can be read as an invariant, provides a simple proof that the plane can be tiled.

Furthermore, we state that the phenomenon demonstrated in Theorem 1 is not specific to Robinson's tileset. Indeed, an analogous semantic statement can be written for every tileset for which one can prove it admits a valid tiling: Propositions 7 and 8. We indeed prove that the most famous tilesets considered in the literature are robust, as they satisfy some invariant holding provably. We do so for Robinson's tileset (Theorem 1), Jeandel-Rao's tileset (Theorem 2) and Kari's tilesets (Proposition 14).

We explain the phenomenon with a parallel made with the theory of Turing machines, for which we can distinguish *robust* Turing machines, where the Halting problem is provably decidable, from *non-robust* ones. A similar statement holds for tilesets: the domino problem for robust tilesets is decidable (Theorem 7). We establish two results with the same flavour: tilesets that encode a Turing machine are robust iff the Turing machine is itself robust. Specifically, we prove this is the case if Turing machines are embedded into a tileset through piecewise affine maps (Sect. 7 with the technique by Kari [12]) or through an adaptation of Robinson's tileset (Sect. 8). As a consequence of our work, we establish a sound and relatively complete method for proving that a tileset can tile the plane. Some experimental facts observed in the systematic computer-assisted exploration of Wang tilesets that have been conducted can be interpreted in a new light thanks to this setting as discussed in Sect. 9.

The paper is organized as follows. Section 2 details how Wang tilesets can be seen as transducers and why this approach is particularly relevant. We propose in Sect. 2.4 a way to use transducers as a logic. We also establish some basic facts about transducers and their relations to Wang tilesets. In Sect. 3 we focus on Robinson's tileset τ_{Robi} to illustrate concepts from Sect. 2. In Sect. 4, we present two notions of robustness for tilesets, distinguishing the cases where an invariant semantically holds and provably holds. We prove in Sect. 4.3 that Robinson's tileset satisfies the strongest of these notions and so does Jeandel-Rao's tileset (Sect. 5). In Sects. 7 and 8 we establish a correspondence between robustness for tilesets and robustness for Turing machines. We conclude with a discussion of our results in Sect. 9. By lack of space, many proofs of statements, as well as various graphical intepretations are missing. More details can be found in the companion article https://arxiv.org/abs/2402.04438.

2 Wang Tilesets as Transducers

2.1 Transducers and Meta-Transducers

Definition 1 (Transducer). *Let Σ be a finite alphabet. A transducer is a triple (Q, Σ, δ) with Q a finite set of states, Σ a finite alphabet and $\delta \subseteq Q \times Q \times \Sigma \times \Sigma$ a set of transitions.*

For example the transducer $\mathcal{T}_1$ with $Q = \{V, B, R\}$, $\Sigma = Q$ is pictured on the right: The fact that $(q, r, w, w') \in \delta$ is represented by an edge between q and r labelled by $w|w'$. To simplify notations and the writing of proofs, it will often be more convenient 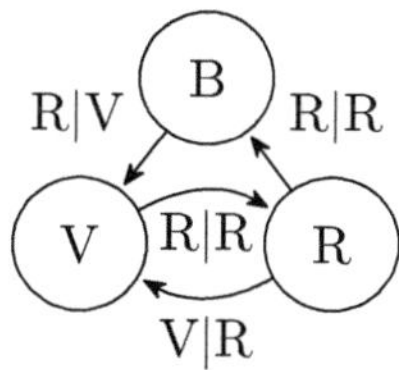 to use *meta-transducers* rather than transducers. Meta-transducers allow us to deal not only with single letters but also with words of arbitrary finite length as labels of transitions. It is more restricted than the usual definition of transducers.

Definition 2 (Meta-transducer). *A meta-transducer is a triple (Q, Σ, δ) with Q a finite set of states, Σ a finite alphabet and $\delta \subseteq Q \times Q \times \Sigma^* \times \Sigma^*$ a set of transitions, or edges, with $|w|, |w'| \geq 1$ where $w, w' \in \Sigma^*$ and $|w| = |w'|$, where $|w|$ denotes the length of w.*

The transition relation δ^* is the smallest relation, containing δ, such that, whenever $(q, r, x, y) \in \delta^*$ and $(r, s, w, w') \in \delta$, then $(q, s, xw, yw') \in \delta^*$. If $\mathcal{T}$ is a (meta-)transducer, for any $w, w' \in \Sigma^*$, we write $w\mathcal{T}w'$ if there exist $q, s \in Q$ and $w, w' \in \Sigma^*$ s.t. $(q, s, w, w') \in \delta^*$. This is equivalent to say that there exists a sequence $q_0, \ldots, q_n$ s.t. $(q_i, q_{i+1}, w_i, w'_i)$ for every $i = 0 \ldots n - 1$, where $w = w_1 \ldots w_n$, $w' = w'_1 \ldots w'_n$ and the w_i, w'_i are words on Σ. By definition, a transducer is a meta-transducer: consider every letter as a word of length 1.

Consider $\mathcal{T} = (Q, \Sigma, \delta)$ a meta-transducer. A transition $(q, q, w, w') \in \delta$ is called a *loop*. Moreover, if the word w' is a cyclic permutation of w, i.e. if there exist two non-trivial words u, v such that $w = uv$ and $w' = vu$, we say that the loop (q, q, w, w') is a *cyclic loop*. Finally, a loop $(q, q, w, w) \in \delta$ is called a *periodic loop*. Note that every periodic loop is a cyclic loop, but the converse is not true. We say that a transducer *contains a loop* if there exists some $h : [[0 \ldots m]] \to Q$ such that $(h(i), h(i + 1), w_i, w'_i) \in \delta$ with $h(m + 1)$ abusively defined as $h(0)$.

2.2 Tiles and Transducers

A *Wang tileset* T is a triple (H, V, τ) where H is a finite set that corresponds to the horizontal colours, V is a finite set that corresponds to vertical colours and $\tau \subseteq H^2 \times V^2$ is the set of tiles. We denote $t = (w, e, s, n) \in \tau$ a *tile* in T where w (resp. e, s, n) is the colour on the left (resp. right, bottom, top) edge. For a tile t, we also write t_w, t_e, t_s and t_n the colours of its edges.

A Wang tileset expressed as triple (H, V, τ) exactly matches the definition of a transducer of Sect. 2.1 where tiles are transitions, horizontal colours H are states Q and vertical colours V are letters of the alphabet Σ (Figs. 1 and 2).

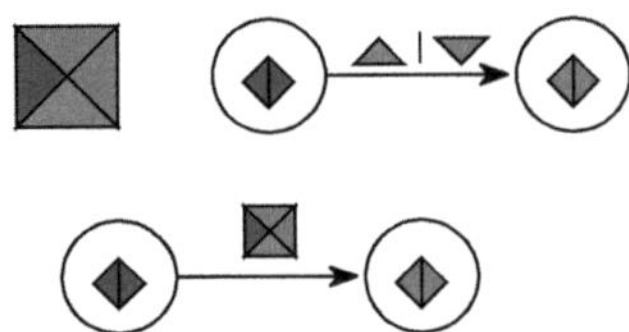

Fig. 1. Graphical representations of a tile as a transition in the associated transducer.

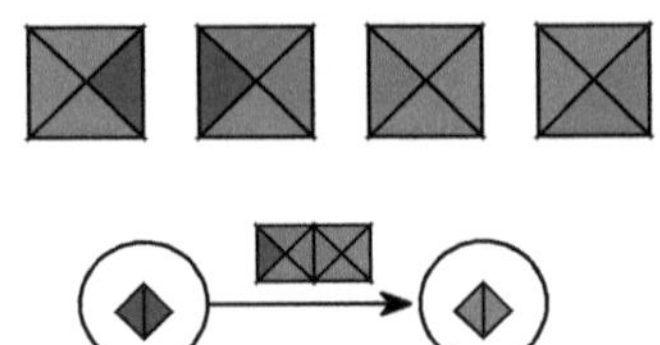

Fig. 2. An example a of meta-transducer derived from the tileset τ composed by the four tiles pictured on the top.

There is a bijection between the tiles of a Wang tileset and the edges of its associated transducer. From now on we will use tileset or transducer interchangeably without any possible confusion. A natural operation on transducers, denoted by $\cup$, is the union: given $\mathcal{T}_1 = (Q_1, \Sigma_1, \delta_1)$ and $\mathcal{T}_2 = (Q_2, \Sigma_2, \delta_2)$, their union $\mathcal{T}_1 \cup \mathcal{T}_2$ is given by (Q, Σ, δ) where $Q = Q_1 \cup Q_2$, $\Sigma = \Sigma_1 \cup \Sigma_2$, and $(q, q', w, w') \in \delta$ iff $(q, q', w, w') \in \delta_1$ or $(q, q', w, w') \in \delta_2$. Any transducer is thus the union of its edges.

A *tiling* x by τ (or by T) is a mapping $x : \mathbb{Z}^2 \to \tau$ such that adjacent edges share the same colour, i.e. for every $(i, j) \in \mathbb{Z}^2$, $x(i, j)_n = x(i, j + 1)_s$ and $x(i, j)_e = x(i + 1, j)_w$. The tiling x is *periodic* with period $(a, b) \in \mathbb{Z}^2 \backslash \{(0, 0)\}$ iff $x(u, v) = x(u + a, v + b)$ for every $(u, v) \in \mathbb{Z}^2$. If there exists a periodic valid tiling with tiles from τ, then there exists a *doubly periodic* valid tiling, i.e. a tiling x such that, for some $a, b > 0, x(u, v) = x(u + a, v) = x(u, v + b)$ for all $(u, v) \in \mathbb{Z}^2$ (see e.g. [11] or [10, Lemma 1]). A tileset τ is called *aperiodic* iff the two following conditions are satisfied (i) there exists a valid tiling and (ii) there does not exist any periodic valid tiling.

A bi-infinite word (or bi-infinite sequence) on the alphabet V is a sequence $(w_i)_{i \in \mathbb{Z}}$ such that, for every $i \in \mathbb{Z}, w_i \in V$. If $w = (w_i)_{i \in \mathbb{Z}} \in V^{\mathbb{Z}}$ is a bi-infinite word, we say $w' = (w'_i)_{i \in \mathbb{Z}} \in V^{\mathbb{Z}}$ is the image of w by the transducer $\mathcal{T}$ if there exists a bi-infinite sequence of states $q = (q_i)_{i \in \mathbb{Z}} \in H^{\mathbb{Z}}$ such $(q_i, q_{i+1}, w_i, w'_i) \in \mathcal{T}$ that for every $i \in \mathbb{Z}$. We denote this by $w \mathcal{T} w'$. This corresponds to the existence of an infinite horizontal stripe of Wang tiles from τ with colours w on the bottom and w' on the top. The relation induced by $\mathcal{T}$ on bi-infinite words is usually non-deterministic, so this is a partial relation, not a function. We say that $\mathcal{T}$ is *compatible* if there exist $\omega, \omega' \in V^{\mathbb{Z}}$ such that $\omega \mathcal{T} \omega'$. The proof of the following is direct from a pumping argument:

Proposition 1. *A tileset τ is compatible if and only if the associated transducer $\mathcal{T}$ contains a loop.*

2.3   Composition

We present here how the notion of composition allows us to make these horizontal patterns grow in the vertical direction. Combining both concepts of meta-transducer and composition, we obtain a formalism in terms of meta-transducers to deal with rectangular patterns valid for τ of arbitrary sizes.

Definition 3 (Notation $\mathcal{T} \circ \mathcal{T}'$). *If (Q, Σ, δ) and (Q', Σ', δ') are two meta-transducers, then the composition of $\mathcal{T}$ and $\mathcal{T}'$ is the meta-transducer $(Q \times Q', \Sigma \times \Sigma', \delta'')$ where $|w| = |w'|$ and $((w, w'), (e, e'), s, n') \in \delta''$ iff we have $(w, e, s, n) \in \delta$ and $(w', e', s', n') \in \delta'$ and $n = s'$.*

Proposition 2. *The composition of two meta-transducers is a meta-transducer.*

As a corollary, if we start with two Wang tilesets, their composition is still a Wang tileset. The states of the compositions are pairs of horizontal colours and the edges correspond to vertical compatibility between the top and bottom vertical colours of tiles associated with the corresponding horizontal colours.

Fix some meta-transducer $\mathcal{T}$. In particular we can compose $\mathcal{T}$ with itself and iterate to get a sequence of transducers $(\mathcal{T}^n)_{n \in \mathbb{N}^*}$ given by: $\mathcal{T}^1 = \mathcal{T}$ and $\mathcal{T}^{n+1} = \mathcal{T}^n \circ \mathcal{T}$ for $n \geq 0$. As a direct consequence of the definition, we relate the compatibility of the composition $\mathcal{T}^n$ with the existence of a valid tiling of a horizontal strip.

Proposition 3. *Consider a transducer $\mathcal{T}$. There exist $w, w' \in V^{\mathbb{Z}}$ such that $w\mathcal{T}^n w'$ iff there exists a valid tiling by τ of $\mathbb{Z} \times [0; n-1]$ the horizontal strip of height n.*

If we start from a Wang tileset, then the set of states of $\mathcal{T}^n$ can be identified with H^n. We say then that $\mathcal{T}^n$ is of *height n* (on Σ^n).

2.4   Inclusion Between Transducers

A meta-transducer can also be seen as a labelled multi-graph whose edges are labelled by pairs of words on Σ. Its vertices are elements of Q (states), and there is an edge between q and r iff $(q, r, w, w') \in \delta$ for some w, w'.

Transducers and meta-transducers are natural points of view for tilings:

Definition 4 (Notation $\mathcal{T} \succ\!\!\prec \mathcal{T}'$). *Given two transducers $\mathcal{T}$ and $\mathcal{T}'$, we write $\mathcal{T} \succ\!\!\prec \mathcal{T}'$ iff every edge of the transducer $\mathcal{T}'$ is an edge of $\mathcal{T}$.*

This just means that the local constraints associated with $\mathcal{T}'$ fulfil all the constraints associated with $\mathcal{T}$. We use a scissor symbol, as this corresponds to deleting (cutting) some edges. As such, when we write $\mathcal{T} \succ\!\!\prec \mathcal{T}'$ and the states of $\mathcal{T}$ are of height n, then $\mathcal{T}'$ is necessarily made of states of height n.

Example 1.

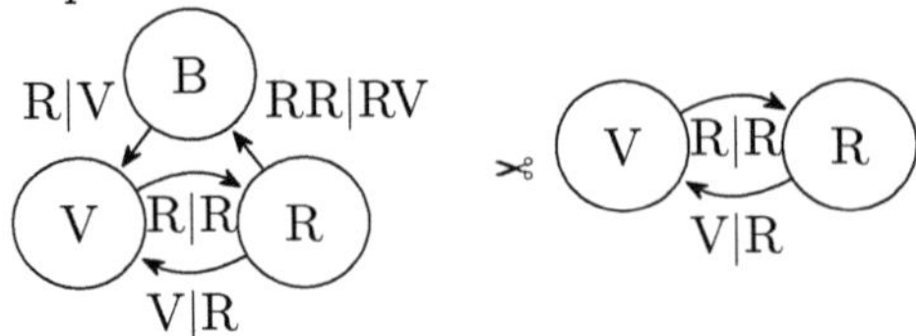

Proposition 4. *Let τ be a tileset and $\mathcal{T}$ the associated transducer. If $\mathcal{T}$ contains a cyclic loop of size (or order) N then $\mathcal{T}^m$ contains a periodic loop for some $1 \leq m \leq N$.*

Proposition 5. *Consider a transducer $\mathcal{T}$. Then $\mathcal{T}^n$ contains a loop iff there exists a valid tiling by τ of $\mathbb{Z} \times [0; n-1]$ the horizontal strip of height n.*

Proposition 6. *Consider a transducer $\mathcal{T}$. Then $\mathcal{T}^n$ contains a loop of order N iff there exists a valid tiling by τ of $\mathbb{Z} \times [0; n-1]$ the horizontal strip of height n using a pattern repeated infinitely with period N.*

3   Robinson's Tileset as a Transducer to Prove Tileability

3.1   Robinson's Tileset

A famous tileset is Robinson's tileset [15] which simplifies the construction of Berger [3]. This aperiodic Wang tileset can be represented more compactly with the geometric tiles pictured in Fig. 3, which may be rotated and reflected. Formally, they are not Wang tiles but they can be transformed in such, using Robinson's arguments [15]. We convert this tileset into a transducer by defining two alphabets for horizontal and vertical colours: $H := \{ \text{⌐}, \text{⌐}, \text{⌐}, \text{⌐}, \text{⌐}, \text{⌐} \}; V := \{ \text{⌐}, \text{⌐}, \text{⌐}, \text{⌐}, \text{⌐}, \text{⌐} \}$. Note that with this choice for H

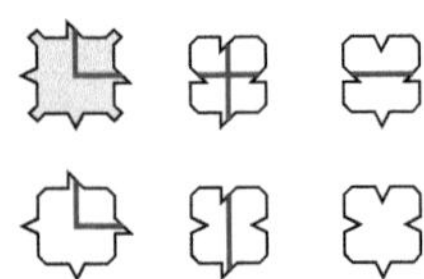

Fig. 3.   Robinson's tileset.

and V, we forget about the corner types (bumpy or dented). However, this will not be an issue later on, since no two tiles are wearing the same colours from H and V with different corner types. Robinson's tileset can be associated with the transducer $\mathcal{T}_{Robi}$. For more readability, in this figure, we label a transition from a state q to a state q' by the corresponding tile from Robinson's tileset with the proper orientation.

3.2   Patterns Inside Robinson's Tilings

We set $g(n) := 2^n - 1$ for every integer $n \geq 0$, so that in the sequel we are only interested in the meta-transducers $\mathcal{T}_n := \mathcal{T}_{Robi}^{g(n)}$. Note that the sequence $(g(n))_{n \in \mathbb{N}}$ satisfies the recurrence relation $g(n) = 2g(n-1) + 1$. We also define notations for some horizontal and vertical patterns

- $u^{\text{Ⓝ}} := \text{⌐}^{g(n-1)} u \text{⌐}^{g(n-1)}$ for $u \in \{ \text{⌐}, \text{⌐}, \text{⌐} \}$;
- $b^{\text{Ⓝ}} := \text{⌐}^{g(n-1)} b \text{⌐}^{g(n-1)}$ for $b \in \{ \text{⌐}, \text{⌐}, \text{⌐} \}$.

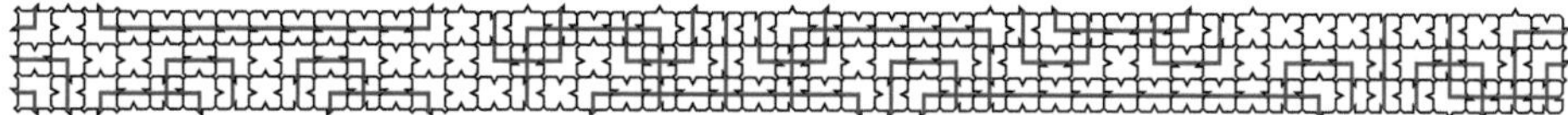

Fig. 4. One of the many possibilities to tile a 46×3 rectangle with Robinson's tileset. All vertical patterns that appear in this pattern are different.

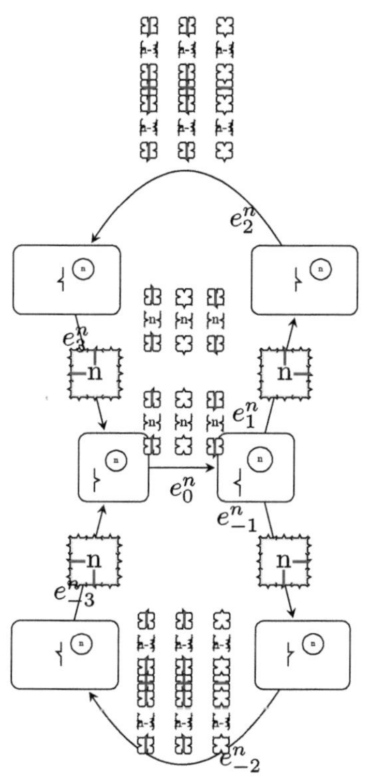

Fig. 5. Graph H_n

We also write $\genfrac{}{}{0pt}{}{a}{\{_n\}}$ (for example, $\genfrac{}{}{0pt}{}{a}{\{_n\}}$) for a pattern made of the tile a repeated $g(n)$ times vertically, where a can be either , , , , or .

Studying the entire transducer $\mathcal{T}_n$ would be too tedious (see Fig. 4 for an insight into the combinatorics of $\mathcal{T}_3$). We rather consider the meta-transducer H_n defined in Fig. 5. As we will see (Theorem 1), this transducer H_n satisfies that $\mathcal{T}_n \succ\!\!\prec H_n$ for every $n \in \mathbb{N}^*$. For the sake of clarity we choose to label the various edges e_j^n with patterns.

For instance the label $\{_n\}$ for edge e_0^n of the Figure 5 corresponds to . In the same spirit, the pattern corre-

sponds to the pair and similarly for other patterns.

3.3 Robinson's Tilings Described by a Recurrence Relation

Theorem 1 (Key property of Robinson tiling). *For all $n \in \mathbb{N}$,*

$$H_n \circ \mathcal{T}_{Robi} \circ H_n \succ\!\!\prec H_{n+1} \tag{1}$$

To prove Theorem 1, we just need to prove that any edge of H_{n+1} is among the edges of $H_n \circ \mathcal{T}_{Robi} \circ H_n$. The proof of Theorem 1 then follows from the following basic steps.

Lemma 1. *For all n, we have* $e_0^n \circ \; \circ e_0^n \succ\!\!\prec \{_{n+1}\}$

122   N. Aubrun et al.

Lemma 2. *For all n, we have $e_0^n \circ \mathcal{T}_{Robi} \circ e_0^n \rightsquigarrow e_0^{n+1}$ and $e_5^n \circ \mathcal{T}_{Robi} \circ e_5^n \rightsquigarrow e_5^{n+1}$.*

Proof. Use exactly the same reasoning, but replacing the role of ▨ by ▨,
▨, ▨, ▨ and ▨ in previous lemma, for the first assertion.
    Replace the role of e_0^n by e_5^n in all previous arguments, to get the second assertion.

Lemma 3. *For all n, we have $e_0^n \circ \mathcal{T}_{Robi} \circ e_0^n \rightsquigarrow e_2^{n+1}$ and $e_0^n \circ \mathcal{T}_{Robi} \circ e_0^n \rightsquigarrow e_{-2}^{n+1}$.*

Lemma 4. *For all n, we have $H_n \circ \mathcal{T}_{Robi} \circ H_n \rightsquigarrow e_1^{n+1}$; $H_n \circ \mathcal{T}_{Robi} \circ H_n \rightsquigarrow e_3^{n+1}$; $H_n \circ \mathcal{T}_{Robi} \circ H_n \rightsquigarrow e_{-1}^{n+1}$; $H_n \circ \mathcal{T}_{Robi} \circ H_n \rightsquigarrow e_{-3}^{n+1}$.*

4   Two Notions of Robustness for Tilesets

Definition 5 (Composition pattern). *Consider a finite set S of transducers. Consider the signature made of the symbols of S (with arity 0) and the symbol $\circ$ (of arity 2). A composition pattern F over S is a term over this signature.*

For example $H \circ \tau \circ H$ is a composition pattern over $\{H, \tau\}$. Such a composition pattern is interpreted as expected: $\circ$ corresponds to composition.

Definition 6. *A tileset τ is **semantically robust** (or* inductive*) if there exists a family of transducers $(\mathcal{T}_n)_n$, of respective heights $(g(n))_n$, with $g : \mathbb{N}^* \to \mathbb{N}^*$ increasing and injective with $g(1) = 1$ and some (fixed) composition pattern $F = F(\mathcal{T}_1, \mathcal{T}_{n-k}, \ldots, \mathcal{T}_n)$ over $\mathcal{T}_1, \mathcal{T}_{n-k}, \ldots, \mathcal{T}_n$ for some integer k, such that:*

1. *Initialisation: $\tau \rightsquigarrow \mathcal{T}_1, \ldots, \tau^{g(k+1)} \rightsquigarrow \mathcal{T}_{k+1}$*
2. *F provides an invariant: $\forall n, \quad F(\mathcal{T}_1, \mathcal{T}_{n-k}, \ldots, \mathcal{T}_n) \rightsquigarrow \mathcal{T}_{n+1}$*
3. *For all n, $\mathcal{T}_n$ contains a loop.*

Proposition 7. *If a tileset τ admits a tiling, then it is semantically robust.*

Proof. Consider $g(n) = n$ and $\mathcal{T}_n = \tau^{g(n)}$ for all n. We have 1) from definition, Condition 2) is satisfied, as we have $\mathcal{T}^{n+1} = \mathcal{T}^n \circ \mathcal{T}$ for all $n \geq 0$. Condition 3) comes from Proposition 5.

Proposition 8. *If a tileset τ is semantically robust, then it admits a tiling.*

Proof. If τ is semantically robust, then by induction and combining conditions 1) and 2, we have that for all n, $\tau^{g(n)} \rightsquigarrow \mathcal{T}_n$. By Proposition 5, we can tile horizontal stripe of height $g(n)$ for arbitrary large n. By a compactness argument, we conclude that τ tile the plane.

4.1 Provably Robust Tilesets

Their might be a strong difference between the fact that $\forall n, F(\mathcal{T}_1, \mathcal{T}_{n-k}, \ldots, \mathcal{T}_n)$ $\rtimes \mathcal{T}_{n+1}$ holds for some composition pattern F, written as $\models \exists F, \forall n, F(\mathcal{T}_1, \mathcal{T}_{n-k}, \ldots, \mathcal{T}_n) \rtimes \mathcal{T}_{n+1}$ and the fact that we can prove it holds, written $\vdash \exists F, \forall n, F(\mathcal{T}_1, \mathcal{T}_{n-k}, \ldots, \mathcal{T}_n) \rtimes \mathcal{T}_{n+1}$. This is for instance the case if a tileset admits a tiling, but we cannot prove the existence of any valid tiling. The same phenomenon happens for Turing machines, for which non-termination may hold while there is no proof of it. In the meantime we introduce the concept of *provably robust* tileset to add this condition.

Definition 7. *A tileset τ is **provably robust** if there exists a family of transducers $(\mathcal{T}_n)_n$, of respective heights $(g(n))_n$, with $g : \mathbb{N}^* \to \mathbb{N}^*$ increasing and injective, $g(1) = 1$ and some (fixed) composition pattern $F = F(\mathcal{T}_1, \mathcal{T}_{n-k}, \ldots, \mathcal{T}_n)$ over $\mathcal{T}_1, \mathcal{T}_{n-k}, \ldots, \mathcal{T}_n$ for some integer k,*

1. *Initialisation: $\tau \rtimes \mathcal{T}_1, \ldots, \tau^{g(k+1)} \rtimes \mathcal{T}_{k+1}$*
2. *F provides an invariant **holding provably**: $\forall n, F(\mathcal{T}_1, \mathcal{T}_{n-k}, \ldots, \mathcal{T}_n) \rtimes \mathcal{T}_{n+1}$*
3. *For all n, $\mathcal{T}_n$ contains a loop.*

Proposition 9. *If a tileset τ is provably robust then it admits a tiling.*

Proof. If we have $\vdash \phi$ for some formula ϕ, then we have $\models \phi$. We did not mention what notion of proof $\vdash$ we use (this could be provability in ZF set theory or provability in first-order logic starting from Peano's axioms of arithmetic), but of course, we expect it to be sound. Consequently, if τ is provably robust it is semantically robust.

Proposition 10. *A periodic tileset is provably robust.*

4.2 On the Role Played by Non-Homogeneous Linear Recurrences

We see that any tileset that admits a tiling necessarily has some invariant and vice-versa. The crucial question is whether this invariant is *simple* or not and whether it is provable. By *simple* we have in mind something similar to the invariant (1) on page 7 satisfied by Robinson's tileset. This invariant involves a meta-transducer with a constant number of edges (7 edges $e^n_{-3}, \ldots, e^n_3$), while in the general case, it may not be the case for the invariant in Definition 6.

In the general case, as the height of $\mathcal{T}_{n+1}$ is $g(n + 1)$, the invariance relation of Definition 6 implies that $g(n + 1)$ is necessarily obtained as the sum of the heights of the $\mathcal{T}_i$ involved in the composition pattern F. Consequently, g must satisfy a (non-homogeneous) linear recurrence of the form $g(n+1) - c_0 y(n) + c_1 g(n-1) + \cdots + c_k g(n-k) + c$ for some non-negative integers $c_0, \ldots, c_k$ and c. As an example $g(n) = n$ from the proof of Proposition 7 satisfies $g(n + 1) = g(n) + 1$. The previous remark, combined with the coming statements explains why tiling discussed in the literature is closely related to non-homogeneous linear recurrences: height $g(n) = 2^n - 1$ appears in Robinson's tiling. Similarly, Fibonacci numbers appear in Jeandel-Rao's tiling [10] discussed in Sect. 5.

4.3 Robinson's Tileset Is Robust

Proposition 11. *For all $n \geq 1$, $\mathcal{T}_n$ contains a loop.*

Proof. An attentive reader will be convinced that the following loop is contained in $e_0^n \cup e_1^n \cup e_2^n \cup e_3^n$ for every $n \in \mathbb{N}$.

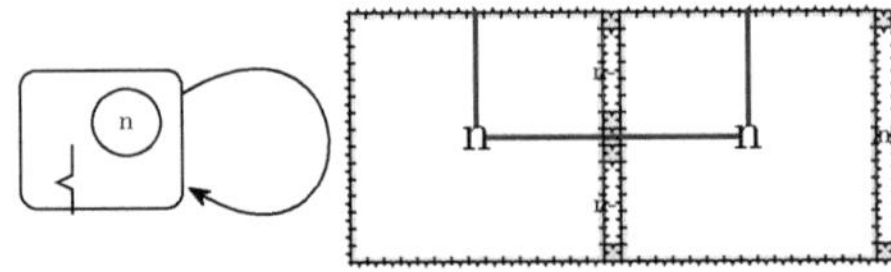

Proposition 12. *Robinson's tileset is provably robust.*

Proof. We prove that the three items of Definition 6 are satisfied: define $\mathcal{T}_1 := \mathcal{T}_{Robi}$ and $\mathcal{T}_n := H_n$. The first two items follow from Theorem 1, considering the pattern $F(H_n, \mathcal{T}_1) = H_n \circ \mathcal{T}_1 \circ H_n$. The last item follows from Proposition 11.

Corollary 1. *Robinson's tileset admits a tiling.*

4.4 Extension: Dealing with More General Shapes

Wang tiles are unit squares and can be arranged, provided local constraints are respected, to compose bigger squared patterns, rectangular patterns or patterns with more general shapes. Reasoning about non-rectangular patterns is sometimes easier. We present here how the formalism of transducers can be adapted.

We choose to define generalised shapes as supports of connected combinations of horizontal patterns. For example, any connected combination of rectangular shapes falls under this definition.

Definition 8 (Generalised shape). *A generalised shape S of height h is a connected subset of $\mathbb{Z} \times [[0 \ldots h - 1]]$, such that for all $0 \leq j < h$, $S_j = \{i : (i,j) \in S\}$ is an non-empty interval $[[\min_j \ldots \max_j]]$.*

The south (resp. north, left, right) frontier of a generalised shape is given by the interval $[[\min_0 \ldots \max_0]]$ (resp. the interval $[[\min_{h-1} \ldots \max_{h-1}]]$, the sequence $\min_0, \ldots, \min_{h-1}$, the sequence $\max_0, \ldots, \max_{h-1}$).

Such a sequence $\min_0, \ldots, \min_{h-1}$ (resp. $\max_0, \ldots, \max_{h-1}$) will be called a left (resp. right) frontier. A left frontier can be encoded, up to a translation, by the sequence of h integers $d_0 = \min_0, d_1 = \min_1 - \min_0, d_2 = \min_2 - \min_1, \ldots, d_{h-1} = \min_{h-1} - \min_{h-2}$. A right frontier can be encoded in a similar way.

A *coloured frontier* is given by a left or right frontier $d_0, \ldots, d_{h-1}$ together with h horizontal colours and $|d_0| + \cdots + |d_{h-1}|$ vertical colours.

We can generalise our concept of transducers so that states are now coloured frontiers. There is an edge between two such states q and q' if there is a valid tiling of the shape whose left frontier is q and the right frontier is q'. This edge is labelled by the word $w|w'$ corresponding to the vertical colours of the south and north frontiers of the shape. We can define a composition operation $\circ$ on transducers, that corresponds to testing the vertical compatibility of edges.

4.5   Extension: Tiling Equivalence

In Sect. 2.1 we defined meta-transducers to simplify notations and proofs. With the same objective in mind, we define a notion of equivalence between tilesets that are not necessarily Wang tilesets. If τ is a tileset, we say that a tileset $\widetilde{\tau}$ is *equivalent* to τ if (1) each tile in $\widetilde{\tau}$ is a valid pattern on τ (2) every tiling x by τ can be decomposed into patterns from $\widetilde{\tau}$.

Definition 9. *A tileset τ is* tiling equivalent *to a tileset τ' if there exist $\widetilde{\tau}$ (resp. $\widetilde{\tau}'$) equivalent to τ (resp. to τ') and a provable bijection $\Phi : \tau \to \tau'$ such that x is a tiling by $\widetilde{\tau}$ iff $\Phi(x)$ is a tiling by $\widetilde{\tau}'$.*

Example 2. The tileset τ_1 made of the geometric tiles with colored decorations is tiling equivalent to the tileset τ_2 pictured on the right, where not two north (resp. south) black decorations have the same width.

Indeed, from a tiling from the first, by ignoring colours we get a tiling from the second. Conversely, a tiling from the second can be transformed into a tiling from the first, as colours can be recovered by only looking at the width of black decorations.

Proposition 13. *Provably robustness and aperiodicity are preserved by tiling equivalence.*

Proof. For provably robustness this follows directly from the definitions, since the bijection involved in tiling equivalence is chosen provable. Assume that x is a periodic tiling by τ which is tiling equivalent to τ' through sets of patterns $\widetilde{\tau}$, $\widetilde{\tau}'$ and a provable bijection Φ. Then $\Phi(x)$ is a tiling by $\widetilde{\tau}'$ that is also periodic. So aperiodicity is also preserved by tiling equivalence.

5   Jeandel-Rao's Tileset

Jeandel Rao's aperiodic tileset [10] $\mathcal{T}$ is the Wang tileset composed with the 11 tiles pictured below.

The two tilesets of Example 2 are tiling equivalent to the Jeandel-Rao's Wang tileset [10].

Theorem 2. *Jeandel Rao's aperiodic tileset $\mathcal{T}$ is provably robust.*

The set of geometric tiles τ_2 from Example 2 is tiling equivalent to Jeandel-Rao's tileset, which is known to be aperiodic. Hence by Proposition 13 τ_2 is also aperiodic. Let us focus on the size of the words that label the edges $\alpha_n, \beta_n, \gamma_n, \delta_n, \epsilon_n, \omega_n$ and $A_n, B_n, C_n, D_n, E_n, O_n$. A careful observation leads to the conclusion that all patterns produced must have a width that satisfies the recurrence relation $g(n+2) = g(n+1) + g(n)$ which determines the Fibonacci numbers with $g(0) = 0$ and $g(1) = 1$. In general a tileset always satisfies some recurrence relation as in Definition 6. If additionally we ask for meta-transducers with a constant number of edges, as it is the case for Jeandel-Rao's tileset, then the same reasoning as above on the widths leads to an analogous conclusion: we get a sequence of arbitrarily large valid patterns whose widths satisfy some (non-homogenous) linear recurrence relation.

6   Robust and Non-Robust Turing Machines

Fix some concept of proof: say, provability in first-order logic starting from Peano's axioms of arithmetic.

On some input w, a Turing machine either halts (in that case the sequence of successive configurations starting on $C[w]$ provides a proof of that fact) or it does not. From Gödel's incompleteness theorem, there is a difference between provable valid statements and valid statements. We thus distinguish the case where there exists a proof of non-termination from the case where there is no proof of it. This leads to the introduction of the following concept:

Definition 10 (Robust Turing machine). *A Turing machine M is* robust *if for all inputs, the machine M halts or there is proof that it does not halt.*

Theorem 3 (A statement for Turing machines). *The halting problem is decidable for robust Turing machines.*

Proof. Given some Turing machine M and some input w, run the two following procedures in parallel

1. simulate M on w: if M accepts w, then halt and accept; if M rejects w, then halt and reject.
2. search for a proof π of the fact that M does not halt on w. If such a proof is found, then halt and reject.

By the robustness hypothesis, this algorithm always halts and is correct.

Definition 10 has similarities with the concept of robustness for Turing machines, and more general dynamical systems, considered in [2]. In [4] the authors extended these concepts to complexity theory [4]. They also relate these concepts to δ-decidability [9], or to statements such as [14]. In all these references, the reasoning for establishing decidability can be interpreted as a variation on the previous proof, by playing on the concept of proof π involved: the problem is proved computably enumarable (c.e.) and co-c.e., hence decidable.

7 Kari's Tilings

In this section, we briefly present a technique due to Kari to encode piecewise rational affine maps computations inside Wang tilings. We encourage the readers to refer to [12] for details about the construction.

7.1 Balanced Representation of Real Numbers

Let $i \in \mathbb{Z}$. We say that a bi-infinite sequence $(x_k)_{k \in \mathbb{Z}}$ of i's and $(i+1)$'s *represents* a real number $x \in [i, i+1]$ if there exists a sequence of intervals $I_1, I_2, \cdots \subseteq \mathbb{Z}$ of increasing lengths $n_1 < n_2 < \ldots$ such that $\lim_{k \to \infty} \frac{\sum_{j \in I_k} x_j}{n_k} = x$, that is to say, the averages of $(x_k)_{k \in \mathbb{Z}}$ over the intervals converge to x. For a real number x, we write $\lfloor x \rfloor$ for the integer part of x, which is the largest integer which is less than or equal to x.

Definition 11 (Balanced representation of real numbers). *Let $x \in \mathbb{R}$. For every $k \in \mathbb{Z}$ define $B_k(x) := \lfloor kx \rfloor - \lfloor (k-1)x \rfloor$. The bi-infinite sequence $(B_k(x))_{k \in \mathbb{Z}}$ is called the* balanced representation *of x.*

$B_k(x) \in \{\lfloor x \rfloor, \lfloor x \rfloor + 1\}$ and $(B_k(x))_{k \in \mathbb{Z}}$ is a representation of x in the sense defined above. A key observation is that balanced representations of irrational $x \notin \mathbb{Q}$ are Sturmian sequences [8], and for rational $x \in \mathbb{Q}$ the sequence is periodic. Definition 11 can be adapted to any $\boldsymbol{x} \in \mathbb{R}^d$: simply define $\lfloor \boldsymbol{x} \rfloor$ or $B_k(\boldsymbol{x})$ coordinate-wise.

7.2 Rational Piecewise Affine Maps

A *rational piecewise affine map* is given by finitely many pairs (U_i, f_i) where U_i are disjoint unit cubes of $\mathbb{R}^d$ with integer corners and f_i are affine maps with rational coefficients. Each set U_i is the domain where f_i can be applied. The system $(U_i, f_i)_{i=1\ldots n}$ determines a function $f : D \longrightarrow \mathbb{R}^d$ whose domain is $D = \bigcup_i U_i$ and $f(\boldsymbol{x}) = f_i(\boldsymbol{x})$ for all $\boldsymbol{x} \in U_i$.

7.3 Tileset Associated with a Rational Piecewise Affine Map

Kari's constructions are based on the idea that with rational (piecewise) affine map f can be associated with some Wang tileset.

Definition 12 (Tileset τ_{f_i} associated with f_i). *The tileset τ_{f_i} corresponding to a rational affine map $f_i(\boldsymbol{x}) = M\boldsymbol{x} + \boldsymbol{b}$ and its domain cube U_i is all the tiles*

$$
\begin{array}{c}
B_k(f_i(\boldsymbol{x})) \\
C_{f_i, k-1}(\boldsymbol{x}) \ \square \ C_{f_i, k}(\boldsymbol{x}) \\
B_k(\boldsymbol{x})
\end{array}
$$

where $k \in \mathbb{Z}$, $\boldsymbol{x} \in U_i$, and $C_{f_i, k}(\boldsymbol{x})$ is a rational number depending on $f_i, \boldsymbol{x}$ and k (details can be found in [12]).

Every tile as pictured above locally computes f_i with some error since we have $f_i\left(B_k(\boldsymbol{x})\right)+C_{f_i,k-1}(\boldsymbol{x}) = B_k\left(f_i\left(\boldsymbol{x}\right)\right)+C_{f_i,k}(\boldsymbol{x})$. Moreover, because f_i is rational, there are only finitely many such tiles (even though there are infinitely many $k \in \mathbb{Z}$ and $\boldsymbol{x} \in U_i$).

Lemma 5 (Main property of τ_{f_i}). *For fixed $\boldsymbol{x} \in U_i$ the tiles for consecutive $k \in \mathbb{Z}$ match so that a horizontal row can be formed whose top and bottom labels bear the balanced representations of $\boldsymbol{x}$ and $f_i(\boldsymbol{x})$.*

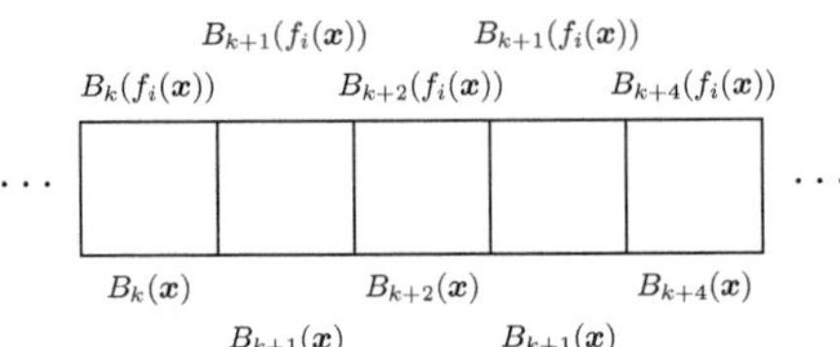

Kari considers the tileset τ_f obtained as the union of the τ_{f_i}, with additional rules to ensure that only tiles from the same τ_{f_i} appear on a given row.

7.4　Tiling Problem and Iterations of Rational Piecewise Map

Let (D, f) be a rational piecewise affine map. A real number $x \in D$ is an *immortal point* for f if for all $k \in \mathbb{N}^*$, $f^{[k]}(x) \in D$. If such an x exists, f is called *immortal*. A non-empty set $I \subseteq D$ is a *rational invariant box* if I is non-empty, $I \cap \mathbb{Q} \neq \emptyset$ and $f(I) \subseteq I$. Every $x \in I$, where I is a rational immortal box, is an immortal point for f.

Example 3. In [11], Kari considers the rational piecewise affine map: $f(x) = 2x$ if $x \leq 1$ and $2x/3$ if $x > 1$, which has $I = [1/2, 2]$ as a rational invariant box.

Lemma 6. *Assume that a rational piecewise affine map f has a (possibly irrational) immortal point $\boldsymbol{x} \in D$. Then for every $k \in \mathbb{N}^*$ there exists a rational $\boldsymbol{x}_k \in D$ such that $\boldsymbol{x}'_k = f^{[k]}(\boldsymbol{x}_k) \in D$ is rational.*

Proposition 14. *Let (D, f) be an immortal rational piecewise affine map. Then τ_f can tile the plane. Moreover if f has a rational invariant box I then τ_f is provably robust.*

The immortality problem for rational piecewise affine maps is undecidable [5]. The proof relies on the encoding of any Turing machine $\mathcal{M}$ inside a rational piecewise map $f_{\mathcal{M}}$, such that $f_{\mathcal{M}}$ is immortal iff $\mathcal{M}$ does not halt. This extends classical ways to simulate a Turing machine using rational piecewise affine maps [13]. We can then consider the associated tiling $\tau_{f_{\mathcal{M}}}$. We obtain:

Theorem 4. *Consider a Turing machine $\mathcal{M}$. Then $\mathcal{M}$ is provably robust and not terminating iff $\tau_{f_{\mathcal{M}}}$ is provably robust.*

8   Turing Machines as Tilesets Through Robinson's Construction

Robinson described in [15, section 7] a method to convert a Turing machine $\mathcal{M}$ into a Wang tileset $\tau_{\mathcal{M}}$ such that $\mathcal{M}$ does not halt on the empty tape iff $\tau_{\mathcal{M}}$ tiles the plane. We call this *Robinson's construction.*

Theorem 5. *If the Turing machine $\mathcal{M}$ is provably robust and not terminating, then the tileset $\tau_{\mathcal{M}}$ is provably robust.*

Theorem 6. *The converse is true: If $\tau_{\mathcal{M}}$ is provably robust, then $\mathcal{M}$ is provably robust and not terminating.*

9   Conclusions and Discussions

Going back to the domino problem, we have the following theorem:

Theorem 7. *The domino problem is decidable for robust tilesets.*

Proof. We prove the problem is decidable by exhibiting two semi-algorithms taking as input a tileset τ. Note that since the set of tilings is a compact set, the fact that a tileset τ cannot tile the plane is equivalent to the existence of an integer n such that τ^n is not compatible. A first semi-algorithm searches for such an integer n: construct successively the transducers $\mathcal{T}^n$, and look for a loop inside $\mathcal{T}^n$. If there exists an n such that $\mathcal{T}^n$ has no loop, then the semi-algorithm stops. Otherwise it keeps running. The fact that a tileset τ can tile the plane is computably enumerable for a robust tileset: from the definition of provably robust, we just need to search for a proof of all the statements of the definition. If we find one, we are sure the plane can be tiled. As the tileset is assumed to be robust, either the search for a n in the first case or of a proof in the second case terminate, hence, in both cases, we can decide effectively which case holds.

The method we described to prove that a given tileset can tile the plane is based on proving invariants of the same forme as in Definition 6. The method of finding an invariant of the proper form for some composition pattern F to prove that a tileset can tile the plane can be seen as a sound and relatively complete method. Jeandel-Rao's tileset $\mathcal{T}$ considered in Sect. 5 arose from a systematic computer-assisted proof of tilesets by considering an increasing number of tiles, conducted in [10]. The authors proved (using the help of about 23 CPU years of computing done on a cluster of parallel machines) that there is no aperiodic Wang set with 10 tiles or less. Another (also provably robust) tileset $\mathcal{T}'$ is considered in [10, Section 7.2]. They mention that some other candidates with 11 tiles could also be aperiodic. Our study provides theoretical explanations for the experimental facts mentioned in [10, Section 8], when the authors discuss why they investigated the tileset $\mathcal{T}$, and not other candidates. They explain that *"it was very easy for a computer to produce the transducer for $\mathcal{T}^k$, even for large values of k ($k \sim 1000$). In contrast, for almost all other tilesets, we were not* *able to reach even $k = 30$"*. As demonstrated in the current article, the possibility of tiling the plane is closely related to the existence of an invariant of the form of Definition 6. When such an invariant exists, then computing $\mathcal{T}^k$ becomes feasible. When it does not hold, then computing $\mathcal{T}^k$ is very costly, as the number of edges is exploding with k. So, their quest can be related to the experimental search of an invariant, and the existence of an invariant, seems to be closely related to the fact, that they mention that *"This suggested this tileset had some particular structure"*. We somehow provide arguments that the involved "structure" is the existence of an invariant for some F.

Our settings pave the way for important and fascinating developments. First, when a tileset is robust, we know it is decidable. But then it makes sense to discuss the complexity of the domino problem and not only its decidability. This can be done by quantifying the robustness in an approach similar to the one done in [4]. Our concept of robustness is inspired by [2,4], but differs from their exact notion of robustness. Furthermore, we think that similar statements can be obtained not only about the fact that the plane can be tiled but also about the fact that the plane can be tiled aperiodically.

References

1. Aanderaa, S.O., Lewis, H.R.: Linear sampling and the ∀∃∀ case of the decision problem. J. Symb. Log. **39**(3), 519–548 (1974). https://doi.org/10.2307/2272894
2. Asarin, E., Bouajjani, A.: Perturbed turing machines and hybrid systems. In: Proceedings of the 16th Annual IEEE Symposium on Logic in Computer Science (LICS-01), Los Alamitos, CA, 16–19 June 2001, pp. 269–278. IEEE Computer Society Press (2001)
3. Berger, R.: The undecidability of the domino problem. Memoirs Am. Math. Soc. **66** (1966)
4. Blanc, M., Bournez, O.: Quantifiying the robustness of dynamical systems. relating time and space to length and precision. In: Computer Science Logic CSL'24, Naples, Italy (2024)
5. Blondel, V.D., Bournez, O., Koiran, P., Tsitsiklis, J.: The stability of saturated linear dynamical systems is undecidable. J. Comput. Syst. Sci. **62**(3), 442–462 (2001). https://doi.org/10.1006/jcss.2000.1737
6. Durand, B., Romashchenko, A., Shen, A.: Fixed-point tile sets and their applications. J. Comput. Syst. Sci. **78**(3), 731–764 (2012). https://doi.org/10.1016/j.jcss.2011.11.001
7. Fernique, T.: Yet another proof of the aperiodicity of robinson tiles. arXiv preprint arXiv:1711.03401 (2017)
8. Fogg, N.P., Berthé, V., Ferenczi, S., Mauduit, C., Siegel, A.: Substitutions in Dynamics, Arithmetics and Combinatorics. Springer, Heidelberg (2002). https://doi.org/10.1007/b13861
9. Gao, S., Avigad, J., Clarke, E.M.: Delta-decidability over the reals. In: 2012 27th Annual IEEE Symposium on Logic in Computer Science (LICS), pp. 305–314. IEEE (2012)
10. Jeandel, E., Rao, M.: An aperiodic set of 11 wang tiles. In: Advances in Combinatorics (2021)
11. Kari, J.: A small aperiodic set of wang tiles. Disc. Math. **160**(1–3), 259–264 (1996)

12. Kari, J.: The tiling problem revisited (extended abstract). In: Durand-Lose, J., Margenstern, M. (eds.) MCU 2007. LNCS, vol. 4664, pp. 72–79. Springer, Heidelberg (2007). https://doi.org/10.1007/978-3-540-74593-8_6
13. Koiran, P., Cosnard, M., Garzon, M.: Computability with low-dimensional dynamical systems. Theor. Comput. Sci. **132**(1–2), 113–128 (1994)
14. Ratschan, S.: Deciding predicate logical theories of real-valued functions. In: Symposium on Mathematical Foundations of Computer Science (MFCS'2023) (2023)
15. Robinson, R.M.: Undecidability and nonperiodicity for tilings of the plane. Invent. Math. **12**, 177–209 (1971)
16. Wang, H.: Proving theorems by pattern recognition - ii. Bell Syst. Techn. J. **40**(1), 1–41 (1961). https://doi.org/10.1002/j.1538-7305.1961.tb03975.x

Creation of Fixed Points in Block-Parallel Boolean Automata Networks

Kévin Perrot[1,2], Sylvain Sené[1,2], and Léah Tapin[1(✉)]

[1] Aix Marseille Univ, CNRS, LIS, Marseille, France
leah.tapin@lis-lab.fr
[2] Université publique, Marseille, France

Abstract. In the context of discrete dynamical systems and their applications, fixed points often have a clear interpretation. This is indeed a central topic of gene regulatory mechanisms modeled by Boolean automata networks (BANs), where a collection of Boolean entities (the automata) update their state depending on the states of others. Fixed points represent phenotypes such as differentiated cell types. The interaction graph of a BAN captures the architecture of dependencies among its automata. A first seminal result is that cycles of interactions (so called feedbacks) are the engines of dynamical complexity. A second seminal result is that fixed points are invariant under block-sequential update schedules, which update the automata following an ordered partition of the set of automata. In this article we study the ability of block-parallel update schedules (dual to the latter) to break this fixed point invariance property, with a focus on the simplest feedback mechanism: the canonical positive cycle. We quantify numerically the creation of new fixed points, and provide families of block-parallel update schedules generating exponentially many fixed points on this elementary structure of interaction.

1 Introduction

A Boolean automata network (BAN) is a finite discrete dynamical system defined by n local functions $f_i : \{0,1\}^n \to \{0,1\}$ for $i \in \{0, \ldots, n-1\}$. A configuration is an element of $\{0,1\}^n$, and an update schedule describes in which order its components (the automata) are updated to compute the next configuration. For example the parallel update schedule corresponds to the fully synchronous application of its local function to each automaton of the network. The structure of interactions is captured by a directed graph G_f on the vertices $\{0, \ldots, n-1\}$, where the arcs represent the effective dependencies among the automata. When G_f is acyclic the dynamics converges to a unique fixed point [16], therefore cycles of interactions are the engines of dynamical asymptotic heterogeneities.

BANs have many applications, in particular they are a classical model of gene regulatory networks since the seminal papers of Kauffman [9] and Thomas [18]. In this context, fixed point configurations correspond to stable patterns of gene expression at the basis of cellular phenotypes [19,20]. From a theoretical standpoint, their quantity is arguably the most studied dynamical property of BANs

© The Author(s), under exclusive license to Springer Nature Switzerland AG 2026
E. Formenti and L. Manzoni (Eds.): UCNC 2025, LNCS 16364, pp. 132–146, 2026.
https://doi.org/10.1007/978-3-032-15641-9_10

[1,2,4,7,14,15,17]. A famous fact is that fixed points are invariant under any block-sequential update schedule, which corresponds to partitionning the set of automata into subsets $W_1, \ldots, W_p$ (called blocks) and updating synchronously the automata in W_1, then the ones in W_2, *etc.*, and eventually the automata in the subset W_p [17]. In this paper we study the block-parallel update schedules, which are dual to the block-sequential, and have the ability to create new fixed points compared to this well established invariance.

Update schedules received a great attention, and debates are still open regarding the most realistic schedule. Indeed, the seminal works [9,18] considered different update schedules. A hierarchy based on a formal notion of simulation has been established, with important members of increasing expressivity behind the parallel update schedule, the sequential ones (permutations of the automata), the block-sequential ones, the periodic ones (any sequence of subsets, repeatedly applied) and the deterministic ones [10]. Block-parallel update schedules are the lowest in this hierarchy with the ability to create fixed points (just above sequential), at a small cost (size at least 5 for a simple cycle, according to our experiments).

Instead of blocks of automata applied in parallel within each block and sequentially the different blocks, block-parallel update schedules apply the local functions sequentially within each block and in parallel the different blocks (called ordered-block, to avoid confusion). The number of substeps, instead of being the number of blocks, becomes the least common multiple of the size of the ordered-blocks (the first time the same set of automata will be updated again). They have been introduced and justified in [5], in relation to the ability of chromatin clocks to influence the speed or rate at which the expression state of genes evolves within the cell nucleus. The chromatin organization (proteins in contact with the DNA) plays an important role in the transcription machinery, as it may disable the accessibility of DNA strands [21]. This process can enforce new attractors to appear (or change in size), as studied in the present work.

Outline. Section 2 presents all the necessary definitions, with an example of block-parallel update schedule creating two new fixed points on the positive cycle of size 5. Section 3 reviews previous works, and Sect. 4 contains our contributions. In Subsect. 4.1 we define a simple family of examples creating new fixed points on disconnected interaction graphs; in Subsect. 4.2 we explain how to efficiently count fixed points on positive cycles from the interaction graph of its parallelization; in Subsect. 4.3 we expose exhaustive numerical experiments quantifying the creation of new fixed points on positive cycles up to size 11; and in Subsect. 4.4 we provide explicit families of block-parallel update schedules creating exponentially many new fixed points on positive cycles. Section 5 concludes and gives perspectives.

2   Definitions

Let $[\![n]\!] = \{0, \ldots, n-1\}$ and $[k] = \{1, \ldots, k\}$. The definitions are illustrated with an example at the end of the section.

Boolean Automata Networks. A *Boolean automata network* (BAN) of size n is a function f on the configuration space $\{0,1\}^n$. Each Boolean component $i \in [\![n]\!]$ is called an *automaton*, and in a configuration $x \in \{0,1\}^n$ it is in a *state* denoted $x_i \in \{0,1\}$. The individual behavior of each automaton $i \in [\![n]\!]$ is described by a *local function* $f_i : \{0,1\}^n \to \{0,1\}$, and a dynamics will be obtained when provided with an order (schedule) on the applications of these local functions to update the states of the automata. A BAN is formally defined as the collection of its n local functions $(f_i)_{i \in [\![n]\!]}$, or equivalently by grouping them into $f : \{0,1\}^n \to \{0,1\}^n$ where $f(x)_i = f_i(x)$ for any x and i.

Interaction Graphs. The architecture of a BAN f is captured in its *interaction graph* denoted $G_f = (V_f, A_f)$, representing the effective dependencies among its automata. We have $V_f = [\![n]\!]$ *i.e.* one vertex per automaton, and an arc from i to j whenever the local function of j depends on the state of i (the in-neighbors of j are the essential variables [3] of the Boolean function $f_j : \{0,1\}^n \to \{0,1\}$).

Cycles in the interaction graph play a central role in automata network theory (discussed in Sect. 3). In this paper, we will explore the simplest form of interaction with feedback and its new capabilities through block-parallel updates: the canonical *positive cycle* of size n. It is the BAN f^{n+} defined by the local functions $f_i^{n+}(x) = x_{i-1 \mod n}$ for $i \in [\![n]\!]$, with $-1 \mod n = n - 1$. Its interaction graph is the simple directed cycle of length n, denoted C^{n+} (on vertex set $[\![n]\!]$ with arcs $\{(i - 1 \mod n, i) \mid i \in [\![n]\!]\}$).

Update Schedules. Local functions provide a static description of some BAN, and its actual dynamics greatly depends on the order of automata updates. This is provided by an *update schedule*, the two most famous being the parallel (fully synchronous update of all states at each step [9]) and asynchronous (non-deterministic update of any, but exactly one, automaton at each step [18]) schedules. More complex schedules mix these two approaches, and in the deterministic landscape the notion of *block* is central: it consists in a subset of automata $W \subseteq [\![n]\!]$ whose states are updated simultaneously during a *substep*. We denote $f_W : \{0,1\}^n \to \{0,1\}^n$ the function such that:

$$f_W(x) = \begin{cases} f_i(x) & \text{if } i \in W, \\ x_i & \text{otherwise.} \end{cases}$$

Block-Sequential Update Schedules. A *block-sequential update schedule* is a sequence of blocks updated sequentially (one block during each substep), such that each automaton is updated exactly once during a step. It is formally given as an ordered partition $\mu = (W_1, \ldots, W_p)$ of $[\![n]\!]$, and its associated *transition function* or *dynamics* on configuration space $\{0,1\}^n$ is defined as $f_{(\mu)} = f_{W_p} \circ \ldots \circ f_{W_1}$, where $\circ$ is the standard composition of functions. The parallel update schedule is a particular case of block-sequential update schedule, namely $\mu = ([\![n]\!])$ with a unique block. Let BS_n denote the set of block-sequential update schedules possible for BANs of size n.

Block-Parallel Update Schedules. *Block-parallel update schedules* are dual of block-sequential schedules: the updates are sequential within one block, and the blocks are updated synchronously. We formally employ the term *o-block* to define them, it is a sequence of automaton, *i.e.* an element of $[\![n]\!]^*$. A partitioned order $\mu = \{S^1, \ldots, S^k\} \subseteq [\![n]\!]^*$ is a finite set of o-blocks (where S^i_j will denote the j-th element of the sequence $S^i \in [\![n]\!]^*$, with $j \in [|S^i|]$), such that the o-blocks have no automaton in common, and the union of their automata is $[\![n]\!]$. The first substep will update the first element of each o-block, the second substep will update the second element of each o-block, *etc.* (considering the automata within each one o-block cyclically), until the last elements of all o-blocks are updated during the same substep, which occurs at the least common multiple of their sizes substep. Formally, it is convenient to "sequentialize" a block-parallel update schedule, via the morphism φ from partitioned orders to sequences of blocks, defined as $\varphi(\mu) = (W_1, \ldots, W_\ell)$ with $\ell = \mathrm{lcm}(|S^1|, \ldots, |S^k|)$ and:

$$W_i = \{S^j_{(i-1 \ \mathrm{mod} \ |S^j|)+1} \mid j \in [k]\} \text{ for } i \in [\ell].$$

The *transition function* or *dynamics* is then defined as $f_{\{\mu\}} = f_{(\varphi(\mu))} = f_{W_\ell} \circ \ldots \circ f_{W_1}$. Let BP_n denote the set of block-parallel udpate schedules possible for BANs of size n. Again, the parallel update schedule is an element of BP_n, corresponding to the set of n singleton sequences $\mu = \{(0), (1), \ldots, (n-1)\}$. The intersection of BP_n and BS_n is characterized in [11].

Fixed Points. Given a BAN f of size n and an update schedule μ, we denote the dynamics $f_{(\mu)}$ when $\mu \in \mathsf{BS}_n$, and $f_{\{\mu\}}$ when $\mu \in \mathsf{BP}_n$. In both cases the dynamics is deterministic, and we abstractely denote it f_μ when μ may be block-sequential or block-parallel. A *fixed-point* is a configuration $x \in \{0,1\}^n$ such that $f_\mu(x) = x$.

Parallelization. Given a BAN f of size n and an update schedule μ, the function $f_\mu : \{0,1\}^n \to \{0,1\}^n$ itself is called their *parallelization*. It can be obtained through the process of computing the substeps, that is for $(W_1, \ldots, W_\ell) \in \mathsf{BS}_n$ or $\varphi(\mu) = (W_1, \ldots, W_\ell)$ where $\mu \in \mathsf{BP}_n$, to consider inductively:

$$f^{(0)}_{\{\mu\}}(x) = \mathrm{x} \quad \text{and} \quad f^{(i)}_{\{\mu\}}(x) = f_{(W_i)}(f^{(i-1)}_{\{\mu\}}(x)) \text{ for } i \in [\ell].$$

Properties on the structure of the sequence of interaction graphs:

$$(G_{f^{(i)}_{\{\mu\}}})_{i \in [\ell]} \text{ with } G_{f^{(\ell)}_{\{\mu\}}} = G_{f_\mu}$$

will be useful to study the dynamics of cycles under block-parallel update schedules.

With block-sequential update schedules, the value of ℓ is capped by n and the parallelization can be computed in polynomial time (given as inputs f, μ and x, output $f_{(\mu)}(x)$) [13], but with block-parallel update schedules the number of substeps may grow exponentially with n and computing the parallelization is hard for PSPACE (even a single bit of $f_{\{\mu\}}(x)$) [12].

Example. Let f^{5+} the canonical positive cycle of size $n = 5$, and the two block-parallel update schedules $\mu_{par} = \{(0), (1), (2), (3), (4)\}$ and $\mu_{new} = \{(0, 1), (2, 3, 4)\}$. Observe that μ_{par} is the parallel update schedule, hence $f^{5+}_{\mu_{par}}$ has two fixed points: 00000 and 11111. The update schedule μ_{new} is more complex: it is not equivalent to a block-sequential update schedule, because it has repetitions (*i.e.*, automata update more than once during one step) in its sequence of blocks form: $\varphi(\mu_{new}) = (\{0, 2\}, \{1, 3\}, \{0, 4\}, \{1, 2\}, \{0, 3\}, \{1, 4\})$.

For the sake of simplicity, we will denote $f^{5+}_{\mu_{new}}$ as g to expose the parallelization process. We start with $g^{(0)}(x) = x$.

$$g^{(1)}(x) : \begin{cases} g_0^{(1)}(x) = x_4 \\ g_1^{(1)}(x) = x_1 \\ g_2^{(1)}(x) = x_1 \\ g_3^{(1)}(x) = x_3 \\ g_4^{(1)}(x) = x_4 \end{cases} \quad g^{(2)}(x) : \begin{cases} g_0^{(2)}(x) = x_4 \\ g_1^{(2)}(x) = x_4 \\ g_2^{(2)}(x) = x_1 \\ g_3^{(2)}(x) = x_1 \\ g_4^{(2)}(x) = x_4 \end{cases} \quad g^{(3)}(x) : \begin{cases} g_0^{(3)}(x) = x_4 \\ g_1^{(3)}(x) = x_4 \\ g_2^{(3)}(x) = x_1 \\ g_3^{(3)}(x) = x_1 \\ g_4^{(3)}(x) = x_1 \end{cases}$$

$$g^{(4)}(x) : \begin{cases} g_0^{(4)}(x) = x_4 \\ g_1^{(4)}(x) = x_4 \\ g_2^{(4)}(x) = x_4 \\ g_3^{(4)}(x) = x_1 \\ g_4^{(4)}(x) = x_1 \end{cases} \quad g^{(5)}(x) : \begin{cases} g_0^{(5)}(x) = x_1 \\ g_1^{(5)}(x) = x_4 \\ g_2^{(5)}(x) = x_4 \\ g_3^{(5)}(x) = x_4 \\ g_4^{(5)}(x) = x_1 \end{cases} \quad g^{(6)}(x) : \begin{cases} g_0^{(6)}(x) = x_1 \\ g_1^{(6)}(x) = x_1 \\ g_2^{(6)}(x) = x_4 \\ g_3^{(6)}(x) = x_4 \\ g_4^{(6)}(x) = x_4 \end{cases}$$

This process is illustrated in Fig. 1, starting with the interaction graph of $g^{(0)}(x)$ which is the identity function. The resulting graph has two connected components, consequently $f^{5+}_{\mu_{new}}$ has two more fixed points compared to $f^{5+}_{\mu_{par}}$: 00111 and 11000.

3　State of the Art

A seminal result of Robert states that cycles in the interaction graph (so called feedbacks) are the engine of dynamical complexity in Boolean automata networks. Indeed, a BAN f of size n with G_f acyclic converges towards a unique fixed point in at most n steps [6,16]. The size of a minimal feedback vertex set in G_f also bounds the maximum number of fixed points obtained under parallel and asynchronous update schedules [1,15].

One easily notices that if a configuration x is a fixed point of f for the parallel update schedule (*i.e.* $f(x) = x$), then x is also a fixed point for any update schedule μ (*i.e.* $f_\mu(x) = x$) [8]. Within block-sequential update schedules the fixed point invariance is stronger: for any $\mu \in \mathsf{BS}_n$ the set of fixed points of $f_{(\mu)}$ is exactly the same [17]. Since the parallel update schedule is an element of BS_n, this can be stated as $f(x) = x$ if and only if $f_{(\mu)}(x) = x$. In the present work we study how this invariance is broken under block-parallel update schedules, where new fixed points can emerge in the dynamics.

Block-parallel update schedules have been introduced in [5], where the authors show examples of fixed points creation (and suggest biological interpretations). This is the track we will develop in the rest of this article. The

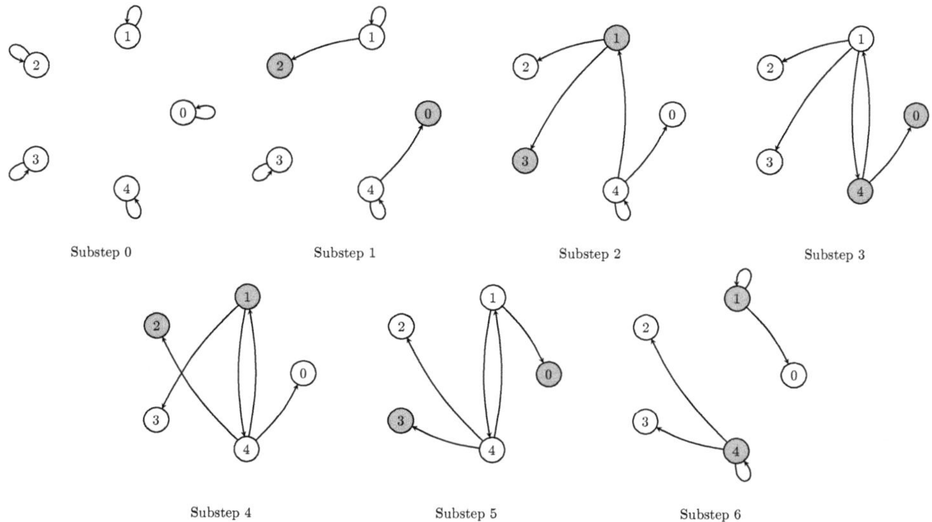

Fig. 1. Substeps of the parallelization of f^{5+} for the block-parallel update schedule μ_{new}. The automata updated at each substep are greyed.

theoretical understanding of block-parallel update schedules has been developed in [11], providing formulas to count them and algorithms to enumerate them (exploited for our numerical simulations in Subsect. 4.3). Update repetitions allow to have an exponential number of substeps within a single iteration (*cf.* our discussion on φ above), and in [12] it is proven that the computational complexity of classical problems tend to be lifted to the level of PSPACE, although this is not a universal rule (unless major complexity collapses).

Let us recall two simple observations. First, the positive cycle BAN f^{n+} of size n whose interaction graph is C^{n+}, has two fixed points in parallel: the two uniform configurations denoted 0^n and 1^n. Second, if we modify f^{n+} to set the local function of one i as $f(x)_i = \neg x_{i-1}$ (instead of $f_i^{n+}(x) = x_{i-1}$), we obtain a negative cycle which has no fixed point in parallel. An exhaustive study of their limit dynamics is exposed in [4].

4   New Fixed Points for Block-Parallel Update Schedules

As an introduction to this section exposing our contributions, we present on Fig. 2 a minimal example of block-parallel update schedule for an automata network of size 3 which creates two new fixed points compared to any block-sequential update schedule. The pitfall of this example is that its interaction graph is disconnected, as the first famility exposed in Subect. 4.1. Subsections 4.2, 4.3 and 4.4 focus on positive cycles, the simplest non-trivial connected structure of interaction for the study of fixed points.

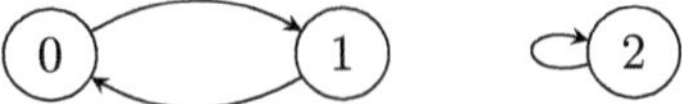

Fig. 2. Example automata network of size $n = 3$ with local functions $f_0(x) = x_1$, $f_1(x) = x_0$ and $f_2(x) = \neg x_2$. It has no fixed point in parallel (hence for any block-sequential update schedule), whereas for the block-parallel update schedule $\mu = \{(0,1),(2)\}$ we have $\varphi(\mu) = (\{0,2\},\{1,2\})$ which admits the four fixed points 000, 001, 110, and 111.

4.1   From Zero to an Exponential Number of Fixed Points

In this section we provide a simple family of Boolean automata networks and block-parallel update schedules, creating arbitrarily many new fixed points (a quantity exponential in the size of the network and in the number of substeps) compared to any block-sequential update schedule.

For any integer $n > 0$, let $\hat{h}^n$ denote the BAN of size $3n$ whose interaction graph is composed of a negative cycle on n automata, plus $2n$ isolated nodes. Formally, its local functions are defined as (interaction graph on Fig. 3):

$$\hat{h}_0^n(x) = \neg x_{n-1},$$
$$\hat{h}_i^n(x) = x_{i-1} \text{ for all } i \in [\![n-1]\!] \setminus \{0\},$$
$$\hat{h}_i^n(x) = 0 \text{ for all } i \in [\![3n-1]\!] \setminus [\![n-1]\!].$$

Under the parallel update schedule, or any block-sequential schedule, it has no fixed point (because of the negative cycle).

Fig. 3. Illustration of $G_{\hat{h}^n}$, the interaction graph of $\hat{h}^n$, for $n = 5$.

Theorem 1. *For any integer $n > 0$, under the block-parallel update schedule:*

$$\hat{\mu}^n = \{(0),(1),(2),\dots,(n-2),(n-1),(n,n+1,n+2,\dots,3n-2,3n-1)\},$$

the dynamics $\hat{h}^n_{\{\hat{\mu}^n\}}$ has 2^n fixed points.

Proof. Automata from the set $[\![3n-1]\!] \setminus [\![n-1]\!]$ have constant 0 local functions, hence their state in any fixed point is 0. The least common multiple of o-block sizes, *i.e.* the number of substeps, is $\ell = 2n$. From any initial configuration $x \in$

$\{0,1\}^{3n}$, after n substeps, each automaton in the cycle has its own initial state negated. For simplicity we denote $\hat{h}^n$ as $\hat{h}$ and $\hat{\mu}^n$ as $\hat{\mu}$, thus for all $i \in [n-1]$, we have $\hat{h}_{\{\hat{\mu}\}}^{(n)}(x)_i = \neg x_i$. It follows that after n additional substeps, these automata get back to their initial state, hence for all $i \in [n-1]$:

$$\hat{h}_{\{\hat{\mu}\}}(x)_i = \hat{h}_{\{\hat{\mu}\}}^{(2n)}(x)_i = x_i.$$

We deduce that any $\tilde{x} \in \{0,1\}^{3n}$ with $\tilde{x}_i = 0$ for all $i \in [\![3n-1]\!] \setminus [\![n-1]\!]$ is a fixed point.

Observe that, redefining $\hat{h}_0^n(x) = x_{n-1}$ (without negation) and removing n constant automata, we obtain a BAN of size $2n$ with two fixed points in parallel (for any $n > 0$), and still 2^n fixed points under $\hat{\mu}^n \in \mathsf{BP}_n$, but now $\hat{\mu}^n$ has only n substeps.

4.2   Counting Fixed Points on Positive Cycles

In this subsection we present two results employed to easily count the number of fixed points obtained through the parallelization of positive cycles (in Subsects. 4.3 and 4.4), whose interaction graphs have in-degree one.

Theorem 2. *Let f be a BAN of size n. If G_f has in-degree one, then for any $\mu \in \mathsf{BP}_n$ the parallelization $G_{f_{\{\mu\}}}$ also has in-degree one.*

Proof. We prove the result by induction on the number of substeps under μ, by noticing that G_f of in-degree one means that every local function is of the form $f_i(x) = x_j$ or $f_i(x) = \neg x_j$ for some $j \in [\![n]\!]$ (hence we abusively say that the local function has in-degree one). Initially $f_{\{\mu\}}^{(0)}$ is the identity, verifying the claim. If $f_{\{\mu\}}^{(s)}$ has in-degree one, automata not udpated will still have the same local function of in-degree one, whereas the update of automaton i of local function $f_i(x) = x_j$ will let:

$$f_{\{\mu\}}^{(s+1)}(x)_i = f_{\{\mu\}}^{(s)}(x)_j$$

which has in-degree one by induction hypothesis.

Theorem 3. *Let f be a BAN of size n. If G_f has in-degree one and only positive arcs, i.e. for all $i \in [\![n]\!]$ the local function is $f_i(x) = x_j$ for some $j \in [\![n]\!]$, then f has 2^c fixed points, with c the number of cycles in G_f (which equals its number of components).*

Proof. Since G_f has in-degree one, each of its component is a cycle with downtrees rooted on it. Components are independent, and each of them has 2 fixed points: the all-0 and the all-1 configurations, because they are the only fixed points on the cycle and spread to the trees (from root to leaves). The number of combinations is 2^c.

n	1 cycle	2 cycles	3 cycles	4 cycles	5 cycles	6 cycles
3	13	0	0	0	0	0
4	67	0	0	0	0	0
5	441	30	0	0	0	0
6	3555	36	0	0	0	0
7	29625	3360	588	0	0	0
8	293091	30552	5400	0	0	0
9	3401113	424278	73296	20700	0	0
10	42263483	4757460	629950	172900	1800	1500
11	551305591	83321513	20529729	7008540	1133550	130680

Fig. 4. Results of the numerical simulations on the number of fixed points, for block-parallel update schedules on canonical positive cycles. For n from 3 to 11, the count in column "c cycles" is the number of different schedules whose parallelization has c cycles (hence 2^c fixed points by Theorem 3).

4.3   Numerical Simulations on Positive Cycles

In this subsection we present numerical simulations to count the number of fixed points that block-parallel update can reach on the primitive structures of canonical positive cycles. More precisely, for each size n, we study all block-parallel update schedules from BP_n that give different sequences of blocks through φ (called representatives for the equivalence classes of $\equiv_0$ in [11]). For each n, we enumerate the schedules and sort them according to the number of cycles of the parallelized interaction graph. From Theorem 3, the number of cycles is the base 2 logarithm of the number of fixed points.

Algorithmically, the enumerator is taken from [11], and the parallelization proceeds substep by substep according to Theorem 2 (updating the in-neighbor of each local function). For in-degree one parallelized interaction graphs, the number of components (equal to the number of cycles) is computed in linear time.

Our code in Python is accessible on the following repository:

https://framagit.org/leah.tapin/blockpargenandcount.

It is archived by Software Heritage at the following permalink:

https://archive.softwareheritage.org/browse/origin/directory/?origin_url=https://framagit.org/leah.tapin/blockpargenandcount.

We have conducted these numerical experiments on a standard laptop (processor Intel-Core^{TM} i7 @ 2.80 GHz), it took around 3 hours for $n = 11$.

The results are presented on Fig. 4. The maximum number of cycles for $n = 10$ and $n = 11$ is 6. One can remark that the increase of the maximum number of cycles $(1, 1, 2, 2, 3, 3, 4, 6, 6)$ is irregular (we got no appropriate record on *OEIS*, because the number of cycles is at most n). We have observed through further experimentations that the maximum number of cycles seems to correspond to update schedules with the largest number of substeps.

4.4 Creation of New Fixed Points on Positive Cycles

Recall that f^{n+} is the BAN whose interaction graph is the cycle of size n with only positive arcs (denoted C^{n+}). In parallel, it has two fixed points (the two uniform configurations). In this section we present two families of update schedules (depending on the parity of n), showing that even on such a primitive architecture of interactions, it is possible to get new fixed points under block-parallel update schedules.

Theorem 4. *Let $n = 2k + 1$ for some $k \geq 2$, and:*

$$\mu_{odd} = \{(0, 1, \ldots, k - 1), (k, k + 1, 2k, 2k - 1, \ldots, k + 2)\}.$$

Then the interaction graph of $f^{n+}_{\{\mu_{odd}\}}$ has k cycles, and 2^k fixed points.

0	1	2	3	$\cdots$	$k-2$	$k-1$	0	1	2	3	4	$\cdots$	$k-2$	$k-1$	0	1	$\cdots$
k	$k+1$	$2k$	$2k-1$	$\cdots$	$k+4$	$k+3$	$k+2$	k	$k+1$	$2k$	$2k-1$	$\cdots$	$k+5$	$k+4$	$k+3$	$k+2$	$\cdots$

$\cdots$	$k-2$	$k-1$	0	1	$\cdots$	$k-3$	$k-2$	$k-1$	0	1	2	$\cdots$	$k-2$	$k-1$
$\cdots$	k	$k+1$	$2k$	$2k-1$	$\cdots$	$k+3$	$k+2$	k	$k+1$	$2k$	$2k-1$	$\cdots$	$k+3$	$k+2$

Fig. 5. Illustration of the "sequentialization" of μ_{odd}, focusing on its beginning and ending. The temporality of substeps flows from left to right, each substep updates two automata: $\{0, k\}$, then $\{1, k + 1\}$, etc. Solid vertical lines emphasize the o-blocks.

The number of substeps under μ_{odd} is $k(k + 1)$, see Fig. 5. For this proof, we will denote the value of automaton i at substep t as x_i^t, with $i \in [\![n]\!]$ and $t \in \{0, \ldots, k(k + 1)\}$ (we have $x^0 = x$). We partition the n automata into three groups: $A = \{0, \ldots, k-1\}$, $B = \{k, k+1\}$, and $C = \{k+2, \ldots, 2k\}$. We will only pay attention to the substeps that are multiples of k (for group A) or $k + 1$ (for groups B and C), which are the substeps at the end of which the entire o-block of the corresponding group has been updated.

The automata of groups A and B are in ascending order. This means that after each automaton in the group has been updated the same number of times in a step (after the entire o-block has been updated) all automata of each of these two groups have the same value.

The automata of group C are in descending order. This means that instead of a single value being propagated as in the previous groups, the values are shifted from an iteration of their o-block to the next one, with the introduction of a new value through automaton $k + 2$. This can be summarized as follows: for any a, b, c such that $a, a - c \in \{k + 2, \ldots, 2k\}$ and $b, b - c \in \{0, \ldots, k\}$, we have $x_a^{(k+1)b} = x_{a-c}^{(k+1)(b-c)}$.

Back to group A, we have (recall Fig. 5):

$$x_0^{ki} = x_{2k}^{(k+1)(i-1)} \text{ for all } i \in \{1, \ldots, k-1\},$$
$$x_0^{k^2} = x_{2k}^{(k+1)(k-2)},$$
$$\text{and } x_0^{k(k+1)} = x_{2k}^{(k+1)(k-1)} = x_{k+2}^{k+1} = x_{k+1}^{k+1} = x_{k-1}^0.$$

We conclude that at the end of a step, every automaton of group A has taken the value of automaton $k-1$ from the previous step.

Now consider group B, we have:

$$x_k^{(k+1)i} = x_{k-1}^{k(i-1)} = x_0^{k(i-1)} \text{ for all } i \in \{1, \ldots, k\},$$

and in particular, $x_k^{(k+1)k} = x_0^{k(k-1)} = x_{2k}^{(k+1)(k-2)} = x_{k+2}^0$ from what precedes.

Regarding group C, we have for any $j \in C$:

$$
\begin{aligned}
x_j^{(k+1)k} &= x_{k+2}^{(k+1)(2k+2-j)} & (2 \leq 2k+2-j \leq k) \\
&= x_{k+1}^{(k+1)(2k+2-j)} & (2 \leq 2k+2-j \leq k) \\
&= x_0^{k(2k+1-j)} & (1 \leq 2k+1-j \leq k-1) \\
&= x_{2k}^{(k+1)(2k-j)} & (0 \leq 2k-j \leq k-2) \\
&= x_j^0.
\end{aligned}
$$

This means that, for the update schedule μ_{odd}, the local function for every automaton of group C in $f_{\{\mu_{odd}\}}^{n+}$ is the identity function.

As a conclusion, the interaction graph of the parallelization $f_{\{\mu_{odd}\}}^{n+}$ is:

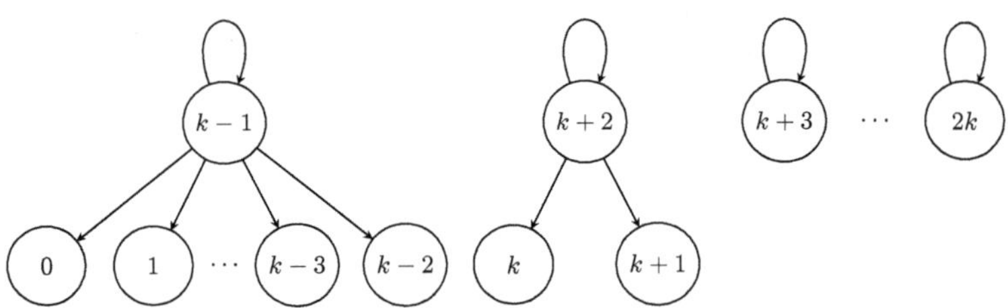

and this graph contains exactly k cycles, hence by Theorem 3 it has 2^k fixed points.

Theorem 5. *Let $n = 2k$ for some $k \geq 4$, and:*

$$\mu_{even} = \{(0), (1, \ldots, k-1), (k, 2k-1, k+1, 2k-2, \ldots, k+2)\}.$$

Then the interaction graph of $f_{\{\mu_{even}\}}^{n+}$ has $k-1$ cycles, and 2^{k-1} fixed points.

0	0	0	0	⋯	0	0	0	0	0	0	0	⋯	0	0	0	0	⋯
1	2	3	4	⋯	$k-2$	$k-1$	1	2	3	4	5	⋯	$k-2$	$k-1$	1	2	⋯
k	$2k-1$	$k+1$	$2k-2$	⋯	$k+4$	$k+3$	$k+2$	k	$2k-1$	$k+1$	$2k-2$	⋯	$k+5$	$k+4$	$k+3$	$k+2$	⋯

⋯	0	0	0	0	⋯	0	0	0	0	0	0	⋯	0	0
⋯	$k-2$	$k-1$	1	2	⋯	$k-3$	$k-2$	$k-1$	1	2	3	⋯	$k-2$	$k-1$
⋯	k	$2k-1$	$k+1$	$2k-2$	⋯	$k+3$	$k+2$	k	$2k-1$	$k+1$	$2k-2$	⋯	$k+3$	$k+2$

Fig. 6. The beginning and the end of the "sequentialization" of μ_{even}.

Proof. The number of substeps under μ_{even} is $(k-1)(k+1)$, see Fig. 6. We proceed as in the proof of Theorem 4, using the notation x_i^t for the value of automaton i at substep t, with $i \in [\![n]\!]$ abd $t \in \{0, \ldots, (k-1)(k+1)\}$. Consider the following partition of the n automata into four groups: $A = \{0\}$, $B = \{1, \ldots, k-1\}$, $C = \{k, k+1\}$, and $D = \{k+2, \ldots, 2k-1\}$. We will only pay attention to the substeps that are multiples of k (for group B) or $k-1$ (for groups C and D) Group A is a particular case since it contains only automaton 0, which is also the only automaton in its o-block, meaning that it is updated at every substep.

Both groups B and C are in ascending order, meaning that they all have the same value after being updated the same amount of times. For group C, the fact that $k+1$ is not updated right after k is not a problem: since k doesn't get updated again in the meantime, it keeps the same value. Group D is in descending order, meaning that we have the same shift of values as with group C in the proof of Theorem 4 for the odd case, with a slightly different formula: for any a, b, c such that $a, a - c \in \{k+2, \ldots, 2k-1\}$ and $b, b - c \in \{0, \ldots, k-1\}$, we have $x_a^{kb} = x_{a-c}^{k(b-c)}$.

For group B, every $k-1$ subsets, automaton 1 takes the value of automaton 0. As we can see in Fig. 6, in all but the two last "cycles" of subsets, 0 has taken the value of $2k-1$ after being updated one time less than 1. In the two last cycles, it has taken the previous value of $2k-1$ (after being updated two times less than 1). We then have:

$$x_1^{k-1} = x_0^0,$$

$$x_1^{(k-1)i} = x_0^{k(i-1)} = x_{2k-1}^{k(i-1)} \text{ for all } i \in \{2, \ldots, k-2\},$$

$$x_1^{(k-1)^2} = x_{2k-1}^{k(k-3)},$$

$$\text{and } x_1^{(k-1)k} = x_{2k-1}^{k(k-2)} = x_{k+2}^k = x_{k+1}^k = x_k^k = x_{k-1}^0.$$

We conclude that at the end of a step, every automaton of group B has taken the value of automaton $k-1$ from the previous step.

Regarding group C, we have:

$$x_k^{ki} = x_{k-1}^{(k-1)(i-1)} \text{ for all } i \in \{1, \ldots, k-1\},$$

and in particular, $x_k^{k(k-1)} = x_{k-1}^{(k-1)(k-2)} = x_1^{(k-1)(k-2)} = x_{2k-1}^{k(k-3)} = x_{2k-1}^{k(k-3)} = x_{k+2}^0.$

That is, at the end of a step, both automata of group C have taken the value of automaton $k+2$ from the previous step.

Considering group D, we have, for any $j \in D \setminus \{2k-1\}$:

$$
\begin{aligned}
x_j^{k(k-1)} = x_{k+2}^{k(2k+1-j)} \qquad & (3 \le 2k+1-j \le k-1) \\
= x_{k+1}^{k(2k+1-j)} \qquad & (3 \le 2k+1-j \le k-1) \\
= x_k^{k(2k+1-j)} \qquad & (3 \le 2k+1-j \le k-1) \\
= x_{k-1}^{(k-1)(2k-j)} \qquad & (2 \le 2k-j \le k-2) \\
= x_1^{(k-1)(2k-j)} \qquad & (2 \le 2k-j \le k-2) \\
= x_{2k-1}^{k(2k-j-1)} \qquad & (1 \le 2k-j-1 \le k-3) \\
= x_j^0 .
\end{aligned}
$$

Except $2k-1$, every automaton of group D regains its original value at the end of the step. Since automaton $2k-1$ is not updated at the last substep, while automaton 0 is, both have the same value at the end of a step:

$$
x_0^{k(k-1)} = x_{2k-1}^{k(k-1)} = x_{k+2}^{k \times 2} = x_{k+1}^{k \times 2} = x_k^{k \times 2} = x_{k-1}^{k-1} = x_1^{k-1} = x_0^0 .
$$

As a conclusion, the interaction graph of the parallelization $f_{\{\mu_{even}\}}^{n+}$ is:

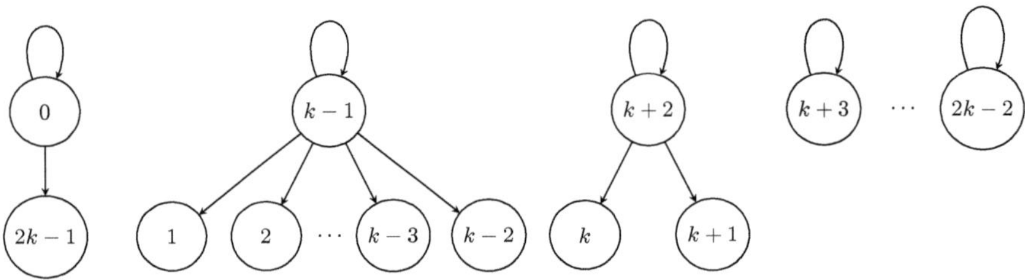

and this graph contains exactly $k-1$ cycles, hence by Theorem 3 it has 2^{k-1} fixed points.

5   Conclusion and Further Work

In this article we have studied the creation of fixed points of Boolean automata networks under block-parallel update schedules:

- we have exhibited a simple family of disconnected (in terms of the interaction graph) BANs that can jump from 0 fixed point in parallel, to an exponential number of fixed points in block-parallel;
- we have conducted numerical experiments on canonical positive cycles, showing that block-parallel schedules are able to create many new fixed points, even on these most primitive BANs with feedback;
- we have identified families of block-parallel update schedules that create exponentially many new fixed points on canonical positive cycles, and proved this behavior through the study of the parallelization process (and its resulting number of cycles).

Theorems 4 and 5 give exactly $2^{\lfloor \frac{n-1}{2} \rfloor}$ fixed points (compared to 2 fixed points in parallel). However, numerical experiments show that this is not the maximum number of fixed points attainable under block-parallel update schedules. Indeed, for $n = 10, 11$ it is possible to have 2^6 fixed points. A perspective would be to characterize the sequence $c(n)$ such that for n automata it is possible to reach a maximum of $2^{c(n)}$ fixed points on a positive cycle (from Theorem 3 it is a power of two). The first terms of $c(n)$ are (starting at $n = 3$):

$$1, 1, 2, 2, 3, 3, 4, 6, 6.$$

There is a trivial bound $c(n) = n$, and obtaining tighter upper bounds would be an interesting start.

A broader continuation would be to consider more complex interaction graphs of BANs, and the ability of block-parallel schedules to create fixed points on them. It suffices to restrict ourselves to connected graphs, because the total number of fixed points of a BAN is the product of the number of fixed points on each of its components. In particular, how does the structure of the interaction graph influence the maximum number of fixed points created by block-parallel update schedules?

Acknowledgments. The authors received support from the projects ANR-24-CE48-7504 ALARICE, HORIZON-MSCA-2022-SE-01 101131549 ACANCOS, and STIC AmSud CAMA 22-STIC-02 (Campus France MEAE).

References

1. Aracena, J.: Maximum number of fixed points in regulatory Boolean networks. Bull. Math. Biol. **70**, 1398–1409 (2008)
2. Aracena, J., Richard, A., Salinas, L.: Number of fixed points and disjoint cycles in monotone Boolean networks. SIAM J. Discret. Math. **31** (2017)
3. Crama, Y., Hammer, P.L.: Boolean Functions: Theory, Algorithms, and Applications. Cambridge University Press, Cambridge (2011)
4. Demongeot, J., Noual, M., Sené, S.: Combinatorics of Boolean automata circuits dynamics. Discret. Appl. Math. **160**(4–5), 398–415 (2012)
5. Demongeot, J., Sené, S.: About block-parallel Boolean networks: a position paper. Nat. Comput. **19**, 5–13 (2020)
6. Gadouleau, M.: Robert's theorem and graphs on complete lattices (2023). Preprint on arXiv:2309.11363
7. Gadouleau, M., Richard, A., Riis, S.: Fixed points of Boolean networks, guessing graphs, and coding theory. SIAM J. Discret. Math. **29** (2015)
8. Goles, E., Martínez, S.: Neural and automata networks: dynamical behavior and applications. In: Mathematics and Its Applications, vol. 58. Kluwer Academic Publishers (1990)
9. Kauffman, S.A.: Metabolic stability and epigenesis in randomly constructed genetic nets. J. Theor. Biol. **22**, 437–467 (1969)
10. Paulevé, L., Sené, S.: Boolean networks and their dynamics: the impact of updates, pp. 173–250. Wiley (2022)

11. Perrot, K., Sené, S., Tapin, L.: Combinatorics of block-parallel automata networks. In: Proceedings of SOFSEM 224. LNCS, vol. 14519, pp. 442–455. Springer, Cham (2024)
12. Perrot, K., Sené, S., Tapin, L.: Complexity of Boolean automata networks under block-parallel update modes. In: Proceedings of SAND 2024. LIPIcs, vol. 292, pp. 19:1–19:19. Dagstuhl (2024)
13. Perrotin, P., Sené, S.: Turning block-sequential automata networks into smaller parallel networks with isomorphic limit dynamics. In: Proceedings of CiE 2023. LNCS, vol. 13967, pp. 214–228. Springer, Cham (2023)
14. Remy, E., Ruet, P., Thieffry, D.: Graphic requirements for multistability and attractive cycles in a Boolean dynamical framework. Adv. Appl. Math. **41**(3), 335–350 (2008)
15. Riis, S.: Information flows, graphs and their guessing numbers. Electron. J. Comb. **14**, 1–17 (2007)
16. Robert, F.: Itérations sur des ensembles finis et automates cellulaires contractants. Linear Algebra Appl. **29**, 393–412 (1980)
17. Robert, F.: Discrete Iterations: A Metric Study. Springer Series in Computational Mathematics, vol. 6. Springer, Cham (1986)
18. Thomas, R.: Boolean formalization of genetic control circuits. J. Theor. Biol. **42**, 563–585 (1973)
19. Thomas, R., d'Ari, R.: Biological Feedback. CRC Press (1990)
20. Wang, R.-S., Saadatpour, A., Albert, R.: Boolean modeling in systems biology: an overview of methodology and applications. Phys. Biol. **9**(5), 055001 (2012)
21. Zhang, S., et al.: TIMELESS regulates sphingolipid metabolism and tumor cell growth through Sp1/ACER2/S1P axis in ER-positive breast cancer. Cell Death Dis. **11**, 892 (2020)

Identification of Cellular Automata on Spaces of Bernoulli Probability Measures

Faizal Hafiz[1]([✉]), Amelia Kunze[2], Enrico Formenti[2], and Davide La Torre[1]

[1] SKEMA Business School, Sophia Antipolis, France
{faizal.hafiz,davide.latorre}@skema.edu
[2] Université Côte d'Azur, CNRS, i3S, Nice, France
{amelia.kunze,enrico.formenti}@univ-cotedazur.fr

Abstract. Classical Cellular Automata (CCAs) are a powerful computational framework for modeling global spatio-temporal dynamics with local interactions. While CCAs have been applied across numerous scientific fields, identifying the local rule that governs observed dynamics remains a challenging task. Moreover, the underlying assumption of deterministic cell states often limits the applicability of CCAs to systems characterized by inherent uncertainty. This study, therefore, focuses on the identification of Cellular Automata on spaces of probability measures (CAMs), where cell states are represented by probability distributions. This framework enables the modeling of systems with probabilistic uncertainty and spatially varying dynamics. Moreover, we formulate the local rule identification problem as a parameter estimation problem and propose a meta-heuristic search based on Self-adaptive Differential Evolution (SaDE) to estimate local rule parameters accurately from the observed data. The efficacy of the proposed approach is demonstrated through local rule identification in two-dimensional CAMs with varying neighborhood types and radii.

Keywords: Cellular Automata · Inverse Problems · Probability Measures · Parameter Estimation · Evolutionary Computation

1 Introduction

Classical Cellular Automata (CCAs) have long been a cornerstone computational framework for modeling and understanding complex systems driven by local interactions. Their simplicity, rooted in a finite set of states and uniform local rules, belies their ability to produce rich and diverse dynamical behaviors. However, a fundamental limitation of CCAs lies in their assumption of complete certainty in the state of all cells. In many real-world systems, uncertainty is inherent in the dynamics, and this deterministic assumption restricts the ability of CCAs to model such dynamics effectively.

© The Author(s), under exclusive license to Springer Nature Switzerland AG 2026
E. Formenti and L. Manzoni (Eds.): UCNC 2025, LNCS 16364, pp. 147–162, 2026.
https://doi.org/10.1007/978-3-032-15641-9_11

To address this limitation, we propose Cellular Automata on spaces of Bernoulli probability Measures (CAMs), a novel generalization of CCAs that incorporates probabilistic uncertainty into the framework. The proposed framework is essentially a special case of the recently proposed Cellular Automata on Measures by the authors in [9]. In our setting, the state of each cell is represented by a Bernoulli probability measure, $\mu(x) \in [0, 1]$, rather than a discrete value, $x \in \{0, 1\}$, and local rules operate on configurations of Bernoulli probability measures. This extension allows for the modeling of systems with spatially varying probabilities and inherent randomness, significantly broadening the scope of cellular automata to mimic complex and realistic phenomena.

Cellular automata were first proposed as an idealized biological system with the aim of modeling biological self-reproduction [6]. They have since been employed as tools in numerous application areas including neural networks, cryptography, and quantum computing (e.g., [7,8,10,12]). Cellular automata have a particularly rich history as discrete models of molecular dynamics for reaction-diffusion equations and fluid dynamics [20] or for studying various forms of population dynamics [3]. These types of applications yield cellular automata which can model the spread of the pollution or the spread of disease in a population. Given this, our work presented below offers a means to successfully construct a cellular automaton which can address meaningful real-world problems.

This study, in particular, focuses on the *inverse* problem of determining the local rule of CAMs, which governs the observed spatio-temporal patterns. This inference process is often referred to as the *identification* of local rules [1,22], and it represents a foundational step in modeling real-world systems [13,14]. The existing investigations often approach the CCA identification as a two-step process, which involves determining the approximate *neighborhood size* followed by further refinements. For instance, Yang and Billings [21] proposed neighborhood detection by selecting cells based on their *contribution* to the updated cell. Often, combinations of candidate neighborhood cells and the updated cell are treated as *patterns* [18,23] or *evidences* [17], which are subsequently used to provide statistics or heuristics to select the appropriate neighborhood. The next identification step involves further refinements, such as removing *redundant* cells [17] and rule refinement using Genetic Algorithms [21]. We refer to [1,22] and the references therein for a detailed treatment of this topic.

It is worth emphasizing that extending the existing CCA-focused identification approaches to CAMs, where a Bernoulli probability measure represents each cell state, may not be trivial. To this end, we demonstrate that the local rule of CAMs can be formulated as a convex combination of neighborhood cell states, which enables us to cast the identification task as an estimation of local rule parameters. Next, we propose a meta-heuristic search approach based on Self-adaptive Differential Evolution (SaDE) [19]. This approach models the local rule parameter estimation as a constrained optimization problem to minimize the *difference* between the observed and simulated states. We show that the distance may be easily computed for Bernoulli probability measures using the Monge-Kantorovich metric. This finding is leveraged to quantify the *difference* between the observed and simulated CAM states, thus driving the optimization process.

The rest of this article is organized as follows: we begin with a discussion of key properties of CCA in Sect. 2, followed by the space of probability measures in Sect. 3 and the framework of Cellular Automata on Bernoulli Probability Measures in Sect. 4. The proposed formulation of CAM identification and the meta-heuristic search approach are provided in Sect. 5. Next, different identification test scenarios and the corresponding results are discussed in Sect. 6. This is followed by the conclusions in Sect. 7.

2 Classical Cellular Automata: Definitions and Main Properties

A cellular automaton consists of a uniform grid of cells, usually infinite in extent. Each cell in the grid takes a value from a discrete set of possible *states*. A particularly simple and classical case is that of binary states, in which every cell in the automaton is either "alive" (in which the state takes value 1) or "dead" (in which the state takes value 0). The automaton evolves at discrete time steps through synchronous updates to the state of all cells according to a local, space-invariant *update rule*.

2.1 Formal Definitions

The cellular grid can be more formally described as a d-dimensional, $d \in \mathbb{N}$, regular lattice Λ. The finite set of states, which indicates all possible values that each cell in the lattice can take, is denoted by $\mathcal{S}$. A *configuration* $x \in \mathcal{S}^\Lambda$ is a mapping that assigns a state $x_i \in \mathcal{S}$ to each cell $i \in \Lambda$. Cells update their state by gathering information about the states of their neighbors. The *neighborhood* $N_r(i)$ of a cell $i \in \Lambda$ is the closed ball of radius r centered at i. The *local rule* is a function $f : \mathcal{S}^{|N_r(i)|} \to \mathcal{S}$ that combines the neighboring states to generate local updates. A *classical cellular automaton* (CCA) is then formally described by the structure $\langle d, \mathcal{S}, r, f \rangle$.

In this work, we focus on two-dimensional cellular automata with lattice $\Lambda = \mathbb{Z}^2$. Different neighborhoods can be obtained by changing how one measures distance from the central cell. We use two different neighborhoods which correspond to using the classical Manhattan and Moore distances:

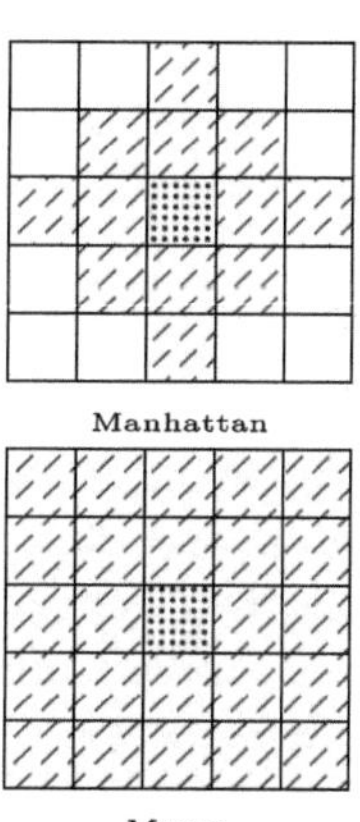

Fig. 1. Illustration of different neighborhoods in two-dimensional cellular automaton for the neighborhood radius, $r = 2$.

$$N_r(i) = \{j \in \Lambda \mid d_1(i,j) \leq r\} \quad \texttt{Manhattan}$$
$$N_r(i) = \{j \in \Lambda \mid d_\infty(i,j) \leq r\} \quad \texttt{Moore} \tag{1}$$

where, d_1 and d_∞ respectively denote Manhattan and Chebyshev distance. Note that the Manhattan neighborhood is quite often referred to as the Von Neumann neighborhood in the CCA literature. The size of neighborhood, K, is controlled by the radius r and it is given by: $K = (2r^2 + 2r + 1)$ (Manhattan) and $K = (2r + 1)^2$ (Moore). The geometry of both neighborhood types is illustrated for $r = 2$ in Fig. 1.

In order to study CCA as dynamical systems, the set of states $\mathcal{S}$ is equipped with the topology induced by the discrete metric d_D defined as $d_D(s,t) = 1$ if $s \neq t$, 0 otherwise, for all $s, t \in \mathcal{S}$. Then, $\mathcal{S}^\Lambda$ is endowed with *Cantor topology*, *i.e.*, the standard product topology induced by the discrete topology on $\mathcal{S}$. Consider the *Cantor distance* defined as

$$d_C(x,y) = \sum_{i \in \Lambda} \frac{d_D(x_i, y_i)}{s^{|i|}}, \tag{2}$$

where s is the size of $\mathcal{S}$ (recall that $\mathcal{S}$ is a finite set here). It is well-known that the Cantor distance induces the Cantor topology on $\mathcal{S}^\Lambda$.

For a fixed $j \in \mathcal{S}$, the *shift map* σ_j is a very well-known and widely studied CA defined as $\sigma_j(x)_i = x_{i+j}, \quad \forall x \in \mathcal{S}^\Lambda, \ \forall i \in \Lambda$. The famous Curtis-Hedlund-Lyndon theorem established that CCAs are exactly those continuous maps commuting with the shift [15]. For an up-to-date bibliography on CAs and their variants, see, for instance [2,4,5,16].

In the next section, we introduce a metric for probability measures to be used in our CAM setting and provide a useful result.

3 The Space of Probability Measures

In the following let us suppose that the state space $S = \{0,1\}$ is endowed with a metric d and that (S,d) is a compact metric space. Let $\mathcal{B}(S)$ be the Borel sigma-algebra defined on S. Let us denote by $\mathcal{M}(S)$ the set of all probability measures defined on S. $\mathcal{M}(S)$ can be endowed with the Monge-Kantorovich metric defined as

$$d_{MK}(\mu, \nu) = \sup_{g \in Lip_1(S)} \int_S g d\mu - \int_S g d\nu \tag{3}$$

where $Lip_1(S)$ is the set of all 1-Lipschitz functions on S, that is

$$Lip_1(S) = \{g : S \to \mathbb{R} \text{ s.t. } |g(z) - g(y)| \leq d(x,y)\}. \tag{4}$$

The Monge-Kantorovic metric arose from a classical problem in transportation of mass [11]. Although it is often used to measure the distance between probability measures, its computation may not appear intuitive. The following result shows that the Monge-Kantorovich metric simplifies in the case of Bernoulli probability measures.

Let μ be a Bernoulli probability measure with parameter p, i.e., $\mu(z) = p^z(1-p)^{(1-z)}$ for $z \in \{0,1\}$. Let ν be a Bernoulli probability measure with parameter q, i.e., $\nu(z) = q^z(1-q)^{(1-z)}$ for $z \in \{0,1\}$.

For the Bernoulli probability measure μ and an arbitrary function $f \in Lip_1(S)$ we have

$$\int_S g d\mu = E[g(x)] = g(1)\mu(1) + g(0)\mu(0) \tag{5}$$

with the analogous equation holding for ν. Their difference is

$$\begin{aligned}
\int_S g d\mu - \int_S g d\nu &= g(1)\mu(1) + g(0)\mu(0) - (g(1)\nu(1) + g(0)\nu(0)) \\
&= g(1)p + g(0)(1-p) - g(1)q - g(0)(1-q) \\
&= (g(1) - g(0))(p - q).
\end{aligned} \tag{6}$$

For any $g \in Lip_1(S)$ we have $|g(1) - g(0)| \leq d(1,0) = 1$ and therefore $\sup_{g \in Lip_1(S)}|g(1) - g(0)| = 1$. In addition, if $g \in Lip_1(S)$ then $-g \in Lip_1(S)$ giving

$$\begin{aligned}
d_{MK}(\mu, \nu) &= \sup_{g \in Lip_1(S)} \left| \int_S g d\mu - \int_S g d\nu \right| = \sup_{g \in Lip_1(S)} |g(1) - g(0)| |p - q| \\
&= |p - q|.
\end{aligned} \tag{7}$$

4 Cellular Automata on Spaces of Bernoulli Probability Measures

In the following, we extend the classical cellular automaton with state space $S = \{0,1\}$. We assume that there is some inherent randomness associated with the realization of the state $x_i = 1$ in a given cell i. The probability of finding a state $x_j \in S$ in cell i is thus described by a Bernoulli random process,

$$\mu_i(x_j) = p^{x_j}(1-p)^{1-x_j} \qquad \text{for } x_j \in S = \{0,1\}. \tag{8}$$

In this framework, the state set S is replaced with $\mathcal{M}(S)$, which denotes the set of Bernoulli probability measures defined on S. In each cell, the Bernoulli random process is described completely by the probability $p_i \in [0,1]$. The local rule thus operates on configurations of Bernoulli probability measures instead of operating directly on the set of states $S = \{0,1\}$. Then the dynamics generated by a spatially-varying Bernoulli distributions on the state space $S = \{0,1\}$ can be completely characterized by the evolution of the values $p_i \in [0,1]$ on the lattice Λ. In other words, our model is equivalent to a cellular automaton with continuous state space $S = [0,1]$. The local rule is then a function $f : [0,1]^{|\mathcal{N}|} \rightarrow [0,1]$ which updates the probability $p_i(t)$ to $p_i(t+1)$ based on the values of its neighbors $p_j(t) \in N_r(i)$. We define a distance between two configurations $x, y \in \mathcal{M}(S)^\Lambda$ in the following way.

Definition 1. *For any pair of configurations* $x, y \in \mathcal{M}(S)^\Lambda$*, the distance* $d_\mathcal{M}$ *is defined as*

$$d_\mathcal{M}(x, y) = \sum_{i \in \Lambda} 2^{-|i|} \cdot d_{MK}(x_i, y_i) = \sum_{i \in \Lambda} 2^{-|i|} \cdot |p_{x_i} - p_{y_i}|. \tag{9}$$

where p_{x_i} *and* p_{y_i} *denote the Bernoulli parameters at the* i^{th} *cells of the config-urations* x *and* y*, respectively, and* $|\cdot|$ *denotes the norm of* i*.*

In the above definition, we have used the fact that in the Bernoulli case, the Monge-Kantorovich metric simplifies to the L^1 distance between the parameters, as shown in Eq. (7). For Bernoulli probabilities, the metric $d_\mathcal{M}$ is a weighted distance which emphasizes agreement between the centrally-located cells in the two configurations.

Observe that the space of configurations $\mathcal{M}(S)^\Lambda$ is closed under convex combinations of its elements. For the proof, it is sufficient to show that the interval $[0, 1]$ is a convex set. This fact motivates a particularly useful choice of local rule, denoted by f, which we first introduced and investigated in [9]. In essence, the local map f takes convex combinations of neighboring cells as explained below.

Let $\{\theta_k\}$ be a set of weights in $[0, 1]$ such that $\sum_{k \in N_r(i)} \theta_k = 1$, for any cell $i \in \Lambda$ and the corresponding neighbors $k \in N_r(i)$, then the action of local rule δ_C on any $x \in \mathcal{M}(S)^\Lambda$ is given by,

$$f(x)_i = \sum_{k=1}^{K} \theta_k \mathcal{N}_i, \quad \text{where,} \quad \mathcal{N}_i = [x_k \mid k \in N_r(i)] \tag{10}$$

$\mathcal{N}_i$ is the vector of neighborhood states of the ith−cell; and K gives the total number of neighbors. The vector $\mathcal{N}_i$ is formed by taking the element of the neighborhood in Marshall order. Note that the set of weights θ_k is only dependent on the position of the cell $k \in N_r(i)$ and is not dependent on cell i.

It is worth highlighting that we introduced CAM as generic framework in [9], where $\mathcal{M}(S)$ is not limited to Bernoulli measures and can represent any arbitrary spatially-varying probability measures. This framework includes the classical setting, if Dirac measures concentrated on specific states, $s \in S$, are employed. In this more general setting, the type of probability measure defined at each cell may also vary. In contrast, the present study deals only with the simple case of spatially-varying Bernoulli probability measures, which is enough for our interests relating to the identification problems presented below. We direct the reader to [9] for a detailed treatment of CAMs.

5   Inverse Problems for CAMs

Identification is the first step of the modeling process, typically involving model inference from observed system behavior. This is of particular interest in practical applications where the local rule governing the observed behavior is unknown.

The identification of such an optimal rule is often non-trivial due to the associated exponential search region, and it is typically approached as an optimization problem [1,22]. To understand this further, consider the proposed CAM framework, which can formally be described as the following structure: $\langle \Lambda, \mathcal{M}(S), N_r, \delta_C \rangle$. Here, Λ denotes a $d-$dimensional cellular lattice, $\mathcal{M}(S)$ gives the Bernoulli probability measures defined on S, N_r denotes the $r-$ radius neighborhood, and δ_C represents a local transition rule. At a particular time step t, each cell $i \in \Lambda$ evolves as the convex combination of its neighbors,

$$x_i(t+1) = f(x(t))_i = \sum_{k=1}^{K} \theta_k \mathcal{N}_i(t) \qquad \forall\, t = 0, \ldots, T-1 \tag{11}$$

where, $\mathcal{N}_i(t)$ is the vector of neighborhood states of the ith cell at time t. Equation (11) represents the *direct problem*, which can be used to study emergent global behavior with a given local rule, *i.e.*, when parameters $\{\theta_k\}$ are known.

In contrast, the identification (or inverse problem) aims to identify the local rule, $f(\cdot)$, from the observed spatio-temporal trajectories of a discrete dynamical system. This can be viewed as a two-step process: the first step involves the selection of an appropriate neighborhood type and radius, N_r, which is followed by the estimation of the corresponding parameters, $\{\theta_k\}$. To understand this further, consider the observed data $\mathcal{O}$ collected over T time steps:

$$\mathcal{O} = \bigcup_{t=0}^{T-1} \{(\mathcal{N}_i(t), x_i(t+1)) \mid i = 1, \ldots, |\Lambda|\} \tag{12}$$

As discussed earlier, the neighborhood type (*i.e.*, Manhattan or Moore, see Fig. 1) and radius r determine the neighborhood size, and, thereby, the total number of parameters, K. This study assumes that such parameters are *a priori* determined either through an information-theoretic (*e.g.*, mutual information) or an empirical trial-and-error criterion [1,17,18,21,23]. This *a priori* step reduces the identification to a parameter estimation problem with the objective of minimizing the distance between the observed states (x_i) and the simulated states $(\hat{x}_i)$ as follows:

$$\Theta^* = \arg\min_{\Theta \in \mathcal{R}_\Theta} \sum_{t=0}^{T-1} d_\mathcal{M}\left(x(t+1), \hat{x}(t+1)\right) \tag{13}$$

$$= \arg\min_{\Theta \in \mathcal{R}_\Theta} \sum_{t=0}^{T-1} \sum_{i \in \Lambda} 2^{-|i|} \cdot d_{MK}\left((x_i(t+1), \sum_{k=1}^{K} \theta_k \mathcal{N}_i(t)\right)$$

$$\text{subject to} \quad \sum_{k=1}^{K} \theta_k = 1,\ 0 \le \theta_k \le 1, \forall k$$

where $\hat{x}_i(t+1) = \sum_{k=1}^{K} \theta_k \mathcal{N}_i(t)$; and $\mathcal{R}_\Theta$ gives the parameter search region as follows: $\mathcal{R}_\Theta = \{\theta \in \mathbb{R}^K \mid \sum_{k=1}^{K} \theta_k = 1,\ 0 \le \theta_k \le 1, \forall k\}$.

5.1 Differential Evolution Based Parameter Estimation Approach

The local rule parameter estimation problem in Eq. 13 essentially represents a constrained numerical optimization problem. For such problems, metaheuristic algorithms offer two critical advantages over classical analytical methods: (1) derivative-free optimization eliminates the computational burden of computing gradients, and (2) global search capability through population-based exploration can avoid local optima inherent in multimodal polynomial landscapes. We, therefore, consider an adaptive variant of Differential Evolution (DE), SaDE [19]. SaDE is a subclass of population-based evolutionary search heuristics, which has been successfully applied to numerical optimization problems across diverse fields. Algorithm 1 outlines the overall steps involved in the proposed SaDE-based parameter estimation approach. In the following, we briefly discuss the key features of SaDE for the sake of completeness, and we refer to [19] for a detailed treatment.

SaDE begins with a population NP of candidate vectors, where each candidate represents a possible solution vector, $e.g.$, $\Theta_i = \{\theta_{i,1}, \theta_{i,2}, \theta_{i,3}, \ldots, \theta_{i,K}\}$. This population is iteratively evolved to identify the $optimal$ solution. This is achieved through two evolutionary operations: Mutation (Line 10–19, Algorithm 1) and Crossover (Line 20–22, Algorithm 1). These evolutionary operators generate new trial candidates, denoted by $\tilde{\Theta}$, at each iteration and represent the core of the search operation. A trial candidate can enter the population if it achieves a better $fitness$ value (to be discussed later), see Line 24–32, Algorithm 1.

It is worth highlighting that the $mutation$ operation explores new trial candidates by sampling the search region $\mathcal{R}_\Theta$, and, therefore, is crucial to successful optimization. For instance, consider a well-known $\mathtt{rand/1}$ mutation strategy, which combines three randomly selected candidates from the population to generate a new trial candidate, as follows: $\tilde{\theta}_{i,k} = \theta_{r_1,k} + F \cdot (\theta_{r_2,k} - \theta_{r_3,k})$ where, $r_1, r_2, r_3 \in [\![1, NP]\!]$ denote randomly selected candidates from the population; F is the scaling factor. It is known that $\mathtt{rand/1}$ strategy demonstrates excellent $exploration$ capability albeit at the expense of slower convergence; such qualities are desirable in multi-modal search regions. Significant efforts have been dedicated to developing different mutation strategies to better adapt to different features of the search region ($e.g.$, uni-modal or multi-modal). However, the nature of the problem landscape is often not known a $priori$, so the selection of the appropriate mutation strategy can be challenging. SaDE [19] was developed to address such issues by maintaining a pool of four strategies, which are probabilistically selected based on their performance over a fixed number of past iterations. Algorithm 1 shows these mutation strategies: $\mathtt{rand/1}$ (Line 12), $\mathtt{rand\text{-}to\text{-}best/2}$ (Line 14), $\mathtt{rand/2}$ (Line 16) and $\mathtt{current\text{-}to\text{-}rand/1}$ (Line 18). We refer to [19] for a detailed discussion on the capabilities of each mutation strategy, the probabilistic update mechanisms of mutation strategies, and the Crossover Rate (CR).

We consider a $solution\text{-}repair$ approach (see Line 23, Algorithm 1) to ensure that all trial candidates are $feasible$ and satisfy the equality constraint: $\sum_k \theta_k =$

Algorithm 1: CA Parameter Identification using SaDE

1: Initialize population $\{\Theta_1, \Theta_2, \ldots, \Theta_{NP}\}$
2: Initialize adaptive crossover rate CR for each strategy
3: `strategy pool` = { `rand/1`, `rand-to-best/2`, `rand/2`, `current-to-rand/1` }
4: Initialize strategy probabilities:
$\quad p_{\text{strategy}} = 1/|\text{strategy pool}|$ for all $\text{strategy} \in \text{strategy pool}$
5: **while** stopping criterion not satisfied **do**
6:    **for** $i = 1$ to NP **do**
7:       Select $\text{strategy} \in \text{strategy pool}$ using Stochastic Universal Sampling [19]
8:       $k_{rand} = \lfloor \text{rand}[0, 1) \times K \rfloor + 1$
9:       Random neighbors sampled without replacement: $\{r_1, r_2, r_3, r_4, r_5\} \subset [1, NP] \setminus \{i\}$,
10:       **for** $k = 1$ to K **do**
11:          **if** strategy = rand/1 **then**
12:             $\hat{\theta}_{i,k} = \theta_{r_1,k} + F(\theta_{r_2,k} - \theta_{r_3,k})$
13:          **else if** strategy = rand-to-best/2 **then**
14:             $\hat{\theta}_{i,k} = \theta_{i,k} + F(\theta_{best,k} - \theta_{i,k}) + F(\theta_{r_1,k} - \theta_{r_2,k}) + F(\theta_{r_3,k} - \theta_{r_4,k})$
15:          **else if** strategy = rand/2 **then**
16:             $\hat{\theta}_{i,k} = \theta_{r_1,k} + F(\theta_{r_2,k} - \theta_{r_3,k}) + F(\theta_{r_4,k} - \theta_{r_5,k})$
17:          **else**
18:             $\hat{\theta}_{i,k} = \theta_{i,k} + \acute{F}(\theta_{r_1,k} - \theta_{i,k}) + F(\theta_{r_2,k} - \theta_{r_3,k})$ {current-to-rand/1}
19:          **end if**
20:          **if** strategy $\neq$ current-to-rand/1 **then**
21:             $\tilde{\theta}_{i,k} = \begin{cases} \hat{\theta}_{i,k}, & \text{if } \text{rand}[0, 1) < CR \text{ or } k = k_{rand} \\ \theta_{i,k}, & \text{otherwise} \end{cases}$
22:          **end if**
23:       **end for**
         {Solution Repair}
         bound repair: $\tilde{\theta}_{i,k} \leftarrow \min\left(\max(\tilde{\theta}_{i,k}, 0), 1\right)$
         sum-to-one: $\tilde{\Theta}_i \leftarrow \tilde{\Theta}_i / \sum_{k=1}^{K} \tilde{\theta}_{i,k}$
24:       Evaluate $J(\tilde{\Theta}_i)$ using CA simulation, see Eq. 14
25:       **if** $J(\tilde{\Theta}_i) \leq J(\Theta_i)$ **then**
26:          $\Theta_i \leftarrow \tilde{\Theta}_i$, record strategy success
27:          **if** $J(\tilde{\Theta}_i) < J(\Theta_{best})$ **then**
28:             $\Theta_{best} \leftarrow \tilde{\Theta}_i$
29:          **end if**
30:       **else**
31:          Record strategy failure
32:       **end if**
33:    **end for**
34:    Update strategy probabilities and CR based on success/failure rates, see [19]
35: **end while**
36: **return** Θ_{best}

1, see Eq. 13. The *fitness* of each trial candidate is calculated next using the distance d_{MK}, as follows:

$$J(\tilde{\Theta}_i) = \sum_{t=0}^{T-1} \sum_{i \in \Lambda} 2^{-|i|} \cdot d_{MK}\left((x_i(t+1), \sum_{k=1}^{K} \tilde{\theta}_{i,k} \mathcal{N}_i(t)\right) \qquad (14)$$

Further, drawing on the recommendations in [19], the search parameters of SaDE are set as follows: population size, $NP \leftarrow 100$; learning period, $LP \leftarrow 50$; scaling factors F which are sampled from a normal distribution with mean $= 0.5$ and variance $= 0.3$, adaptive crossover rate CR (initial value 0.5), and four strategies with equal initial probabilities $p_{\text{strategy}} = 0.25$. A total of 20 independent runs of SaDE are carried out, where each run is terminated upon reaching the maximum number of iterations $(G = 500)$ to account for the stochastic nature of the algorithm.

6   Results

We consider several identification scenarios to emulate distinct local behaviors. This is achieved by variations in the local neighborhood of a two-dimensional CAM: *two neighborhood types* (Manhattan/von Neumann and Moore) and *three neighborhood radii* $(r = 1, 2,$ and $3)$. Moreover, two sets of parameters are generated for each neighborhood: the first set is generated *randomly*, *i.e.*, each neighbor is assigned a random parameter $\theta_k \in [0, 1]$, see Fig. 2a and 2c. In contrast, the other set (denoted by *distance-based*) assigns parameter values that are inversely proportional to distance, θ_k, see Fig. 2b and 2d. This leads to a total of 12 different identification scenarios (2 neighborhood types $\times$ 3 radii $\times$ 2 parameter sets). For convenience, we split these scenarios into two cases: *Case-I* includes all the scenarios where θ_k values are randomly assigned, whereas *Case-II* encompasses the remaining scenarios. Note that the parameter assignments under both cases ensure the sum of parameters $\sum_k \theta_k = 1$.

In each scenario, the identification data $\mathcal{O}$ is collected on a two-dimensional rectangular lattice $(\Lambda = 51 \times 51$ cells) over $T = 10$ time steps. Note that this temporal horizon is selected through trial-and-error to balance computational complexity with parameter identifiability. A Gaussian white noise $(SNR = 40 \ dB)$ is added to the observed states to simulate the real-world measurement noise. Each scenario begins with the same randomly assigned $\in [0, 1]$ initial states at $T = 0$, which ensures that any change in dynamic behavior across multiple scenarios can be ascribed to the local rule. This is further illustrated in Fig. 3, which demonstrates CAM evolution with the Manhattan topology $(r = 2)$ over four different time steps under two scenarios: Fig. 3a with *randomly* assigned parameters shown in Fig. 2a, and Fig. 3b with *distance-based* parameters shown in Fig. 2b.

6.1   Parameter Estimation Results

Given that the identification data is collected using the known local rule (and thus the parameter θ_k), we use the Normalized Root Mean Square Error (NRMSE) metric to gauge the accuracy of the estimates $\hat{\theta}_k$, as follows:

$$NRMSE = \frac{\sqrt{\frac{1}{K} \sum_{k=1}^{K} (\hat{\theta}_k - \theta_k)^2}}{\max(\Theta) - \min(\Theta)} \times 100 \tag{15}$$

It is worth noting that the equality constraint, $\sum_k \theta_k = 1$, dictates that overall parameter values decrease with the increase in the neighborhood radius, which can skew error metrics. NRMSE avoids numerical instability with near-zero parameters in such scenarios and provides consistent comparisons across different neighborhood sizes and types.

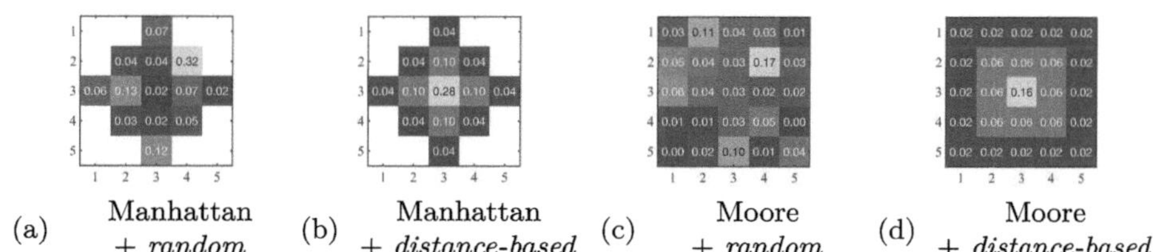

(a) Manhattan + *random* (b) Manhattan + *distance-based* (c) Moore + *random* (d) Moore + *distance-based*

Fig. 2. Parameter assignment for $r = 2$.

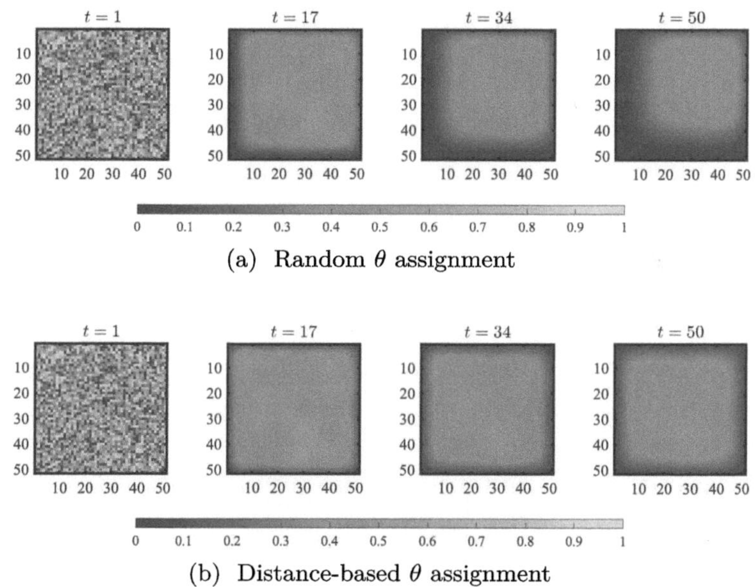

(a) Random θ assignment

(b) Distance-based θ assignment

Fig. 3. Evolution of CAMs with Manhattan neighborhood and with $r = 2$

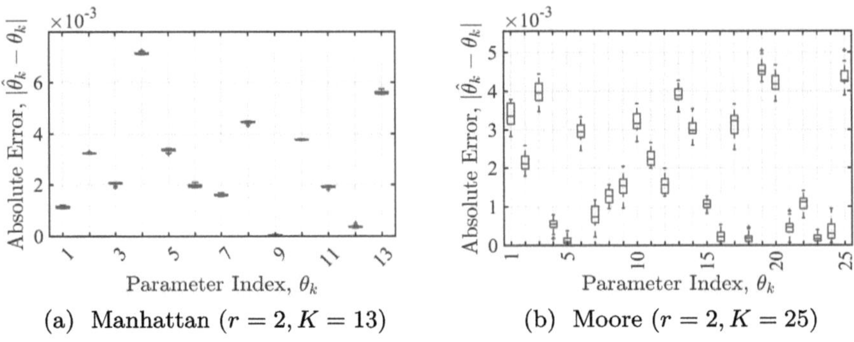

(a) Manhattan ($r = 2, K = 13$)   (b) Moore ($r = 2, K = 25$)

Fig. 4. Parameter Estimation Error with random θ assignments

To evaluate the efficacy of the proposed approach, the parameter estimation is carried out on all identification scenarios (Sect. 6) following the search setup described in Sect. 5.1. A total of 20 independent runs of SaDE are carried out for each identification scenario due to its stochastic nature. At the end of each run, the estimated parameters, $\hat{\Theta} = \{\hat{\theta}_1, \hat{\theta}_2, \ldots, \hat{\theta}_K\}$, are recorded for further analysis. Figure 4 depicts such variations in the absolute parameter estimation error, $|\hat{\theta}_k - \theta_k|$, over 20 independent runs of SaDE for the identification scenarios with $r = 2$ and random parameter assignments.

For the sake of brevity, we summarize the parameter estimation results of all identification scenarios over 20 independent runs of SaDE in terms of NRMSE in Table 1. The corresponding variations in the fitness function, $J(\cdot)$, are also shown in Table 2. The results indicate an excellent parameter recovery with NRMSE $\leq$ 2% across all identification scenarios. As expected, the neighborhood size impacts the estimation performance; the quadratic increase in parameter count (K) with increasing radius (r) slightly affects parameter recovery. For example, the mean NRMSE values of Manhattan neighborhoods rise from 0.8–1.3% (with $r = 1$) to 1.7–2.2% (with $r = 3$). Moreover, the fitness function results in Table 2 further demonstrate that the proposed approach can accurately capture spatio-temporal dynamics. For all identification scenarios, the average value of the fitness function $J(\cdot) \equiv d_\mathcal{M} \leq 0.22$, which indicates the predicted states $\hat{x}$ closely simulate the observed states x (see Eq. 13 and 14), even in the presence of measurement noise. Further, the results show consistently better performance for the Manhattan neighborhoods irrespective of the neighborhood radius, r. This, in part, can be explained by its relatively compact structure, which leads to fewer parameters (K) for a particular radius compared to the Moore neighborhood. Additionally, distance-based parameter assignments yield marginally better fitness values than random assignments, particularly for Manhattan neighborhood, indicating that parameter distribution patterns influence identification performance.

Table 1. Parameter estimation errors over 20 independent runs of SaDE

NRMSE	$r = 1$		$r = 2$		$r = 3$	
	Manhattan, $K = 5$	Moore, $K = 9$	Manhattan, $K = 13$	Moore, $K = 25$	Manhattan, $K = 25$	Moore, $K = 49$
Case-I: Random θ Assignment						
Best	0.784	0.497	1.123	1.451	1.676	1.722
Mean	0.784	0.497	1.128	1.548	1.736	2.042
SD	5.52E−11	5.70E−06	3.05E−03	4.75E−02	3.92E−02	2.01E−01
Max	0.784	0.497	1.133	1.631	1.830	2.396
Case-II: Distance-based θ Assignment						
Best	1.314	1.301	1.603	1.774	1.947	1.774
Mean	1.314	1.301	1.605	1.861	2.185	1.972
SD	6.85E−12	5.25E−06	2.29E−03	5.02E−02	1.22E−01	9.03E−02
Max	1.314	1.301	1.612	1.948	2.362	2.100

Table 2. Variations in the fitness function $J(\cdot)$ over 20 independent runs of SaDE

$J(\Theta_{best})$	$r = 1$		$r = 2$		$r = 3$	
	Manhattan, $K = 5$	Moore, $K = 9$	Manhattan, $K = 13$	Moore, $K = 25$	Manhattan, $K = 25$	Moore, $K = 49$
Case-I: Random θ Assignment						
Best	0.16	0.27	0.12	0.23	0.14	0.21
Mean	0.16	0.27	0.12	0.23	0.14	0.21
SD	1.13E−15	7.08E−09	1.31E−06	2.01E−05	7.17E−05	1.00E−03
Max	0.16	0.27	0.12	0.23	0.14	0.22
Case-II: Distance-based θ Assignment						
Best	0.14	0.26	0.12	0.24	0.12	0.22
Mean	0.14	0.26	0.12	0.24	0.12	0.22
SD	8.52E−17	3.56E−09	1.45E−06	4.41E−05	1.73E−04	4.75E−04
Max	0.14	0.26	0.12	0.24	0.12	0.22

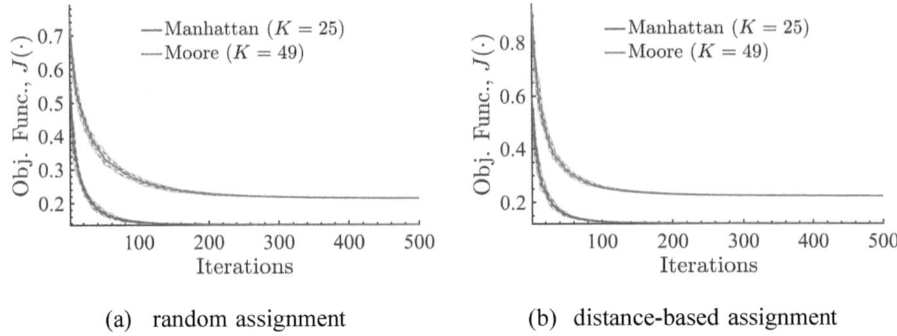

(a) random assignment   (b) distance-based assignment

Fig. 5. Convergence behavior of SaDE over 20 independent runs for different topologies with $r = 3$. Note that y-axis is zoomed and begin from $J(\cdot) \approx 0.14$ to appreciate the difference in $J(\cdot)$ obtained on Manhattan and Moore topologies.

6.2   SaDE: Convergence and Runtime Analysis

Finally, we focus on the mutation strategies of SaDE and their impacts on the proposed parameter estimation problem. To this end, we observed the iterative variations in the selection probabilities, p_{strategy}, of all strategies being considered: {rand/1, rand-to-best/2, rand/2, current-to-rand/1}, see Algorithm 1. Figure 6a depicts these results for the Manhattan topology with $r = 3$ and *random* θ assignments. The results clearly demonstrate that two strategies dominate throughout the search process: rand-to-best/2 and current-to-rand/1. Similar patterns are also observed in the remaining identification scenarios (not shown here for brevity). This may be ascribed to a key challenge of the proposed parameter estimation problem: the equality constraint, $\sum_k \theta_k = 1$. The evolutionary operators (such as mutation and crossover) associated with a particular strategy are likely to generate *infeasible* solutions that violate this constraint.

It is worth emphasizing that both *dominant* strategies use the current parameter vector $\theta_{i,k}$ as their *mutation* basis to generate the trial candidate, see Line 14 (rand-to-best/2) and Line 18 (current-to-rand/1), Algorithm 1. This key distinction is likely to reduce the disruption arising from the binomial crossover operation (see Line 20-22, Algorithm 1), and thereby conserve parameter relationships and limit constraint violations. Additionally, current-to-rand/1 replaces binomial crossover with an arithmetic recombination, possibly leading to reductions in infeasible solutions. To test this hypothesis, in each iteration, we calculated the ratio between infeasible and total solutions (denoted as the Constraint Violation Rate) generated by a particular strategy, before the solution repair is applied (see Line 23, Algorithm 1). Figure 6b depicts the results of this analysis, which confirm the hypothesis: current-to-rand/1 maintains near-zero violations throughout, while rand-to-best/2 reduces violations faster than the remaining strategies.

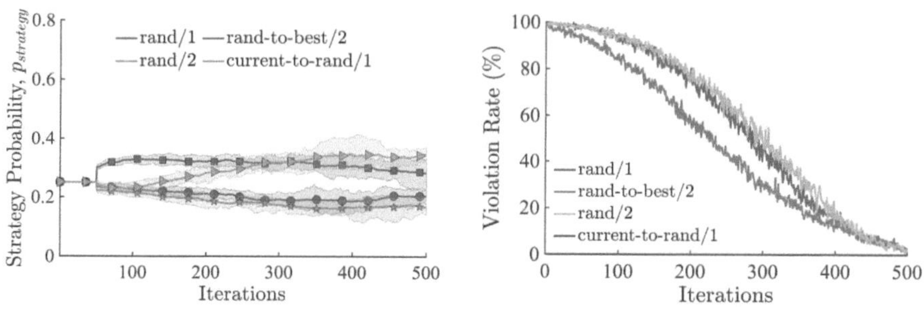

(a)   Iterative Strategy Probability   (b)   Iterative Infeasible Solution Rate

Fig. 6. Comparison of SaDE mutation strategies over 20 independent runs on Manhattan topology with $r = 3$ and *random* θ assignment.

7   Conclusions

A new approach to identifying local rules from the observed data of Cellular Automata on spaces probability measures (CAMs) was proposed. To this end, a special class of CAMs was considered, where cell states are represented by Bernoulli probability distributions and local rules act on such probabilistic configurations. We leveraged the fact that the local rules can be formulated as a convex combination of neighborhood states in this framework. Building upon this notion, it was demonstrated that the identification can be approached as a constrained optimization task involving the estimation of local rule parameters, assuming that the neighborhood size and type are *a priori* determined. Next, a meta-heuristic identification approach was proposed that can estimate the local rule parameters directly from the observed CAM states using Self-adaptive Differential Evolution (SaDE). The performance of the proposed approach was evaluated by considering 12 distinct identification scenarios that involved varied combinations of neighborhood types, neighborhood radii and parameter assignments. The results demonstrate that the known local rule parameters of the test scenarios could be accurately estimated even in the presence of measurement noise. These results are encouraging and demonstrate the potential of the proposed approach to accurately identify local rules governing the observed spatio-temporal data of real-world systems.

Finally, while this study assumes the *a priori* knowledge of neighborhood size and type, this issue remains one of the fundamental identification challenges, and it will be the topic of our further investigation. Furthermore, the runtime analysis of SaDE underlines the equality constraint, $\sum_k \theta_k = 1$, as the key optimization challenge. In this context, some of the existing mutation and crossover strategies were found to be disruptive and led to infeasible solutions. A further investigation into the design or selection of new evolutionary operators to address such issues seems promising.

References

1. Adamatzky, A.I.: Identification of Cellular Automata. CRC Press (2018)
2. Bhattacharjee, K., Naskar, N., Roy, S., Das, S.: A survey of cellular automata: types, dynamics, non-uniformity and applications. Nat. Comput. **19**(2), 433–461 (2020)
3. Cattaneo, G., Dennunzio, A., Farina, F.: A full cellular automaton to simulate predator-prey systems. In: El Yacoubi, S., Chopard, B., Bandini, S. (eds.) ACRI 2006. LNCS, vol. 4173, pp. 446–451. Springer, Heidelberg (2006). https://doi.org/10.1007/11861201_52
4. Cattaneo, G., Dennunzio, A., Margara, L.: Chaotic subshifts and related languages applications to one-dimensional cellular automata. Fundam. Inform. **52**(1–3), 39–80 (2002)
5. Cattaneo, G., Dennunzio, A., Margara, L.: Solution of some conjectures about topological properties of linear cellular automata. Theor. Comput. Sci. **325**(2), 249–271 (2004)
6. Delorme, M.: An Introduction to Cellular Automata, pp. 5–49. Springer, Cham (1999)
7. Farina, F., Dennunzio, A.: A predator-prey cellular automaton with parasitic interactions and environmental effects. Fundam. Inform. **83**(4), 337–353 (2008)
8. Farrelly, T.: A review of quantum cellular automata. Quantum **4**, 368 (2020)
9. Formenti, E., Hafiz, F., Kunze, A., La Torre, D.: Cellular automata on probability measures (2025). https://arxiv.org/abs/2503.15086
10. Formenti, E., Imai, K., Martin, B., Yunès, J.B.: Advances on Random Sequence Generation by Uniform Cellular Automata, pp. 56–70. Springer, Cham (2014)
11. Formenti, E., La Torre, D.: Cellular automata on probability measures (2023). http://dx.doi.org/10.2139/ssrn.4645495
12. Gilpin, W.: Cellular automata as convolutional neural networks. Phys. Rev. E **100**(3) (2019)
13. Hafiz, F., Swain, A., Mendes, E.: Orthogonal floating search algorithms: from the perspective of nonlinear system identification. Neurocomputing **350**, 221–236 (2019)
14. Hafiz, F., Swain, A., Mendes, E.: Two-dimensional (2D) particle swarms for structure selection of nonlinear systems. Neurocomputing **367**, 114–129 (2019)
15. Hedlund, G.A.: Endomorphisms and automorphisms of the shift dynamical system. Math. Syst. Theory **3**(4), 320–375 (1969)
16. Kari, J.: Theory of cellular automata: a survey. Theor. Comput. Sci. **334**(1–3), 3–33 (2005)
17. Maeda, K., Sakama, C.: Identifying cellular automata rules. J. Cell. Autom. **2**(1) (2007)
18. Mei, S., Billings, S.A., Guo, L.: A neighborhood selection method for cellular automata models. Int. J. Bifurc. Chaos **15**(02), 383–393 (2005)
19. Qin, A.K., Huang, V.L., Suganthan, P.N.: Differential evolution algorithm with strategy adaptation for global numerical optimization. IEEE Trans. Evol. Comput. **13**(2), 398–417 (2009)
20. Wolfram, S.: Cellular automaton fluids 1: Basic theory. J. Stat. Phys. **45** (1986)
21. Yang, Y., Billings, S.: Neighborhood detection and rule selection from cellular automata patterns. IEEE Trans. Syst. Man Cybern. **30**(6), 840–847 (2000)
22. Zhao, Y., Billings, S.: The identification of cellular automata. J. Cell. Autom. **2**, 47–65 (2007)
23. Zhao, Y., Billings, S.A.: Neighborhood detection using mutual information for the identification of cellular automata. IEEE Trans. Syst. Man Cybern. **36**(2), 473–479 (2006)

Symport/Antiport P Systems with Membrane Separation Characterize $\mathbf{P}^{(\#\mathbf{P})}$

Vivien Ducros[1]([✉]) and Claudio Zandron[2]

[1] ENS Paris-Saclay, Université Paris-Saclay, Gif-sur-Yvette, France
`vivien.ducros@ens-paris-saclay.fr`
[2] Dipartimento di Informatica, Sistemistica e Comunicazione, Università degli Studi di Milano-Bicocca, Milan, Italy

Abstract. Membrane systems represent a computational model that operates in a distributed and parallel manner, inspired by the behavior of biological cells. These systems feature objects that transform within a nested membrane structure. This research concentrates on a specific type of these systems, based on cellular symport/antiport communication of chemicals.

Results in the literature show that systems of this type that also allow cell division can solve **PSPACE** problems. In our study, we investigate systems that use membrane separation instead of cell division, for which only limited results are available. Notably, it has been shown that any problem solvable by such systems in polynomial time falls within the complexity class $\mathbf{P}^{(\#\mathbf{P})}$.

By implementing a system solving **MIDSAT**, a $\mathbf{P}^{(\#\mathbf{P})}$-complete problem, we demonstrate that the reverse inclusion is true as well, thus providing an exact characterization of the problem class solvable by P systems with symport/antiport and membrane separation.

Moreover, our implementation uses rules of length at most three. With this limit, systems were known to be able to solve **NP**-complete problems, whereas limiting the rules by length two, they characterize **P**.

Keywords: Membrane Computing · Symport/Antiport P System · Membrane Separation · Computational Complexity

1 Introduction

Membrane systems (or P systems) were introduced in [8]. They are a computational model inspired by the operations made by biological cells on different types of chemicals. The structure of these systems comprises a hierarchy of nested membranes, each defining distinct regions. A system is enclosed in an outermost membrane called the skin, isolating it from the external environment. Within these regions, objects undergo transformations based on specific evolution rules, enabling their movement and communication across membranes between different regions.

© The Author(s), under exclusive license to Springer Nature Switzerland AG 2026
E. Formenti and L. Manzoni (Eds.): UCNC 2025, LNCS 16364, pp. 163–178, 2026.
https://doi.org/10.1007/978-3-032-15641-9_12

An interesting and deeply investigated feature of such systems was proposed shortly after in [9], where the concept of *active membranes* was introduced: membranes can be *divided* by evolutionary rules, duplicating their contents in both obtained copies, thus allowing for the generation of an exponential amount of resources within a polynomial time frame. An alternative approach, later considered, is *separation* of membranes [1], where membranes are still duplicated but the contents of the original membrane are distributed between the two resulting membranes rather than being duplicated.

Since then, many studies have investigated the complexity classes defined by different types of P systems with active membranes, employing various features and constraints. These studies aim to understand how specific properties influence the development of time-efficient systems that can solve computationally difficult problems in polynomial time while using exponential space, in contrast to less efficient systems. In this paper, we continue this line of research by concentrating on a variant of membrane systems known as symport/antiport P systems, introduced in [7]. In this model, objects can be communicated between two regions in one of two ways: either by moving multiple objects simultaneously from the same region to an adjacent one (symport) or by exchanging two or more objects between two adjacent regions (antiport).

In [14], authors proved that symport/antiport P systems with membrane division and rules of length (the number of objects involved in the rule) not exceeding 3 can uniformly solve the well-known **PSPACE**-complete problem **QSAT**, in polynomial time and exponential space. In [15], the authors conjectured that the reverse inclusion also holds, which would provide a characterization of **PSPACE** in terms of these systems.

Regarding symport/antiport P systems with membrane separation, only partial results have been obtained so far. For example, they are known to solve the **SAT** problem [17], indicating that **NP** is a lower bound for these systems. The influence of the environment has been studied in [6], showing that while it does not affect the computational power of symport/antiport P systems with membrane division, removing the environment in the case of membrane separation reduces the class of solvable problems to **P**. When the length of the rules is restricted to 1 (resp. 2), symport/antiport P systems with membrane division (resp. separation) are inefficient and can only solve problems in **P** [3] (resp. [4]).

In all these instances, the membranes are arranged in a nested structure. However, another variant, known as *tissue P systems*, arranges the membranes as a directed graph. When fission (division or separation) rules are allowed, these systems characterize the class $\mathbf{P}^{(\#\mathbf{P})}$ of problems solved by polynomial time Turing Machine using $\#\mathbf{P}$ oracles [2]. By exploiting this last result and the relations between cell-like and tissue-like P systems making use of membrane separation, we first show in this work that $\mathbf{P}^{(\#\mathbf{P})}$ represents an upper bound for P systems with symport/antiport and membrane separation.

Moreover, we prove that the opposite inclusion also holds by solving the $\mathbf{P}^{(\#\mathbf{P})}$-complete problem **MIDSAT**, thus obtaining a precise characterization of the class of problems solved by such kind of P systems. The **MIDSAT** problem takes a boolean formula $\varphi(x_1, \ldots, x_n)$ as input, and decides whether the variable x_n is assigned to 1 in the lexicographically middle satisfying assignment of φ. The system solving **MIDSAT** we designed is based on the one solving **SAT** from [17]. We provide an implementation of the system in the `MeCoSim` application, a simulator of P systems [12], with the specific model for symport/antiport systems [5].

The rest of the paper is organized as follows: Sect. 2 is dedicated to the definitions of symport/antiport P systems and their complexity classes. In Sect. 3, we give some results towards the characterization (Corollary 1). The system solving **MIDSAT** is described in Sect. 4 and we prove it is correct. Finally, we add a few comments about our implementation in `MeCoSim` in Sect. 5.

2   Definitions

We give in this section the usual definitions of P systems and complexity classes of problems solved by P systems. For more details, our reader can refer to the introduction and to the handbook of membrane computing [10,11]. The notions of complexity theory for membrane systems are treated in [13].

The set of natural integers is written $\mathbb{N}$ and includes 0. $[\![i,j]\!]$ denotes the set of all integers from i to j.

A *multiset* m over a set A is a pair (f, A) where $f : A \longrightarrow \mathbb{N}$. For $a \in A$, $f(a)$ is the number of occurrences of a in m. Its *support* is $\mathrm{Supp}(m) = \{a \in A : f(a) > 0\}$. If the support is finite, its cardinal $|m| = \sum_{a \in A} f(a)$ is finite as well. We write $a \in m$ when $a \in \mathrm{Supp}(m)$. The sum $m_0 + m_1$ of two multisets $m_0 = (f_0, A)$ and $m_1 = (f_1, A)$ is the multiset $m = (f, A)$ where $f(a) = f_0(a) + f_1(a)$ for all $a \in A$. m_0 is included in m_1, written $m_0 \sqsubseteq m_1$, when $f_0(a) \leqslant f_1(a)$ for all $a \in A$.

2.1   Symport/Antiport P Systems

Definition 1. *A membrane structure is a rooted tree in which nodes are labeled by integers from 1 to q, where $q > 0$ is the size of the membrane structure. The nodes are called membranes. The depth $d(h)$ of a membrane h is its distance from the root. The root is called the* skin. *A leaf is called an* elementary *membrane. The parent of a membrane h is denoted by $p(h)$.*

Definition 2. *A symport/antiport P system with membrane separation rules is a tuple*

$$\Pi = (\Gamma, (\Gamma_0, \Gamma_1), \Sigma, \mathcal{E}, \mu, \mathcal{M}_1, \ldots, \mathcal{M}_q, \mathcal{R}_1, \ldots, \mathcal{R}_q, h_{in}, h_{out})$$

where:

- Γ *is a finite alphabet of objects with a partition $\{\Gamma_0, \Gamma_1\}$: $\Gamma_0 \uplus \Gamma_1 = \Gamma$,*
- $\Sigma \subset \Gamma$ *is the input alphabet,*
- $\mathcal{E} \subseteq \Gamma \backslash \Sigma$ *is the set of objects present in the environment, each one available in arbitrary number of copies,*
- μ *is a membrane structure of size q,*

- $\mathcal{M}_h$ is the finite multiset of objects of $\Gamma \setminus \Sigma$ initially located in membrane $h \in [\![1, q]\!]$,
- $\mathcal{R}_h$ is the finite set of evolution rules of the membrane $h \in [\![1, q]\!]$ of type:
  - (u, out) or (u, in) called the symport rules,
  - $(u, out \mid v, in)$ called the antiport rules,
  - $[a]_h \to [\Gamma_0]_h [\Gamma_1]_h$, if h is an elementary membrane distinct from the skin and the output membrane h_{out}, called the separation rules,

  where u and v are non-empty finite multisets of Γ and $a \in \Gamma$,
- $h_{in} \in [\![1, q]\!]$ is the index of the input membrane,
- $h_{out} \in [\![0, q]\!]$ is the index of the output membrane.

How Does a Symport/Antiport P System Compute? Let Π be a symport/antiport P system with membrane separation rules defined as above. As P systems are models of computation, they evolve step by step from an initial configuration. Our definition allows to give a multiset m over the input alphabet Σ as an input to the system in the membrane h_{in}. We describe here the evolution of $\Pi(m)$. Note that the computation is non-deterministic, but we will later restrict ourselves to *confluent* systems, where all possible computations give the same result (see Remark 1).

Each membrane $h \in \mu$ of the system delimits a *region* in which objects of Γ lie. A region is the area between h and its children (membranes h' such that $p(h') = h$). Objects go through membrane h, when communicated between region h and $p(h)$, according to symport and antiport rules in $\mathcal{R}_h$. Elementary membranes (the one having no children) can split according to the separation rules. We distinguish two regions: the *environment* is the region outside the system, containing in arbitrary many occurrences the objects of $\mathcal{E}$. The *skin* is the limit between the system and the environment, its index is the root of μ. It cannot separate, however, it is the only membrane able to communicate with the environment, so we consider the environment as the parent of the skin.

The *configuration* $\mathcal{C}_n = (\mu_n, (m_n(h))_{h \in \mu_n})$ of the system after step $n \in \mathbb{N}$ is a pair where μ_n is the membrane structure of the system and $(m_n(h))_{h \in \mu_n}$ is the family of all multisets corresponding to the objects lying in each region after step n. The *initial configuration* is $\mathcal{C}_0 = (\mu_0, (m_0(h))_{h \in \mu_0})$ where $\mu_0 = \mu$, $m_0(h) = \mathcal{M}_h$ for all $h \in [\![1, q]\!]$, $h \neq h_{in}$, and $m_0(h_{in}) = \mathcal{M}_{h_{in}} + m$.

The symport/antiport P system evolves according to the *maximal parallel mode*, that is: all rules that can be applied are applied in parallel as long as they can. Objects can be used by only one rule at a given step, and the objects communicated can't be used again in the same step. If an object can be used by more than one rule, the applied rule is chosen non-deterministically. If a separation rule is applied, it is the only one applied to this membrane.

During step $n > 0$, the symport rule $(u, out) \in \mathcal{R}_h$ can be used if the objects of u are in region h: $u \sqsubseteq m_{n-1}(h)$. It sends the objects of u outside region h (through membrane h to the region $p(h)$). The symport rule $(u, in) \in \mathcal{R}_h$ can be used if $u \sqsubseteq m_{n-1}(p(h))$. It sends u from region $p(h)$ inside region h. The antiport rule $(u, out \mid v, in) \in \mathcal{R}_h$ can be used if $u \sqsubseteq m_{n-1}(h)$ and $v \sqsubseteq m_{n-1}(p(h))$. Then u is sent outside h whereas v is sent inside h.

The length of a rule is $|u|$ for symport rules and $|u| + |v|$ for antiport rules.

The membrane separation rule $[a]_h \to [\Gamma_0]_h[\Gamma_1]_h \in \mathcal{R}_h$ can be used when the object a is in h. The object a is consumed, and the membrane h is split into two membranes, the first one containing the objects of Γ_1 in h and the second the objects of Γ_2 in h. The initial membrane h is deleted and the new membranes keep the index h so the rules of $\mathcal{R}_h$ can be applied to both at the next step. By definition, only elementary membranes different from the output membrane can be separated.

Starting with the initial configuration $\mathcal{C}_0$ and applying the rules in the maximal parallel order, we obtain a sequence $\mathcal{C}_0, \mathcal{C}_1, \ldots$ of configurations called computation, representing the state of the system after each step. When no rules can be applied to a configuration, the computation halts. In this case, the result of the computation is the multiset of objects in the membrane h_{out} of the final configuration. If a computation never halts, then it produces no output.

2.2   Decision Problems and Complexity Classes

Definition 3. *A decision problem X is a pair (I_X, θ_X) where I_X is the set of instances and $\theta_X : I_X \longrightarrow \{0, 1\}$. An instance $x \in I_X$ is said to be accepted by the problem if $\theta_X(x) = 1$, else it is rejected.*

Definition 4. *A recognizer symport/antiport P system with membrane separation rules is a symport/antiport P system with membrane separation Π where:*

- **yes**, **no** $\in \Gamma \setminus \Sigma$ *are the answer objects,*
- *they are initially present in the system:* $\exists h, h' \in [\![1, q]\!], \textbf{yes} \in m_h, \textbf{no} \in m_{h'}$,
- *the output membrane is the environment:* $h_{out} = 0$,

and such that:

- *every computation halts,*
- *for all computations of Π on the same input, either* **yes** *(the computation is said to be accepting) or* **no** *(the computation is said to be rejecting) must be sent to the environment* **only at the last step**.

We denote by **CSC** *the class of all recognizer symport/antiport P systems with separation rules. When all rules have length at most $k \in \mathbb{N}$, the recognizer P system is in* **CSC**(k).

Definition 5. *Let $\mathcal{D}$ be a class of recognizer P systems. A family* $\mathbf{\Pi} = (\Pi_n)_{n \in \mathbb{N}}$ *of P systems is $\mathcal{D}$-consistent if $\Pi_n \in \mathcal{D}$ for all $n \in \mathbb{N}$. Moreover, it is said to be polynomially uniform if there exists a deterministic Turing Machine that computes, from the unary representation of $n \in \mathbb{N}$, the P system Π_n in polynomial time of n.*

Definition 6. *A decision problem $X = (I_X, \theta_X)$ is decidable in a uniform way by a polynomially uniform family $\mathbf{\Pi}$ of recognizer P systems if there exist two polynomially computable functions s (size) and cod (encoding) such that for all $x \in I_X$:*

- $s(x) \in \mathbb{N}$,
- $cod(x)$ is a multiset over $\Sigma_{s(x)}$, the input alphabet of $\Pi_{s(x)} \in \mathbf{\Pi}$,
- if $\theta_X(x) = 1$ then all computations of $\Pi_{s(x)}(cod(x))$ are accepting,
- if there exists one accepting computation of $\Pi_{s(x)}(cod(x))$, then $\theta_X(x) = 1$.

Remark 1. The last two properties are called *soundness* and *completeness*. They imply *confluence*, which is that the computations of $\Pi_{s(x)}(cod(x))$, $x \in I_X$ are all accepting or all rejecting. Such a P system is said to be confluent.

Definition 7. *Let $\mathcal{D}$ be a class of recognizer P systems. A decision problem X is in* $\mathbf{PMC}_{\mathcal{D}}$ *if there exists a polynomially uniform $\mathcal{D}$-consistent family $\mathbf{\Pi}$ deciding X in a uniform way, such that for all $x \in I_X$, each computation of $\Pi_{s(x)}(cod(x))$ halts in a polynomial number of steps relatively to $s(x)$.*

Proposition 1. *Let $\mathcal{D}$, $\mathcal{D}'$ be two classes of recognizer P systems such that $\mathcal{D} \subseteq \mathcal{D}'$. The following statement holds by definition:* $\mathbf{PMC}_{\mathcal{D}} \subseteq \mathbf{PMC}_{\mathcal{D}'}$.

3   Characterization of $\mathbf{P}^{(\#\mathbf{P})}$

The main contribution of this article is the proof that the class $\mathbf{P}^{(\#\mathbf{P})}$ characterizes exactly the problems solved by symport/antiport P systems with membrane separation rules, see Corollary 1.

This section provides all the tools and the results needed in the proof of this characterization. First, we recall a previous result from [2] characterizing $\mathbf{P}^{(\#\mathbf{P})}$ for tissue-like symport/antiport P systems, another type of systems where membranes are organized in a graph structure (instead, we use a more specific definition with a hierarchy structure between the membranes). Then we state our principal theorem: we can solve the $\mathbf{P}^{(\#\mathbf{P})}$-complete problem **MIDSAT** with a symport/antiport P system with membrane separation using rules of length not exceeding 3 (Theorem 1).

Combining these results, we prove the characterization (Corollary 1).

Definition 8. *A tissue P system is a tuple*

$$\Pi = (\Gamma, (\Gamma_0, \Gamma_1), \Sigma, \mathcal{E}, \mathcal{M}_1, \ldots, \mathcal{M}_q, \mathcal{R}, h_{in}, h_{out})$$

The difference with the cell-like P systems we defined (Definition 2) is that we do not have a membrane structure anymore. Every membrane can separate (except the output membrane) and can communicate objects by symport/antiport rules to any other membrane, including the environment. The rules are all grouped in the rule set $\mathcal{R}$. Symport rules are written $[u]_h \leftrightarrow [\ ]_{h'}$, and antiport rules are written $[u]_h \leftrightarrow [v]_{h'}$ (u, v being non-empty finite multisets over Γ; h, $h' \in [\![0, q]\!]$ being the communicating membranes, where 0 identifies the environment).

To define **TSC**, *the class of recognizer tissue symport/antiport P systems with membrane separation, we ask for the same properties as for the cell-like systems (* **yes** *and* **no** *objects, $h_{out} = 0$, computations halt when releasing* **yes** *or* **no** *in the environment, see Definition 4).*

Proposition 2. $\mathbf{PMC_{CSC}} \subseteq \mathbf{PMC_{TSC}}$

Proof. A recognizer (cell-like) P system is, in fact, a recognizer tissue P system. We just need to translate symport and antiport rules in $\mathcal{R}_h$ as follows: (u, out), (u, in), and $(u, \text{out} \mid v, \text{in})$ become $[u]_h \leftrightarrow [\,]_{p(h)}$, $[u]_{p(h)} \leftrightarrow [\,]_h$, and $[u]_h \leftrightarrow [v]_{p(h)}$.
    Then $\mathbf{CSC} \subseteq \mathbf{TSC}$ and $\mathbf{PMC_{CSC}} \subseteq \mathbf{PMC_{TSC}}$ by Proposition 1.

Proposition 3 (Leporati et Al., [2]). $\mathbf{PMC_{TSC}} = \mathbf{P}^{(\#\mathbf{P})}$

Definition 9. *The* **MIDSAT** *decision problem is defined as follows:*
    Instance: *A satisfiable boolean formula $\varphi(x_1, \ldots, x_n)$ in conjunctive normal form.*
    Question: *Does the least significant bit of the middle (according to lexicographic order) satisfying assignment of φ equal 1?*

The question is to determine the truth value of x_n in the assignment which is the median among all the satisfying assignments of φ, ordered by the lexicographic order. This is one of the problems complete for $\mathbf{P}^{(\#\mathbf{P})}$ listed by Toda in [16].

Proposition 4 (Toda, [16]). MIDSAT *is $\mathbf{P}^{(\#\mathbf{P})}$-complete.*

Example 1. With the boolean formula $\varphi(x_1, x_2, x_3) = (x_1 \vee x_2 \vee \neg x_3) \wedge (x_1 \vee \neg x_2 \vee x_3) \wedge (\neg x_1 \vee x_2 \vee x_3)$, the satisfying assignments are $(0, 0, 0)$, $(0, 1, 1)$, $(1, 0, 1)$, $(1, 1, 0)$, $(1, 1, 1)$, ordered in the lexicographic order. Then the median is $(1, 0, \mathbf{1})$ and the answer to the problem is **yes**.
    When an even number of assignments satisfy the formula, the lower of the two median assignments is chosen as the median.

Theorem 1. MIDSAT $\in \mathbf{PMC_{CSC(3)}}$

We prove this result in the following section by implementing a family of recognizer symport/antiport P systems with membrane separation rules and rules of length at most 3 solving **MIDSAT** in a uniform way and in polynomial time.

Corollary 1 (Characterization of $\mathbf{P}^{(\#\mathbf{P})}$). $\mathbf{PMC_{CSC(3)}} = \mathbf{P}^{(\#\mathbf{P})}$

Proof. By definition, $\mathbf{PMC_{CSC(3)}} \subseteq \mathbf{PMC_{CSC}}$, and from Propositions 2 and 3, we obtain $\mathbf{PMC_{CSC(3)}} \subseteq \mathbf{P}^{(\#\mathbf{P})}$.
    By Theorem 1 and because **MIDSAT** is $\mathbf{P}^{(\#\mathbf{P})}$-complete (Proposition 4), the following inclusion holds: $\mathbf{P}^{(\#\mathbf{P})} \subseteq \mathbf{PMC_{CSC(3)}}$.
    The corollary comes out from these inclusions.

4   A System Solving MIDSAT

This section is devoted to the proof of Theorem 1.

The P system we create is an evolution of the one used to solve the **SAT** problem in [17]. We first explain briefly how their system works and what changes we made. Then we provide an overview of the computation. Finally, we give a detailed proof of the theorem.

The complexity of the system arises principally from the constraint to have rules of length at most three. This requires us to add steps when we want four objects to interact at a precise step or when we want to exchange an object with n copies of another one.

4.1   Some Modifications to the CSC(3) SAT Solver

Our work is based on the system implemented by [17] solving **SAT**.

This one runs in three phases: the generation phase, where 2^n membranes are created with the separation rule, each one representing an assignment of the n variables of the formula; the checking phase consists in the evaluation of the formula in each membrane; finally, the output phase answers **yes** or **no** depending on whether a membrane satisfies the formula.

The instance is a boolean formula $\varphi(x_1, \ldots, x_n) = \bigwedge_{j=1}^{m} C_j$ in CNF, where $C_j = l_{i_1,j} \vee \cdots \vee l_{i_{r_j},j}$ and $l_{i_k,j} = x_{i_k}$ or $\neg x_{i_k}$, $j \in [\![1, m]\!]$, $k \in [\![1, r_j]\!]$. They encode it by $cod(\varphi) = \{x_{i_k,j} : l_{i_k,j} = x_{i_k,j}\} + \{\bar{x}_{i_k,j} : l_{i_k,j} = \neg x_{i_k,j}\}$, and $s(\varphi) = \langle m, n \rangle$, where $\langle m, n \rangle = ((n+m)(n+m+1)/2) + n$ is the common bijection between $\mathbb{N}$ and $\mathbb{N}^2$.

We denote by $\Pi_{\langle m,n \rangle}^{\mathbf{SAT}}$ the systems of the family they provide.

$$\Pi_{\langle m,n \rangle}^{\mathbf{SAT}} = (\Gamma, (\Gamma \setminus \Gamma_1, \Gamma_1), \Sigma, \mathcal{E}, [[\;]_2[\;]_3]_1, \mathcal{M}_1, \mathcal{M}_2, \mathcal{M}_3, \mathcal{R}_1, \mathcal{R}_2, \mathcal{R}_3, 1, 0)$$

The system of [17] has initially three membranes. The skin (membrane 1) is the input membrane. It contains: membrane 2, that will separate to obtain 2^n membranes 2 (one for each truth assignment), and membrane 3, used during the output phase. The rules have a maximum length of 3 objects.

At the end of the checking phase, that is after $\Delta_{SAT} = 3n + 2m$ steps, every membrane 2 corresponding to a satisfying assignment of φ contains an object $e_{i,m}$ or $\bar{e}_{i,m}$ for $i \in [\![1, n]\!]$.

In our implementation, we re-use this system until the end of the checking phase. We delete the membrane 3 and the rules associated with it, because it was used only during the output phase. We remove the objects $\{f_r\}_{r \in [\![0, \Delta_{SAT}]\!]} \cup \{f'_p\}_{p \in [\![0, \Delta_{SAT}+1]\!]} = D$. We also remove the following rules: $\{(E_0 f_{\Delta_{SAT}} \mathbf{yes}, out), (f_{\Delta_{SAT}} \mathbf{no}, out)\} = D_1 \subseteq \mathcal{R}_1$, and $\{(e_{i,m} E_0, out), (\bar{e}_{i,m} E_0, out)\}_{i \in [\![1,n]\!]} = D_2 \subseteq \mathcal{R}_2$, so that at this point, we obtain a halting configuration after Δ_{SAT} steps.

Then, we add three phases so the computation solves **MIDSAT**:

- The *activation phase* makes the membranes 2 corresponding to a satisfying assignment able to interact during the next phases.

- The *search phase* searches for the lexicographically middle satisfying assignment, determining the values of the variables, one by one. The determination of the k-th variable needs three sub-phases ($k \in [\![1, n]\!]$):
  - *Preparation sub-phase*: the preparation cannot be done previously in parallel because it needs the result of the previous determination phase.
  - *Comparison sub-phase*: all the satisfying assignments are compared to a well-chosen assignment, see the next paragraph giving the main idea.
  - *Determination sub-phase*: depending on the number of assignments lower or greater than the one they have been compared with, the k-th variable of the median assignment is determined.
- The *output phase* returns **yes** or **no** whether $x_n = 1$ or $x_n = 0$ in the assignment found previously.

We describe all these phases more precisely in the following subsection.

The Key Idea. To determine the median $x_1 \cdots x_n$ of satisfying assignments, the search phase does the same thing as a research of the median of a set of values between 0 and $2^n - 1$. This works by finding the bits of the median, from the most to the least significant bit, by counting the number of values greater and lower than a current well-chosen number. This process is very close to a binary search: we first consider $10 \cdots 0 = 2^{n-1}$, if there are strictly more numbers greater or equal than $10 \cdots 0$, than numbers less than $10 \cdots 0$, then the most significant bit x_1 of the median is 1, else 0. To find the k-th most significant bit x_k, we must consider $x_1 \cdots x_{k-1} 10 \cdots 0$. After n iterations, we obtain the median.

For the sake of a better understanding of the different phases, we define a few constants for the number of steps each (sub-)phase lasts for and when it begins:

Sub-phases duration	Phases duration
$\Delta_{Prep} = n + 5$	$\Delta_{SAT} = 3n + 2m$
$\Delta_{Comp} = 2n + 1$	$\Delta_{ACT} = n + 2$
$\Delta_{Dete} = 3$	$\Delta_{SEA} = n \cdot \Delta_{Loop}$
$\Delta_{Loop} = \Delta_{Prep} + \Delta_{Comp} + \Delta_{Dete}$	$\Delta_{OUT} = 1$
Initial step of (sub)-phases	
$I_{Prep}(k) = \Delta_{SAT} + \Delta_{ACT} + (k - 1) \cdot \Delta_{Loop}, \ k \in [\![1, n]\!]$	
$I_{Comp}(k) = I_{Prep}(k) + \Delta_{Prep}, \ k \in [\![1, n]\!]$	
$I_{Dete}(k) = I_{Comp}(k) + \Delta_{Comp}, \ k \in [\![1, n]\!]$	
$I_{OUT} = \Delta_{SAT} + \Delta_{ACT} + \Delta_{SEA}$	

Thus, the family $\mathbf{\Pi} = (\Pi_{\langle m,n \rangle})_{m,n}$ we provide (described earlier), extends the system $\Pi_{\langle m,n \rangle}^{\mathbf{SAT}}$. It is a polynomially uniform family of recognizer symport/antiport P systems with membrane separation and rules of length at most 3, defined as follows:

$$\Pi_{\langle m,n \rangle} = (\Gamma', (\Gamma' \setminus \Gamma_1, \Gamma_1), \Sigma, \mathcal{E}', [[\ ]_2]_1, \mathcal{M}_1', \mathcal{M}_2, \mathcal{R}_1', \mathcal{R}_2', 1, 0)$$

Membrane 3 has been removed, together with its rule set $\mathcal{R}_3$, and initial multiset $\mathcal{M}_3$. While we deleted objects and rules D, D_1, and D_2 (see previous page), we added the objects of $\mathcal{E}''$ in the environment, the rules of $\mathcal{R}_1''$ to membrane 1, and the rules of $\mathcal{R}_2''$ to membrane 2:

- $\mathcal{R}_1' = (\mathcal{R}_1 \setminus D_1) \cup \mathcal{R}_1''$,
- $\mathcal{R}_2' = (\mathcal{R}_2 \setminus D_2) \cup \mathcal{R}_2''$,
- $\mathcal{E}' = (\mathcal{E} \setminus D) \cup \mathcal{E}''$,
- $\Gamma' = (\Gamma \setminus D) \cup \mathcal{E}'' \cup \{\xi_{i,0}\}_{i \in [\![1,n]\!]} \cup \{\omega_0\}$,
- $\mathcal{M}_1' = \mathcal{M}_1 - \{f_0\} - \{f_p'\}_{p \in [\![1,\Delta_{SAT}+1]\!]} + \{\mathbf{no}, \omega_0\} + \{\xi_{i,0}\}_{i \in [\![1,n]\!]}$.

Our reader will find the complete definition of the system in this document. The appendix contains the full description of $\mathcal{E}''$, $\mathcal{R}_1''$, and $\mathcal{R}_2''$.

4.2   Computation Overview

Here we present a detailed overview of the computation of the system $\Pi_{\langle m,n \rangle}$ on input $cod(\varphi)$. As mentioned in the previous subsection, the system solves **SAT** until step $\Delta_{SAT} = 3n + 2m$, where the membranes 2 corresponding to a satisfying assignment of φ contain an object $e_{i,m}$ or $\bar{e}_{i,m}$, $i \in [\![1,n]\!]$. We explain the effect of activation, search, and output phases.

Remind that the step s applies between configurations $\mathcal{C}_{s-1}$ and $\mathcal{C}_s$.

A Few Words About Counters. To control the timing of the system, we use counters situated in membrane 1. These objects keep the information of which is the current step and allow some rules to be used at a precise step. Their objective is to bring other objects into the system. The counters are the objects $\xi_{i,s}$ and ω_s, they appear only in rules of $\mathcal{R}_1'$.

Counters increment their index each step with rules like $(\xi_{i,t}, \mathrm{out} \mid \xi_{i,t+1}, \mathrm{in})$ and $(\omega_t, \mathrm{out} \mid \omega_{t+1}, \mathrm{in})$. They verify the following property: after step s (that is, in configuration $\mathcal{C}_s$), the only ξ and ω objects present in the system are $\xi_{i,s}$, $\omega_s \in m_s(1)$, in membrane 1, $i \in [\![1,n]\!]$.

We have n variants of the ξ object thanks to the index $i \in [\![1,n]\!]$. Sometimes, they bring together with their successor $\xi_{i,s+1}$ a different object so that, at a precise step, we can make appear n (possibly distinct) objects in membrane 1:

- $\nu_{i,0}$, at step $\Delta_{SAT} + 1$, $i \in [\![1,n]\!]$,
- $\zeta_{k,i}'$, at step $I_{Prep}(k)$, $k \in [\![1,n]\!]$, $i \in [\![1,n]\!]$,
- $\bar{\mu}_k$, at step $I_{Prep}(k+1) + 1$, $k \in [\![1,n]\!]$.

ω objects are present in the system in 2^n occurrences. They are useful to bring into the system a lot of objects in only one step. The rules $(\omega_s, \mathrm{out} \mid \omega_{s+1}^2, \mathrm{in})$, $s \in [\![0, n-1]\!]$ apply in the first n steps of the computation (in parallel to **SAT** computation) to duplicated ω objects until they are in 2^n occurrences. Then, as for ξ, ω counters bring the following objects in the system within 2^n occurrences:

- $\omega_{\Delta_{SAT}-1}'$, at step $\Delta_{SAT} - 1$,
- σ, from step $\Delta_{SAT} + 1$ to step $\Delta_{SAT} + n$,

- $\theta_{I_{Comp}(k)-1}$, at step $I_{Comp}(k)-1$, $k \in [\![1,n]\!]$,
- $\omega'_{I_{Comp}(k)+2i}$, at step $I_{Comp}(k)+2i$, $k \in [\![1,n]\!]$, $i \in [\![0,n-2]\!]$,
- $p'_{k,i+1}$, at step $I_{Comp}(k)+2i+1$, $k \in [\![1,n]\!]$, $i \in [\![0,n-2]\!]$,
- κ'''_k, at step $I_{Comp}(k)+2n$, $k \in [\![1,n]\!]$,
- χ_k, at step $I_{Dete}(k)$, $k \in [\![1,n]\!]$.

Activation Phase. This phase lasts for $\Delta_{ACT} = n+2$ steps and brings in the membranes 2 corresponding to a satisfying assignment of φ the objects Λ_i (the assignment makes x_i true) and Ψ_i (the assignment makes x_i false) that will interact in the next phases ($i \in [\![1,n]\!]$).

In membrane 1, during steps $\Delta_{SAT} + s$, $s \in [\![0,n-2]\!]$, objects $\omega'_{\Delta_{SAT}+s}$ bring the objects σ'_s in 2^n occurrences, and objects $\nu_{i,s}$ duplicated. At step $\Delta_{SAT} + n - 1$, 2^n objects σ'_{n-1} enter the system and every $\nu_{i,n-1}$ ($i \in [\![1,n]\!]$, in 2^n occurrences) bring objects Λ_i and Ψ_i from the environment. In parallel, from steps $\Delta_{SAT} + 1$ to $\Delta_{SAT} + n$, 2^n objects σ enter membrane 1 with the help of the counter ω.

At step $\Delta_{SAT} + 1$, each membrane 2 corresponding to a satisfying assignment of φ exchanges its object $e_{i,m}$ (or $\bar{e}_{i,m}$) for σ'_0 and T_i (or F_i). Then during each step $\Delta_{SAT} + i + 2$, σ'_i is exchanged with σ'_{i+1} and one object σ (for $i \in [\![0,n-2]\!]$), and at step $\Delta_{SAT} + n + 1$, an other object σ enters those membranes 2. Finally, at step $\Delta_{SAT} + n + 2 = \Delta_{SAT} + \Delta_{ACT}$, the rules $(\sigma, T_i, \text{out} \mid \Lambda_i, \text{in})$ and $(\sigma, F_i, \text{out} \mid \Psi_i, \text{in}) \in \mathcal{R}'_2$ are used so the objects T_i, F_i are replaced by Λ_i and Ψ_i in the activated membranes.

Search Phase. This phase determines the values of the variables $x_1, \ldots, x_n$ of the lexicographically middle satisfying assignment. The P system will repeat n times the three sub-phases, finding the value of one variable at each loop. The variables are determined in order, from x_1 to x_n.

Let's describe the k-th iteration of the sub-phases, in which x_k is determined ($k \in [\![1,n]\!]$). Preparation, comparison, and determination sub-phases last for $\Delta_{Prep} = n + 5$, $\Delta_{Comp} = 2n + 1$, and $\Delta_{Dete} = 3$ steps. Thus, the search phase lasts for $n \cdot \Delta_{Loop} = 3n^2 + 9n$ steps.

Preparation Sub-phase k. When this sub-phase begins, the values $x_1, \ldots, x_{k-1}$ are coded by the presence in membrane 1 of μ_j, if $x_j = 1$, or $\bar{\mu}_j$, if $x_j = 0$ ($j \in [\![1, k-1]\!]$). The objective of this phase is to bring 2^n occurrences of the objects $\ddot{\epsilon}^N_{k,1} \cdots \ddot{\epsilon}^N_{k,k-1} \, \epsilon_{k,k} \, \bar{\epsilon}^N_{k,k+1} \cdots \bar{\epsilon}^N_{k,n}$, where $\ddot{\epsilon}^N_{k,j} = \epsilon^N_{k,j}$ or $\bar{\epsilon}^N_{k,j}$ whether $x_j = 1$ or 0, $j \in [\![1, k-1]\!]$.

Here, we use ϵ for l, g, p; and N for Λ, Ψ, so that in fact $6n \cdot 2^n$ objects must appear (6 possible combinations for each $i \in [\![1,n]\!]$, in 2^n occurrences). The presence/absence of the bar on these objects represents $x_1 \cdots x_{k-1} 10 \cdots 0$, the reference assignment we want to compare to all satisfying assignments. Moreover, $n2^n$ occurrences of φ, φ' will appear in the system at the end of this step.

This sub-phase uses only rules from the rule set $\mathcal{R}'_1$ of membrane 1. One step before the preparation sub-phase, objects $\zeta'_{k,i}$ are created with the $\xi_{i,I_{Prep}(k)}$

counters. Next step is $I_{Prep}(k) + 1$, and $\zeta'_{k,i}$ objects become either $\zeta_{k,i}$ or $\bar{\zeta}_{k,i}$, with a bar iff the i-th value of the reference is a zero. For $i < k$, the choice is done by using μ_i or $\bar{\mu}_i$. For $i = k$, it will be $\zeta_{k,k}$, and for $i > k$, $\bar{\zeta}_{k,i}$. Then $\zeta_{k,i}/\bar{\zeta}_{k,i}$ are exchanged for $\eta_{k,i,0}/\bar{\eta}_{k,i,0}$, and if $i < k$ then $\mu_i/\bar{\mu}_i$ re-enter membrane 1. Objects $\eta_{k,i,j}/\bar{\eta}_{k,i,j}$ are duplicated at each of the n next steps, until $\eta_{k,i,n}/\bar{\eta}_{k,i,n}$ are present in membrane 1 in 2^n occurrences. Objects $\eta_{k,i,n}/\bar{\eta}_{k,i,n}$ are exchanged with $\lambda_{k,i}/\bar{\lambda}_{k,i}$ and $\pi_{k,i}/\bar{\pi}_{k,i}$, the formers then become $l_{k,i}/\bar{l}_{k,i}$ and $g_{k,i}/\bar{g}_{k,i}$, whereas the laters become $p_{k,i}/\bar{p}_{k,i}$ and π'_k. Finally, objects $\epsilon^N_{k,i}/\bar{\epsilon}^N_{k,i}$, appear in 2^n occurrences ($N \in \{\Lambda, \Psi\}$, $i \in [\![1, n]\!]$) and objects φ, φ' appear in $n2^n$ occurrences.

Comparison Sub-phase k. During this phase, every active membrane 2 will compare their assignment $N_1 \cdots N_n$ (stored by the objects Λ_i/Ψ_i, $i \in [\![1, n]\!]$) to the reference assignment $\nu = x_1 \cdots x_{k-1}10\cdots0$ coded in membrane 1. An object $\epsilon^N_{k,i}/\bar{\epsilon}^N_{k,i}$ is sent in the membrane 2 to indicate if the number given by the first i bits of the assignment is lower ($\epsilon = l$), equal ($\epsilon = p$) or greater ($\epsilon = g$) than the one given by the first i bits of the reference.

The following property holds during the comparison sub-phase:

For $i \in [\![1, n-1]\!]$, the object $\epsilon^N_{k,i}/\bar{\epsilon}^N_{k,i}$ present in a membrane 2 after step $I_{Comp}(k) + 2i - 1$ verifies: (1) $\epsilon = l$ if $N_1 \cdots N_i < \nu_1 \cdots \nu_i$, $\epsilon = p$ if $N_1 \cdots N_i = \nu_1 \cdots \nu_i$, or $\epsilon = g$ if $N_1 \cdots N_i > \nu_1 \cdots \nu_i$; (2) $N = \Lambda$ if $N_i = \Lambda_i$ or $N = \Psi$ if $N_i = \Psi_i$; and (3) there is a bar over ϵ iff $\nu_i = 0$. For the case $i = n$, the property asks to consider configuration $I_{Comp}(k) + 2n$ instead of $I_{Comp}(k) + 2n - 1$.

The exponent $N = \Lambda$ or Ψ is used to know which object among Λ_i and Ψ_i must be re-sent to the membrane 2 in the next step.

First, we explain what happens between the environment and membrane 1 during this sub-phase. At step $I_{Comp}(k)+2i+1$, $i \in [\![0, n-2]\!]$, object $\omega'_{I_{Comp}(k)+2i}$ that barely appeared in membrane 1, will be exchanged with objects $l'_{k,i+1}$ and $g'_{k,i+1}$. At the same time, object $p'_{k,i+1}$ enters the system. These three objects are present in 2^n occurrences. Still during this first step, object φ'_k becomes φ''_k. In the next step, objects φ_k and φ''_k will leave the system together. Independently, object $\theta_{I_{Comp}(k)+j}$ (created before by the counter ω) will duplicate until $\theta_{I_{Comp}(k)+2n-2}$ being in $6 \cdot 2^n$ occurrences in configuration $\mathcal{C}_{I_{Comp}(k)+2n-2}$. Then it is exchanged with κ_k and κ'_k. At step $I_{Comp}(k)+2n$, κ'_k becomes κ''_k and some occurrences of κ_k leave the system together with all objects $\epsilon^N_{k,n}/\bar{\epsilon}^N_{k,n}$ present in membrane 1. Such objects are still in the system, but in membranes 2. To avoid them to be pulled out from the system after they leave membranes 2, the objects κ'_k are sent out by objects κ''_k. We recall that 2^n objects κ'''_k appear in membrane 1 at step $I_{Comp}(k) + 2n$ with the counter ω.

Secondly, for activated membranes 2, the comparison sub-phase decomposes into $n - 1$ groups of two steps and one group of three steps, each one extending the comparison by one bit. First (steps $I_{Comp}(k) + 1$, $I_{Comp}(k) + 2$), an object $\epsilon^N_{k,1}/\bar{\epsilon}^N_{k,1}$ enters the membrane together with φ_k, pulling out N_1, with respect to the properties (1), (2) and (3). Then N_1 can re-enter the membrane 2 with $\epsilon'_{k,1}$ to keep the result of the comparison. Steps $I_{Comp}(k) + 2(i-1) + 1$, $I_{Comp}(k) + 2(i-1) + 2$, $i \in [\![2, n-1]\!]$ do the same things, but there is no need of φ_k, and they use $\epsilon'_{k,i-1}$ to preserve the properties. Finally, the group of three steps begins by $I_{Comp}(k) + 2n - 1$ which is similar to the others, nothing occurs at step $I_{Comp}(k) + 2n$ (at that time, in membrane 1, objects $\epsilon^N_{k,n}/\bar{\epsilon}^N_{k,n}$ go out of the system). Last step of the subphase is $I_{Comp}(k) + 2n + 1$ where $\epsilon^N_{k,n}/\bar{\epsilon}^N_{k,n}$ leaves membrane 2 and brings back N_n from membrane 1 to 2, together with κ'''_k.

The value of ϵ in this last object keeps the result of the comparison of the reference and the assignment of the membrane 2 considered.

Determination Sub-phase k. This sub-phase will count the number of satisfying assignments greater or equal than the reference. If they are in majority (strictly more than the ones strictly lower than the reference), then the system must contain the object μ_k at the end of this phase, else it will be $\bar{\mu}_k$.

To count, we first exchange the objects $\bar{g}^N_{k,n}$, $\bar{p}^N_{k,n}$ (or $g^N_{n,n}$, $p^N_{n,n}$ when $k = n$) with new objects G_k, and the objects $\bar{l}^N_{k,n}$ (or $l^N_{n,n}$ when $k = n$) with objects L_k. This is done with the help of objects χ_k, created at the previous step with the counter $\omega_{I_{Dete}(k)}$. At the next step, objects G_k and L_k leave the system by pairs with the rule $(L_k, G_k, \text{out}) \in \mathcal{R}_1$ and the object $\bar{\mu}_k$ enters membrane 1 with the counter $\xi_{1,I_{Dete}(k)+2}$. Finally, if there are objects G_k remaining, the rule $(G_k, \bar{\mu}_k, \text{out} \mid \mu_k, \text{in})$ applies, bringing μ_k in the system. If not, $\bar{\mu}_k$ stays into membrane 1.

Preparation sub-phase $k + 1$ (or the output phase if $k = n$) can now begin.

Output Phase. At the end of the search phase (step I_{OUT}), membrane 1 contains μ_n or $\bar{\mu}_n$ depending on whether $x_n = 1$ or 0 in the lexicographically middle satisfying assignment. This phase lasts for only one step, where one of these two rules $(\mu_n \text{ **yes** } \xi_{1,I_{OUT}}, \text{out})$, $(\bar{\mu}_n \text{ **no** } \xi_{1,I_{OUT}}, \text{out}) \in \mathcal{R}_1$ is applied. The object $\xi_{1,I_{OUT}}$ has appeared in membrane 1 during the previous step, if it was present previously, the rules could have been applied before. The objects **yes** and **no** are present in membrane 1 since the beginning of the computation.

After this step, no rules can be applied yet, and the environment contains either **yes** or **no**, according to the value of x_n in the lexicographically middle satisfying assignment.

4.3   Proof of Theorem 1

We prove here that **MIDSAT** $\in$ **PMC**$_{\mathbf{CSC(3)}}$. Previous subsection gives us the following Lemma:

Lemma 1. *When $x_n = 1$ in the lexicographically middle satisfying assignment of φ, then the object **yes** is released in the environment.*

Proof (of Theorem 1). First, the family $\mathbf{\Pi} = (\Pi_{\langle m,n \rangle})$ is **CSC**(3)-consistent: for all $n, m \in \mathbb{N}$, $\Pi_{\langle m,n \rangle} \in \mathbf{CSC}(3)$.

Secondly, it is polynomially uniform since the alphabet Γ' has a polynomial number of objects, initial multisets have a polynomial cardinal, and there is a polynomial number of rules of constant length.

Then, $\mathbf{\Pi}$ recognizes $\mathbf{MIDSAT}$ in a uniform way: the functions $s(\varphi) = \langle m, n \rangle$ and $cod(\varphi) = \{x_{i_k,j} : l_{i_k,j} = x_{i_k,j}\} \cup \{\bar{x}_{i_k,j} : l_{i_k,j} = \neg x_{i_k,j}\}$ are polynomially computable, and our system is confluent: the sets of rules we added to the P system solving $\mathbf{SAT}$ are deterministic. Lemma 1 gives the last argument.

Finally, every computation of $\Pi_{\langle m,n \rangle}(cod(\varphi))$ halts after $\Delta_{SAT} + \Delta_{ACT} + \Delta_{SEA} + \Delta_{OUT} = 3n^2 + 13n + 2m + 3$ steps, a polynomial in $\langle m, n \rangle$.

Thus, we have $\mathbf{MIDSAT} \in \mathbf{PMC}_{\mathbf{CSC}(3)}$. $\qquad\qquad\square$

5 Implementation in P-Lingua

We implemented the family $\mathbf{\Pi}$ solving $\mathbf{MIDSAT}$ in the P system simulator `MeCoSim` [5,12]. Our implementation is available on the Inria gitlab instance.

The authors of [17] have provided the implementation of their system solving $\mathbf{SAT}$ on the MeCoSim website. We based our implementation on it.

During the computation, some objects stay in the system even if they won't be used anymore. Then it is possible to add garbage rules to make them exit the system. This has been done in [17] and we also added such rules. After some tests, it appears that the simulation is slower when the garbage rules are added, meaning that it is better to let objects lie in the system in a computer simulation. This has no influence on the number of steps of the computation.

The results of the simulation are gathered in the following Table 1:

Table 1. Simulation time of the P system with/without garbage rules.

$\langle n,m \rangle$	Formula φ	Middle satisfying assignment	Simulation time with garbage rules (seconds)	Simulation time without garbage rules (seconds)
$\langle 3,3 \rangle$	$(x_1 \vee x_2 \vee \neg x_3)$ $\wedge(x_1 \vee \neg x_2 \vee x_3)$ $\wedge(\neg x_1 \vee x_2 \vee x_3)$	$(1,0,1)$	0.990	0.944
$\langle 5,4 \rangle$	$(x_1 \vee x_2 \vee \neg x_5)$ $\wedge(\neg x_2 \vee x_3)$ $\wedge(\neg x_1 \vee x_3 \vee x_4)$ $\wedge(x_1 \vee x_4 \vee \neg x_5)$	$(1,0,0,1,1)$	16.242	15.458
$\langle 6,5 \rangle$	$(x_1 \vee x_2 \vee x_5 \vee x_6)$ $\wedge(\neg x_2 \vee x_3 \vee x_4)$ $\wedge(\neg x_1 \vee \neg x_4 \vee x_6)$ $\wedge(x_1 \vee \neg x_3 \vee \neg x_5)$ $\wedge(x_3 \vee \neg x_6)$	$(1,0,0,0,1,0)$	55.108	53.904
$\langle 7,7 \rangle$	$(x_1 \vee x_2 \vee x_3 \vee x_4)$ $\wedge(x_5 \vee x_6 \vee x_7)$ $\wedge(\neg x_1 \vee x_2 \vee x_5)$ $\wedge(x_3 \vee \neg x_5 \vee \neg x_6)$ $\wedge(x_3 \vee \neg x_4 \vee x_7)$ $\wedge(\neg x_1 \vee \neg x_7)$ $\wedge(\neg x_2 \vee x_6)$	$(0,1,0,0,0,1,1)$	145.825	138.009

6   Conclusion

In this article, we improved the system solving **SAT** from [17] to get a cell-like symport/antiport P system with membrane separation and rules of length at most 3 solving the **MIDSAT** problem. Because this problem is $\mathbf{P}^{(\#\mathbf{P})}$-complete, we obtained the inclusion $\mathbf{P}^{(\#\mathbf{P})} \subseteq \mathbf{PMC}_{\mathbf{CSC}(3)}$.

We noticed that these systems are indeed a specific case of tissue P system, already proved to characterize $\mathbf{P}^{(\#\mathbf{P})}$ in [2]. This leads us to characterize $\mathbf{P}^{(\#\mathbf{P})}$ by the set of all problems solved in polynomial time by a polynomially uniform family of cell-like symport/antiport P systems with membrane separation and rules of length at most 3.

This result implies that the computational powers of cell-like and tissue-like symport/antiport P systems with membrane separation running in polynomial time are the same. Moreover, it is known that when cell-like symport/antiport systems use membrane division instead of membrane separation, they can solve problems beyond **PSPACE** [14]. Then, our characterization reveals the computational power of membrane division against membrane separation (as long as $\mathbf{P}^{(\#\mathbf{P})} \subset \mathbf{PSPACE}$).

Disclosure of Interests. The authors have no competing interests to declare that are relevant to the content of this article.

References

1. Alhazov, A., Ishdorj, T.O.: Membrane operations in P systems with active membranes. In: Păun, G., Riscos-Núñez, A., Romero-Jimenez, A., Sancho-Caparrini, F. (eds.) Proceedings of the Second Brainstorming Week on Membrane Computing, pp. 37–44. University of Sevilla (2004)
2. Leporati, A., Manzoni, L., Mauri, G., Porreca, A.E., Zandron, C.: Characterising the complexity of tissue P systems with fission rules. J. Comput. Syst. Sci. **90**, 115–128 (2017). https://doi.org/10.1016/j.jcss.2017.06.008
3. Macías Ramos, L., Song, B., Song, T., Pan, L., Pérez Jiménez, M.d.J.: Limits on efficient computation in P systems with symport/antiport rules. In: BWMC 2017: 15th Brainstorming Week on Membrane Computing, pp. 147–160 (2017)
4. Macías-Ramos, L., Song, B., Valencia-Cabrera, L., Pan, L., Pérez-Jiménez, M.: Membrane fission: a computational complexity perspective. Complexity **21** (2015). https://doi.org/10.1002/cplx.21691
5. Macías-Ramos, L., Valencia-Cabrera, L., Song, B., Song, T., Pan, L., Pérez-Jiménez, M.: A P-lingua based simulator for P systems with symport/antiport rules. Fundamenta Informaticae **139**, 211–227 (2015). https://doi.org/10.3233/FI-2015-1232
6. Orellana-Martín, D., Martínez-del-Amor, M.A., Valencia-Cabrera, L., Song, B., Pan, L., Pérez-Jiménez, M.J.: P systems with symport/antiport rules: when do the surroundings matter? Theor. Comput. Sci. **805**, 206–217 (2020). https://doi.org/10.1016/j.tcs.2018.04.052
7. Păun, A., Păun, G.: The power of communication: P systems with symport/antiport. New Generation Comput. **20**, 295–306 (2002). https://doi.org/10.1007/BF03037362

8. Păun, G.: Computing with membranes. J. Comput. Syst. Sci. **61**(1), 108–143 (2000). https://doi.org/10.1006/jcss.1999.1693
9. Păun, G.: P systems with active membranes: attacking **NP** complete problems. J. Automata Lang. Comb. **6**, 75–90 (2001)
10. Păun, G.: Introduction to Membrane Computing, pp. 1–42. Springer, Heidelberg (2006). https://doi.org/10.1007/3-540-29937-8_1
11. Păun, G., Rozenberg, G., Salomaa, A.: The Oxford Handbook of Membrane Computing. Oxford University Press Inc., Oxford (2010)
12. Pérez-Hurtado, I., Valencia-Cabrera, L., Pérez-Jiménez, M.J., Colomer, M.A., Riscos-Núñez, A.: Mecosim: a general purpose software tool for simulating biological phenomena by means of P systems. In: IEEE Fifth International Conference on Bio-inpired Computing: Theories and Applications (BIC-TA 2010), vol. I, pp. 637–643 (2010). http://www.cs.us.es/~marper/investigacion/mecosim.pdf
13. Pérez–Jiménez, M.J.: A computational complexity theory in membrane computing. In: Păun, G., Pérez-Jiménez, M.J., Riscos-Núñez, A., Rozenberg, G., Salomaa, A. (eds.) WMC 2009. LNCS, vol. 5957, pp. 125–148. Springer, Heidelberg (2010). https://doi.org/10.1007/978-3-642-11467-0_10
14. Song, B., Pérez-Jiménez, M.J., Pan, L.: Efficient solutions to hard computational problems by P systems with symport/antiport rules and membrane division. Biosystems **130**, 51–58 (2015). https://doi.org/10.1016/j.biosystems.2015.03.002
15. Sosík, P.: P systems attacking hard problems beyond **NP**: a survey. J. Membrane Comput. **1** (2019). https://doi.org/10.1007/s41965-019-00017-y
16. Toda, S.: Simple characterizations of **P**(#**P**) and complete problems. J. Comput. Syst. Sci. **49**(1), 1–17 (1994). https://doi.org/10.1016/S0022-0000(05)80082-0
17. Valencia-Cabrera, L., Song, B., Macías-Ramos, L., Pan, L., Riscos-Núñez, A., Pérez-Jiménez, M.J.: Computational efficiency of P systems with symport/antiport rules and membrane separation. In: Thirteenth Brainstorming Week on Membrane Computing, vol. 1, pp. 325–370 (2015). http://www.cs.us.es/~marper/investigacion/proceedings-13th-BWMC.pdf

Cellular Automata on Spaces
of Probability Measures

Amelia Kunze[1], Enrico Formenti[1(✉)], Faizal Hafiz[2], and Davide La Torre[2]

[1] Université Côte d'Azur, CNRS, i3S, Nice, France
enrico.formenti@univ-cotedazur.fr
[2] SKEMA Business School, Sophia Antipolis, France

Abstract. Classical Cellular Automata (CCAs) are a powerful computational framework widely used to model complex systems driven by local interactions. Their simplicity lies in the use of a finite set of states and a uniform local rule, yet this simplicity leads to rich and diverse dynamical behaviors. CCAs have found applications in numerous scientific fields, including quantum computing, biology, social sciences, and cryptography. However, traditional CCAs assume complete certainty in the state of all cells, which limits their ability to model systems with inherent uncertainty. This paper introduces a novel generalization of CCAs, termed Cellular Automata on spaces of probability Measures (CAMs), which extends the classical framework to incorporate probabilistic uncertainty. In this setting, the state of each cell is described by a probability measure, and the local rule operates on configurations of such probability measures. This study lays the groundwork for future exploration of CAMs, offering a flexible and robust framework for modeling uncertainty in cellular automata and opening new directions for both theoretical analysis and practical applications.

Keywords: Cellular Automata · Spaces of Probability Measures · Convex combination maps

1 Introduction

Classical Cellular Automata (hereafter referred to as CCAs) are extensively used as formal tools to model various complex phenomena arising from local interactions and a finite number of states (for other bio-inspired models [12]). CCAs find applications in numerous scientific domains, including quantum computing, biology, social sciences, and more [1,3,27,29,31]. In computer science, they are particularly valuable in cryptography, where they are employed for tasks such as pseudo-random number generation, secret sharing schemes, and the design of S-boxes [13,14,22,23].

Informally, a CCA consists of an infinite collection of automata (referred to as *cells*), each of which assumes a *state* (or a value from a finite alphabet) drawn from a finite set (the *set of states*). These cells are arranged on a regular

© The Author(s), under exclusive license to Springer Nature Switzerland AG 2026

E. Formenti and L. Manzoni (Eds.): UCNC 2025, LNCS 16364, pp. 179–193, 2026.
https://doi.org/10.1007/978-3-032-15641-9_13

grid (the *lattice*), typically indexed by $\mathbb{Z}^d$, where d denotes the dimensionality of the CCA. Each cell updates its state synchronously based on its own state and the states of neighboring cells (its *neighborhood*), using a uniform *local rule* that applies to all cells. The overall state of the automata at a given moment is referred to as a (instantaneous) configuration.

Despite the simplicity of the CCA model, it exhibits a remarkable diversity of dynamical behaviors. To study these behaviors, the set of configurations is equipped with the product topology (as detailed in the next section). This area is an active research field that has led to numerous publications in international conferences and journals [5,6,8,21,24].

Over time, researchers have proposed several variants of CCAs [4,9,10], some of which have inspired this work. For example, Cattaneo et al. introduced a model of cellular automata within fuzzy frameworks to better analyze the chaotic behavior of CCAs [7]. This foundational work has spurred a series of follow-up studies with practical applications, such as modeling crowd dynamics [15].

In [7], the authors justify their approach by noting that in a configuration only a finite subset of cells has a known state, while the states of the other cells are subject to uncertainty. Our paper proposes an alternative method to incorporate uncertainty into the local state. Specifically, we assume that the state of each cell is described by a probability measure which may vary from cell to cell. By imposing additional constraints on the structure of the local rule, we generalize CCAs to work with configurations of probability measures on the set of states. We term this new model Cellular Automata on probability Measures (CAMs).

This approach generalizes the conventional scenario in CCAs, where each cell is associated with the same measure, and then the product measure is taken. For a comprehensive survey of the results in this classical setting, we refer the reader to [24]. Our paper aims to provide a foundation for much more general scenarios, some of which will be discussed in the last section and developed in a forthcoming paper.

We recall that probabilistic CAs is another way to incorporate uncertainty in CCAs. In this case, time and states are exactly like in CCAs but each cell is updated by randomly choosing (independently from the other cells) the new state accordingly to a distribution (which might depend on some local neighborhood). This is a different approach from ours, which has produced an extensive literature and is also prone to numerous applications. We refer to [21] as a starting point on this subject.

We stress that, for lack of space, in this introduction we cited only a few pointers to the huge existing literature. The interested reader is invited to expand on the subject starting from the references provided.

The paper is structured as follows. The next section recalls the basic definitions about CCAs together with the famous Curtis-Hedlund-Lyndon theorem. Section 3 introduces the space of probability measures and describes some of its interesting properties. Section 4 provides the formal definition of CAMs and some interesting results about their dynamics. In Sect. 5 we study a particular family of CCAs. In the last section we draw our conclusions and propose some perspectives.

For lack of space some proofs have been removed. They will appear in the long version of the paper.

2   Classical Cellular Automata: Definitions and Main Properties

Formally, a *classical cellular automaton* (CCA or simply CA) is a structure $\langle d, r, S, \delta \rangle$ where d is the *dimension*, i.e., the dimension of the regular lattice L (in this work we set either $L = \mathbb{N}^d$ or $L = \mathbb{Z}^d$); r is the *radius*; S is the finite set of *states*; and $f \colon S^{(2r+1)^d} \to S$ is the *local rule*. We denote by $N_r(i)$ the closed ball of center $i \in L$ and radius $r \in \mathbb{N}$ according to the standard Manhattan distance on L. We call $N_r(i)$ the *neighborhood* of the cell i (on L). When we will write, for example, $N_r(i) = (-1, 0, 1)$ it means that we are considering the cells with index $i-1, i, i+1$ in this order. Any element of $x \in S^L$ is called a *configuration* and it represents the overall state of the CA at a given generic time step. For any $i \in L$, denote by $x_i = x(i)$ the element of x contained in the cell $i \in L$. This notation can be extended to (ordered) sequences of elements of L in the obvious way. In particular, $x_{N_r(i)} = x(N_r(i))$ denotes the content of the neighborhood of cell i for the configuration x. Any local rule f induces a *global rule* G_f as follows

$$ G_f(x)_i = f\left(x_{N_r(i)}\right), \quad \forall x \in S^L, \ \forall i \in L. $$

Hence, the global rule illustrates the overall evolution of current configuration of the cellular automaton after one unity of time. A space-time diagram is a convenient way to represent the evolution of a CCA which can help to grasp some initial guess on their behavior. More formally, a space time diagram is a function $ST_f : L \times \mathbb{N} \times S^L \to S$ such that $\forall x \in S^L, \forall (i, t) \in L \times \mathbb{N}, ST_f(i, t, x) = G_f^t(x)_i$.

In order to study CCA as dynamical systems, the set of states S is equipped with the topology induced by the discrete metric $d_D(s, t) = 1$ if $s \neq t, 0$ otherwise for all $s, t \in S$. Then, S^L is endowed with *Cantor topology* i.e. the standard product topology induced by the discrete topology on S. Consider the *Cantor distance* defined as:

$$ d_C(x, y) = \sum_{k \in L} \frac{d_D(x_k, y_k)}{s^{|k|}} $$

where s is the size of S (recall that S is a finite set here). It is well-known that the Cantor distance induces the Cantor topology on S^L.

For a fixed $j \in L$, the *shift map* σ_j is a very well-known and widely studied CA defined as follows $\sigma_j(x)_i = x_{i+j}, \quad \forall x \in S^L, \ \forall i \in L$.

The famous Curtis-Hedlund-Lyndon theorem established that CCAs are exactly those continuous maps commuting with the shift and can be formally stated as follows.

Theorem 1. *[16, Th. 3.4 p. 324] The class of CCAs is exactly the class of continuous mappings $f\colon S^L \to S^L$ such that for all $j \in \mathbb{N}$, $f \circ \sigma^j = \sigma^j \circ f$.*

In Sect. 4 we will provide a new version of this theorem adapted to our context.

3   The Space of Probability Measures

In the following let us suppose that the state space S is endowed with a metric d and that (S, d) is a compact metric space. The space S can be discrete as $\{0, 1\}$ or continuous as $[0, 1]$. Let $\mathcal{B}(S)$ be the Borel sigma-algebra defined on S. Let us denote by $\mathcal{M}(S)$ the set of all probability measures defined on S. $\mathcal{M}(S)$ can be endowed with the Monge-Kantorovich metric defined as:

$$d_{\mathrm{MK}}(\mu, \nu) = \sup_{g \in \mathrm{Lip}_1(S)} \left\{ \int_S g\, d\mu - \int_S g\, d\nu \right\} \tag{1}$$

where $\mathrm{Lip}_1(S)$ is the set of all 1-Lipschitz functions on S, that is

$$\mathrm{Lip}_1(S) = \{g : S \to \mathbb{R} \text{ s. t. } |g(x) - g(y)| \le d(x, y)\}. \tag{2}$$

The Monge-Kantorovich metric arose from a classical problem in transportation of mass [20]. Since then it has been employed in many applications areas, several of which are stochastic in nature and require a means to measure the distance between two probability measures. Indeed, the set of probability measures can be equipped with the Monge-Kantorovich metric in order to yield the topology of weak convergence [11]. A particularly important feature of the Monge-Kantorovich distance is that it is tied to the underlying metric d on the space S. A simple example illustrating this link is provided below.

Proposition 1. *Let us suppose that S is a finite and discrete set. Let d the discrete metric defined on S, and let us consider the set $D \subseteq \mathcal{M}(\mathcal{S})$ composed by all Dirac measures on S, that is $\mathcal{D} = \{\delta_s, s \in S\}$. Then the metric d_{MK} collapses to the discrete metric d on $\mathcal{D}$.*

Proof. Let us consider two points s and t in S and the associated Dirac probability measures $x = \delta_s$ and $y = \delta_t$. Computing $d_{\mathrm{MK}}(x, y)$ yields

$$d_{\mathrm{MK}}(x, y) = \sup_{g \in \mathrm{Lip}_1(S)} g(s) - g(t) = d_D(s, t), \tag{3}$$

that is, the d_{MK} metric collapses to the discrete metric d.   ∎

The following result shows that the metric d_{MK} may be easily computed for real-valued probability measures.

Proposition 2. *[28] Let μ and ν be two probability measures on $S \subset \mathbb{R}$ and denote their cumulative distribution functions by F_μ and F_ν, respectively. Then the metric d_{MK} can be expressed as*

$$d_{MK}(\mu, \nu) = \int_{\mathbb{R}} |F_\mu - F_\nu|\, dx. \tag{4}$$

Note that d_{MK} simplifies even further in the case of Bernoulli probability measures. Let μ be a Bernoulli measure with parameter q and ν be a Bernoulli measure with parameter p (we recall that the value of the parameter is the probability of getting a 1). Then,

$$\begin{aligned}
d_{\mathrm{MK}}(\mu, \nu) &= \int_{\mathbb{R}} |F_\mu - F_\nu|\, dx \\
&= \int_{-\infty}^{0} |0 - 0|\, dx + \int_{0}^{1} |(1 - q) - (1 - p)|\, dx + \int_{1}^{\infty} |1 - 1|\, dx \\
&= |p - q|.
\end{aligned} \tag{5}$$

In words, we obtain the desirable property that the Monge-Kantorovich distance simplifies to the L^1 distance between the parameters p and q.

Remark 1. Let us notice that for $g \in \mathrm{Lip}_1(S)$ and $0 \in S$ it holds

$$\int_S g\, d\mu - \int_S g\, d\nu \leq \int_S (g(0) + d(x, 0))\, d\mu - \int_S (g(0) - d(x, 0))\, d\nu \leq 2\, K \tag{6}$$

where $K = \max_{x \in S} d(x, 0)$. In other words, $d_{\mathrm{MK}}(\mu, \nu) \leq 2K$ for any $\nu, \mu \in \mathcal{M}(S)$, and hence $(\mathcal{M}(S), d_{\mathrm{MK}})$ is a bounded metric space.

Remark 2. From Sect. 4 on, the set of states considered is either $S = \{0, 1\}$ or $S = [0, 1]$. Therefore, the distance d_{MK} is bounded by 1 for the former set (see Eq. (5)) or by 2 for the latter (see the previous remark).

Proposition 3. *[17] If (S, d) is a compact metric space then the metric space $(\mathcal{M}(S), d_{MK})$ is compact and, therefore, complete.*

Let us notice that the space $\mathcal{M}(S)$ is not a group but it is closed with respect to the convex combination of probability measures (see [2,17,19] for more details).

Proposition 4. *[2] Let $\mu_1, \mu_2, ..., \mu_n \in \mathcal{M}(S)$ and $\nu_1, \nu_2, ...\nu_n \in \mathcal{M}(S)$ be two sets of probability measures and $\lambda_1, \lambda_2, ..., \lambda_n$ a set of weights such that $\sum_{j=1}^{n} \lambda_j = 1$. Then, the following inequality holds:*

$$d_{MK}\left(\sum_{j=1}^{n} \lambda_j \mu_j, \sum_{j=1}^{n} \lambda_j \nu_j\right) \leq \sum_{j=1}^{n} \lambda_j d_{MK}(\mu_j, \nu_j) \tag{7}$$

4   Cellular Automata on Probability Measures

A **proto Cellular Automaton on probability Measures** (pCAM) is defined on configurations that have probability measures in their cells. In other words, in this new setting, the standard state set S is replaced with $\mathcal{M}(S)$. More formally, a pCAM is a structure $\langle d, r, \mathcal{M}(S), f \rangle$ where d and r have the same meaning as for CCAs, $\mathcal{M}(S)$ is the set of all probability measures on the finite set of classical states $S = \{0, 1\}$ or on $S = [0, 1]$ and the local function f goes from $\mathcal{M}(S)^{(2r+1)^d}$ to $\mathcal{M}(S)$.

Let us notice that the definition of pCAM collapses to the classical one when the probability measure in the cell i coincides with the Dirac measure on a specific state $s \in S$ (and S is finite). In other words, a string of states for a CCA can be embedded into this new stochastic framework by associating with each cell element $s \in S$ the probability mass at s.

We added the word *proto* to the new class of pCAMs since it is not exactly the generalization of CCAs that we wanted, but a kind of prototype of it that needs further refinements. In order to understand these necessary refinements, we need to introduce a topology for pCAMs (which will be also used later for CAMs).

Definition 1. *For any pair of elements $x, y \in \mathcal{M}(S)^L$, let us define the distance*

$$d_{\mathcal{M}}(x, y) = \sum_{i \in L} \frac{d_{MK}(x_i, y_i)}{2^{|i|}} \tag{8}$$

where $|\cdot|$ is the L^1-norm of i.

We equip $\mathcal{M}(S)^L$ with the topology induced by the $d_{\mathcal{M}}$ distance. It is not difficult to see that this topology coincides with the product topology of $\mathcal{M}(S)$ equipped with the distance d_{MK} [25]. The following simple example illustrates the new framework.

Example 1. Consider $S = \{0, 1\}$ and $\mathcal{M}(S)$, the set of all possible discrete probability measures on S. For any $\mu \in \mathcal{M}(S), \mu$ must be a Bernoulli probability measure, say $B(p)$, and so the outcomes 0 and 1 have probability $1 - p$ and p, respectively. Let us now consider a set of possible configurations in $S^{\mathbb{Z}}$ and, for the sake of simplicity, consider the cells x_{-1}, x_0 and x_1 at time t. We now consider the transition map that associates each triple as above with a new probability measure defined by the convex combinations of x_{-1}, x_0, and x_1 with weights $\lambda_{-1} = 0.2$, $\lambda_0 = 0.4$, and $\lambda_1 = 0.4$, respectively.

Figure 1 provides space-time diagrams illustrating the dynamical behavior of this rule on Bernoulli measures starting from (a) a random initial configuration and (b) a particular configuration in which three central cells contain $p = 0.5$ and all others contain $p = 0.01$.   ∎

The following propositions shed some light on some topological properties of the space $(\mathcal{M}(S)^L, d_{\mathcal{M}})$ that will be useful later on.

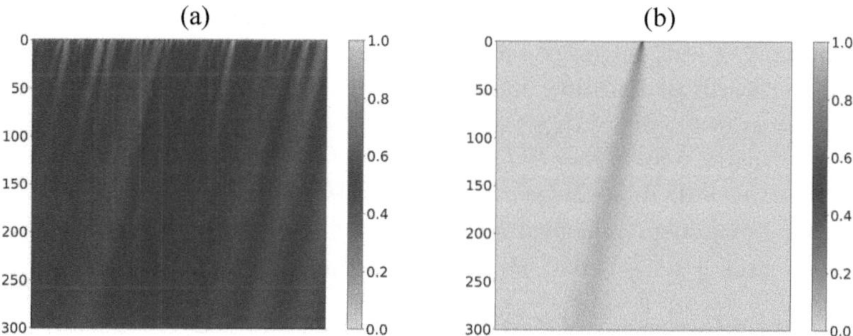

Fig. 1. Some examples of dynamics of CAMs on Bernoulli measures. Here only the value of the parameter p is plotted. (a) and (b) use local rule $f(x_i) = 0.2x_{i-1}+0.4x_i+0.4x_{i+1}$, where (a) has an initial random choice of the parameters and (b) has a specific initial pattern.

Proposition 5. *Suppose that (S, d) is compact. Then, $(\mathcal{M}(S)^L, d_{\mathcal{M}})$ is compact and, therefore, complete.*

Proof. The compactness of S implies that the above function $d_{\mathcal{M}}$ is well-defined and bounded. Furthermore, $d_{\mathcal{M}}$ generates the product topology [25]. The remainder of the proof follows from Tychonoff's theorem which states that any countable product of compact metric spaces is compact in the product topology [18, Th. 13 p. 143]. Completeness follows immediately from compactness. ∎

Proposition 6. $\mathcal{M}(S)^L$ *is a convex space.*

Proof. To show the convexity property, let $\lambda \in [0, 1]$ and $x, y \in \mathcal{M}(\mathcal{S})^L$. By trivial calculations it is easy to prove that $\lambda x + (1 - \lambda)y \in \mathcal{M}(\mathcal{S})^L$. ∎

The following proposition provides a first hint that the behavior of CAs in this new setting is original and cannot be easily deduced by the knowledge acquired about CCAs.

Proposition 7. *Let S be a compact set, and $F : (\mathcal{M}(S)^L, d_{\mathcal{M}}) \to (\mathcal{M}(S)^L, d_{\mathcal{M}})$ be a continuous function. Then F has at least one fixed point.*

Proof. By Propositions 5 and 6, $(\mathcal{M}(S)^L, d_{\mathcal{M}})$ is compact and convex. Since F is continuous, the thesis follows from Schauder's fixed point theorem [26,30]. ∎

From the next proposition one can deduce that the classical theorem of Curtis-Hedlund-Lyndon (Theorem 1) does not hold any more in this new setting.

Proposition 8. *On $(\mathcal{M}(S)^L, d_{\mathcal{M}})$, there exist shift-commuting non-continuous pCAMs.*

Proof. In this proof we consider $L = \mathbb{Z}$, the generalization to $\mathbb{Z}^D$ for $D \in \mathbb{N}$ is straightforward. First, let us consider the case $S = \{0, 1\}$. For $\theta \in (0, 1)$, denote $\mathfrak{B}(\theta)$ the Bernoulli probability measure on $\{0, 1\}$ of parameter θ.

Then, for $p, q \in (0, 1)$, $d_{\mathrm{MK}}(\mathfrak{B}(p), \mathfrak{B}(q)) = |p - q|$. Now, let x (resp., y) be a configuration which has $\mathfrak{B}(p)$ (resp., $\mathfrak{B}(q)$) in each cell of the lattice L. Hence, $d_{\mathcal{M}}(x, y) = 3 \cdot d_{\mathrm{MK}}(\mathfrak{B}(p), \mathfrak{B}(q)) = 3 \cdot |p - q|$ (see Eq. (5)). For $\varepsilon > 0$, let $0 < \delta < \varepsilon < 1$. Choose the parameters p and q such that $|p - q| < \frac{\delta}{3}$ by letting $p = \frac{1}{2} + \frac{\delta}{12}$ and $q = \frac{1}{2} - \frac{\delta}{12}$. Hence, with these values, $d_{\mathrm{MK}}(\mathfrak{B}(p), \mathfrak{B}(q)) = |\frac{1}{2} + \frac{\delta}{12} - \frac{1}{2} + \frac{\delta}{12}| = \frac{\delta}{6}$ and $d_{\mathcal{M}}(x, y) = 3 \cdot \frac{\delta}{6} = \frac{\delta}{2}$. Let us define a local rule $f : \mathcal{M}(S) \to \mathcal{M}(S)$ of radius 0 such that for all $\zeta \in \mathcal{M}(S)$, $f(\zeta) = \zeta'$ where

$$
\zeta' = \begin{cases} \mathfrak{B}(2h) & \text{if } \zeta = \mathfrak{B}(h) \text{ and } h < 1/2, \\ \mathfrak{B}(2h - 1) & \text{if } \zeta = \mathfrak{B}(h) \text{ and } h > 1/2, \\ \zeta & \text{otherwise.} \end{cases}
$$

It is easy to see that G_f is shift-commuting. However, it is not continuous at x. In fact, for how we chose x and y, we have $d_{\mathcal{M}}(x, y) = \frac{\delta}{2} < \delta$ but $d_{\mathcal{M}}(G_f(x), G_f(y)) = 3 \cdot |2p - 1 - 2q| = 3 \cdot |\frac{\delta}{6} - 1 + \frac{\delta}{6}| = 3 - \delta > \varepsilon$.

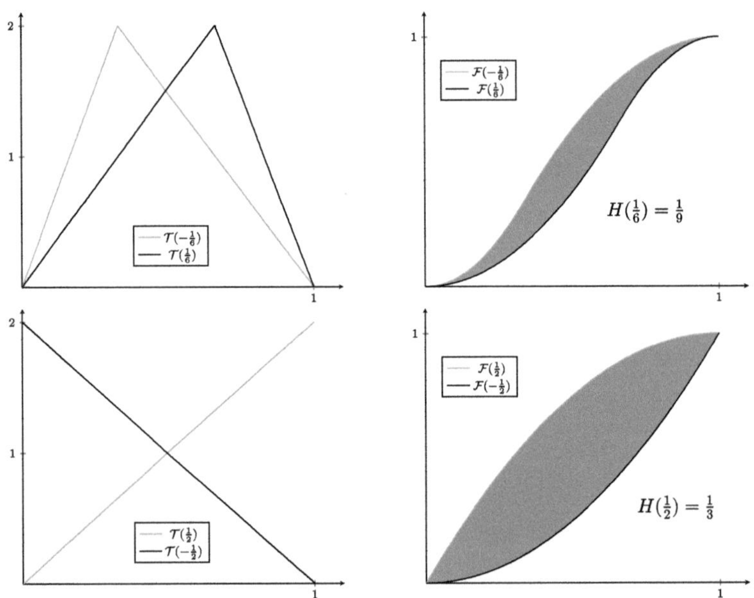

Fig. 2. Some values of the function $H(\alpha)$ used in the proof of Proposition 8: $\alpha = 1/6$ (above) and $\alpha = 1/2$ (below). H is given by the surface of the orange regions. Here, $\tau(\alpha)$ (resp., $\mathcal{F}(\alpha)$) is the density (resp., the cumulative) function of $\mathfrak{T}(\alpha)$.

Now, for the case $S = [0, 1]$. Denote $\mathfrak{T}(\theta)$ the triangle probability distribution with support $[0, 1]$ and parameter $\frac{1}{2} + \theta$ with $\theta \in [-\frac{1}{2}, \frac{1}{2}]$. Let

$H(\alpha) = d_{\mathrm{MK}}(\mathfrak{T}(\alpha), \mathfrak{T}(-\alpha))$ for $\alpha \in [0, \frac{1}{2}]$. Figure 2 illustrates the behavior of the function H for some interesting values of α. It is not difficult to see that H is continuous and monotonically increasing in $[0, \frac{1}{2}]$ with maximum 1 at $\alpha = 1/2$ and minimum 0 at $\alpha = 0$. For $0 < \varepsilon < 1/2$, choose $\delta > 0$ and $\alpha > 0$ such that $3H(\alpha) < \delta < \varepsilon$. Consider a configuration x (resp., y) in which each cell contains $\mathfrak{T}(\alpha)$ (resp., $\mathfrak{T}(-\alpha)$). Then, $d_{\mathcal{M}}(x, y) = 3 \cdot d_{\mathrm{MK}}(\mathfrak{T}(\alpha), \mathfrak{T}(-\alpha)) = 3 \cdot H(\alpha) < \delta$. Similarly to the case $S = \{0, 1\}$, let us define a local rule $f \colon \mathcal{M}(S)^L \to \mathcal{M}(S)^L$ such that for any $\zeta \in \mathcal{M}(S)^L$, $f(\zeta) = \zeta'$ with

$$
\zeta' = \begin{cases} \mathfrak{T}(-1/2) & \text{if } \zeta = \mathfrak{T}(\alpha) \text{ and } \alpha < 0, \\ \mathfrak{T}(1/2) & \text{if } \zeta = \mathfrak{T}(\alpha) \text{ and } \alpha > 0, \\ \zeta & \text{otherwise.} \end{cases}
$$

Since f has radius zero and does not depend on the cell of the lattice (but only on its content), G_f is shift-commuting but we have

$$
\begin{aligned}
d_{\mathcal{M}}(G_f(x), G_f(y)) &= 3 \cdot d_{\mathrm{MK}}(f(\mathfrak{T}(\alpha)), f(\mathfrak{T}(-\alpha))) = \\
&= 3 \cdot d_{\mathrm{MK}}(\mathfrak{T}(1/2), \mathfrak{T}(-1/2)) = \\
&= 3 \cdot H(1/2) = 1 > \varepsilon.
\end{aligned}
$$

∎

The following proposition is essentially due to the compactness of $\mathcal{M}(S)$ and $\mathcal{M}(S)^L$. It is inspired by a similar result in [16].

Proposition 9. *Continuous functions F from $\mathcal{M}(S)^L$ to itself are induced by local functions i.e. for all $i \in L$ there exist a finite ordered subset $N_i \subset L$ and a function $f_i \colon \mathcal{M}(S)^{|N_i|} \to \mathcal{M}(S)$ such that $F(x)_i = f_i(x|_{N_i})$. If F is shift-commuting then f_i is unique i.e. for all $i, j \in L$, $f_i = f_j$.*

Proof. We provide the proof for $L = \mathbb{Z}$, the generalization to higher dimensions is straightforward. Consider an open cover $O_1, O_2, \ldots$ of $\mathcal{M}(S)$. Since $\mathcal{M}(S)^L$ is compact, then we can extract a finite sub-cover $O_1, \ldots, O_k$. For $i \in \mathbb{Z}$ and $j \in \{1, \ldots, k\}$, define $C_{i,j} = \{x \in \mathcal{M}(S)^L \mid x_i \in O_j\}$. Clearly, the $C_{i,j}$ form a sub-base for the product topology on $\mathcal{M}(S)^L$.

Since F is continuous, then for a fixed $i \in \mathbb{Z}$ and for all $j \in \{1, \ldots, k\}$, $F^{-1}(C_{i,j})$ is an open cover of $\mathcal{M}(S)^L$. Any $F^{-1}(C_{i,j})$ is the union of a set of finite intersections of elements of the sub-base. By compactness, this set of finite intersections can be chosen to be finite. So considering all the $F^{-1}(C_{i,j})$ for all possible j, we must find two integers $m_i, n_i \in \mathbb{N}$ such that no set $F^{-1}(C_{i,j})$ depends on an element $C_{i,t}$ for $t < -m_i$ or $t > n_i$. In other words, the value of F at $i \in L$ depends only on a finite number of neighboring cells. Hence, there exists a local function $f_i \colon \mathcal{M}(S)^{m_i + n_i + 1} \to \mathcal{M}(S)$ such that for any $x \in \mathcal{M}(S)^L$ and for this precise $i \in L$, $F(x)_i = f_i(x_{i-m_i}, \ldots, x_i, \ldots, x_{i+n_i})$. If F is shift-commuting i.e. for any $x \in \mathcal{M}(S)^L$, $F(\sigma(x)) = \sigma(F(x))$, then we have

$$
F(\sigma(x))_i = f_i(x_{i-m_i+1}, \ldots, x_{i+n_i+1})
$$

and

$$\sigma(F(x))_i = F(x)_{i+1} = f_{i+1}(x_{i-m_{i+1}+1}, \ldots, x_{i+n_{i+1}+1}).$$

Hence, we must have $f_i = f_{i+1}$, $n_i = n_{i+1}$ and $m_i = m_{i+1}$. ∎

Looking at the obstructions to continuity used in the proof of Proposition 8, one can adopt the following. A function $F\colon \mathcal{M}(S)^L \to \mathcal{M}(S)^L$ is **projection continuous** if for all $i \in L$ the projections $F_i\colon \mathcal{M}(S)^L \to \mathcal{M}(S)$ are continuous w.r.t. $(\mathcal{M}(S), d_{\mathrm{MK}})$. A local function is projection continuous if it induces a projection continuous global function. Projection continuity is the key concept that will allow us to prove a result similar to the classical Curtis-Hedlund-Lyndon theorem for CCAs.

We are ready to provide the desired definition of CAMs.

Definition 2. *A **Cellular Automaton on probability Measures** (CAM) is a pCAM that has a unique projections continuous local rule.*

Theorem 2 (Curtis-Hedlund-Lyndon theorem revisited). *The class of continuous shift-commuting functions from $\mathcal{M}(S)^L$ to itself coincides with the class of CAMs.*

Proof. Since continuity implies projection continuity, then one direction is given by Proposition 9. For the opposite direction, we prove the case $L = \mathbb{Z}$, the generalization to higher dimensions is straightforward.

First of all, we remark that a global rule induced by a local rule always commutes with the shift map whenever this same local rule is applied to all cells of the lattice and it does not depend on the position of the cell. Now, let us prove that if the local rule f is projections continuous and unique, then the global rule G_f is continuous. In fact, for any $x \in \mathcal{M}(S)^L$, fix some $\varepsilon > 0$. Consider some integer $n \in \mathbb{N}$ larger than the radius r of f (the exact value will be made precise later) and let $\varepsilon' = \frac{\varepsilon}{2(3-2^{1-n})}$. Define V_n as $\{-n, -n+1, \ldots, 0, \ldots, n\}$. Since f is projections continuous, there exists $\delta_0 > 0$ such that $y \in B_{\delta_0}(x)$ implies $d_{\mathrm{MK}}(G_f(x)_0, G_f(y)_0) < \varepsilon'$. Similarly, for $i \in V_{n+r}$, by the same argument, we can find a $\delta_i > 0$ such that $y \in B_{\delta_i}(x)$ implies $d_{\mathrm{MK}}(G_f(x)_i, G_f(y)_i) < \varepsilon'$. Let $\bar{\delta} = \min_{i \in V_{n+r}} \delta_i$. Of course, $\bar{\delta}$ is different from 0 since all $\delta_i \neq 0$ and it is well-defined since the set V_{n+r} is finite. For any $y \in B_{\bar{\delta}}(x)$ and for the configuration x we fixed above, we have

$$d_{\mathcal{M}}(G_f(x), G_f(y)) = d_{\mathcal{M}}(G_f(x), G_f(y))|_{\mathbb{Z}\setminus V_n} + d_{\mathcal{M}}(G_f(x), G_f(y))|_{V_n}$$

Let us compute separately the two addends of the right-hand side of the equation above

$$d_{\mathcal{M}}(G_f(x), G_f(y))|_{V_n} = \sum_{i=-n}^{n} \frac{d_{\mathrm{MK}}(G_f(x)_i, G_f(y)_i)}{2^{|i|}} <$$

$$< \varepsilon' \sum_{i=-n}^{n} \frac{1}{2^{|i|}} = \varepsilon' \left[\left(2 \sum_{i=0}^{n} \frac{1}{2^i} \right) - 1 \right] = (3 - 2^{1-n})\varepsilon' = \frac{\varepsilon}{2}.$$

In the computation of the other addend, we can assume that the distance between two cells d_{MK} is bounded by 2 (see Remark 2), and hence

$$d_{\mathcal{M}}(G_f(x), G_f(y))|_{\mathbb{Z}\setminus V_n} \leq 2\left(\sum_{i=-\infty}^{\infty} \frac{1}{2^{|i|}} - \sum_{i=-n}^{n} \frac{1}{2^{|i|}}\right) = 2^{2-n}$$

Now to grant continuity, we must choose an integer n such that $2^{2-n} + \frac{\varepsilon}{2} < \varepsilon$ that is to say $2^{3-n} < \varepsilon$. Clearly, any $n > 3 - \log_2 \epsilon$ is enough to satisfy the last inequality. ∎

The following proposition analyzes the special case of the *shift* map on $\mathcal{M}(S)^L$. By Theorem 2, it easy to see that it is continuous. However, the proposition proves a stronger property, namely that it is Lipschitz.

Proposition 10. *For any $j \in L$, the shift map $\sigma_j(x)_i = x_{i+j}$ is Lipschitz continuous on $(\mathcal{M}(\mathcal{S})^L, d_{\mathcal{M}})$.*

It is easy to prove that the shift map σ_j has an uncountable set of fixed points, namely all constant sequences $x \in \mathcal{M}(S)^L$ such that $x_i = \mu$ for a probability measure $\mu \in \mathcal{M}(S)$. It is not clear if almost all CAMs have an uncountable number of fixed points. The family of CAMs presented in the next section might be a good starting point.

5   The Convex Combination Maps

In this section we explore the properties of a specific family of rules f_C which are defined as follows. For any $i \in L$ and all $j \in N_r(i)$, let $\{\lambda_j\}$ be a set of weights in $[0, 1]$ such that $\sum_{j \in N_r(i)} \lambda_j = 1$. Let us notice that the set of weights λ_j is only dependent on the position of the element j in $N_r(i)$ and it does not depend on i. Let $T : \mathcal{M}(S) \to \mathcal{M}(S)$ be a map defined on the space of probability measures $\mathcal{M}(S)$. For any $x \in \mathcal{M}(S)^L$ the action of f_C on x is defined as

$$(f_C(x))_i = \sum_{j \in N_r(i)} \lambda_j T(x_j) \tag{9}$$

Note that the shift map $\sigma_j(x)_i$ is contained in the set of possible convex combination rules expressible in the form of Eq. (9). To see this, take the identity map $T(x) = x$ and then set $N_r(i) = (j)$ and $\lambda_1 = 1$. We also remark that when T is the identity map, any CCA induced by a convex combination local rule f_C has an uncountable number of fixed points, namely any constant configuration x such for all $i \in L$, $x_i = \mu$ for $\mu \in \mathcal{M}(S)$. In fact, we have

$$\mu = (f_C(x))_i = \sum_{j \in N_r(i)} \lambda_j x_j = \mu \sum_{j \in N_r(i)} \lambda_j = \mu.$$

An interesting consequence of this framework is that the Lipschitz property of the map f_C depends entirely on the Lipschitz property of the map T. The remainder of this section establishes such Lipschitz properties and provides convergence results in the case of contractivity.

190   A. Kunze et al.

Proposition 11. *Suppose that the map T satisfies the following average Lipschitz property: There exists a constant c such that for any configurations $x, y \in \mathcal{M}(S)^L$ and any $i \in L$ it holds:*

$$\sum_{j \in N_r(i)} \lambda_j d_{MK}(T(x_j), T(y_j)) \leq c d_{MK}(x_i, y_i) \tag{10}$$

Then the map $f_C : \mathcal{M}(S)^L \to \mathcal{M}(S)^L$ is a Lipschitz map with respect to $d_{\mathcal{M}}$ with Lipschitz factor c.

Remark 3. Let us notice that the above average Lipschitz property collapses to the classical notion whenever $r = 0$ and $N_r(i) = \{0\}$ for all $i \in L$.

Proposition 12. *Suppose that the map T satisfies the above average Lipschitz property with Lipschitz factor $c \in [0, 1)$. Then the map $f_C : \mathcal{M}(S)^L \to \mathcal{M}(S)^L$ is a contraction with contractivity factor c. Furthermore, there exists a unique invariant configuration $\bar{x}$ such that $f_C(\bar{x}) = \bar{x}$ and, for any configuration x_0, the orbit $f_C^n(x_0)$ converges to $\bar{x}$ whenever $n \to +\infty$.*

Proposition 13. *Suppose that the map T satisfies the following cell-dependent Lipschitz property: There exists a constant c such that for any $i \in L$ the following condition holds:*

$$d_{MK}(T(x_i), T(y_i)) \leq \frac{c}{2^{|i|}} d_{MK}(x_i, y_i) \tag{11}$$

Then, the map $f_C : \mathcal{M}(S)^L \to \mathcal{M}(S)^L$ is Lipschitz with respect to $d_{\mathcal{M}}$ with Lipschitz constant $\sum_{i \in L} \frac{c}{2^{|i|}}$.

Proposition 14. *Suppose that the map T satisfies the above cell-dependent Lipschitz property with constant $\sum_{i \in L} \frac{c}{2^{|i|}} < 1$. Then, the map $f_C : \mathcal{M}(S)^L \to \mathcal{M}(S)^L$ is a contraction with contractivity factor c. Furthermore, there exists a unique invariant configuration $\bar{x}$ such that $f_C(\bar{x}) = \bar{x}$ and, for any configuration x_0, the orbit $f_C^n(x_0)$ converges to $\bar{x}$ whenever $n \to +\infty$ in the $d_{\mathcal{M}}$ metric.*

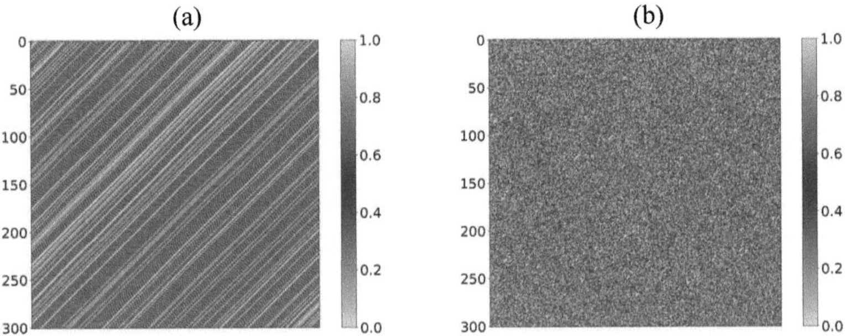

Fig. 3. Some examples of dynamics of CAMs on Bernoulli measures. Here only the value of the parameter p is plotted. (a) uses $f(x_i) = 2x_{i+1} \mod 1$ and (b) uses $f(x_i) = 2x_i + 3x_{i+1} \mod 1$, both having random initial configurations.

6   Conclusions

We have extended the notion of CCA to work on spaces of probability measures. This extension is motivated by the idea of introducing uncertainty in cellular automata by acting on the status of each cell rather than on the transition map. This is the main difference between our approach and those currently existing in the literature. We have proved a characterization of CAMs in terms of properties that allows us to obtain a result similar to the Curtis-Hedlund-Lyndon theorem and proved that all these properties are necessary. We also showed that in this new setting any CAM has at least a fixed point. And, from there we started by analyzing a family of CAMs which admit either one or uncountably many fixed points. Therefore, a first interesting future research direction consists in trying to find which property controls the number of fixed points for generic CAMs.

Our work is just a first exploration in the domain of CAMs, it would be very interesting to study and characterize interesting dynamical behaviors (stability, sensitivity to initial conditions, *etc..*). Our initial study has shown that convex combination rules with sufficiently nice maps T generate regular dynamics on CAMs. Different local rules can generate complex dynamics on CAMs, as shown in Fig. 3. In Fig. 3 (a) a periodic pattern emerges from an initially random configuration on Bernoulli measures, while Fig. 3 (b) maintains 'noisy' configurations. Similarly, the effects of the map T in the convex combination rule on the behavior of the CAM should be studied.

We are actively working in concrete applications of the theory developed so far. A first application develops a connection between random graphs and CAMs. Recall that a random graph can be described by an adjacency matrix in which the entry $x_{ij} \in [0, 1]$ indicates the probability of an edge between the nodes i and j. A particular realization of the random graph is obtained by sampling edges from the corresponding Bernoulli distributions. In this light, the adjacency matrix can be viewed as a two-dimensional CAM on Bernoulli probabilities. Applying the local rule will then generate dynamics on the random graph. If we choose S different from the Bernoulli case, for example $S = \{0, 1, 2, 3, 4\}$ or $S = [0, 1]$, then $\mathcal{M}(S)$ describes the probability of realizing a *weighted* edge where the possible weights correspond to all values $s \in S$.

The CAM framework can be extended to simulate classes of neural networks using CAMs. Recall that an ideal neural network is characterized by a single output neuron and infinitely many input neurons. In our setting, this structure can be represented by a CAM with lattice $\mathbb{Z}^1$ and $S = [0, 1]$. The entry in cell i is a probability measure over the set of possible weights $s \in S$ linking the ith input neuron to the output neuron. Once again, application of the local rule produces dynamics on the ideal neural network. Since the weights of the network are also updated during training, different CAMs can simulate this process. Networks with multiple output neurons can be described by two-dimensional CAMs, and networks with multiple layers can be described by higher-dimensional CAMs.

Another future avenue includes the analysis of inverse problems for CAMs since they can be used in applications. An inverse problem consists of constructing a CAM which approximates the evolution of an observed real-world system.

A possible class of phenomena are those described by diffusion models; for example, we may wish to describe the spread of disease in a geographical area using the CAM framework. Given data describing the evolution of the system of interest, we then infer the local rule of the CAM that best describes the observed trajectory. Specifically, inferring the local rule amounts to solving a multi-criteria optimization problem. We have already obtained successful results in this direction. Early experiments infer not only the form of the local rule f, but also the size and structure of the neighborhood with good accuracy. Our study relating to inverse problems has also been submitted to this conference.

Acknowledgements. This study was partially supported by the HORIZON-MSCA-2022-SE-01 project 101131549 "Application-driven Challenges for Automata Networks and Complex Systems (ACANCOS)" and the ANR project ANR-24-CE48-0335-01 "Ordinal Time Computations" (OTC). Moreover, we acknowledge the support from OTC project of RISE Academy of Université Côte d'Azur.

Disclosure of Interests. The authors have no competing interests to declare that are relevant to the content of this article.

References

1. Arrighi, P.: An overview of quantum cellular automata. Nat. Comput. **18**(4), 885–899 (2019)
2. Barnsley, M.F.: Fractals Everywhere. Academic Press (2014)
3. Cattaneo, G., Dennunzio, A., Farina, F.: A full cellular automaton to simulate predator-prey systems. In: El Yacoubi, S., Chopard, B., Bandini, S. (eds.) ACRI 2006. LNCS, vol. 4173, pp. 446–451. Springer, Heidelberg (2006). https://doi.org/10.1007/11861201_52
4. Cattaneo, G., Dennunzio, A., Formenti, E., Provillard, J.: Non-uniform cellular automata. In: Dediu, A.H., Ionescu, A.M., Martín-Vide, C. (eds.) LATA 2009. LNCS, vol. 5457, pp. 302–313. Springer, Heidelberg (2009). https://doi.org/10.1007/978-3-642-00982-2_26
5. Cattaneo, G., Dennunzio, A., Margara, L.: Chaotic subshifts and related languages applications to one-dimensional cellular automata. Fundam. Informaticae **52**(1–3), 39–80 (2002)
6. Cattaneo, G., Dennunzio, A., Margara, L.: Solution of some conjectures about topological properties of linear cellular automata. Theor. Comput. Sci. **325**(2), 249–271 (2004)
7. Cattaneo, G., Flocchini, P., Mauri, G., Vogliotti, C.Q., Santoro, N.: Cellular automata in fuzzy backgrounds. Physica D **105**(1), 105–120 (1997)
8. Dennunzio, A.: From one-dimensional to two-dimensional cellular automata. Fund. Inform. **115**(1), 87–105 (2012)
9. Dennunzio, A., Formenti, E., Manzoni, L.: Computing issues of asynchronous CA. Fund. Inform. **120**(2), 165–180 (2012)
10. Dennunzio, A., Guillon, P., Masson, B.: Sand automata as cellular automata. Theor. Comput. Sci. **410**(38–40), 3962–3974 (2009)
11. Durrett, R.: Probability metrics and the stability of stochastic models. SIAM Rev. **34**(1), 147 (1992)

12. Ehrenfeucht, A., Rozenberg, G.: Reaction systems. Fundam. Informaticae **75**(1–4), 263–280 (2007)

13. Farina, F., Dennunzio, A.: A predator-prey cellular automaton with parasitic interactions and environmental effects. Fund. Inform. **83**(4), 337–353 (2008)

14. Formenti, E., Imai, K., Martin, B., Yunès, J.-B.: Advances on random sequence generation by uniform cellular automata. In: Calude, C.S., Freivalds, R., Kazuo, I. (eds.) Computing with New Resources. LNCS, vol. 8808, pp. 56–70. Springer, Cham (2014). https://doi.org/10.1007/978-3-319-13350-8_5

15. Gerakakis, I., Gavriilidis, P., Dourvas, N.I., Georgoudas, I.G., Trunfio, G.A., Sirakoulis, G.C.: Accelerating fuzzy cellular automata for modeling crowd dynamics. J. Comput. Sci. **32**, 125–140 (2019)

16. Hedlund, G.A.: Endomorphisms and automorphisms of the shift dynamical system. Math. Syst. Theory **3**(4), 320–375 (1969)

17. Hutchinson, J.E.: Fractals and self similarity. Indiana Univ. Math. J. **30**(5), 713–747 (1981)

18. Kelley, J.L.: General Topology. Graduate Texts in Mathematics. Springer, Cham (1975)

19. Kunze, H.E., La Torre, D., Mendivil, F., Vrscay, E.R.: Fractal-Based Methods in Analysis. Springer, Cham (2011)

20. La Torre, D., Mendivil, F.: The Monge–Kantorovich metric on multimeasures and self–similar multimeasures. Set-Valued Variational Anal. **23** (2015)

21. Louis, P.Y., Nardi, F.R. (eds.): Probabilistic Cellular Automata, Emergence, Complexity and Computation, vol. 27. Springer, Cham (2018)

22. Mariot, L., Gadouleau, M., Formenti, E., Leporati, A.: Mutually orthogonal Latin squares based on cellular automata. Des. Codes Cryptogr. **88**(2), 391–411 (2020)

23. Picek, S., Mariot, L., Yang, B., Jakobovic, D., Mentens, N.: Design of s-boxes defined with cellular automata rules. In: Proceedings of the Computing Frontiers Conference, CF 217, Siena, Italy, 15–17 May 2017, pp. 409–414. ACM (2017)

24. Pivato, M.: The ergodic theory of cellular automata. Int. J. Gen. Syst. **41**(6), 583–594 (2012)

25. Sagar, G., Ravi, D.: Compactness of any countable product of compact metric spaces in product topology without using Tychonoff's theorem. arXiv preprint arXiv:2111.02904 (2021)

26. Schauder, J.: Der fixpunktsatz in funktionalraümen. Stud. Math. **2**(1), 171–180 (1930)

27. Sipahi, R., Zupanc, G.K.: Stochastic cellular automata model of neurosphere growth: roles of proliferative potential, contact inhibition, cell death, and phagocytosis. J. Theor. Biol. **445**, 151–165 (2018)

28. Vallender, S.S.: Calculation of the Wasserstein distance between probability distributions on the line. Theory Probab. Appl. **18**(4), 784–786 (1974)

29. Van Scoy, G.K., George, E.L., Opoku Asantewaa, F., Kerns, L., Saunders, M.M., Prieto-Langarica, A.: A cellular automata model of bone formation. Math. Biosci. **286**, 58–64 (2017)

30. Zeidler, E.: Applied Functional Analysis: Applications to Mathematical Physics, vol. 108. Springer, Cham (2012)

31. Zhao, R., Zhai, Y., Qu, L., Wang, R., Huang, Y., Dong, Q.: A continuous floor field cellular automata model with interaction area for crowd evacuation. XXPhys. A **575**, 126049 (2021)

The Morita Gate is Universal

Matthew Cook[1,2(✉)] and Ethan Palmiere[2]

[1] University of Groningen, Groningen, The Netherlands
cook@ini.phys.ethz.ch
[2] Institute of Neuroinformatics University of Zurich and ETH Zurich,
Zurich, Switzerland

Abstract. We consider 2-state reversible gates that route tokens from input wires to output wires. This can be thought of as a model that reduces computation to just flow control, with wires representing execution paths, and the token representing the current execution point. Memory and computation are strictly co-localized, in the sense that the state of a gate is only changeable or observable by a token that passes through it. Also known as Reversible Logic Elements with Memory (RLEMs), such gates have been studied extensively by Kenichi Morita and his co-authors. Through a series of surprising results, they have shown that *all* non-degenerate 2-state RLEMs with 3 or more inputs are universal, meaning any of them can be used to build any other RLEM of any size and number of states. They also proved that all *smaller* (2 or fewer inputs) 2-state RLEMs are *not* universal, with the exception of one gate that defied analysis. This one remaining gate, known as 2-17, was conjectured to also be non-universal, but this has remained an open question since 2012. Here we resolve this open question by showing that this extremely simple gate is in fact universal. This makes it the smallest universal reversible gate, and we name it the Morita gate in honor of Morita's extensive foundational work in this area.

1 What is the Morita Gate?

In this paper we look at computation performed by circuits of stateful routing elements, which we refer to as gates. These routing elements are reversible finite automata that receive tokens on input lines and route them to output lines. Our primary interest is in the Morita gate, shown in Fig. 1.

In Fig. 1(A), the two input lines are labeled "A" and "B", and the two output lines are labeled "S" and "T". In state 0, the Morita gate routes tokens from A to S and from B to T, while in state 1 this mapping is flipped, so A routes to T and B routes to S. The Morita gate flips its state every time a token passes through it.

In Fig. 1(B), we see a device that implements a Morita gate using a rotor whose position gets toggled when a marble passes through it. For example, in state 0, if a marble comes in at A, it will hit the rotor, moving it slightly counterclockwise, and exit at S, leaving the rotor in state 1. If another marble then comes in at A, the rotor will now send it to the right, so the marble will exit at

© The Author(s), under exclusive license to Springer Nature Switzerland AG 2026
E. Formenti and L. Manzoni (Eds.): UCNC 2025, LNCS 16364, pp. 194–213, 2026.
https://doi.org/10.1007/978-3-032-15641-9_14

T, with the rotor having been pushed back to the state 0 position. We can see that this routing behavior from input lines to output lines is exactly the same as that shown in Fig. 1(A).

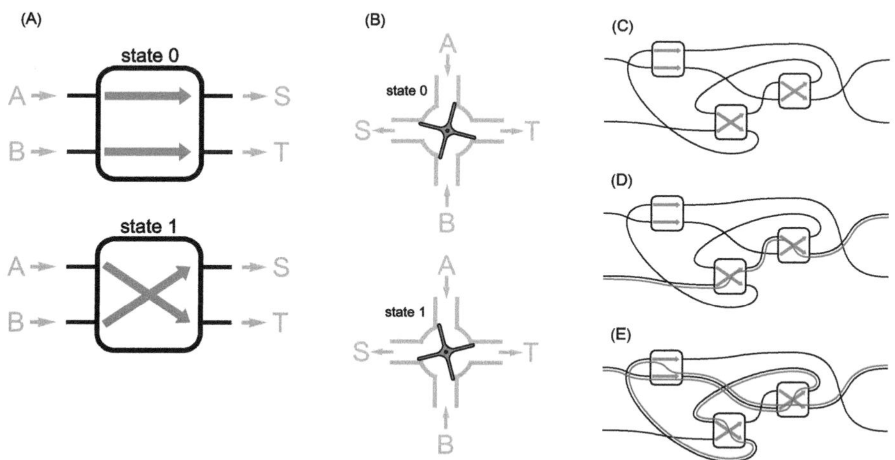

Fig. 1. (A) The two states of a Morita gate. Morita gates alternate between their two states, sending the token straight through (state 0), or switching the token to the other side (state 1). The state changes each time the token passes through. **(B) Another view of a Morita gate.** If we think of the gate as routing a marble, a Morita gate can take a simple physical form. Langton's Ant [1] corresponds to inserting a marble into an infinite grid of these devices, all starting in state 0. **(C) An example of a circuit of Morita gates.** A token can enter the circuit on the left, and exits on the right. **(D) An example path through the circuit.** If the token enters this circuit on the lower input wire, it will take the path shown, exiting on the upper output wire. After the token has passed, all three gates will be in state 0. **(E) Another path example.** If the token instead enters the circuit on the upper input wire, then it travels around a bit before exiting on the upper output wire. Note that when it visits a gate for the second time, the gate is in the opposite state. After the token has gone through the circuit, two of the three gates have been visited twice and thus been put back into their original state.

In Fig. 1(C), we see a little circuit of Morita gates. In such circuits, a single token is fed into one of the input lines of the circuit, and it travels around until it exits on an output line. In this example, there are two input lines, shown on the left. There are also two output lines, shown on the right.

Such circuits, being reversible, do not have fan-in or fan-out. A wire's starting and ending points are not shared with any other wires: each wire goes directly from a unique starting point to a unique endpoint without any merging or branching. The wires provide a bijective mapping from the circuit inputs and gate outputs to the circuit outputs and gate inputs. When not at a gate, the token follows this bijective mapping.

In Fig. 1(D), we see the path that a token would take if it arrives on the lower input wire. Each of the two gates it comes to is in state 1, and thus routes it to the opposite side, leading the token to exit on the upper output wire. Both of the gates get flipped from state 1 to state 0 as the token passes through it.

In Fig. 1(E), we see the path that the token would take if it instead arrives on the upper input wire. This path is more contorted than the previous one, and some gates are visited twice, but the token eventually exits on the upper output wire. At the first gate, which is in state 0, the token is routed straight through, flipping that gate into state 1. When the token comes back to that gate later in its journey, it gets routed according to the state 1 mapping, and the gate gets flipped again, back to state 0. We only consider one token at a time in a circuit.

We can see that the Morita gate is reversible. For example, if a token is exiting on wire T, and the gate is in state 1, then we know the token entered on wire B with the gate in state 0. This is easiest to see in the form shown in Fig. 1(B), where watching the operation in reverse looks just the same as watching it forwards. At any time during the marble's travels we can reverse the marble and start rolling it backwards, and it will trace out the entire path that it took through the circuit, in reverse, including repeated uses of gates and wires (marble tracks).

In the form shown in Fig. 1(A), however, the arrows show how the *next* token will be routed, not how the *previous* token was routed. So when considering the activity in reverse, crossed arrows unintuitively mean that a token going in reverse will go straight through (for example from S to A), and parallel arrows mean the token going in reverse will cross to the other side! One way to think of this is that the gate switches state right when the token crosses the *output* side of the gate. So going forwards, the state switches as the token leaves the gate, after the token has followed its arrow, but when going backwards, the state switches as the token enters the gate, before the token follows the arrow in reverse.

We will not usually be considering the token going in reverse, and we will use arrow markings as shown in Fig. 1(A) to indicate the state. Even though we almost never explicitly run the system in reverse, the mere fact that the system is reversible will imply many other properties, as we will see below.

1.1   Contents of this Paper

The bulk of this paper is Sects. 2 and 3.

In Sect. 2, we introduce several RLEMs of interest, and recall Morita's results demonstrating universality of all non-degenerate gates with 3 or more inputs, and non-universality of all gates with 2 or fewer inputs (except for 2-17). Gate 2-17, which we refer to as the Morita gate, was the only gate not known to be universal or non-universal. It was conjectured to be non-universal, like the other gates of its size. The results mentioned in Sect. 2 all come from the many papers of Morita et al.

In Sect. 3, we present the new material, consisting of gadgets that show how the Morita gate can be used. We go on to show that the Morita gate is in fact universal, meaning that for any functionality that any circuit of reversible gates can provide, the same functionality can be provided by a circuit of Morita gates. This makes the Morita gate the simplest universal reversible routing element, and the only one with just two states and two inputs.

2   Reversible Logic Elements with Memory (RLEMs)

This section introduces a number of relevant RLEMs, and concludes by recalling a few of Morita's results that relate to the hierarchy given in Sect. 4.

2.1   RLEM Definition

A reversible logic element with memory (RLEM) has n input wires and n output wires, and has m states, for some $n \in \mathbb{N}^{0+}$ and $m \in \mathbb{N}^{+}$.

In this paper, when we refer to a "gate", we are talking about an RLEM. Note that an RLEM is a completely different kind of gate from a digital logic gate. Digital logic circuits assign Boolean values to wires, whereas the wires in RLEM circuits are simply paths that a token may travel on. We can see that RLEM gates are in fact slightly closer to the original meaning of how gates have been used for millennia to route people or animals.

When a token arrives on one of the n input wires, then, according to its current state, the RLEM routes the token to one of its n output wires, and perhaps switches to a new state. Its behavior is thus defined by its routing function, which maps $\{1..m\} \times \{1..n\} \mapsto \{1..m\} \times \{1..n\}$. Since the RLEM is reversible, this mapping must be a bijection. (This is why the number of outputs must equal the number of inputs). Furthermore, since a circuit only contains a single token travelling around (analogous to an execution path), we do not need to consider the case of multiple tokens arriving simultaneously.

It is also possible to consider *non-reversible* logic elements with memory, such as the well-known balancer [2–4], but we will not consider such elements in this paper.

Following convention, we refer to a gate's inputs as $A, B, \ldots$, and we refer to its outputs as $S, T, \ldots$. Occasionally, we will refer to all the input wires and output wires of a gate as the *legs* of the gate. For the names of the m states, we use the numbers from 0 to $m - 1$. In this paper we will usually be considering 2-state RLEMs, which have states 0 and 1 (Fig. 2).

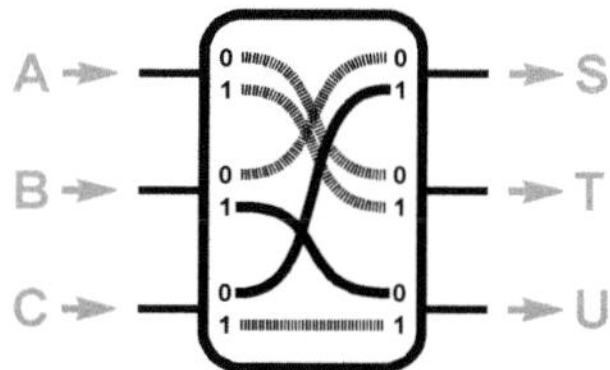

Fig. 2. An example gate with 3 inputs and 2 states. An arriving token will be routed depending on the state of the gate (0 or 1) and on the input where it arrives (A, B, or C). The lines drawn inside the gate, which are fixed and define the gate, show which output the token will be routed to, as well as what state the gate will be in afterwards. Following convention, transitions that flip the state of the gate are shown as solid lines, while transitions that leave the gate in the same state are shown as dotted lines. Thus lines are drawn dotted iff the endpoint states match.

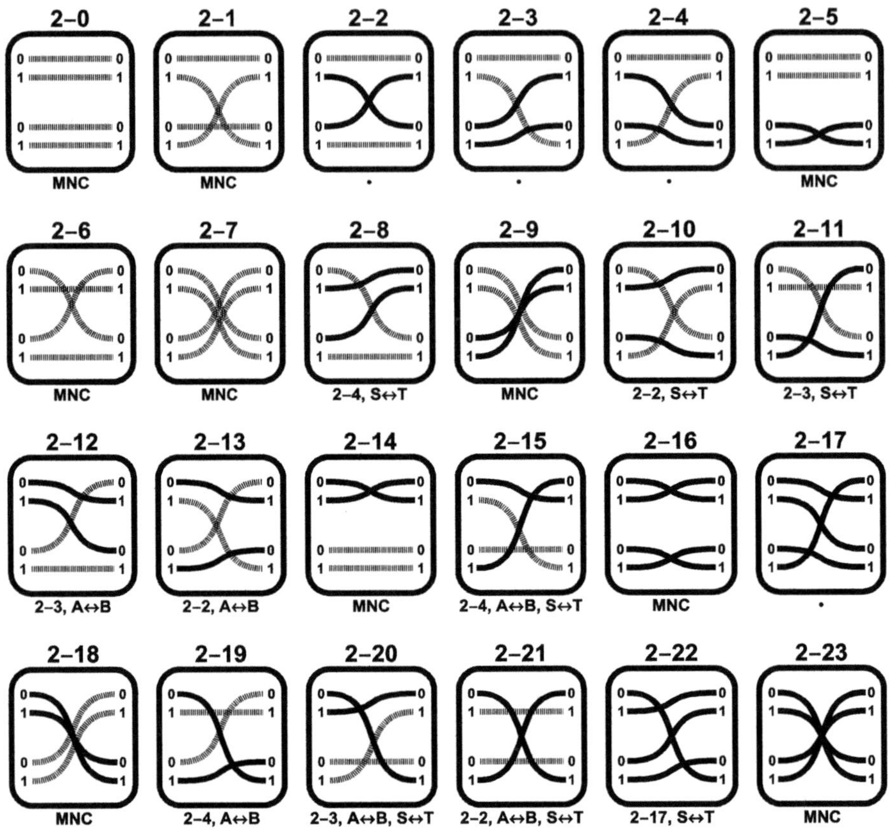

Fig. 3. All twenty-four RLEMs having two states and two inputs. Each is labeled on top with its numeric code, while the bottom text indicates equivalences. Transitions are shown with curved lines, which for additional visual clarity are drawn solid if the state changes (0→1 or 1→0) and dotted if the state stays the same (0→0 or 1→1). The pre-transition state is shown at the left end of the line and the post-transition state is shown at the right end. "MNC" stands for "mapping never changes", meaning that tokens from any given input always get routed to the same output, either because the RLEM cannot change its state (all lines are dotted: 2-0, 2-1, 2-6, and 2-7), and/or because the routing is the same regardless of the state (2-0, 2-5, 2-7, 2-9, 2-14, 2-16, 2-18, and 2-23). If the mapping never changes, then the functionality of the RLEM is trivial, and it can be replaced by just wires. The four basic nontrivial gates (2-2, 2-3, 2-4, and 2-17, discussed in Sect. 2.2) are indicated by a "·". The other gates are equivalent to one of these four basic gates as indicated by the text below the gate, simply by swapping the input wires (A↔B) and/or the output wires (S↔T). Some equivalences are obtainable in more than one way (perhaps also by swapping the state names 0↔1), but just one way is shown. Note that in 2-9, 2-18, and 2-23, we see solid lines cross at a steep angle. Visually it might not be clear whether they cross or not, but they necessarily cross, since solid lines always change the state, from 0→1 and 1→0. If the lines do not cross, then they are dotted and clearly do not cross, as can be seen in 2-7, 2-9, or 2-18.

A two-state, two-input RLEM is therefore given by a bijective mapping between the sets $\{0A, 0B, 1A, 1B\} \mapsto \{0S, 0T, 1S, 1T\}$. There are $4! = 24$ such mappings, and if we list them in lexicographic order as in Fig. 3 then we get their standard [5] numeric names, from 2-0 to 2-23. The "2" prefix in the names indicates that there are two inputs.

For completeness, an RLEM with m states and n inputs has a numeric identifier of the form m-n-i, where i indicates which of the $(mn)!$ gates (having m states and n inputs) it is [5]. However, we will only use numeric identifiers for 2-state RLEMs in this paper, so since m always equals 2, we will just use the simplified numeric identifier n-i.

When we speak of a gate receiving an input, we think of "the input" as being the information of which of the n input wires the token arrived on. Similarly, when we say that a gate generates an output, it refers to which of the n output wires the gate routes the token to. So if we say that a gate receives input B and produces output S, we mean that a token arrives on leg B and the gate routes it to leg S.[1]

Due to reversibility, it is not possible for an RLEM to get stuck in a subset of states by making a "bad choice" transition. If some sequence of inputs takes the RLEM from state s_1 to state s_2, then there must also be a sequence of inputs that takes the RLEM back from state s_2 to s_1, as can be seen from a simple pigeonhole argument due to the bijective mapping.

More generally, any partitioning of the state transition graph (the multigraph having states as vertices and directed edges each labeled with an input leg name at its source and an output leg name at its destination) will necessarily consist of parts that have as many transitions entering the part as exiting, since the graph is regular: Each state vertex has one outgoing edge for each of the gate's input legs, and one incoming edge for each output leg.

However, it is possible for the state transition graph to be disconnected, in which case only a subset of the states will be reachable. Gates 2-0, 2-1, 2-6, and 2-7 are of this form, for example, because they cannot switch between state 0 and state 1. Any use of such a gate is equivalent to a simpler gate with fewer states, consisting of only the part of the state transition graph that contains the current state.

2.2   Getting to Know Some Key Gates

Any gate with fewer than two states (i.e., with just one state), or with fewer than two inputs (i.e., with just zero or one inputs), cannot provide any variation in how the token is routed. Thus, all such gates are trivial and can be replaced

---

[1] With this perspective on inputs and outputs, an RLEM is similar to a Mealy finite state machine. However, in contrast to the intuition behind Mealy machines, for an RLEM we see that inputs arrive from different sources, and the input symbol corresponds to the source sending the token, rather than being a symbol chosen by the source. Outputs are likewise sent to different destinations, with the output symbol indicating the destination the token is routed to, rather than being a symbol readable by the destination. The input alphabet and output alphabet are thus disjoint sets of wires whose cardinality must be identical due to reversibility, and the token carries no information beyond its presence. .

by just wires. However, with two states and two inputs, interesting behavior becomes possible. All of the two-state, two-input gates are shown in Fig. 3, and there are (up to symmetries) four nontrivial gates among them.

The 2-2 (Delay) Gate. The input to a 2-2 does not immediately affect its output, which is determined solely by the state. In state 0, the output will be S, and in state 1 the output will be T, regardless of where the input arrives. However, the input determines the new state: An input on A sets the state to 0, while an input on B sets the state to 1. Thus, the input determines the output that will be produced the *next* time the gate is used. Since the output on the nth use is given by the input on the $n-1$st use, the output sequence exactly matches the input sequence except that it is delayed by one usage step, and so we call this a delay gate. Note that there is no temporal delay during any individual use.

The 2-3 (Integrator) Gate. Like the 2-2, the output of a 2-3 is determined by the state, not by the current input. An input at A leaves the state unchanged, while an input at B toggles the state. Thus, we can think of the state (and therefore the output) as counting the number of previous B inputs, mod 2. If we think of A as 0 and B as 1, we see the gate is keeping track of the total sum so far, mod 2.

It might seem unintuitive that the output is the *previous* sum (prior to adding the current input) rather than the new sum, but we can see that this is required for reversibility: If both the state and the output indicate only the new sum, then in reverse there would be no way to recover the input or previous sum. For the correct intuition, think of *"output, state"* as the sequence of the two latest sums.

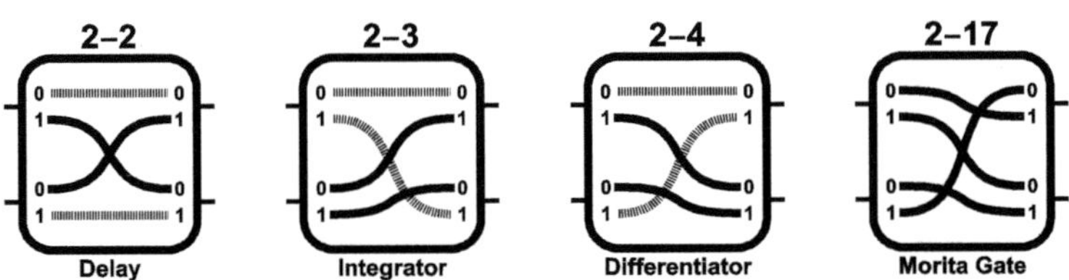

Fig. 4. The four nontrivial gates on two inputs. All other two-input gates are either trivial (equivalent to just wires), or are equivalent to one of these four gates, as indicated in Fig. 3. Any gate with fewer than two inputs or fewer than two states is clearly trivial, so these four gates are the smallest nontrivial gates. The first three gates will be shown to be non-universal in Sect. 2.3, and the Morita gate will be shown to be universal in Sect. 3.4.

The 2-4 (Differentiator) Gate. The 2-4 is a time-reversed 2-3, as can be seen from their transitions as shown in Fig. 4, which are mirror images of each other. The token is output on S whenever the current input matches the previous input, and on T when the current input differs from the previous input. Thus the output is the derivative (mod 2) of the input. The state simply stores the most recent input.

The 2-17 (Morita) Gate. This gate sends the token "straight through" (A→S and B→T) in state 0, and "crossed" (A→T and B→S) in state 1. It changes state every time it is used, so it just alternates between "straight through" and "crossed" with every use. This gate is also the subject of Fig. 1, and will be the main topic of discussion in Sect. 3.

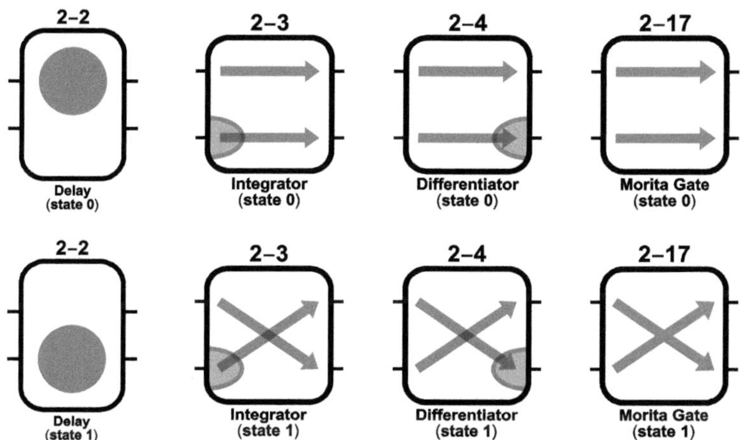

Fig. 5. The four nontrivial gates on two inputs (same as Fig. 4), shown in an iconic form in states 0 (top row) and 1 (bottom row). The **Delay** gate's icon shows a token that is "stuck" in the gate, either on the upper path ($A \rightarrow S$) or on the lower path ($B \rightarrow T$). When the next token arrives (on either the upper or the lower path), it will knock this token out, and the incoming token will get stuck. Thinking of it like this, in terms of tokens stuck on straight paths, we see that tokens never change paths. The **Integrator** and **Differentiator** are shown with bubble notation, explained in the text. The state changes when a token passes through the bubble. The **Morita** gate shows its state with arrows that indicate how the next token will pass through the gate, as in Fig. 1(A).

Bubble Notation. Another way to think of the 2-3 is shown in Fig. 5. The arrows indicate that in state 0 the token goes straight through the gate, while in state 1 the token crosses vertically to the other side. If the token passes through the "bubble" in the lower left, then the gate flips to the other state *while the token is in the bubble.* For example, in state 0, if the token enters on the lower input wire, then it immediately hits the bubble, switching the gate to state 1, with crossed arrows. As the token continues past the bubble, it follows the now-crossed arrow and exits on the upper output wire.

The 2-4 is shown similarly, but with the bubble on the lower right, indicating that the state will only change *after* the token has travelled along an arrow ending at the lower right.

Larger Gates. The number of possible gates grows very quickly with the number of inputs. We saw that there are 24 gates with 2 inputs, reducing to only 4 nontrivial gates after accounting for symmetries. For 3 inputs, there are $6! = 720$ gates, reducing to 14 new gates. For 4 inputs, there are $8! = 40320$ gates, reducing to 55 new gates. For the purposes of this paper, we do not need to be familiar with all of these gates. Here we will focus only on the 3-input **Shifter** and the 4-input **Rotating Element**.

The 3-10 (Shifter) Gate. Figure 6 shows the Shifter. The Shifter is a three-input gate that shifts the token to one side or the other, depending on the state. In state 0, it shifts the token upwards, and in state 1, it shifts the token downwards. If the token is already at the edge of the gate and cannot shift further to that side, then it just goes straight, and the gate changes to the other state. In Fig. 6(B) we see the Shifter in its iconic form, where a diagonal line indicates the direction the token will be shifted in.

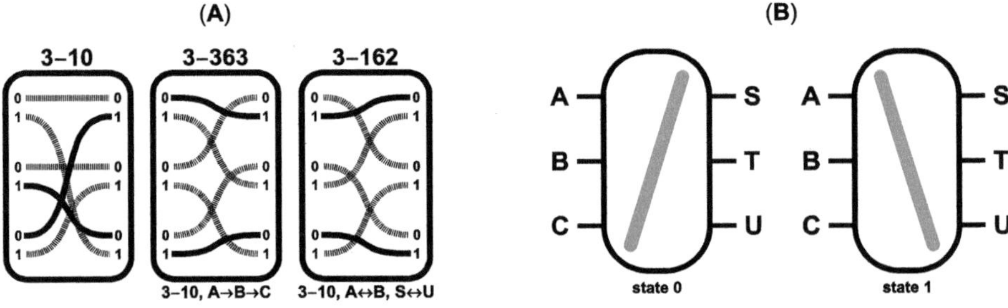

Fig. 6. The Shifter. **(A)** These are three of the 36 equivalent forms of the Shifter. In general we think of the Shifter as being in the 3-363 form. The mirror symmetry of the 3-363 form and the 3-162 form shows that this gate is equivalent to its time-reversed version. Panel B shows an iconic notation for the 3-363 form. **(B)** We draw a Shifter like this to indicate the direction of the shift. In the first form (state 0), tokens are shifted up: B→S and C→T, but a token at A cannot be shifted up, so it goes straight through to S, changing the state to the second form. In the second form (state 1), tokens are shifted down: A→T and B→U, but a token at C cannot be shifted farther down, so it goes straight through to U, toggling the gate back to the first form.

The 4-289 (RE) (Rotating Element) Gate. The Rotating Element, or RE, is shown in Fig. 7. The state is indicated by the orientation of the rotor: horizontal for state 0, and vertical for state 1.

If the input arrives parallel to the rotor, then it continues straight through to the other side without touching the rotor. But if the input arrives perpendicularly to the rotor, then it hits the rotor, causing the rotor to rotate by 90 degrees, at which point the rotor knocks the token back out of the gate just on the other side of the corner where the token came in. For example, an input on I_N in state 0 will hit the rotor and exit at O_W, leaving the RE in state 1.

The four-way rotational symmetry of its operation and its geometrical nature make the RE very intuitive to work with. It is what we will use to implement arbitrary RLEMs in Sect. 2.3.

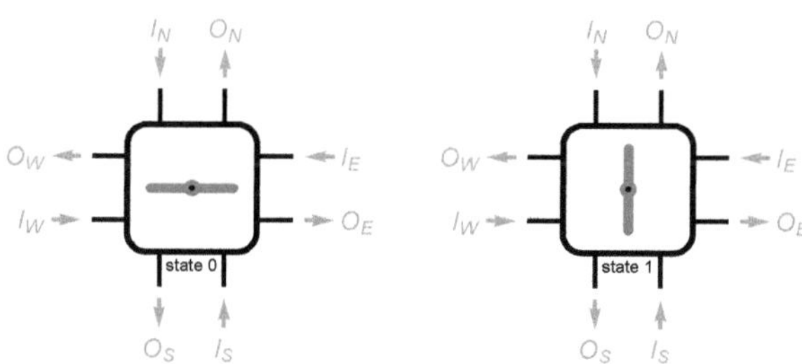

Fig. 7. The Rotating Element (RE). A token entering parallel to the rotor passes straight across to the opposite side, while a token entering at right angles to the rotor will bounce off the rotor and exit from the same corner where it entered, leaving the rotor in the opposite state.

2.3 Apart from the Morita Gate, All Gates with ≤2 Inputs are Non-universal, and All Non-degenerate Gates with ≥3 Inputs are Universal

Apart from the first classic result, all the results listed here are from papers with Kenichi Morita as first, last, or sole author.

Lemma 1 (Bennett). *[6] A reversible Turing machine can calculate anything that an ordinary Turing machine can calculate.*

Lemma 2. *[7, 8] An RE can implement any reversible Turing machine (RTM).*

Lemma 3. *[9] An RE can implement any RLEM, and is therefore universal.*

Lemma 4. *[10] A Shifter can implement an RE, and is therefore universal.*

Lemma 5. *[11] An Integrator (2-3) together with a Differentiator (2-4) can implement a Shifter, and are therefore jointly universal.*

Lemma 6. *[10] All non-degenerate gates with 3 inputs can implement both the Integrator and the Differentiator, and are therefore universal.*

Lemma 7. *[10] Any non-degenerate gate with 3 or more inputs can implement (in a circuit with only one gate) a non-degenerate gate with one fewer inputs.*

Theorem 1. *[10] All non-degenerate gates with ≥3 inputs are universal, by using Lemma 7 zero or more times until Lemma 6 applies.*

Lemma 8. *[12] The Delay (2-2) cannot implement the Integrator or Differentiator or Morita gate.*

Lemma 9. *[12] The Integrator and Differentiator cannot implement each other, nor can they implement the Morita gate.*

Lemma 10. *[13] The Integrator, Differentiator, and Morita gate can all implement the Delay.*

All that is missing from these impressive results, spanning many years of work, is: What exactly can the Morita gate do? In particular, if it can implement the Integrator, then the reversed circuit, which is also a circuit of Morita gates, implements a Differentiator, so in this case it would be universal, by Lemma 5. On the other hand, if it cannot implement the Integrator, then it cannot implement the Differentiator (because if it could then the reversed circuit would implement an Integrator), and so in light of Lemma 9 it would generate a third incomparable class, alongside the classes generated by the Integrator and the Differentiator. It is known that any two of these three elements are universal together [11,13]. So the complete classification of two-state gates is waiting only for the determination of the Morita gate's power.

3   Circuits of Morita Gates

In this section we present the new results of this paper, showing how to build circuits with Morita gates. We will see that the Morita gate is universal, disproving a long-standing conjecture [10,18] that 2-input gates cannot be universal. As we will discuss in Sect. 4, this result also finishes the classification of all 2-state RLEMs (of all sizes), regarding which RLEMs or groups of RLEMs are able to implement which others.

3.1   What can a Morita Gate Easily do?

The only way to become familiar with the Morita gate is to start playing around with it.

A single Morita gate has a couple of simple capabilities. For example, if inputs are repeatedly received at a single input, then the outputs alternate. Conversely, if the inputs alternate, then the output is constant.

Moving from single Morita gates to circuits of Morita gates might appear to be a daunting step, but we will start with an extremely simple circuit, where we simply chain two Morita gates together, as in Fig. 8A. If we just feed tokens into the top input, we see that they will be fed into the two internal wires in alternation, and then the second Morita gate will merge these paths back together, so the tokens always arrive at the top output. (Recall that only one token at a time is in the circuit). Another way to see it is that the first token passes straight across, as shown in Fig. 8B, flipping both gates into state 1, shown in Fig. 8C. Later, when the token arrives for the second time, it gets diverted to the lower path, but then the second gate sends it back to the upper path, as shown in Fig. 8D, after which both gates are back in state 0, as in Fig. 8A again.

This circuit can be thought of as a "Do-Half" gadget. If we imagine some subcircuit being placed along the upper internal wire, then if $2n$ tokens arrive to the input, the subcircuit will be executed only n times.

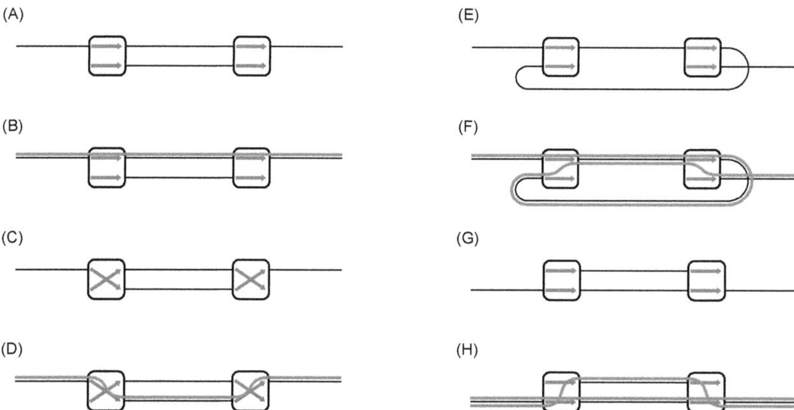

Fig. 8. **(A)** The **Do-Half** gadget. Each internal wire receives half of the tokens fed to the circuit. **(B)** The first of a pair of tokens travels across the upper internal wire. **(C)** After the first token has passed, the gates are both in state 1. **(D)** The second of a pair of tokens travels across the lower internal wire, leaving the circuit in the original state shown in (A), ready for the next pair of tokens. **(E)** The **Do-Twice** gadget. As in other diagrams, never-used wires are not shown. **(F)** A token coming into the Do-Twice gadget is fed to the internal wire twice, leaving the circuit in the same state as how it started. **(G)** The Do-Half gadget shown in (A) also works if used from the lower input and output. **(H)** A sequence of two tokens arriving to the circuit again results in each wire being used once. The usage order is reversed relative to the order shown in (B) and (D).

For our next circuit, we try routing the upper output back around to the other input, as shown in Fig. 8E. Now we see that when a token enters the circuit, it travels straight across the top, and is then sent back to the bottom input of the first gate, where it gets diverted to the top again and then it gets diverted back to the bottom by the second gate, where it exits, as shown in Fig. 8F. After all of this, the circuit is back in the same state as it started, so the next token will behave in exactly the same way. We can think of this circuit as a "Do-Twice" circuit, because the upper path is traversed twice by every token.

Note that even if a circuit has feedback loops, the token can't enter an infinite loop (which would become a periodic loop as soon as a circuit state is repeated), because in a reversible system any periodic loop is also a periodic loop in reversed time. If we run the circuit forwards until the token is cycling in the periodic loop, and then we reverse time, then the token will be cycling backwards in the periodic loop, and the token will not suddenly be able to exit the loop to go back to the circuit input. If we put a Do-Half gadget (as in Fig. 8A) inside a Do-Twice gadget (as in Fig. 8E), the result is a bit anti-climactic, a "Do-Once" gadget, basically as powerful as a single wire. But the Do-Half gadget (shown in Fig. 8A) is usable not only from its top legs, but also from its bottom legs (as shown in Fig. 8G), meaning that if the token is sent twice to the bottom input (as shown in Fig. 8H), then it also traverses the top internal wire just once, the same as for the top input. (Well, almost the same – when used from the top input, it will call the subroutine the first time around, but when used from the bottom input, it will call the subroutine the second time around. But this does not matter).

So we can actually attach two different Do-Twice gadgets to the same Do-Half gadget, one connected to the upper legs of the Do-Half gadget, and one connected to its lower legs. This will allow *either* of the Do-Twice gadgets to send a token to the *same* Do-Half gadget, before returning the token. In effect, this allows two different execution paths to access the same subcircuit. This is exactly the concept of a subroutine, so we call this agglomeration a Subroutine Gadget (SUBG). This idea is shown in Fig. 9.

We think of this as a gadget rather than an RLEM, because the subroutine is external to the gadget, so the token enters the subroutine from an output wire of the gadget (labelled *CallSub* in Fig. 9), and the wire where it returns is an input wire of the gadget (labelled *SubRet* in Fig. 9). The gadget thus has 3 inputs and 3 outputs, and to use it correctly you must alternate sending the token to input 1 or 2 (which the gadget will route to *CallSub*), and sending the token to *SubRet* (which the gadget will route to circuit output 1 or 2, depending on which input was used most recently). Unlike an RLEM, the abstract gadget is not defined for other types of input sequences.

Usually we intuitively think of the subroutine wires as being sort of internal to the gadget's operation, rather than as an input or output of the gadget, even though the subroutine is typically drawn far outside the gadget. This is exactly like how a subroutine in standard programming languages is thought of as being sort of internal to the code that calls it, even though the subroutine's code is often far from the caller's code.

The SUBG only allows two callers. To allow three callers, use two SUBGs, with the subroutine of the first SUBG being one of the callers of the second SUBG. In general, to allow n callers, use a stack or tree of $n - 1$ SUBGs, as shown in Fig. 9.

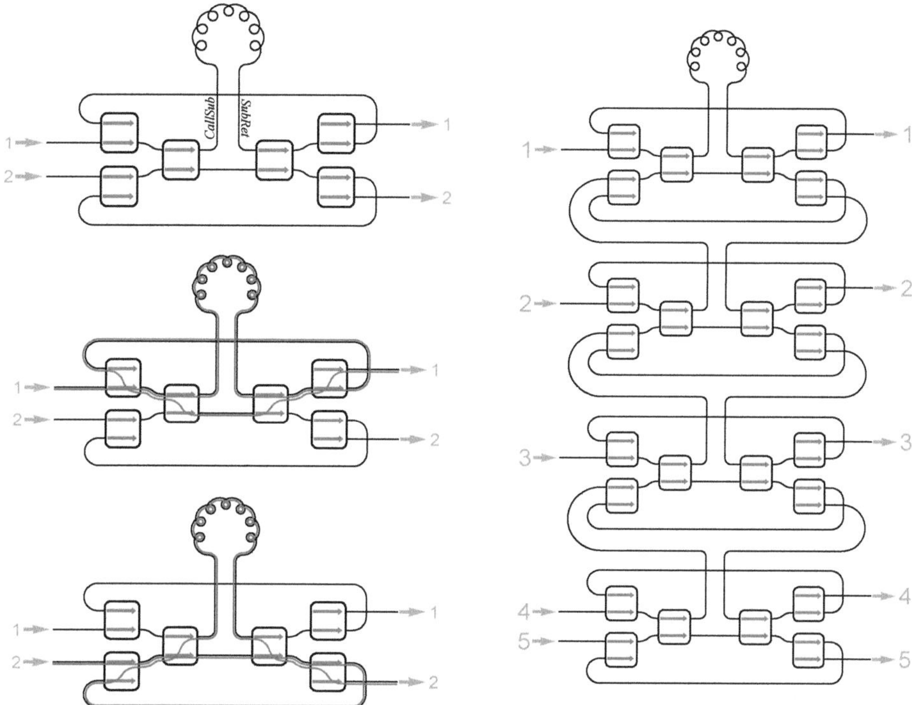

Fig. 9. A Subroutine Gadget (SUBG) implementation, built from Morita gates. **(upper left)** This gadget allows two different "callers", shown as 1 and 2, to call a single "subroutine", shown here as the tree (or light bulb filament) sticking up in the middle of the circuit. **(center left)** If a token enters on input line 1, it will pass through the subroutine and then exit on output line 1, leaving all of the circuit's gates in state 0, the same as how they started. **(lower left)** If a token enters on line 2, it will also pass through the subroutine, but then it will exit on output line 2, also leaving the overall circuit state the same as before it entered. **(right)** To support more callers, a stack of SUBGs can be used.

3.2   Gadgets vs. RLEMs

A gadget is like an RLEM that has restrictions on how you use it. For example, the wires *CallSub* and *SubRet* in Fig. 9 are supposed to be connected to a subroutine such that whenever a token is sent on *CallSub*, it comes back on *SubRet* before either of the inputs 1 or 2 are used again. So we can see that this gadget is always in one of three states: "waiting for an input", "waiting for the subroutine to return for caller 1", and "waiting for the subroutine to return for caller 2". This gadget is shown in Fig. 10.

If a token is received at an input where the gadget is not defined, the resulting behavior is undefined. This means that for an implementation, any behavior is ok—it is even ok if it destroys the gadget. The gadget is only required to work

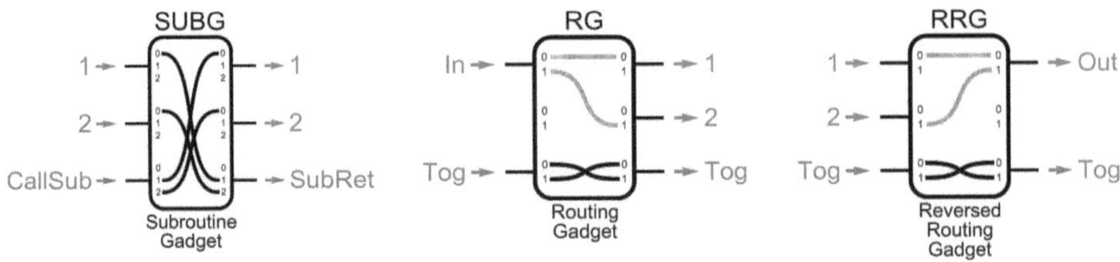

Fig. 10. The SUBG, RG, and RRG gadgets. Unlike RLEMs, gadgets can have missing transitions. Tokens should not arrive to inputs where there is no transition defined for the current state. That could break the gadget.

when the inputs follow the correct protocol, meaning only defined transitions are used.

Thus, a gadget is not an actual object or circuit, it is simply a specification that an actual object or circuit might satisfy. In an abuse of terminology, circuits satisfying the specification are referred to as being such a gadget, with context making it clear whether the reference is to a specific circuit or to the abstract specification. (This is like the class vs. instance distinction).

Another gadget that will be relevant for the next two sections is the Routing Gadget (RG), also shown in Fig. 10. It routes the incoming input (labeled "In") to one of two outputs, depending on the state. Another input, labeled *Tog*, toggles the state. This is like a railroad track that can be switched to one of two continuing tracks. This gadget cleanly separates the toggling action from the use of the toggled path. To merge two tracks together, we can use the reversed form of this gadget, also shown in Fig. 10. In reversible systems, it is not possible for two paths to simply merge into one. This lack of merging trajectories is the essence of reversibility. And this is why the Reversed Routing Gadget (RRG) is useful. It lets two paths merge into one, because the information about the source is clearly represented in the state, so at any given time actually only one source can lead to the output, preserving reversibility.

3.3   The Morita Gate can Implement a Routing Gadget

The circuit shown in Fig. 11 was found by a search program that ran for 2 weeks, exhaustively enumerating all ten-gate circuits (up to symmetries). It was looking for any 3-input circuit that implements a 2-state gadget when just two of the inputs are used and it is started in an appropriate state. It only found two results that were not equivalent to a 2-input gate: the ten-gate circuit shown here, and a six-gate circuit equivalent to a 3-input, 3-output Delay gate. A rough estimate is that the same program would take five years to search all eleven-gate circuits, so we were lucky that a ten-gate circuit was waiting to be found. The ten gate circuit that was found actually alternated its usage of the output lines, and so adding an eleventh Morita gate to the end of the circuit made the circuit perfectly implement a Routing Gadget (RG), and this eleven gate circuit is what is shown in Fig. 11.

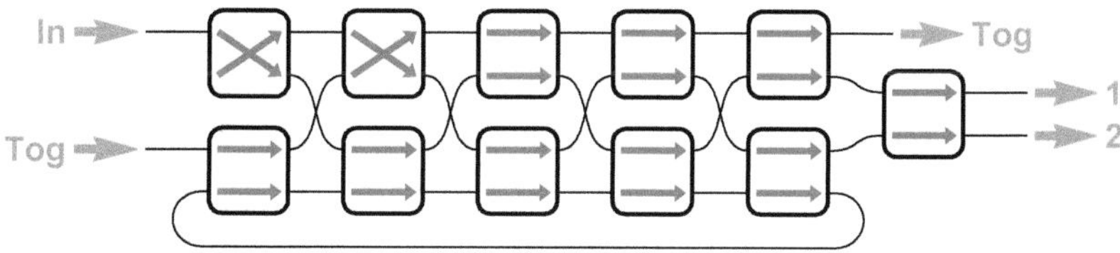

Fig. 11. The key circuit for showing that the Morita gate is universal. This circuit of Morita gates implements a Routing Gadget (RG). Tokens sent to the "In" input are routed to output 1. But after a token is sent to the "Tog" (Toggle) input (which always routes its token to the Toggle output), then tokens sent to the "In" input will be routed to output 2 instead. The "Tog" and "In" input lines can freely be used in any order. Each time a token is sent to "Tog", it flips the output line that the tokens sent to "In" get routed to. This is exactly the behavior of an RG. The circuit has a very regular wiring pattern, which is mostly feed-forward, except for the single long feedback wire at the bottom of the circuit. Despite its regular structure, we do not have any explanation (not even a heuristic one) for why this circuit acts correctly as a Routing Gadget. It simply happens to operate correctly.

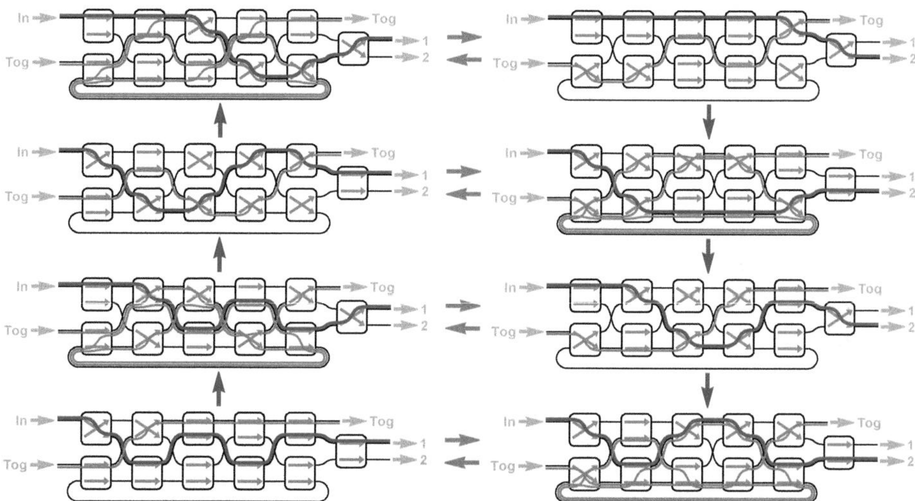

Fig. 12. Sending tokens to the "In" input causes the circuit to cycle through the states shown in a single column, as shown by the blue arrows. (The wrap-around arrow is not shown, but each column's cycle does wrap around). Sending tokens to the "Tog" input causes the circuit to move to the other column, as shown by the red arrows. The first column routes "In" tokens to output "1", while the second column routes "In" tokens to output "2". Tokens sent to the "Tog" input are always routed to the "Tog" output. If the circuit starts in the lower left state (the state shown in Fig. 11), then the n'th token sent to the circuit will use the bottom feedback wire twice to pass through the main part of the circuit three times before exiting, if n is even and the token was sent to the "Tog" input.

If a token is sent to the unused input (the lower input of the upper left Morita gate), then the circuit state becomes corrupted and it stops working as an RG.

The circuit shown in Fig. 11 actually uses four states that correspond to RG state 0, and another four states that correspond to RG state 1. Every time the "In" line is used, it cycles to another of the four states. Figure 12 shows all eight states, and the blue and red arrows show how the state changes when a token is sent to the "In" or "Tog" input.

Note that if we do a left-right flip of this circuit, so outputs become inputs and vice versa, then we get a Reversed Routing Gadget (RRG). The flipped Morita gates are still Morita gates (although the states also need to be flipped from $0 \leftrightarrow 1$), so the flipped circuit is an implementation of an RRG from Morita gates. The RRG feeds tokens from the correct input line (1 or 2) to a single output line. Sending a token on the toggle line toggles the correct input line between 1 and 2. Sending a token on the wrong input line corrupts the gadget. The RRG may not appear at first to be useful, since the token needs to enter on the correct input wire and is then always routed to the same output wire regardless of the RRG's state. But the RRG is just as useful as the RG: Just like how the RG provides controlled fan-out, the RRG allows controlled fan-in, so two inputs can be routed to the same output. Uncontrolled fan-in is not possible in reversible systems.

3.4   RG, RRG and SUBG Together can Implement Any RLEM

The Routing Gadget is a very powerful building block. Figure 13 shows how we can use the RG, RRG, and SUBG to implement an arbitrary target RLEM.

The idea, which follows a similar strategy to how Morita [9] implemented an arbitrary target RLEM using REs, is to use horizontal wires running from the inputs to the outputs, passing through a column for each state. Each column has the horizontal wire pass through an RG and then an RRG. When the circuit is in its "quiescent state", meaning all the RGs and RRGs are in state 0, any incoming token would simply pass straight across horizontally, from the input to the output.

Of course the quiescent state is not a useful state for the circuit to be in, since then the circuit would be equivalent to just wires, never changing its state. Instead, the circuit represents the state of the simulated RLEM by having all the RGs and RRGs in the corresponding column being in state 1. We refer to that condition as the column being "turned on", while if all the RGs and RRGs in that column are in state 0, we say the column is "turned off". So to represent state s, column s is turned on while all the other columns are turned off.

When the circuit is representing state s and a token arrives to the circuit on input wire i, the token will continue horizontally until it reaches column s, at which point it will be diverted by the RG onto the lower wire.

On this lower wire, the first thing it will do (indicated by a * next to the wire in Fig. 13) is toggle all the RGs and RRGs in that column, turning the column off,

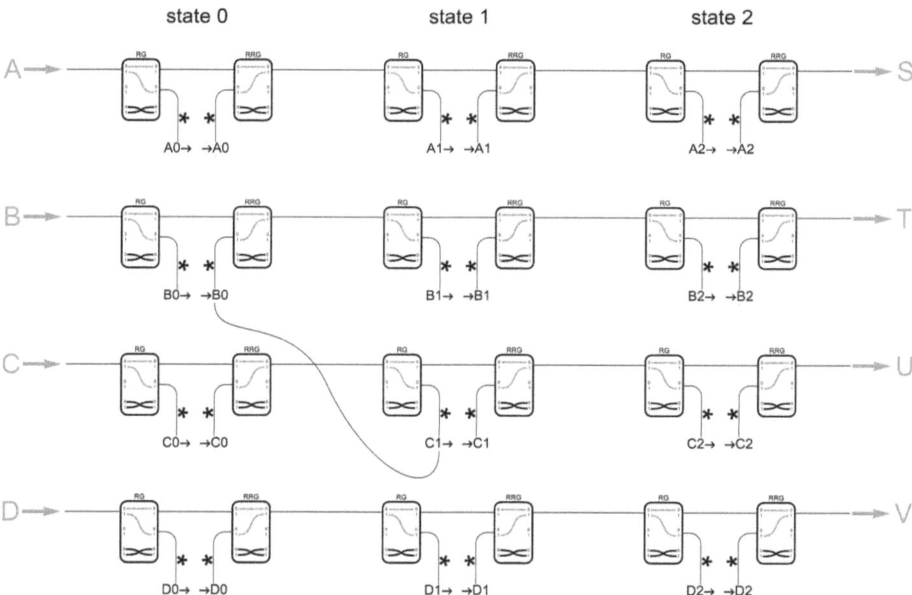

Fig. 13. A circuit of RG, RRG, and SUBG elements can implement any RLEM with the method shown here. The RGs and RRGs are all set to state 0, so an incoming token just moves along horizontally, except for the column corresponding to the currently simulated state, where all the RGs and RRGs are set to state 1. All of the dangling wires are paired up according to the transitions of the RLEM being implemented. However, for clarity, only one transition wire is shown, implementing the transition C1→B0. It goes from the wire labeled C1→ to the wire labeled →B0. If the simulated RLEM is in state 1, then a token entering the circuit at C will go straight until it gets diverted towards this transition wire. As it passes the * it will turn off the column for state 1, via a SUBG stack providing access to the toggling path for that column (not shown). Then it will travel along the transition wire, and when it arrives to the * at the end, it will turn the new column on, representing the new simulated state. This allows it to use the RRG to merge onto the horizontal wire there, whence it will continue straight to the right, exiting at T. Time-reversing the whole process yields the same flow logic.

thereby putting the circuit into the quiescent state. This can be done by a single long path that passes through the *Tog* legs (lower legs) of all the RGs and RRGs in the column. However, since *all* of the wires marked with * in that column need to access this toggling path, they actually use a SUBG stack to share access to the toggling path. For clarity, each column's toggling path and SUBG stack are not explicitly shown in Fig. 13, but rather the * notation is used.

After turning off the column, thereby putting the circuit into its quiescent state, the token continues on a transition wire. In Fig. 13 only one transition wire is shown, going from "C1→" to "→B0", representing the transition C1→B0.

(If the figure would show all the transition wires, it would visually become an intimidating mess).

Since the transitions of an RLEM form a bijection between all (input wire, current state) pairs and all (output wire, new state) pairs, the transition wires completely connect all of the dangling ends in Fig. 13 without duplication.

The transition wire carries the token to the position of the new state and output, where there is another *, meaning the token toggles the new column, turning it on. This allows it to use the RRG to merge onto the correct output wire. The token continues horizontally, exiting the circuit on the correct output, and leaving the circuit in the correct state.

Since the Morita gate can implement RG, RRG, and SUBG, we see that with this construction it can implement any target RLEM with any number of states and inputs. Therefore the Morita gate is universal.

4   Conclusion

We have shown that the Morita gate is the smallest universal RLEM. This completes the classification of all sets of reversible 2-state elements.

Given some building block RLEMs (called *generators*), the set of all 2-state RLEMs constructable from these generators is always a *closed* set, meaning it already includes among its members all the 2-state RLEMs that can be built from any members of the set. Since the Morita gate is universal, there are just 5 such closed sets: (It happens to be the case that each set can be generated by a single 2-state generator of size 2, or even, in the simplest case, by zero generators).

- {2-17: two-state RLEMs implementable with Morita gates (i.e., all RLEMs)}
- {2-4: two-state RLEMs implementable with Differentiator gates}
- {2-3: two-state RLEMs implementable with Integrator gates}
- {2-2: two-state RLEMs implementable with Delay gates}
- {just wires: two-state RLEMs whose mapping never changes}

Each class is strictly larger (more powerful) than the classes listed below it, except that the 2-3 and 2-4 classes are incomparable (neither is larger; they are equivalent under a time reversal symmetry).

Open questions include the general analysis of gates with more than 2 states. In particular, are there additional closed classes of gates when 3 or more states are permitted? For example, is there any gate, with any number of states, that can implement more than just wires but *cannot* implement a Delay (2-2) gate?

Additionally, the question of non-intrinsic (rather, Turing machine style) universality remains open for the 2-2, 2-3, and 2-4 classes. That is, even though they cannot directly simulate all RLEMs, perhaps one or more of them could still simulate a Turing machine via an appropriate encoding. We note that Morita [7] has shown that an RE can very nicely (much more nicely than Boolean logic gates) implement a reversible Turing machine (see Lemma 2), so by the results in Sect. 3, the 2-17 class is also Turing universal.

References

1. Langton, C.G.: Studying artificial life with cellular automata. Physica **22**(D), 120–149 (1986)
2. Dowd, M., Perl, Y., Rudolph, L., Saks, M.: The periodic balanced sorting network. J. Assoc. Comput. Mach. **36**(4), 738–757 (1989)
3. Cormen, T., Leiserson, C., Rivest, R.: Introduction to Algorithms. McGraw Hill and MIT Press (1990)
4. Aspnes, J., Herlihy, M., Shavit, N.: Counting networks. J. Assoc. Comput. Mach. **41**(5), 1020–1048 (1994)
5. Morita, K., Ogiro, T., Tanaka, K., Kato, H.: Classification and universality of reversible logic elements with one-bit memory. In: Margenstern, M. (ed.) MCU 2004. LNCS, vol. 3354, pp. 245–256. Springer, Heidelberg (2005). https://doi.org/10.1007/978-3-540-31834-7_20
6. Bennett, C.H.: Logical reversibility of computation. IBM J. Res. Dev. **17**, 525–532 (1973)
7. Morita, K.: A simple universal logic element and cellular automata for reversible computing. In: Margenstern, M., Rogozhin, Y. (eds.) MCU 2001. LNCS, vol. 2055, pp. 102–113. Springer, Heidelberg (2001). https://doi.org/10.1007/3-540-45132-3_6
8. Morita, K.: Constructing a reversible turing machine by a rotary element, a reversible logic element with memory. Technical report, Hiroshima University Institutional Repository (2010)
9. Morita, K.: A new universal logic element for reversible computing. In: Martin-Vide, C., Mitrana, V. (eds.) Grammars and Automata for String Processing: From Mathematics and Computer Science to Biology, and Back: Essays in Honour of Gheorghe Paun, pp. 285–294. Taylor & Francis (2003)
10. Morita, K., Ogiro, T., Alhazov, A., Tanizawa, T.: Non-degenerate 2-state reversible logic elements with three or more symbols are all universal. J. Multiple-Valued Logic Soft Comput. **18**(1), 37–54 (2012)
11. Lee, J., Peper, F., Adachi, S., Morita, K.: An asynchronous cellular automaton implementing 2-state 2-input 2-output reversed-twin reversible elements. In: Umeo, H., Morishita, S., Nishinari, K., Komatsuzaki, T., Bandini, S. (eds.) ACRI 2008. LNCS, vol. 5191, pp. 67–76. Springer, Heidelberg (2008). https://doi.org/10.1007/978-3-540-79992-4_9
12. Mukai, Y., Morita, K.: Hierarchy of reversible logic elements with memory. In: LA Winter Symposium, Kyoto (2012). (in Japanese)
13. Mukai, Y., Morita, K.: Universality of 2-symbol reversible logic elements with memory. In: LA Summer Symposium, Kosai, Sizuoka Prefecture (2011). (in Japanese)
14. Morita, K.: Reversible computing and cellular automata-a survey. Theor. Comput. Sci. **395**, 101–131 (2008)
15. Mukai, Y., Morita, K.: Realizing reversible logic elements with memory in the billiard ball model. Int. J. Unconv. Comput. **8**(1), 47–59 (2012)
16. Morita, K.: Reversible logic elements with memory and their universality. In: Neary, T., Cook, M. (eds.) Machines, Computations and Universality (MCU 2013), vol. 128, pp. 3–14. Electronic Proceedings in Theoretical Computer Science (EPTCS) (2013)
17. Morita, K.: Theory of Reversible Computing. Monographs in Theoretical Computer Science, an EATCS Series. Springer, Cham (2017)
18. Mukai, Y., Ogiro, T., Morita, K.: Universality problems on reversible logic elements with 1-bit memory. Int. J. Unconv. Comput. **10**(5), 353–373 (2014)

Bridging Chaos Game Representations and k-mer Frequencies of DNA Sequences

Haoze He[1(✉)], Lila Kari[2], and Pablo Millan Arias[2]

[1] École Polytechnique Fédérale de Lausanne (EPFL), Lausanne, Switzerland
`haoze.he@epfl.ch`
[2] University of Waterloo, Waterloo, Canada
`{lila,pmillana}@uwaterloo.ca`

Abstract. This paper establishes formal mathematical foundations linking Chaos Game Representations (CGR) of DNA sequences to their underlying k-mer frequencies. We prove that the Frequency CGR (FCGR) of order k is mathematically equivalent to a discretization of CGR at resolution k, and its vectorization corresponds to the k-mer frequencies of the sequence. Additionally, we characterize how symmetry transformations of CGR images correspond to specific nucleotide permutations in the originating sequences. Leveraging these insights, we introduce an algorithm that generates synthetic DNA sequences from prescribed k-mer distributions by constructing Eulerian paths on De Bruijn multigraphs. This enables reconstruction of sequences matching target k-mer profiles with arbitrarily high precision, facilitating the creation of synthetic CGR images for applications such as data augmentation for machine learning-based taxonomic classification of DNA sequences. Numerical experiments validate the effectiveness of our method across both real genomic data and artificially sampled distributions. To our knowledge, this is the first comprehensive framework that unifies CGR geometry, k-mer statistics, and sequence reconstruction, offering new tools for genomic analysis and visualization. The web application implementing the reconstruction algorithm is available at https://tinyurl.com/kmer2cgr.

Keywords: DNA sequence · Genomic signature · Chaos Game Representation CGR · Frequency Chaos Game Representation FCGR · k-mer frequency vector · synthetic DNA

1 Introduction

The increased availability of complete genome sequences has motivated a paradigm shift in comparative genomics, from homology-based to whole-genome analyses based on sequence composition patterns [8,34]. The observation of different structural patterns in DNA sequences dates back to 1990, when Jeffrey applied concepts from chaotic dynamics to DNA sequences and introduced

© The Author(s), under exclusive license to Springer Nature Switzerland AG 2026
E. Formenti and L. Manzoni (Eds.): UCNC 2025, LNCS 16364, pp. 214–229, 2026.
https://doi.org/10.1007/978-3-032-15641-9_15

Chaos Game Representations (CGR) of DNA sequences [14]. A CGR is visualized within a unit square, with each of the four vertices labelled by one of the nucleotides (A, C, G, and T). The plotting process follows a simple iterative procedure: the first nucleotide in the sequence is plotted at the midpoint between the center of the square and the vertex corresponding to that nucleotide. Each subsequent nucleotide is then plotted at the midpoint between the previously plotted point and the vertex representing the current nucleotide.

The appearance of interesting geometric patterns in CGRs of real DNA sequences, such as fractals and parallel lines, motivated further research in the field [7,9,12,15]. Notably, Oliver *et al.* [26] computed the first discretization of CGRs. In these representations, referred to hereafter as Frequency Chaos Game Representations (FCGR), the unit square is divided into a $2^k \times 2^k$ grid, and the CGR is discretized by counting the number of points in a given cell. Each cell is associated with a particular subword of length k (k-mer), and in [26] it was suggested that the count in each cell must be equal to the frequency of the corresponding k-mer in the DNA sequence. Similarly, Hao introduced k-frames [10], a class of self-similar and self-overlapping fractals mapping each k-mer to a cell inside the unit square using the Kronecker product.

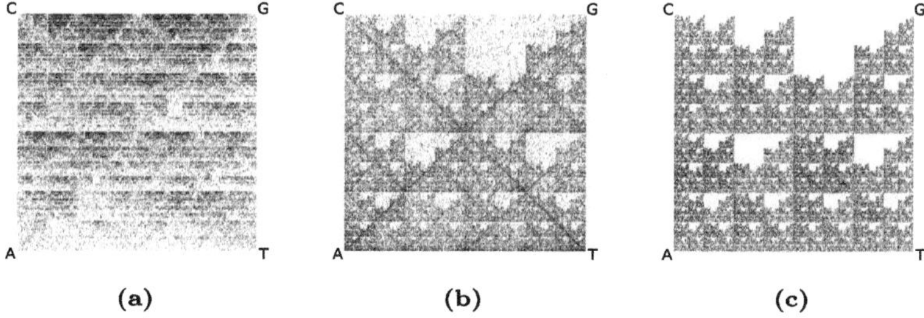

Fig. 1. (a) CGR of a 100,000 bp DNA sequence randomly extracted from the complete genome of *Pseudomonas aeruginosa* strain PAO1 (RefSeq NC_002516.2); (b) CGR of a 100,000 bp DNA sequence randomly selected from human chromosome 4 (GRCh38.p14 primary assembly, RefSeq NC_000004.12); (c) CGR of a computer-generated DNA sequence that is random in all other aspects except that the dinucleotide GC is absent.

Independent of CGR research, in 1995, Karlin and Burge introduced the notion of a *genomic signature* [18], as an umbrella term for any numerical quantity that shows greater similarity among DNA sequences of closely related organisms compared to those of more distantly related organisms. In that work, dinucleotide relative abundance profiles (DRAP) were proposed as genomic signatures, as they were effective in capturing expected variations between some species and similarities within the genome of a single species. Building on this concept, in 1999, Deschavanne *et al.* [4], characterized both CGR and FCGR as genomic signatures and showed that the variation between FCGR images of sequences along a genome was smaller than the variation between FCGR images of sequences taken from different genomes. For example, as seen in Figs. 1a and

1b, the CGR of a DNA sequence from the genome of *P. aeruginosa* exhibits visual patterns that are significantly different from patterns in the CGR of a DNA sequence from human chromosome 4. Subsequently, these CGR studies lead to the generalization of genomic signatures across various orders k, introduced in 2005 by Wang *et al.* [32] where, e.g., $k = 2$ corresponds to DRAP. In [32], it was shown that higher-order FCGRs capture sequence features that are not encoded by the DRAPs that had initially been proposed as genomic signatures.

The concepts of CGR, FCGR of order k, and k-mer frequency vector of a sequence (the latter comprising the counts of all its k-mers) collectively referred to here as *genomic signature*, have been widely used in comparative genomics as an alternative or a complement to alignment-based methods. For example, CGRs have inspired numerous alignment-free methods for taxonomic classification [16,17,21,28,29], clustering [1,23], and phylogenetic analyses [13,20,30]. Similarly, distance measures between k-mer frequency profiles serve as robust proxies for evolutionary relatedness, making them the core component in similar applications [5,31,33,34].

Despite the interchangeable use of these representations of DNA sequences in many applications [2,8,23], a formalization of their interconnections is still missing. In this work, we introduce a theoretical and algorithmic framework that rigorously links these genomic signatures. The main contributions of this paper are that it: *(i)* formally establishes two-way connections between a symmetry transformation of the CGR of a sequence and a morphism applied to the underlying sequence (Sect. 2), *(ii)* demonstrates the mathematical equivalence between a CGR of resolution $2^k \times 2^k$ of a DNA sequence s, and the Frequency Chaos Game Representation (FCGR) of order k of that sequence, as well as k-mer frequency vector of the sequence (Sect. 3), *(iii)* provides an algorithm and a software tool that computes a CGR and the corresponding synthetic DNA sequence, from a target k-mer frequency vector (Sect. 4). The proofs of all the results in this paper can be found in [11].

1.1 Notation

Throughout this paper, Σ will denote the DNA alphabet, namely the set $\Sigma = \{A, C, G, T\}$. The cardinality of a set A will be denoted by $\mathrm{card}(A)$. A non-empty word (string) over Σ, $w = a_1 a_2 ... a_n$, $n \geq 1$, is a concatenation of letters $a_i \in \Sigma$, $1 \leq i \leq n$. The empty word is denoted by λ, by Σ^+ we denote the set of all non-empty words over Σ, and $\Sigma^* = \Sigma^+ \cup \{\lambda\}$, while for $k \geq 1$ we have that Σ^k is the set of words of length k over Σ. The length of a word w is denoted by $|w|$, and $|\lambda| = 0$. For a given $k \geq 1$, a word $w \in \Sigma^k$ will be called a k-mer. The number of occurrences of the k-mer w in the DNA sequence s will be denoted by $occ(s, w)$. For a word $w \in \Sigma^+$, we will denote by $\mathrm{sub}_k(w)$ the set of all subwords of length k in the word w. We use Δ^{n-1} to denote the standard simplex $\{(x_1, \ldots, x_n) : x_i \geq 0 \text{ for all } i = 1, \ldots, n, \sum_{i=1}^{n} x_i = 1\}$ in $\mathbb{R}^n$.

2   CGR Symmetries and DNA Letter Permutations

This section first recalls the formal definition of the Chaos Game Representation (CGR) of a DNA sequence [14] (Definition 2) and introduces several other definitions and notations. Subsequently, Theorem 1 shows that, if a CGR is the image of another CGR via a symmetry transformation of the square, then their originating DNA sequences are connected by a certain letter permutation, and viceversa. In addition, Theorem 2 establishes the correspondence between the k-mers avoided in each of the sequences underlying two CGRs that are obtained from one another by a symmetry transformation of the square.

Definition 1 (CGR square). *A CGR square is the square centred at the origin*

$$\{(x, y)|\ -1 < x < 1, -1 < y < 1\}$$

with corners $(-1, -1), (-1, 1), (1, 1), (1, -1)$, *each labeled by the labeling function* $label : \Sigma \longrightarrow \mathbb{Z}^2$

$$label(A) = (-1, -1), label(C) = (-1, 1), label(G) = (1, 1), label(T) = (1, -1)$$

Note that the CGR square defined above has a side length of 2, whereas the original CGR square defined in [15] is a unit square with side length 1.

Definition 2 (CGR representation of a DNA sequence). *Let $n \geq 1$ and let $s = a_1 a_2 ... a_n$ be a sequence of length n over the DNA alphabet Σ. The CGR representation of the sequence s is the set of points $CGR(s) = \{p_0, p_1, .., p_n\} \subseteq \mathbb{Q}^2$ whose coordinates are defined recursively by*

$$p_0 = (x_0, y_0) = (0, 0), \ \ and \ p_i = \frac{p_{i-1} + label(a_i)}{2} \ for \ all \ 1 \leq i \leq n.$$

The *dihedral group* [6] of degree 4 and order 8 is the symmetry group of a square $D_8 = \{e, r, r^2, r^3, s, sr, sr^2, sr^3\}$, and it comprises rotations, reflections across the horizontal and vertical axis, as well as reflections across the diagonals. The symmetries of the axis-aligned CGR square, centered at the origin, can be represented by 2×2 permutation matrices, acting on the plane by multiplication on column vectors of coordinates $\begin{bmatrix} x \\ y \end{bmatrix}$. The group composition operation is represented as matrix multiplication.

Definition 3. *For two words $u, w \in \Sigma^n$ and a symmetry $h \in D_8$, we will say that $CGR(u) = h \cdot CGR(w)$ if and only if, for all $1 \leq i \leq n$, we have that $q_{u,i} = h \cdot p_{w,i}$ where $q_{u,i}$ is the ith point in the generation of $CGR(u)$ and $p_{w,i}$ is the ith point in the generation of $CGR(w)$.*

Definition 4. *Let $CGR(w)$ be the CGR of a DNA sequence $w \in \Sigma^*$. The image of a $CGR(w)$ via a transformation $h \in D_8$ is defined as $\{h \cdot x|\ x \in CGR(w)\}$, the set obtained by multiplying the matrix h with each point in $CGR(w)$.*

Consider a permutation $\sigma : \Sigma \to \Sigma$ of the letters in the DNA alphabet Σ, and consider that $\sigma(\lambda) = \lambda$, where λ denotes the empty word. Such a permutation σ can be extended to a morphism $\sigma : \Sigma^* \to \Sigma^*$ by the morphism property whereby $\sigma(uv) = \sigma(u)\sigma(v)$, for all $u, v \in \Sigma^*$. We now explore the relationships between such morphisms applied to a DNA sequence and symmetry transformations in D_8 applied to its CGR.

Definition 5. *Let S be the following set of permutations extended to morphisms on Σ^*:*

$$S = \{(), (A,T,G,C), (A,G)(C,T),$$

$$(A,C,G,T), (A,C)(G,T), (C,T), (A,T)(C,G), (A,G)\}$$

In Definition 5, in the customary cyclic notation, () denotes the identity permutation on $\{A,C,G,T\}$, and, e.g., (A,T,G,C) denotes a circular permutation that maps A to T, T to G, G to C, and C to A. Note also that S is a subset of the S_4, set of all 24 permutations of $\{A,C,G,T\}$.

Recall that, [6, Section 1.2], the dihedral group D_8 is connected to a subgroup of the permutation group S_4 of all permutations of four elements, by the mapping $f : D_8 \to S_4$ defined as $f(e) = ()$, $f(r) = (A,T,G,C)$, $f(r^2) = (A,G)(C,T)$, $f(r^3) = (A,C,G,T)$, $f(s) = (A,C)(G,T)$, $f(sr) = (A,G)$, $f(sr^2) = (A,T)(C,G)$, $f(sr^3) = (C,T)$.

The following theorem proves that, if a CGR is observed to be the image of another CGR via one of the symmetry transformations in D_8, then their originating DNA sequences are connected by one of the letter permutations in S (see Definition 5).

Theorem 1. *Let $u, w \in \Sigma^n$ be two DNA sequences of length n, and let $\sigma \in S$ be one of the morphisms in S. Then $u = \sigma(w)$ if and only if $CGR(u) = f^{-1}(\sigma) \cdot CGR(w)$.*

In other words, Theorem 1 states that, given two DNA sequences u and w and a letter permutation $\sigma \in S$, then $u = \sigma(w)$ iff $CGR(u)$ can be obtained from $CGR(w)$ via the symmetry $f^{-1}(\sigma)$ in D_8. A direct consequence of this result is that the image of a CGR of a DNA sequence under some symmetry transformation $h \in D_8$ is still a CGR of another DNA sequence, i.e., the set of CGRs of DNA sequences is closed under transformations in D_8.

Corollary 1. *Given a word $u \in \Sigma^n$ and a symmetry transformation $h \in D_8$, there exists a word $w \in \Sigma^n$ such that $CGR(w) = h \cdot CGR(u)$. Constructively, this word can be computed as $w = f(h) \cdot u$, with $f : D_8 \to S_4$ as defined above.*

A DNA sequence s is said to *avoid a word w* if s does not contain any occurrence of w as a subword. We now investigate the mathematical connection between k-mer avoidance in DNA sequences and the family of CGR images related by the symmetry group D_8. This question is of interest because the CGR of a DNA sequence that avoids a k-mer w has a similar visual appearance as the CGR of a real DNA sequence in which the word w is under-represented, but not completely absent. For instance, Figs. 1b and 1c illustrate that avoiding the dinucleotide GC produces the same "double-scoop" CGR pattern as that observed in CGRs of human DNA sequences (characterized by an under-representation of GC). Given the CGR of a DNA sequence that avoids a certain k-mer, and another CGR obtained from the first through a symmetry transformation in D_8, the following theorem identifies the k-mer avoided by the underlying sequence of the second CGR.

Theorem 2. *Let $w, u \in \Sigma^n$ be two DNA words of length n, and let $\alpha, \beta \in \Sigma^k$ be two k-mers, $k < n$, such that $sub_k(w) = \Sigma^k \backslash \{\alpha\}$ and $sub_k(u) = \Sigma^k \backslash \{\beta\}$. Then, for any permutation morphism $\sigma \in S$, we have that $CGR(u) = f^{-1}(\sigma) \cdot CGR(w)$ implies $\beta = \sigma(\alpha)$.*

3   Frequency CGR, and Equivalence to k-mer Frequencies

In this section, we first recall the definition of $FCGR_k(s)$ [4], the frequency CGR of order k, of a sequence s, which is discretized version of $CGR(s)$ at resolution $2^k \times 2^k$. We formalize the discretization process of $CGR(s)$ (Proposition 1), and establish several properties of this discretized version of $CGR(s)$. Lastly, we prove that $FCGR_k(s)$ can be equivalently computed either by discretizing $CGR(s)$ at resolution $2^k \times 2^k$, or by directly counting the occurrences of all k-mers in the sequence s (Theorem 3 and its Corollary 2).

We start by describing the discretizing process of the CGR of a DNA sequence, by subdividing the CGR square (of size 2×2) into $2^k \times 2^k$ equal sub-squares of size $1/2^{k-1}$.

Definition 6 (grid cell of order k). *Given a CGR square, $k \geq 1$, and indices $0 \leq i, j \leq 2^k - 1$, the grid cell (i, j) of order k, denoted by $cell_k(i, j)$, is the region of the square defined by*

$$cell_k(i, j) = \{(x, y) : x_i - \frac{1}{2^k} < x < x_i + \frac{1}{2^k}, y_j - \frac{1}{2^k} < y < y_j + \frac{1}{2^k}\}$$

where the center of the grid cell $cell_k(i, j)$ is the point

$$(x_j, y_i) = \left(-\frac{2^k - 1}{2^k} + \frac{j}{2^{k-1}}, \frac{2^k - 1}{2^k} - \frac{i}{2^{k-1}} \right). \tag{1}$$

Note that the size of each grid cell $cell_k(i, j)$ of order k is $\frac{1}{2^{k-1}} \times \frac{1}{2^{k-1}}$. Observe also that each grid cell $cell_k(i, j)$ (without boundary) $i = 0, \ldots, 2^k - 1, j = 0, \ldots, 2^k - 1$, is uniquely determined by its center (x_j, y_i), where the indices i and j are swapped so as to match the indexing convention of matrices. Lastly, note that grid cells of order k correspond to an image resolution of $2^k \times 2^k$.

We can now define the Frequency Chaos Game Representation of order k of a DNA sequence.

Definition 7 (FCGR of order k, of a sequence s). *Let $k \geq 1$ and $n \geq k$. A Frequency Chaos Game Representation of order k, of a sequence $s \in \Sigma^n$, is a matrix $FCGR_k(s) \in \mathbb{N}^{2^k \times 2^k}$ whose entries are defined as*

$$FCGR_k(s)(i, j) = card(CGR(s) \cap cell_k(i, j)) \text{ for all } 0 \leq i, j \leq 2^k - 1.$$

The $FCGR_k(s)$ matrix provides a compressed representation of the sequence s. Definition 7 can be viewed as a discretization of $CGR(s)$ into $2^k \times 2^k$ cells, each counting the number of points of $CGR(s)$ that fall inside that cell. We will show later that counting the points in a grid cell of order k is equivalent to calculating the number of occurrences of a specific k-mer in the sequence s. For that, we need to define the notion of a *CGR cell associated with a k-mer*.

Definition 8. *Let $n \geq 1$ and $s \in \Sigma^n$. The last point of the CGR representation of s, $CGR(s) = \{p_0, p_1, \ldots, p_n\}$, is defined as $p_{last}(s) = p_n$.*

Let $p_{last}(w) = (x_w, y_w)$ be the last point in $CGR(w)$ of a word $w = a_1 a_2 \ldots a_k$, where $k \geq 1$, and assume that $label(a_l) = (x_l, y_l)$ for all $1 \leq l \leq k$. By the definition of $CGR(s)$, it is easy to see that

$$x_w = \frac{\Sigma_{l=1}^{k} x_l \cdot 2^{l-1}}{2^k}, \quad y_w = \frac{\Sigma_{l=1}^{k} y_l \cdot 2^{l-1}}{2^k}. \tag{2}$$

As observed in [13], distinct sequences $s \in \Sigma^+$ have distinct last points $p_{last}(s)$ in their respective CGR representations.

Definition 9 (CGR cell associated with a k-mer). *Let $k \geq 1$, let w be a k-mer in Σ^k, and let $p_{last}(w) = (x_w, y_w)$ be the last point of $CGR(w)$. The CGR cell associated with the k-mer w is*

$$c(w) = \{(x,y)\mid\ x_w - \frac{1}{2^{|w|}} < x < x_w + \frac{1}{2^{|w|}}, y_w - \frac{1}{2^{|w|}} < y < y_w + \frac{1}{2^{|w|}}\}.$$

Figure 2a illustrates $c(ACG)$, the CGR cell associated with the 3-mer ACG. It is easy to see that, for a k-mer w, the last point $p_{last}(w)$ of $CGR(w)$ is the center of the sub-square $c(w)$. Note also that for two different k-mers w_1 and w_2, the cells $c(w_1)$ and $c(w_2)$ do not intersect.

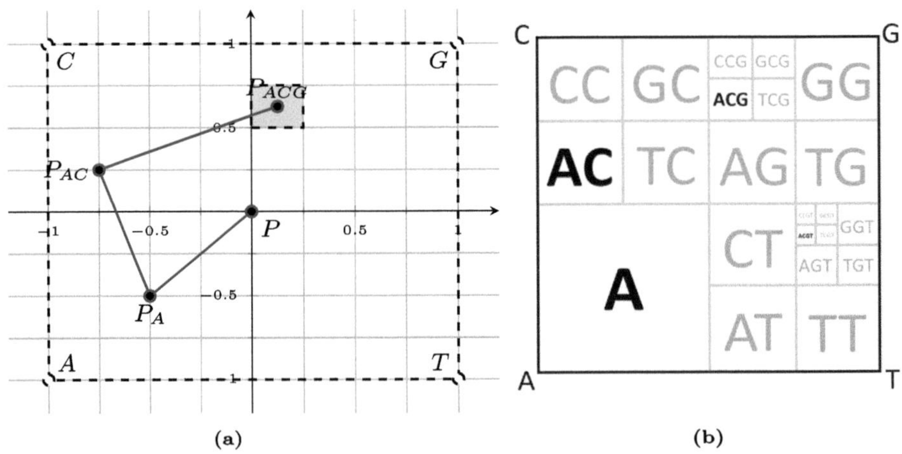

Fig. 2. (a) The cell $c(ACG)$ associated with the 3-mer ACG is marked in green. Its center P_{ACG} coincides with $p_{last}(ACG)$, the last point in the CGR representation of ACG. (b) Illustration of the hierarchical structure of CGR cells described by Lemma 2 and Proposition 3 (adapted from [21]). (Color figure online)

The following result now establishes the correspondence between the grid cells of order k (Definition 6) and the CGR cells associated with a k-mer (Definition 9).

Proposition 1. *Let $k \geq 1$, let $w = a_1 \cdots a_k$ be a k-mer, and assume that $label(a_l) = (x_l, y_l)$ where $x_l, y_l \in \{-1, 1\}$, for all $1 \leq l \leq k$. Then $c(w) = cell_k(i, j)$, that is, the CGR cell associated with the k-mer w equals the grid cell (i, j) of order k with indices*

$$j = \frac{2^k - 1 + \Sigma_{l=1}^k x_l \cdot 2^{l-1}}{2}, \quad i = \frac{2^k - 1 - \Sigma_{l=1}^k y_l \cdot 2^{l-1}}{2}. \tag{3}$$

For a given k, Proposition 1 establishes a bijection between the set of CGR cells associated with k-mers (Definition 9) and the set of grid cells of order k (Definition 6). Observe that, in both Definition 6 and Definition 9, the respective cells do not include their boundaries. The next proposition, aided by the following lemma, proves that no point of $CGR(s)$ falls on the boundary of a cell associated with a k-mer, for any k with $1 \leq k \leq n$, where $|s| = n$ (ensuring thus that the $FCGR_k(s)$ as defined in Definition 7 does not miscount).

Lemma 1. *Let $n \geq 1, k \geq 1$, and let $u \in \Sigma^n, w \in \Sigma^k$ be two DNA sequences. Let $p_{last}(u), p_{last}(w)$ be the last points of $CGR(u), CGR(w)$ respectively. Then, the last point of $CGR(uw)$ is*

$$p_{last}(uw) = \frac{p_{last}(u)}{2^{|w|}} + p_{last}(w).$$

Proposition 2. *Let s be a sequence of length n over Σ, and let $1 \leq k \leq n$. The last point $p_{last}(s)$ of $CGR(s)$ is not on the boundary of a cell $c(w)$ associated with any k-mer w in Σ^k.*

Lemma 1 also serves as an aide to showing the hierarchical nested structure of the CGR cells associated with a k-mer w, as defined in Definition 9. The next result shows that the CGR cell associated with a given k-mer w includes all the CGR cells associated with words ending in w.

Lemma 2. *Let w_1, w_2 be words in Σ^+. Then $c(w_1 w_2) \subseteq c(w_2)$.*

The *closed CGR cell associated with a k-mer w* is denoted by $\overline{c(w)}$ and is defined as the union between $c(w)$ and its boundaries (sides of the square). The next result proves that, given $n \geq 1$, the closed CGR cell associated with a k-mer w equals the union of all closed CGR cells associated with words $w'w$ with $|w'| = n$. This hierarchical structure of the closed cells associated with k-mers is illustrated in Fig. 2b.

Proposition 3. *Let $k \geq 1$ and $n \geq 1$. For a given k-mer $w \in \Sigma^k$ we have that $\overline{c(w)} = \bigcup_{w' \in \Sigma^n} \overline{c(w'w)}$.*

Lemma 1 can now be used to show that the $FCGR_k(s)$ matrix defined in Definition 7 through discretizing $CGR(s)$ into grid cells of order k (i.e., at resolution $2^k \times 2^k$), can also be obtained directly by counting the number of occurrences of all k-mers in the sequence s.

Theorem 3. *Given a DNA sequence $s \in \Sigma^n$, and a k-mer $w \in \Sigma^k$, where $1 \leq k \leq n$, we have that*

$$occ(s, w) = card(CGR(s) \cap c(w))$$

where $occ(s, w)$ denotes the number of occurrences of the k-mer w in s.

Corollary 2. *Let $s \in \Sigma^n$ be a sequence, and let $1 \leq k \leq n$ be a k-mer length. Computing $FCGR_k(s)$ by discretizing $CGR(s)$ into grid cells of order k as defined in Definition 7 (i.e., at resolution $2^k \times 2^k$) is equivalent to counting the number of occurrences of k-mers in the sequence s.*

Importantly, we now observe that in [10] one finds a definition of "FCGR of order k of a sequence s" via Kronecker products that is *different* from our Definition 7. The advantage of the concept as defined in Definition 7 is that, besides being equivalent to computing the counts of k-mers, $FCGR_k(s)$ in Definition 7 is a discretization of $CGR(s)$, at resolution $2^k \times 2^k$, and it is thus both visually similar and obtainable from $CGR(s)$. In contrast, the FCGR of order k of a sequence s as defined in [10] is not connected to $CGR(s)$, and it is visually different from $CGR(s)$.

4   From k-mer Distributions to Synthetic DNA

In this section, we explore the concept of empirical k-mer distribution of a sequence s, and formalize the correspondence with its $FCGR_k(s)$. We exploit this correspondence to generate synthetic DNA sequences and their respective CGRs by tracing an Eulerian path on the De Bruijn multigraph built from a target k-mer distribution. The main result (Theorem 5) shows that, for any target distribution on the standard probability simplex satisfying the structural constraints imposed by the linear nature of DNA sequences, we can construct a synthetic sequence s whose empirical k-mer distribution matches the target with arbitrarily high precision. A computational tool implementing the algorithm is available at https://tinyurl.com/kmer2cgr, and is used to empirically evaluate the algorithm's practical reconstruction accuracy through computational experiments using both real and synthetic target distributions.

Definition 10. *Given the DNA alphabet $\Sigma = \{A, C, G, T\}$ and a k-mer $w = a_1 \ldots a_k \in \Sigma^k$, we define the index function $idx \colon \Sigma^k \to \{0, 4^k - 1\}$ as:*

$$idx(w) = \sum_{t=1}^{|w|} 4^{|w|-t} \times \xi(a_t), \tag{4}$$

where the labelling function $\xi : \Sigma \to \{0,1,2,3\}$ *assigns each nucleotide to an integer as follows:*

$$\xi(A) = 0, \ \xi(C) = 1, \ \xi(G) = 2, \ \xi(T) = 3.$$

The previous definition interprets each k-mer w as a base-4 numeral via the digit assignment $\xi(\cdot)$. Note that $\mathrm{idx}(w)$ is a bijection between Σ^k and $\{0, 4^k - 1\}$, with $k = |w|$. This bijection induces a natural ordering on Σ^k, where for $u, v \in \Sigma^k$, we write $u < v$ if and only if $\mathrm{idx}(u) < \mathrm{idx}(v)$. Based on the label assignment $\xi(\cdot)$, this ordering corresponds to the lexicographic order given by the nucleotide ordering $A < C < G < T$.

Definition 11 (k-mer frequency vector). *Given a DNA sequence s, its k-mer frequency vector $F_k(s)$ is a vector in $\mathbb{N}^{4^k}$ defined as:*

$$F_k(s) = (occ(s, w_0), \ occ(s, w_1), \ldots, \ occ(s, w_{4^k - 1}))$$

where $w_i = idx^{-1}(i)$ *is the k-mer corresponding to index i under the mapping in* (4).

We now formalize the intuition that the k-mer frequency vector $F_k(s)$ corresponds to a vectorization of the $FCGR_k(s)$ matrix and compute the mapping between the matrix coordinates and the vector positions. For that purpose, we use the *label($\cdot$)* function (Definition 1) that links the coordinates of each corner in the CGR square with each nucleotide. Recall that for each k-mer $w = a_1 \cdots a_k$, $label(a_l) = (x_l, y_l)$, with $x_l, y_l \in \{-1, 1\}$ for all $1 \leq l \leq k$. Consequently, $label^{-1}(x_l, y_l) = a_l$.

Lemma 3. *Let* $k \geq 1$, *and* $0 \leq i, j \leq 2^k - 1$. *If* $i = \sum_{l=0}^{k-1} \alpha_l 2^l$ *and* $j = \sum_{l=0}^{k-1} \beta_l 2^l$, *with* $\alpha_l, \beta_l \in \{0,1\}$ *are the binary expansions of i and j, then* $cell_k(i,j) = c(a_1 \cdots a_k)$ *where $c(a_1...a_k)$ is the cell associated with $a_1...a_k$ constructed as*

$$a_l = label^{-1}(1 - 2\alpha_{l-1}, 2\beta_{l-1} - 1), \ for \ 1 \leq l \leq k \tag{5}$$

Proposition 4. *Let* $k \geq 1$, $n \geq 1$, $s \in \Sigma^n$ *and* $0 \leq i, j \leq 2^k - 1$. *If* $i = \sum_{l=0}^{k-1} \alpha_l 2^l$ *and* $j = \sum_{l=0}^{k-1} \beta_l 2^l$, *with* $\alpha_l, \beta_l \in \{0,1\}$ *(their binary expansions), then* $FCGR_k(s)(i,j) = occ(s, w_\tau)$, *where*

$$\tau = \sum_{l=0}^{k-1} 4^{k-l-1} \times \xi(label^{-1}(1 - 2\alpha_{l-1}, 2\beta_{l-1} - 1)) \tag{6}$$

Proposition 4 confirms the intuition that the k-mer frequency vector $F_k(s)$ corresponds to a vectorization of the $FCGR_k(s)$ matrix. Furthermore, we now have a connection between each word w, its corresponding point in the CGR ($p_{last}(w)$), the indices (i, j) in the FCGR matrix (Definition 7) and its position in the k-mer frequency vector given by the lexicographic order, $\mathrm{idx}(w)$.

We will now proceed to define a probabilistic framework for comparing the k-mer composition of sequences of different lengths, with the ultimate purpose of generating synthetic DNA sequences from given k-mer frequency vectors.

Definition 12 (k-mer distribution). *For a sequence s of length n, the empirical k-mer distribution of s is defined as the probability vector*

$$\boldsymbol{\theta}_s = (\theta_0^s, \theta_1^s, \ldots, \theta_{4^k-1}^s) = \frac{1}{n-k+1} \cdot F_k(s).$$

In other words, $\theta_i^s = \frac{occ(s,w_i)}{n-k+1}$ represents the normalized frequency of the k-mer w_i in s, where w_i denotes the i-th k-mer in the lexicographic order. Given $\boldsymbol{\theta} \in \Delta^{4^k-1}$ and a k-mer w, we denote the component $\theta_{\text{idx}(w)}$ by θ_w.

Note that any $\boldsymbol{\theta}_s$ is a point in the 4^k-dimensional probability simplex Δ^{4^k-1} and represents the empirical distribution of k-mers in s. However, not all points $\boldsymbol{\theta} \in \Delta^{4^k-1}$ correspond to an empirical k-mer distribution; in particular, the points with at least one irrational coordinate do not correspond to an empirical k-mer distribution.

Besides the normalization constraint, if $\boldsymbol{\theta}_s \in \Delta^{4^k-1}$ is a valid k-mer distribution, every internal $(k-1)$-mer is shared by two overlapping k-mers. For example, every occurrence of a k-mer $w = a_1 a_2 \ldots a_k$ must be followed sequentially by a k-mer $w' = a_2 a_3 \ldots a_{k+1}$. This can be formalized by the requirement $\text{SUFFIX}(w) = \text{PREFIX}(w')$, where $\text{PREFIX}(w)$ (respectively $\text{SUFFIX}(w)$) is defined as being the word consisting of the first (respectively last) $|w| - 1$ letters in w. This condition imposes a marginalization constraint on all the k-mers, e.g.

$$\theta_{AXY} + \theta_{CXY} + \theta_{GXY} + \theta_{TXY} = \theta_{XYA} + \theta_{XYC} + \theta_{XYG} + \theta_{XYT} = \theta_{XY}.$$

Thus, for each $(k-1)$-mer $v \in \Sigma^{k-1}$ (except edge cases in finite sequences), the sum of the frequencies of k-mers with v as a prefix must equal the sum of the frequencies of k-mers with v as a suffix. When considered across all k-mers, this condition imposes a set of additional marginal constraints that imply that $\boldsymbol{\theta}_s$ does not occupy the full 4^k-dimensional simplex but rather a lower-dimensional sub-polytope [19,25]. Formally, for every $v \in \Sigma^{k-1}$,

$$\sum_{a \in \Sigma} \theta_{va}^s = \sum_{a \in \Sigma} \theta_{av}^s = \theta_v^s. \tag{7}$$

Note that the constraints in (7) represent necessary conditions for a point $\boldsymbol{\theta} \in \Delta^{4^k-1}$ to be the k-mer distribution of a valid DNA sequence. We will later show that these constraints are a sufficient condition for sequence reconstruction.

Given a k-mer distribution $\boldsymbol{\theta}$, our objective is to reconstruct a sequence s, whose empirical k-mer frequency vector approximates the target distribution $\boldsymbol{\theta}$ as closely as possible. Then, we can compute the CGR according to Definition 2 to visualize the patterns corresponding to specific k-mer over- or under-representation. A similar problem has been studied in the field of genome assembly [24], where De Bruijn-graph–based methods have proven effective for assembling genomes from fragmented reads [3,27]. We will show that a suitable traversal of an appropriately constructed De Bruijn graph, combined with corrections for rounding error and violation of (7), produces a sequence that satisfies an approximation criterion.

Definition 13. *A directed multigraph M is a pair (V, E), where V is a set of nodes, and E is a multiset of ordered pairs from $V \times V$, called edges. Loops and parallel edges are permitted.*

Definition 14. *Let $k \geq 2$, $n > k$, and $s \in \Sigma^n$. The De Bruijn multigraph, denoted $\textsc{De Bruijn}_k^+(s)$, is the directed multigraph $\bigl(V_{k-1}(s), E_k(s)\bigr)$, where*

$$V_{k-1}(s) = \bigl\{\, v \in \Sigma^{k-1} \mid v \text{ appears as a substring of } s \,\bigr\},$$

$$E_k(s) = \{(u, v) \in V_{k-1}(s)^2 \mid \text{there exists a } k\text{-mer } w \in \Sigma^k \text{ that appears in } s$$
$$\text{with } \textsc{Prefix}(w) = u \text{ and } \textsc{Suffix}(w) = v\}.$$

In particular, for each $(u, v) \in V_{k-1}(s)^2$, the multiplicity of the edge (u, v) in $E_k(s)$ is exactly $occ(s, w)$.

Based on Definition 14, $\textsc{De Bruijn}_k^+(s)$ is completely determined by the collection of k-mers of s, so it can be constructed directly from a k-mer frequency vector without explicit knowledge of the sequence s. With this framework in place, assume a target distribution $\boldsymbol{\theta}_s$. It is then possible to rescale $\boldsymbol{\theta}_s$ by $n - k + 1$ to obtain an array of pseudo counts $\boldsymbol{c}_s$ such that each component is defined by $c_w^s = (n - k + 1) \cdot \theta_w^s$. Consequently, the problem of string reconstruction reduces to finding a path in the De Bruijn multigraph (constructed from these pseudo-counts) that traverses every edge exactly once, i.e., an Eulerian path. Euler's theorem provides the necessary and sufficient conditions for the existence of such a path.

Theorem 4. *(Euler) Let $G = (V, E)$ be a directed graph, and for any vertex $v \in V$ let $\textsc{in}(v)$ and $\textsc{out}(v)$ denote its indegree and outdegree, respectively. Then G contains an Eulerian cycle if and only if: (1) G is strongly connected, i.e., there exists a directed path between any pair of vertices in G, and (2) for every vertex $v \in V$, it holds that $\textsc{IN}(v) = \textsc{OUT}(v)$.*

Thus, for a directed graph G to contain an Eulerian cycle, it must be strongly connected and balanced at every vertex. In our setting, this implies that the hypothetical counts c_w must satisfy the marginal constraint on the de Bruijn graph to yield a valid DNA sequence s. In particular, if we associate each $(k-1)$-mer v with the k-mers that have v as a prefix or suffix, then the following condition in (7) must hold for every vertex v, except for the unique vertices corresponding to the start and end of an Eulerian path. We now formalize these conditions in the following theorem:

Theorem 5. *Let $1 < k < n \in \mathbb{Z}$, $\epsilon \in [0, 1]$, and let $\boldsymbol{\theta}$ be a point in $\Delta^{4^k - 1}$ satisfying the marginal constraint in (7) for every vertex v. If $n > \frac{2(k-1) \cdot 4^k}{\epsilon} + k - 1$, then there exists an algorithm that constructs a sequence s whose empirical k-mer distribution $\hat{\boldsymbol{\theta}}_s = \frac{F_k(s)}{\|F_k(s)\|_1}$ satisfies $\|\hat{\boldsymbol{\theta}}_s - \boldsymbol{\theta}\|_1 \leq \epsilon$.*

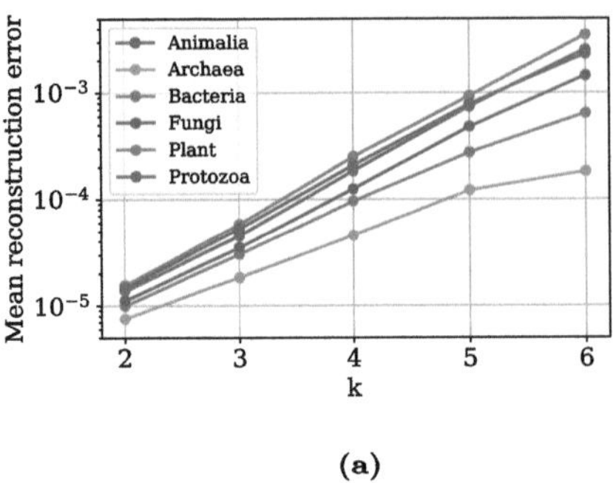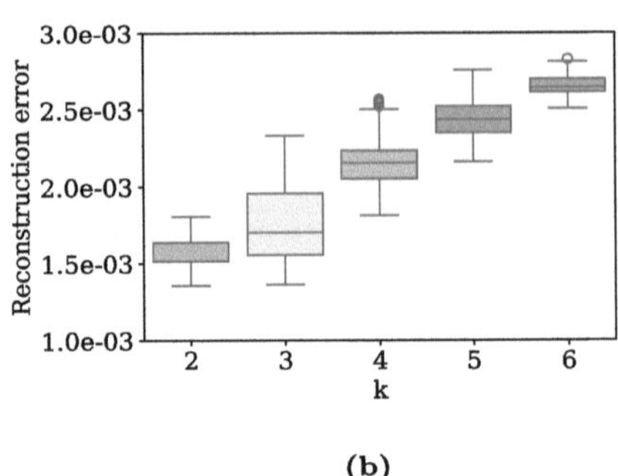

(a)   (b)

Fig. 3. (a) Reconstruction of sequences from k-mer distributions of real DNA sequences. The reconstruction is computed across different values of k, and sequences spanning species from each of the six kingdoms of life. For each kingdom, 100 genomes were selected, and a single 100,000-long DNA fragment was randomly selected from each genome. The mean error was computed for each pair (k, kingdom). **(b)Reconstruction of sequence from arbitrary k-mer distributions that satisfy the marginal constraints.** Each point is sampled from the standard probability simplex $\Delta^{4^{k}-1}$ subject to the constraints in Eq. 7 for different values of $k \in \{2, \ldots, 6\}$. The sequence length $n_{\min}$, calculated for $\epsilon = 0.01$, ranges from 3,200 for $k = 2$, to 819,200 for $k = 6$. The reconstruction error is less than ϵ for all the values of k.

We validate the theoretical construction through numerical experiments, using two different approaches. First, a fragment of 100,000 base pairs (bp) is sampled from the reference genome of 100 different species, each representing a distinct kingdom of life. For each sampled fragment, empirical k-mer distributions are computed for $k \in \{2, \ldots, 6\}$, and the average reconstruction error, measured by the total variation of k-mer distances, is determined for each kingdom. These results are summarized in Fig. 3a. Second, we sample arbitrary distributions from the relevant sub-polytope of $\Delta^{4^{k}-1}$. To this end, we employ a variation of the hit–and–run sampling procedure [22], which operates within the null-space defined by the linear constraints in (7). Here, the sequence length is fixed at $n_{\min} = 2 \cdot 4^{k}/\epsilon + k - 1$ for $\epsilon = 0.01$, and sampling is performed for $k \in \{2, \ldots, 6\}$. Figure 3b illustrates the reconstruction error for different values of k over points sampled from the desired sub-polytope.

The results of our experiments indicate that the reconstruction algorithm is capable of recovering sequences whose k-mer distributions closely approximate the target distributions, regardless of whether the target is sampled from a reference DNA sequence or sampled from the simplex. A computational tool implementing the algorithm is available at https://tinyurl.com/kmer2cgr (see Fig. 4).

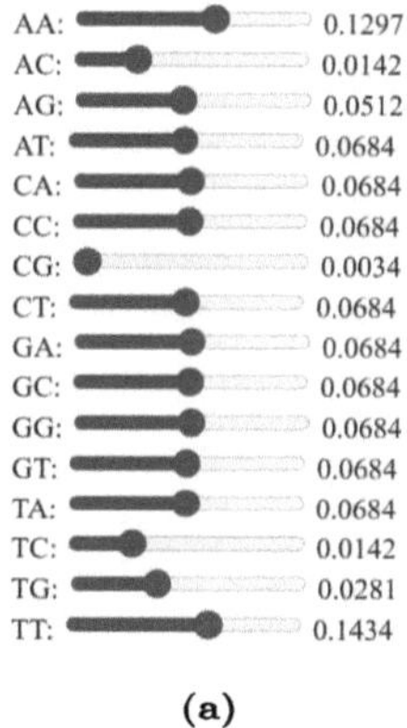

(a)

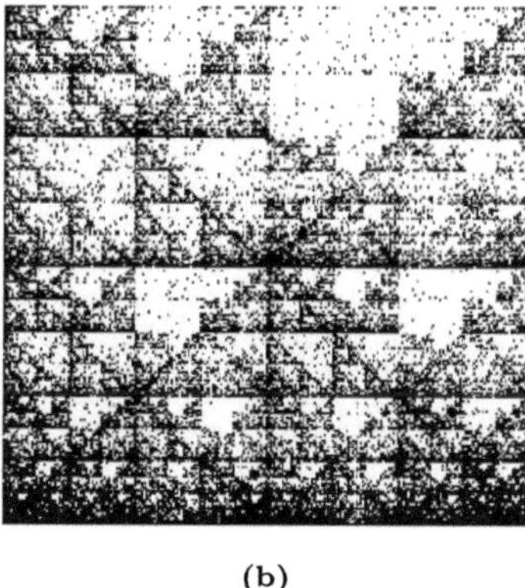

(b)

Fig. 4. Snapshot of the functionality of the computational tool: **(a)** Input: Dinucleotide ($k = 2$) distribution selected by a user via interactively adjusting 16 different sliders, each corresponding to one dinucleotide. **(b)** Output: CGR of a reconstructed DNA sequence whose k-mer distribution closely approximates the k-mer distribution in (a).

5   Conclusions and Future Work

This work formalizes the connections between CGR, $FCGR_k$, and k-mer frequency vectors as genomic signatures. We note, however, that each genomic signature has distinct features and that each has proven effective in specific applications, depending on various factors such as sequence length or sequence dataset size. The aforementioned demonstrated connections between these types of genomic signatures could potentially aid in the determination of their suitability for particular practical applications. We also present an algorithm that generates synthetic DNA sequences and their corresponding CGRs from target k-mer frequency vectors. This methodology opens a new avenue for synthetic DNA data generation in contrastive machine learning pipelines for taxonomic classification applications, whereby data augmentations (training samples artificially generated from existing data), are used to learn meaningful representations of the DNA sequences. Future work could explore the applicability of k-mer sampling for data augmentation and the effect of less stringent constraints, e.g., information-theoretic, on reconstruction guarantees.

Acknowledgments. The authors acknowledge the support of the Natural Sciences and Engineering Research Council of Canada (NSERC), [RGPIN-2023-03663] to L.K., and thank Niousha Sadjadi for generating Fig. 2b.

References

1. Alipour, F., Hill, K.A., Kari, L.: CGRclust: chaos game representation for twin contrastive clustering of unlabelled DNA sequences. BMC Genom. **25**, 1–17 (2024)
2. Avila Cartes, J. et al.: Accurate and fast clade assignment via deep learning and frequency chaos game representation. GigaScience **12**, giac119 (2023)
3. Bankevich, A., et al.: SPAdes: a new genome assembly algorithm and its applications to single-cell sequencing. J. Comput. Biol. **19**(5), 455–477 (2012)
4. Deschavanne, P.J., et al.: Genomic signature: characterization and classification of species assessed by chaos game representation of sequences. Mol. Biol. Evol. **16**(10), 1391–1399 (1999)
5. Dexter Dyer, B., Kahnand, M.J., LeBlanc, M.D.: Classification and regression tree (CART) analyses of genomic signatures reveal sets of tetramers that discriminate temperature optima of archaea and bacteria. Archaea **2**, 159–167 (2008)
6. Dummit, D.S., Foote, R.M.: Abstract Algebra, 3rd edn. Wiley, Hoboken (2004)
7. Dutta, C., Das, J.: Mathematical characterization of chaos game representation: new algorithms for nucleotide sequence analysis. J. Mol. Biol. **228**(3), 715–719 (1992)
8. De la Fuente, R. et al.: Genomic signature in evolutionary biology: a review. Biology **12**(2) (2023)
9. Goldman, N.: Nucleotide, dinucleotide and trinucleotide frequencies explain patterns observed in chaos game representations of DNA sequences. Nucleic Acids Res. **21**(10), 2487–2491 (1993)
10. Hao, B.L.: Fractals from genomes – exact solutions of a biology-inspired problem. Phys. A: Stat. Mech. Appl. **282**(1), 225–246 (2000)
11. He, H., Kari, L., Arias, P.M.: Bridging CGR and k-mer frequencies of DNA sequences (2025). https://arxiv.org/abs/2506.22172
12. Hill, K.A., Schisler, N.J., Singh, S.M.: Chaos game representation of coding regions of human globin genes and alcohol dehydrogenase genes of phylogenetically divergent species. J. Mol. Evol. **35**, 261–269 (1992)
13. Hoang, T., Yin, C., Yau, S.S.T.: Numerical encoding of DNA sequences by chaos game representation with application in similarity comparison. Genomics **108**(3–4), 134–142 (2016)
14. Jeffrey, H.J.: Chaos game representation of gene structure. Nucleic Acids Res. **18**(8), 2163–2170 (1990)
15. Jeffrey, H.: Chaos game visualization of sequences. Comput. Graph. **16**(1), 25–33 (1992)
16. Joseph, J., Sasikumar, R.: Chaos game representation for comparison of whole genomes. BMC Bioinform. **7**, 1–10 (2006)
17. Karamichalis, R., et al.: An investigation into inter-and intragenomic variations of graphic genomic signatures. BMC Bioinform. **16**(1) (2015)
18. Karlin, S., Burge, C.: Dinucleotide relative abundance extremes: a genomic signature. Trends Genet. **11**(7), 283–290 (1995)
19. Kislyuk, A., et al.: Unsupervised statistical clustering of environmental shotgun sequences. BMC Bioinform. **10**(1), 316 (2009)
20. Lichtblau, D.: Alignment-free genomic sequence comparison using FCGR and signal processing. BMC Bioinform. **20**, 1–17 (2019)
21. Löchel, H.F., Heider, D.: Chaos game representation and its applications in bioinformatics. Comput. Struct. Biotechnol. J. **19**, 6263–6271 (2021)

22. Lovász, L., Vempala, S.: Hit-and-run from a corner. SIAM J. Comput. **35**(4), 985–1005 (2006)
23. Millán Arias, P.A., et al.: DeLUCS: deep learning for unsupervised clustering of DNA sequences. PLoS ONE **17**(1), e0261531 (2022)
24. Miller, J.R., Koren, S., Sutton, G.: Assembly algorithms for next-generation sequencing data. Genomics **95**(6), 315–327 (2010)
25. Nissen, J.N., et al.: Improved metagenome binning and assembly using deep variational autoencoders. Nat. Biotechnol. **39**(5), 555–560 (2021)
26. Oliver, J., et al.: Entropic profiles of DNA sequences through chaos-game-derived images. J. Theor. Biol. **160**(4), 457–470 (1993)
27. Pevzner, P.A., Tang, H., Waterman, M.S.: An Eulerian path approach to DNA fragment assembly. PNAS **98**(17), 9748–9753 (2001)
28. Randhawa, G.S., et al.: Machine learning using intrinsic genomic signatures for rapid classification of novel pathogens: COVID-19 case study. PLoS ONE **15**(4), e0232391 (2020)
29. Rizzo, R., et al.: Classification experiments of DNA sequences by using a deep neural network and chaos game representation. In: Proceedings of the 17th International Conference on Computer Systems and Technologies, pp. 222–228 (2016)
30. Sengupta, D.C., et al.: Similarity studies of corona viruses through chaos game representation. Comput. Mol. Biosci. **10**(3), 61 (2020)
31. Solis-Reyes, S., et al.: An open-source k-mer based machine learning tool for fast and accurate sub-typing of HIV-1 genomes. PLoS ONE **13**(11), e0206409 (2018)
32. Wang, Y., et al.: The spectrum of genomic signatures: from dinucleotides to chaos game representation. Gene **346**, 173–185 (2005)
33. Zielezinski, A., et al.: Benchmarking of alignment-free sequence comparison methods. Genome Biol. **20**(1), 144 (2019)
34. Zielezinski, A., et al.: Alignment-free sequence comparison: benefits, applications, and tools. Genome Biol. **18**(1), 186 (2017)

Machine Learning by Adiabatic Evolutionary Quantum Systems
(Preliminary Report)

Tomoyuki Yamakami[(✉)]

Faculty of Engineering, University of Fukui, 3-9-1 Bunkyo, Fukui 910-8507, Japan
tomoyukiyamakami@gmail.com

Abstract. A computational model of adiabatic evolutionary quantum system (or AEQS, pronounced "eeh-ks") was introduced in [26] as a sort of quantum annealing and its underlying input-driven Hamiltonians are generated quantum-algorithmically by various forms of quantum automata families (including 1qqaf's). We study an efficient way to accomplish certain machine learning tasks by training these AEQSs quantumly. When AEQSs are controlled by 1qqaf's, it suffices in essence to find an optimal 1qqaf that approximately solves a target relational problem. For this purpose, we develop a basic idea of approximately utilizing well-known quantum algorithms for quantum counting, quantum amplitude estimation, and quantum approximation. We then provide a rough estimation of the efficiency of our quantum learning algorithms for AEQSs.

Keywords: adiabatic quantum computing · quantum automata family · optimization problem · quantum counting · quantum amplitude estimation · quantum finding of maxima

1 A Historical Account and an Overview of Contributions

We briefly state the background of the current work and provide a quick overview of our major contributions in this work.

1.1 Adiabatic Evolutionary Quantum Systems

Quantum physics have offered a new computing device, known as quantum computers. In theoretical fields, various models of quantum computing have been proposed, including quantum finite(-state) automata, quantum Turing machines, quantum circuits, and quantum recursion schemata [25]. In many models, at each step of computation applies a unitary operation (unless making measurement). In the 1980s, Benioff [5] and Deutsch [11,12] initiated a study on quantum computability based on some of these computation models. The models that were later formulated as quantum Turing machines and quantum circuits essentially execute computations by sequentially conducting unitary operations one by one.

© The Author(s), under exclusive license to Springer Nature Switzerland AG 2026

E. Formenti and L. Manzoni (Eds.): UCNC 2025, LNCS 16364, pp. 230–244, 2026.
https://doi.org/10.1007/978-3-032-15641-9_16

Unlike such unitary-operation-based quantum computing, *adiabatic quantum computing* is a physical model of performing quantum computation induced by an appropriately defined Schrödinger equation. A time-depending quantum state $|\psi(t)\rangle$ evolves according to the equation $i\hbar\frac{d}{dt}|\psi(t)\rangle = H(t)|\psi(t)\rangle$ for a specific time-dependent Hamiltonian[1] $H(t)$, where $\hbar$ is the *reduced Plank constant* and i denotes $\sqrt{-1}$. It was proven in [2] that adiabatic quantum computers are as powerful as quantum Turing machines in the polynomial-time setting. Adiabatic computing in fact has taken advantages over solving a combinatorial optimization problem, whose goal is to find a solution that maximizes a given objective function under a certain set of constraints. There have been numerous works demonstrating the use of adiabatic quantum computing for solving such optimization problems, e.g., [14].

Lately, in [26], the computational model of so-called *adiabatic evolutionary quantum systems* (or AEQS, pronounced as "eeh-ks") was introduced based on the Schrödinger equation. The main target of [26] was numerous decision problems and its purpose was to determine computational resources necessary for solving the problems on AEQSs. Notice that each Schrödinger equation is controlled by an appropriate choice of a Hamiltonian. In other words, the evolution of adiabatic quantum computation is controlled by underlying Hamiltonians. The importance of AEQSs was demonstrated in [26], which shows that every decision problem (or equivalently, language) is solvable by an appropriate choice of AEQS with accuracy exactly 1 (see also Theorem 1). This suggests an alternative way to categorize all decision problems according to the computational complexity of AEQSs that solve them.

From a practical viewpoint, it is desirable to make those Hamiltonians be "generated" by running families of machines. In [26], nonetheless, those machines are taken as restricted forms of quantum automata families. The initial goal of [26] was to solve numerous decision problems (identified with languages) using AEQSs under various constraints. It has been open to develop a coherent theory on search and optimization problems. In this work, we investigate another field of machine learning and try to use AEQSs as a new leaning model.

1.2   Quantum Machine Learning

We wish to seek a direct application of AEQSs to quantum machine learning. A study of machine learning has made a crucial influence on developing artificial intelligence (AI), which is rapidly changing the society's underlying conceptual mind-set regarding intellectual activities. A widely used model for machine learning consists of various forms of neural networks. In real applications, however, a large set of parameters is required to control these neural networks for better learning consequences.

If learning a "concept" represented by a set of data is to find an optimal algorithm whose outcomes match the "concept", then our task is to search for such an algorithm approximately within a reasonable amount of time. Thus, learning

---

[1] A Hamiltonian is a complex Hermitian matrix.

can be viewed as a process of approximating optimal solutions of some forms of optimization problems. In this work, we take this view and study efficient learning procedures in the quantum setting.

Lately, quantum machine learning has been making a crucial progress on promoting the understandings of learning process and power of quantum computing as well. For instance, Aimeur, Brassard, and Gambs [1] analyzed a quantum algorithm for clustering. Moreover, Beer et al. [4] studied a quantum analogue of neural network to accomplish given learning tasks. Our intention of this work is to provide an alternative quantum learning model to enrich our understandings of machine learning in general. More precisely, our goal is to train an AEQS to learn a target "concept" by finding its underlying 1qqaf that appropriately expresses the concept.

1.3 Major Contributions

This work will provide only a set of preliminary results and leave more advanced results to an updated version of this work. Through this work, nonetheless, we wish to demonstrate the following. Here, we focus only on simple tasks of learning certain types of "concepts". For simplicity, a "concept" is assumed to be represented as a predetermined relation (or property) on an alphabet Σ.

We wish to study how to learn a given relation (presented by a so-called *supervisor*) using the model of AEQSs as accurate as possible. As a simple, concrete example, let us consider the binary relation $EQ(x, y)$, which asserts the identity relation between two instances x and y of a fixed length. We must train an AEQS $\mathcal{S}$ to "learn" EQ in the sense that $\mathcal{S}$ behaves approximately like EQ; in other words, $\{(x, y) \mid \mathcal{S} \text{ accepts } (x, y)\}$ is a good approximation of $\{(x, y) \mid EQ(x, y)\}$. To simplify the basis setting further, we formulate our learning task in the following simple way. When instances are limited to a fixed length n, we tend to write $\mathcal{S}^{(n)}$ to denote $\mathcal{S}$ that takes only length-n instances. We now restrict our attention to a simpler unary relation.

Simple Learning Task:

- instance: a positive integer n.
- target: a relation $R_n \subseteq \{0, 1\}^n$, which is provided by a supervisor through queries.
- task: train an AEQS $\mathcal{S}^{(n)}$ to learn R_n; that is, the set $\{x \in \{0, 1\}^n \mid \mathcal{S}^{(n)} \text{ accepts } x\}$ is a "good" approximation of R_n with "high" accuracy.

Since a 1qqaf controls an AEQS, it suffices to construct a suitable 1qqaf so that the AEQS can learn a given relation properly. Hence, a key ingredient of our strategy is how to approximate a given relation R by "quantum automata," in particular, 1qqsf's. An underlying idea of our approximation procedure is an application of Grover's quantum search algorithm [15,16]. Remember that Grover's discovery of fast quantum search algorithm has found important applications to quantum amplitude amplification [8], quantum phase/amplitude estimation [21], quantum counting [7], and quantum searching of minima [13,19,23].

Our learning algorithm for the above simple learning task utilizes the underlying ideas behind those quantum algorithms.

The fundamental notions and notation regarding AEQSs and 1qqaf's are provided in Sect. 2. To accomplish the above simple learning task, we make a paradigm shift from learning by AEQSs to approximating 1qqaf's in Sect. 3. The desired quantum algorithms for approximating 1qqaf's are presented in Sect. 4.

2 Fundamental Notions and Notation

Let us explain the basic notions and notation necessary for the reader to read through the rest of this work. We loosely follow [26] for the description of these notions and notation. For the underlying reasoning behind their introduction, the reader may refer to [26].

2.1 Numbers, Languages, and Quantum States

The sets of all integers, of all natural numbers (including 0), and all real numbers are denoted $\mathbb{Z}$, $\mathbb{N}$, and $\mathbb{R}$, respectively. We further set $\mathbb{N}^+$ to be $\mathbb{N} - \{0\}$. Given two integers m and n with $m \leq n$, the notation $[m,n]_{\mathbb{Z}}$ expresses the *integer interval* $\{m, m+1, m+2, \ldots, n\}$. This is compared to the standard notation $[s,t]$ for the real interval $\{r \in \mathbb{R} \mid s \leq r \leq t\}$. We further set $[n] = [1,n]_{\mathbb{Z}}$ for each number $n \in \mathbb{N}^+$. For simplicity, we define $\mathrm{ilog}(0) = 0$ and $\mathrm{ilog}(x) = \lceil \log_2 x \rceil$ for any real number $x > 0$. Moreover, we write $\mathbb{C}$ for the set of all complex numbers. We write $\imath$ for $\sqrt{-1}$. Given a number $\alpha \in \mathbb{C}$, the notation α^* denotes its *complex conjugate*. The *commutator* $[A,B]$ between two square matrices A and B is defined to be $AB - BA$.

An *alphabet* is a finite nonempty set of "symbols" or "letters". Given such an alphabet Σ, a *string* over Σ means a finite sequence of symbols taken from Σ. The notation Σ^* denotes the set of all strings over Σ. The *length* of a string x, denoted $|x|$, is the total number of symbols in x. A subset of Σ^* is treated as a *(unary) relation* over Σ in this work whereas it is usually called a *language*. Given such a relation R and an input $x \in \Sigma^*$, we often say that $R(x)$ (resp., $\neg R(x)$) is *true* if $x \in R$ (resp., $x \notin R$).

In this work, we consider only quantum states in finite-dimensional Hilbert spaces. To express such quantum states, we use Dirac's ket notation. The *norm* (i.e., ℓ_2-norm) $\| |\phi\rangle \|$ of a quantum state $|\phi\rangle$ is set to be $\sqrt{\langle \phi | \phi \rangle}$. A *quantum bit* (or a *qubit*) is a unit-norm quantum state in the Hilbert space spanned by two basis vectors $|0\rangle$ and $|1\rangle$ (which form the *computational basis*). In contrast, the set $\{|\hat{0}\rangle, |\hat{1}\rangle\}$ is called the *Hadamard basis*, where $|\hat{0}\rangle = \frac{1}{\sqrt{2}}(|0\rangle + |1\rangle)$ and $|\hat{1}\rangle = \frac{1}{\sqrt{2}}(|0\rangle - |1\rangle)$. For two quantum states $|\phi\rangle, |\psi\rangle$ of the same dimension, $|\psi\rangle$ is said to be *ε-close* to $|\psi\rangle$ if $\| |\phi\rangle - |\psi\rangle \| \leq \varepsilon$ holds. A complex Hermitian matrix is called a *Hamiltonian*. Given a Hamiltonian H, the *spectral gap* of H, denoted $\Delta(H)$, is the difference between the lowest eigenvalue and the second lowest eigenvalue of H. We also use the following notion of "closeness" between two quantum states. Given two quantum states $|\phi\rangle, |\psi\rangle$ in the same Hilbert space, let $dis(|\phi\rangle, |\psi\rangle) = |\langle\phi|\psi\rangle|$.

For convenience, we use the notation I for the identity operator acting on a Hilbert space of an "arbitrary" dimension. Moreover, the notation $\mathbf{0}$ is used to express the *null vector*. The *Walsh-Hadamard transform* WH is defined as $\frac{1}{\sqrt{2}}\begin{bmatrix} 1 & 1 \\ 1 & -1 \end{bmatrix}$. For any $a, b \in \{0, 1\}$, we set $CNOT(|a\rangle|b\rangle) = |a\rangle|b\rangle$ if $a = 0$, and $CNOT(|a\rangle|b\rangle) = |a\rangle|\bar{b}\rangle$ otherwise, where $\bar{b} = 1 - b$. It is well-known that the set of $CNOT$ and all single-qubit gates forms a universal gate set (see, e.g., [17, 24]). For later use, we expand the scope of $CNOT$ from two qubits to the n input qubits by simply defining $CNOT_{i,j}$ to express the application of $CNOT$ to the ith and jth qubits of the n input qubits. In particular, when $i = j$, $CNOT_{i,j}$ equals I. Each single-qubit gate can be in general expressed as

$$e^{\imath\psi}\begin{bmatrix} e^{-\imath\beta} & 0 \\ 0 & e^{\imath\beta} \end{bmatrix}\begin{bmatrix} \cos\theta & -\sin\theta \\ \sin\theta & \cos\theta \end{bmatrix}\begin{bmatrix} e^{-\imath\alpha} & 0 \\ 0 & e^{\imath\alpha} \end{bmatrix}$$

using four parameters $\alpha, \beta, \theta, \psi \in [0, 2\pi)$, where $e^{\imath\psi}$ is a *(global) phase shift*. We write $U_{(\psi,\alpha,\theta,\beta)}$ for this matrix. For example, if $\alpha = \beta = \psi = 0$, then $U_{(\psi,\alpha,\theta,\beta)}$ simply represents the rotation at angle θ. We further expand $U_{(\psi,\alpha,\theta,\beta)}$ to handle all n qubits by setting $\hat{U}^{(i)}_{(\psi,\alpha,\theta,\beta)} = I^{\otimes(i-1)} \otimes U_{(\psi,\alpha,\theta,\beta)} \otimes I^{\otimes(n-i)}$ for any number $i \in [n]$. We assert that any quantum transform can be approximated with arbitrary precision by an appropriate quantum circuit composed only of $CNOT_{i,j}$s and $\hat{U}^{(i)}_{(\psi,\alpha,\theta,\beta)}$s. The *quantum Fourier transform* $\mathcal{F}_k$ is defined as

$$\mathcal{F}_k(|z\rangle) = \frac{1}{\sqrt{k}}\sum_{d=0}^{k-1} e^{2\pi\imath \cdot num(z)d/k}|d\rangle,$$

where $num(z)$ is the natural number whose binary representation matches a binary string z (by ignoring the leading 0s).

Given a Hamiltonian, its eigenvalue is particularly called an *energy* and its eigenvector is called an *eigenstate*. The lowest energy is called the *ground energy* and its associated eigenstate is called the *ground state*.

2.2   Adiabatic Evolutionary Quantum Systems or AEQSs

We quickly review the basic definition of *adiabatic evolutionary quantum system* or AEQS (pronounced "eeh-ks"). Formally, an AEQS $\mathcal{S}$ is a septuple $(m, \Sigma, \varepsilon, \{H^{(x)}_{ini}\}_{x\in\Sigma^*}, \{H^{(x)}_{fin}\}_{x\in\Sigma^*}, \{S^{(n)}_{acc}\}_{n\in\mathbb{N}}, \{S^{(n)}_{rej}\}_{n\in\mathbb{N}})$, where $m : \Sigma^* \to \mathbb{N}$ is a size function (called the *system size*), Σ is an (input) alphabet, ε is an *accuracy bound* in $[0, 1)$, both $H^{(x)}_{ini}$ and $H^{(x)}_{fin}$ are Hamiltonians acting on the Hilbert space of $2^{m(x)}$ dimension (where this space is called the system's *evolution space*), and both $S^{(n)}_{acc}$ and $S^{(n)}_{rej}$ are subsets of $\{0, 1\}^n$. The pair $(S^{(n)}_{acc}, S^{(n)}_{rej})$ is collectively called an *acceptance/rejection criteria pair*. A typical example of

m is given as $m(x) = |x|$ for all strings $x \in \Sigma^*$. We always demand that $H_{ini}^{(x)}$ and $H_{fin}^{(x)}$ should have *unique* ground states.

The Hilbert spaces spanned by the basis vectors in $\{|u\rangle \mid u \in S_{acc}^{(n)}\}$ and those in $\{|u\rangle \mid u \in S_{acc}^{(n)}\}$ are respectively denoted $QS_{acc}^{(n)}$ and $QS_{rej}^{(n)}$ and also called the *accepting space* and the *rejecting space*. Intuitively, the AEQS $\mathcal{S}$ is said to *accept* (resp., *reject*) x if the ground state of $H_{fin}^{(x)}$ is "sufficiently" close to a certain normalized vector in $QS_{acc}^{(m(x))}$ (resp., $QS_{rej}^{(m(x))}$).

Given an input $x \in \Sigma^*$, T_x denotes the minimum evolution time of $\mathcal{S}$ on x. By setting $H^{(x)}(t) = (1 - \frac{t}{T_x})H_{ini}^{(x)} + \frac{t}{T_x}H_{fin}^{(x)}$, $\mathcal{S}$ is thought to run in time $T_x \cdot \max_{t \in [0,T_x]} \|H^{(x)}(t)\|$. The ground state of $H^{(x)}(t)$ is expressed as $|\psi_g^{(x)}(t)\rangle$. In particular, $|\psi_g^{(x)}(0)\rangle$ and $|\psi_g^{(x)}(T_x)\rangle$ respectively match the ground states of $H_{ini}^{(x)}$ and $H_{fin}^{(x)}$.

A lower bound of T_x is given by the *adiabatic theorem* [20,22]. For any two constants $\varepsilon, \delta > 0$, if

$$T_x \geq \Omega\left(\frac{\|H_{fin}^{(x)} - H_{ini}^{(x)}\|}{\varepsilon^\delta \min_{t \in [0,T_x]}\{\Delta(H^{(x)}(t))^{2+\delta}\}}\right),$$

then $|\psi(T_x)\rangle$ is ε-close to the ground state $|\psi_g(T_x)\rangle$ of $H_{fin}^{(x)}$, as stated in [2].

When we need to restrict our interest of $\mathcal{S}$ within inputs of length n, we use the special notation $\mathcal{S}^{(n)}$ in place of $\mathcal{S}$. Yamakami [26] formulated the acceptance/rejection criteria of $\mathcal{S}$ as follows. Given a relation $R \subseteq \{0,1\}^*$ and a constant $\eta \in (1/2, 1]$, $\mathcal{S}$ *solves* (or *computes*) R *with accuracy at least* η if, for any $x \in \Sigma^*$, (1) there exists two unique ground states $|\psi_g^{(x)}(0)\rangle$ of $H_{ini}^{(x)}$ and $|\psi_g^{(x)}(T_x)\rangle$ of $H_{fin}^{(x)}$, (2) if $R(x)$ is true, then $|\psi_g^{(x)}(T_x)\rangle$ is $\sqrt{2(1-\eta)}$-close to a normalized quantum state in $QS_{acc}^{(m(x))}$, and (3) if $\neg R(x)$ is true, then $|\psi_g^{(x)}(T_x)\rangle$ is $\sqrt{2(1-\eta)}$-close to a normalized quantum state in $QS_{rej}^{(m(x))}$.

The computational power of AEQSs is exemplified in the following theorem of [26].

Theorem 1 [26]. *For any decision problem (or equivalently, any language) L over alphabet Σ, there exists an AEQS $\mathcal{S}$ of system size 1 for which $\mathcal{S}$ solves L with accuracy 1.*

This theorem supports the idea that AEQSs are useful tool for categorizing all decision problems in terms of the computational complexity of AEQSs that solve them.

2.3 Quantum Quasi-automata Families for AEQSs

In [26], certain AEQSs are "controlled" by machine families and the computational complexity of AEQSs is measured in terms of the complexity of those machine families. To accomplish the simple learning task described in Sect. 1.3, the core part of the rest of this work is therefore to search for a suitable machine family that controls an AEQS. See, e.g., [3] for the general information on quantum finite automata.

We use the notion of quantum (finite) automata to control the Hamiltonians of the Schrödinger equation. A *one-way quantum quasi-automata family* (abbreviated as 1qqaf) is a family $\mathcal{M} = \{M_n\}_{n \in \mathbb{N}}$ of machines M_n of the form $(Q^{(n)}, \Sigma, \{\triangleright, \triangleleft\}, \{A_\sigma^{(n)}\}_{\sigma \in \check{\Sigma}}, \Lambda_0^{(n)}, Q_0^{(n)})$, where $Q^{(n)}$ is a finite set of inner states, Σ is an (input) alphabet, $\triangleright, \triangleleft$ are endmarkers, each $A_\sigma^{(n)}$ is a quantum operator, $\Lambda_0^{(n)}$ is an initial mixture of the form $\sum_{u \in Q^{(n)}} \gamma_u |u\rangle\langle u|$ for real numbers $\gamma_u \geq 0$, and $Q_0^{(n)}$ is a subset of $Q^{(n)}$. Each $A_\sigma^{(n)}$ is defined on the Hilbert space of linear operators acting on the configuration space spanned by the vectors in $\{|q\rangle \mid q \in Q^{(n)}\}$. The *projective measurement* $\Pi_0^{(n)}$ is the projection onto the Hilbert space $span\{|u\rangle \mid u \in Q^{(n)} - Q_0^{(n)}\}$.

Notice that each $A_\sigma^{(n)}$ has a *Kraus representation* $\mathcal{K}_\sigma = \{K_{\sigma,j}\}_{j \in [k]}$ composed of k Kraus operators (or operation elements) for a constant $k \in \mathbb{N}^+$; namely, $A_\sigma^{(n)}(H) = \sum_{j=1}^{k} K_{\sigma,j} H (K_{\sigma,j})^\dagger$ for any linear operator H. Here, the number k is referred to as the *Kraus number* of $A_\sigma^{(n)}$. We always demand that $\mathcal{K}_\sigma$ satisfies the completeness relation: $\sum_{j=1}^{k} (K_{\sigma,j})^\dagger K_{\sigma,j} = I$. We further assume that all entries of $K_{\sigma,j}$ are indexed by the elements of $Q^{(n)} \times Q^{(n)}$.

Each 1qqaf produces a series of Kraus operators. We further expand $A_\sigma^{(n)}$ by setting inductively $A_{z\sigma}^{(n)}(H) = A_\sigma^{(n)}(A_z^{(n)}(H))$ for $\sigma \in \Sigma \cup \{\lambda\}$ and $z \in \{\triangleright, \lambda\}\Sigma^* - \{\lambda\}$. Each 1qqaf M_n is said to *generate* (or *produce*) a quantum operator $E_{\triangleright x \triangleleft}^{(n)} = \Pi_0^{(n)} A_{\triangleright x \triangleleft}^{(n)}(\Lambda_0^{(n)}) \Pi_0^{(n)}$.

A *selector* refers to a function from Σ^* to $\mathbb{N}$. For instance, the standard selector has the form $\mu(x) = |x|$. A selector μ naturally induces the *input domain* $\Delta_n = \{x \in \Sigma^* \mid \mu(x) = n\}$. A family $\{H^{(x)}\}_{x \in \Sigma^*}$ of Hamiltonians is *generated by 1qqaf's* if there are a polynomially-bounded selector μ and a 1qqaf $\mathcal{M} = \{M_n\}_{n \in \mathbb{N}}$ such that, for any $x \in \Sigma^*$, M_n generates $E_{\triangleright x \triangleleft}^{(\mu(x))}$ and $E_{\triangleright x \triangleleft}^{(\mu(x))}$ coincides with $H^{(x)}$.

We remark that the time-dependent Hamiltonian $H^{(x)}(t)$ of the Schrödinger equation depends on each input x. The use of a 1qqaf $\mathcal{M} = \{M_n\}_{n \in \mathbb{N}}$, whose elements depend only on n, provides a means of controlling an AEQS on "all" inputs x of length $m(x)$.

Example 2 [26]. Let *Equal* denote the language $\{w \mid \#_0(w) = \#_1(w)\}$ over the alphabet $\Sigma = \{a, b\}$, where $\#_b(w)$ is the total number of occurrences of b in w. It is known that no one-way finite automata solve *Equal*. However, this language *Equal* is solved by an AEQS controlled by a certain 1moqqaf, where 1moqqaf refers to "measure-once" 1qqaf (which corresponds to the case that the Kraus number is 1).

3   A New Setting for Machine Learning by AEQSs

To complete the simple learning task described in Sect. 1.3, we use the computational model of AEQS to train it appropriately so that it learns a given relation correctly. When the AEQS is controlled by a 1qqaf, it is possible to focus on constructing such a 1qqaf. In this section, we explain how to make a paradigm shift from "learning by AEQSs" to "optimizing 1qqaf's."

3.1   From Learning by AEQSs to Optimizing 1qqaf's

Since each AEQS is controlled by an appropriate 1qqaf, it is possible to focus on our attention only on the approximation of 1qqaf's instead of directly designing AEQSs. In [26], several formal languages are shown to be decidable by AEQSs with appropriate choices of underlying 1qqaf's. Those constructions actually show similar traits. By exploiting such traits, in this work, we intend to restrict the framework of AEQSs as follows.

A 1qqaf $\mathcal{M} = \{M_n\}_{n\in\mathbb{N}}$ consists of machines M_n of the form $(Q^{(n)}, \Sigma, \{\triangleright, \triangleleft\}, \{A_\sigma^{(n)}\}_{\sigma\in\check{\Sigma}}, \Lambda_0^{(n)}, Q_0^{(n)})$. To simplify the description of the quantum learning algorithms in Sect. 4, we intend to consider a restricted case by fixing some of the parameters of M_n. We fix a designated element, say, q_0 in $Q^{(n)}$ and then set $Q_-^{(n)} = Q^{(n)} - \{q_0\}$. Moreover, we set $\Lambda_0^{(n)} = \sum_{u\in Q_-^{(n)}} |u\rangle\langle u|$. For each $x \in \Sigma^*$, we define $m(x) = \mathrm{ilog}|Q^{(\mu(x))}|$ and informally treat $Q^{(\mu(x))}$ as $\{0,1\}^{m(x)}$. For each symbol $\sigma \in \check{\Sigma}$, $A_\sigma^{(n)}$ is a square matrix indexed by $Q^{(n)}\times Q^{(n)}$ and we focus on the case where the Kraus number of $A_\sigma^{(n)}$ is 1. In other words, $A_\sigma^{(n)}(H)$ has the form $U_\sigma^{(n)} H (U_\sigma^{(n)})^\dagger$ for an appropriate unitary operator $U_\sigma^{(n)}$, where H is an arbitrary linear operator. In analogy to the expansion of $A_\sigma^{(n)}$ to $A_z^{(n)}$ for any string z, we naturally expand $U_\sigma^{(n)}$ to $U_z^{(n)}$. Notice that $E_{\triangleright x\triangleleft}^{(n)}$ expresses $\Pi_0^{(n)} A_{\triangleright x\triangleleft}^{(n)}(\Lambda_0^{(n)})\Pi_0^{(n)}$. Furthermore, we set $Q_0^{(n)} = \varnothing$ for all $n \in \mathbb{N}$. This makes $\Pi_0^{(n)}$ be identified with I.

Next, we consider an AEQS $\mathcal{S}$ of the form $(m, \Sigma, \varepsilon, \{H_{ini}^{(x)}\}_{x\in\Sigma^*}, \{H_{fin}^{(x)}\}_{x\in\Sigma^*}, \{S_{acc}^{(n)}\}_{n\in\mathbb{N}}, \{S_{rej}^{(n)}\}_{n\in\mathbb{N}})$. Here, we force $S_{acc}^{(m(x))}, S_{rej}^{(m(x))} \subseteq Q^{(\mu(x))}$ for all $x \in \Sigma^*$. We further define $H_{ini}^{(x)}$ to be $WH^{\otimes m(x)}\Lambda_0^{(\mu(x))}(WH^{\otimes m(x)})^\dagger$ and set $H_{fin}^{(x)} = E_{\triangleright x\triangleleft}^{(\mu(x))}$.

Lemma 3. *For any $x \in \Sigma^*$, the ground state $|\phi_x\rangle$ of $H_{fin}^{(x)}$ is of the form $U_{\triangleright x\triangleleft}^{(\mu(x))}|q_0\rangle$.*

Proof. Fix $x \in \Sigma^*$ and set $|\phi_x\rangle = U_{\triangleright x\triangleleft}^{(\mu(x))}|q_0\rangle$. Let us calculate $H_{fin}^{(x)}|\phi_x\rangle$ to see that $|\phi_x\rangle$ is indeed the ground state of $H_{fin}^{(x)}$. Since $\Pi_0^{(n)} = I$ for any $n \in \mathbb{N}$, it follows that $H_{fin}^{(x)}|\phi_x\rangle = U_{\triangleright x\triangleleft}^{(\mu(x))}\Lambda_0^{(\mu(x))}(U_{\triangleright x\triangleleft}^{(\mu(x))})^\dagger|\phi_x\rangle = U_{\triangleright x\triangleleft}^{(\mu(x))}\Lambda_0^{(\mu(x))}(U_{\triangleright x\triangleleft}^{(\mu(x))})^\dagger U_{\triangleright x\triangleleft}^{(\mu(x))}|q_0\rangle$. The last term is further simplified to $U_{\triangleright x\triangleleft}^{(\mu(x))}\Lambda_0^{(\mu(x))}|q_0\rangle$. Since $\Lambda_0^{(\mu(x))} = \sum_{u\in Q_-^{(\mu(x))}} |u\rangle\langle u|$, we instantly obtain

$\Lambda_0^{(\mu(x))}|q_0\rangle = 0$. Therefore, $H_{fin}^{(x)}|\phi_x\rangle = 0$ follows. This indicates that $|\phi_x\rangle$ is an eigenvector of $H_{fin}^{(x)}$ and its eigenvalue is 0. $\square$

The outcome of $\mathcal{S}$ on input x depends on whether its ground state $|\phi_x\rangle$ is close enough in distance to $QS_{acc}^{(m(x))}$ or $QS_{rej}^{(m(x))}$. As shown below, this is determined by simply making a measurement of $|\phi_x\rangle$ in the basis of $\{|u\rangle \mid u \in S_{acc}^{(m(x))}\}$ or $\{|u\rangle \mid u \in S_{rej}^{(m(x))}\}$. We use the notation $(S_{acc}^{(m(x))})^{\perp}$ to denote the set $Q^{(\mu(x))} - S_{acc}^{(m(x))}$.

Lemma 4. *Let $|\phi_x\rangle$ denote the ground state of $H_{fin}^{(x)}$ and let $\eta \in (1/2, 1]$. The following two statements are logically equivalent. (1) There exists a unit-norm quantum state $|\xi\rangle \in QS_{acc}^{(m(x))}$ such that $\||\phi_x\rangle - |\xi\rangle\| \le \sqrt{2(1-\eta)}$. (2) After a measurement of $|\phi_x\rangle$ in the basis of $\{|u\rangle \mid u \in S_{acc}^{(m(x))}\}$, we observe u in $S_{acc}^{(m(x))}$ with probability at least η. The same also holds for $S_{rej}^{(m(x))}$.*

Proof. $(1) \Rightarrow (2)$. We express the ground state $|\phi_x\rangle$ as $|\psi_1\rangle + |\psi_2\rangle$, where $|\psi_1\rangle = \sum_{v \in S_{acc}^{(m(x))}} \alpha_v |v\rangle$ and $|\psi_1\rangle = \sum_{v \in (S_{acc}^{(m(x))})^{\perp}} \beta_v |v\rangle$ with $\alpha_v, \beta_v \in \mathbb{C}$. We define $|\tilde{\psi}_1\rangle$ to be the normalized $|\psi_1\rangle$, that is, $\frac{1}{\||\psi_1\rangle\|}|\psi_1\rangle$. Consider the value $\||\phi_x\rangle - |\xi\rangle\|$ for any unit-norm quantum state $|\xi\rangle \in QS_{acc}^{(m(x))}$. Since this value becomes the minimum when $|\xi\rangle = |\tilde{\psi}_1\rangle$, by (1), it follows that $\||\phi_x\rangle - |\tilde{\psi}_1\rangle\| \le \sqrt{2(1-\eta)}$.

After making a measurement of $|\phi_x\rangle$ in the basis of $\{|u\rangle \mid u \in S_{acc}^{(m(x))}\}$, we obtain $|\psi_1\rangle$. Now, we wish to show that $\||\psi_1\rangle\| \ge \eta$. We first calculate the value $\langle\tilde{\psi}_1|\phi_x\rangle$. It follows that $\langle\tilde{\psi}_1|\phi_x\rangle = \langle\tilde{\psi}_1|(|\psi_1\rangle + |\psi_2\rangle) = \langle\tilde{\psi}_1|\psi_1\rangle = \frac{1}{\||\psi_1\rangle\|}\langle\psi_1|\psi_1\rangle = \||\psi_1\rangle\|$. Notice also that $\langle\phi_x|\psi_1\rangle = \||\psi_1\rangle\|$. From these equalities, we obtain $\||\phi_x\rangle - |\tilde{\psi}_1\rangle\|^2 = \||\tilde{\psi}_1\rangle\|^2 + \||\phi_x\rangle\|^2 - (\langle\phi_x|\tilde{\psi}_1\rangle + \langle\tilde{\psi}_1|\phi_x\rangle) = 2 - 2\||\psi_1\rangle\|$. Thus, $2(1 - \||\psi_1\rangle\|) \le 2(1-\eta)$ follows. We therefore conclude that $\||\psi_1\rangle\| \ge \eta$, as requested.

$(2) \Rightarrow (1)$. It follows from (1) that $|\langle\xi|\phi_x\rangle|^2 \ge \eta$ holds for a certain unit-norm quantum state $|\xi\rangle \in QS_{acc}^{(m(x))}$. Since $|\langle\tilde{\psi}_1|\phi_x\rangle| \ge |\langle\xi|\phi_x\rangle|$ holds, we obtain $|\langle\tilde{\psi}_1|\phi_x\rangle| \ge \eta$. As shown before, $\||\phi_x\rangle - |\tilde{\psi}_1\rangle\|^2 = 2(1 - \langle\tilde{\psi}_1|\phi_x\rangle)$ follows. Hence, we conclude that $\||\phi_x\rangle - |\tilde{\psi}_1\rangle\| \le \sqrt{2(1-\eta)}$. $\square$

To make our later analysis simpler, we set $\mu(x) = |x|$ for all $x \in \Sigma^*$. Note that $m(x) = m(y)$ holds for all pairs $x, y \in \Sigma^*$ with $|x| = |y|$ because of $m(x) = \mathrm{ilog}|Q^{(\mu(x))}|$. For simplicity, we write $\bar{n}$ for $\mathrm{ilog}|Q^{(n)}|$.

For each machine M_n, we modify it into another machine $\hat{M}_n$ by changing its behavior as follows. The machine $\hat{M}_n$ starts with the initial inner state $|q_0\rangle$ and simulates M_n as reading input x of length n. When M_n halts, $\hat{M}_n$ observes its inner state. This machine $\hat{M}_n$ is said to *accept* x if its observed inner state is in $S_{acc}^{(\bar{n})}$ and to *reject* x if the observed inner state is in $S_{rej}^{(\bar{n})}$. We briefly call $\hat{M}_n$ an *acceptor version* of M_n (with respect to $\mathcal{S}$). In what follows, we automatically set $S_{rej}^{(\bar{n})} = Q^{(n)} - S_{acc}^{(\bar{n})}$.

Although the above framework looks too restrictive, this in fact makes it possible to shift from "learning by AEQSs" to "optimizing 1qqaf's".

3.2 Matrix-Decomposition of Quantum Automata

Now, we shift our attention from "learning by AEQSs" to "optimizing 1qqaf's". More precisely, we search for the best possible 1qqaf that correctly computes a given relation on most instances.

Unfortunately, there are *uncountably many* possible 1qqaf's usable for AEQSs. Hence, it is necessary to restrict our attention to a "small" set of 1qqaf's in search of an "optimal" 1qqaf. To achieve this goal for a given 1qqaf $\mathcal{M} = \{M_n\}_{n\in\mathbb{N}}$, we hereafter focus on an acceptor version $\hat{M}_n$ of each machine M_n in $\mathcal{M}$. To treat such $\hat{M}_n$ in later quantum algorithms, we require its appropriately defined encoding $\langle \hat{M}_n \rangle$.

Firstly, we set $\Sigma = \{0, 1\}$. We demand that $Q^{(n)} \subseteq [0, n+1]_{\mathbb{Z}}$, and then define its encoding $\langle Q^{(n)} \rangle$ to be the sequence $(i_1, i_2, \ldots, i_t)$ of numbers if $Q^{(n)} = \{i_1, i_2, \ldots, i_t\}$ with $i_1 < i_2 < \cdots < i_k$. Since $S_{acc}^{(\bar{n})} \subseteq Q^{(n)}$, we further define $\langle S_{acc}^{(\bar{n})} \rangle$ to be $(j_1, j_2, \ldots, j_s)$ if $S_{acc}^{(\bar{n})} = \{j_1, j_2, \ldots, j_s\}$ with $j_1 < j_2 < \cdots < j_s$.

Next, we focus on each quantum transition matrix $A_\sigma^{(n)}$ of M_n for a symbol $\sigma \in \check{\Sigma}$. Since $A_\sigma^{(n)}$ is unitary, we wish to "decompose" it using its quantum circuit representation given in the following way. Let us recall $CNOT_{i,j}$ and $\hat{U}_{(\psi,\alpha,\theta,\beta)}^{(i)}$ working on n input qubits. Letting $\tau = ((i_1, \psi_1, \alpha_1, \theta_1, \beta_1), (i_2, \psi_2, \alpha_2, \theta_2, \beta_2), \ldots, (i_k, \psi_k, \alpha_k, \theta_k, \beta_k), (i, j))$, we define $V_\sigma^{(\tau)}$ as $\hat{U}_{(\psi_1,\alpha_1,\theta_1,\beta_1)}^{(i_1)} \hat{U}_{(\psi_2,\alpha_2,\theta_2,\beta_2)}^{(i_2)} \cdots \hat{U}_{(\psi_k,\alpha_k,\theta_k,\beta_k)}^{(i_k)} CNOT_{i,j}$. We call this tuple τ the *design* of $V_\sigma^{(\tau)}$. Since $CNOT$ and all single-qubit gate form a universal gate set, $A_\sigma^{(n)}$ can be expressed as $V_\sigma^{(\tau_1)} V_\sigma^{(\tau_2)} \cdots V_\sigma^{(\tau_m)}$ for an appropriate choice of $(\tau_1, \tau_2, \ldots, \tau_m)$. The sequence $(\tau_1, \tau_2, \ldots, \tau_m)$ is called the *(design) encoding* of $A_\sigma^{(n)}$ and it is denoted by $\langle A_\sigma^{(n)} \rangle$. Finally, we define the encoding $\langle \hat{M}_n \rangle$ of $\hat{M}_n$ by setting it to be the sequence $(\langle Q^{(n)} \rangle, \langle S_{acc}^{(\bar{n})} \rangle, \langle A_{\sigma_1}^{(n)} \rangle, \langle A_{\sigma_2}^{(n)} \rangle, \ldots, \langle A_{\sigma_t}^{(n)} \rangle)$.

To limit the number of possible encodings, we restrict the choice of $(\tau_1, \tau_2, \ldots, \tau_m)$ by allowing only a polynomial number of tuples $(i, \psi, \alpha, \theta, \beta)$ in each τ_k (such $(\tau_1, \tau_2, \ldots, \tau_m)$'s are called "admissible") and thus by allowing only polynomially-growing matrix-decomposition complexity. We say that $\hat{M}_n$ is *admissible* if its encoding $\langle \hat{M}_n \rangle$ consists of only admissible numbers. This restriction helps us search for $\hat{M}_n$ from only a pool of exponentially many encodings instead of a pool of uncountably many possible encodings.

Here, we remark that we generally assume an efficient encoding and decoding of each series of "numbers" so that we can freely run $\hat{M}_n$ on any input whenever $\langle \hat{M}_n \rangle$ is given.

4 How to Train AEQSs to Learn Relations

This section presents how to accomplish the simple learning task described in Sect. 1.3. Meanwhile, we fix n ($\in \mathbb{N}^+$) arbitrarily and focus on a target relation R_n, which is simply a subset of $\{0, 1\}^n$. To learn R_n by AEQSs, we wish to find a good approximation of R_n. Since an AEQS is essentially controlled by its underlying 1qqaf, we only deal with 1qqaf's. In the rest of this section, the notation M_n generally refers to the n-th machine of an arbitrarily fixed 1qqaf $\mathcal{M} = \{M_n\}_{n \in \mathbb{N}}$ and $\hat{M}_n$ refers to an acceptor version of M_n. We also fix a constant $\eta \in (1/2, 1]$.

To simplify the notation, as long as there is no confusion, we write "$\hat{M}_n(x) =_\eta R_n(x)$" to mean that (i) $\hat{M}_n(x)$ accepts x with probability at least η if R_n is true and (ii) $\hat{M}_n(x)$ rejects x with probability at least η if $\neg R_n$ is true.

4.1   Simple Case

We intend to present quantum algorithms to optimize $\hat{M}_n$ that solves R_n. As a starter, we consider a simple case where

(*) there exists an admissible machine $\hat{M}_n$ such that $\hat{M}_n(x) =_\eta R_n(x)$ for all instances $x \in \{0, 1\}^n$.

In this simple case, our task is to find such a machine $\hat{M}_n$ (actually its encoding $\langle \hat{M}_n \rangle$) with the help of R_n. We begin with sketching the basic quantum algorithm $\mathcal{A}$, which is a backbone of the subsequent quantum algorithm.

Basic Quantum Algorithm $\mathcal{A}$:

1. Starting with the quantum state $|0^m\rangle|0^n\rangle$, generate $\frac{1}{\sqrt{s}} \sum_{m_{(n)}} |m_{(n)}\rangle|0^n\rangle$, where $m_{(n)}$ means the admissible encoding $\langle \hat{M}_n \rangle$ of a machine M_n and s is the total number of all possible such machines.
2. Generate $\frac{1}{\sqrt{2^n}} \sum_{x:|x|=n} |x\rangle$ from $|0^n\rangle$ by applying $WH^{\otimes n}$.
3. For each $m_{(n)} = \langle M_n \rangle$, run M_n on x to obtain $\hat{M}_n(x)$.
4. Query to the relation R and change the qubits $|m_{(n)}\rangle|x\rangle$ to $-|m_{(n)}\rangle|x\rangle$ if $\hat{M}_n(x) =_\eta R_n(x)$, and $|m_{(n)}\rangle|x\rangle$ otherwise. The obtained quantum state is of the form $\frac{1}{\sqrt{s}} \sum_{m_{(n)}} |m_{(n)}\rangle \otimes \frac{1}{\sqrt{2^n}} \sum_{x:|x|=n} \xi_{m_{(n)},x} |x\rangle |r_{m_{(n)},x}\rangle$ with $\xi_{m_{(n)},x} \in \{\pm 1\}$ and $r_{m_{(n)},x} \in \{0, 1\}$.
5. Transform $\frac{1}{\sqrt{2^n}} \sum_{x:|x|=n} |x\rangle$ to $|0^n\rangle$ by applying $WH^{\otimes n}$.

Let $|\Psi\rangle$ denote the quantum state obtained by running $\mathcal{A}$, that is, $|\Psi\rangle = \mathcal{A}(|0^m\rangle|0^n\rangle)$. We can split this quantum state $|\Psi\rangle$ into two parts $|\Psi_0\rangle + |\Psi_1\rangle$, where $|\Psi_1\rangle$ has the form $\sum_{m_{(n)}} \alpha_{m_{(n)},n} |m_{(n)}\rangle|0^n\rangle$ for appropriate amplitudes $\alpha_{m_{(n)},n} \in \mathbb{C}$ and $|\Psi_0\rangle$ has the form $\sum_{m_{(n)}} \sum_{y:|y|=n \wedge x \neq 0^n} \beta_{m_{(n)},n,x} |m_{(n)}\rangle|x\rangle$. From our assumption, there exists an admissible machine M_n with $m_{(n)} = \langle \hat{M}_n \rangle$ satisfying $\hat{M}_n(x) =_\eta R_n(x)$ for all $x \in \{0, 1\}^n$. It thus follows that $\sum_{m_{(n)}} |\alpha_{m_{(n)},n}|^2 \geq 1/s$.

First Quantum Algorithm:

1. Apply $\mathcal{A}$ to $|0^m\rangle|0^n\rangle$ and obtain $|\Psi\rangle$. We split $|\Psi\rangle$ into two parts $|\Psi_0\rangle + |\Psi_1\rangle$, as stated above.

2. Let $\zeta = \|\langle \Psi_1 | \Psi_1 \rangle\|$ and take $\theta \in [0, \pi/2]$ satisfying $\zeta = \sin^2 \theta$. In the case where θ is known, with additional qubits, there is a way to increase the success probability over $1 - \varepsilon$ for a small constant ε [10]. Since θ is unknown in general, we first need to estimate θ by running a quantum algorithm for quantum amplitude estimation. Let $\tilde{\theta}$ denote the obtained estimation of θ and set $\tilde{\zeta} = \sin^2 \tilde{\theta}$.

3. Using the value $\tilde{\zeta}$, we amplify the amplitude of $|\Psi_1\rangle$.

4. Perform a measurement on the first register and obtain a desired machine $\hat{M}_n$ with $m_{(n)} = \langle \hat{M}_n \rangle$.

In what follows, we give a series of detailed explanation of each step of the first quantum algorithm stated above.

(Step 2) In this step, we conduct quantum amplitude estimation of [8] in the following way. Instead of calculating the value η, we here intend to estimate the value θ. Let the controlled-$\mathcal{A}$ operator be defined as $\mathcal{O}[\mathcal{A}](|j\rangle|z\rangle) = |j\rangle \otimes \mathcal{A}^j(|z\rangle)$ for any $j \in \mathbb{N}$ and $|z\rangle = |m_{(n)}\rangle|0^n\rangle$. Recall the quantum Fourier transform $\mathcal{F}_k$. We apply $(\mathcal{F}_k^{-1} \otimes I)\mathcal{O}[\mathcal{A}](\mathcal{F}_k \otimes I)$ to $|0\rangle \otimes \mathcal{A}(|0^m\rangle|0^n\rangle)$. We then obtain an approximate value $\tilde{\theta} = \frac{\pi}{k}z$, where z is the value obtained by performing measurement on the first register. Finally, we obtain the desired $\tilde{\zeta}$ by calculating $\sin^2 \tilde{\theta}$.

(Step 3) We then amplify the amplitude of $|\Psi_1\rangle$ using $\tilde{\zeta}$. For this purpose, we set $C_{\tilde{\zeta}} = \frac{2}{1-\tilde{\zeta}}|\Psi_0\rangle\langle\Psi_0| - I$ and $C_{\mathcal{A}} = 2|\Psi\rangle\langle\Psi| - I$, which equals $\mathcal{A}(2|0^m, 0^n\rangle\langle 0^m, 0^n| - I)\mathcal{A}^\dagger$. We then define $Q = C_{\mathcal{A}}C_{\tilde{\eta}}$ and apply Q^l to $|0^m\rangle|0^n\rangle$, where l is $\lfloor \pi/4\tilde{\theta} \rfloor$. We then obtain $\frac{1}{\sqrt{\tilde{\zeta}}}\sin((2l+1)\tilde{\theta})|\Psi_1\rangle + \frac{1}{\sqrt{1-\tilde{\zeta}}}\cos((2l+1)\tilde{\theta})|\Psi_0\rangle$. The probability of observing $|m_{(n)}\rangle|0^n\rangle$ is at least $\sin^2((2l+1)\tilde{\theta}) \geq 1 - \tilde{\zeta}$.

4.2 Slightly More General Case

In the previous subsection, we have considered the simple case in which the aforementioned condition (*) actually holds. Here, we study a slightly more general case where this condition (*) is not assumed. Hence, we do not use Step 5 of the basic quantum algorithm $\mathcal{A}$.

Let us consider an admissible machine $\hat{M}_n$ and set $m_{(n)} = \langle \hat{M}_n \rangle$. We define $r_{m_{(n)},x}$ as $r_{m_{(n)},x} = 1$ if $\hat{M}_n(x) =_\eta R_n(x)$, and 0 otherwise. As noted in Sect. 4.1, if the condition (*) holds, then it follows that $dis(\frac{1}{\sqrt{2^n}}\sum_{x:|x|=n}|x\rangle|r_{m_{(n)},x}\rangle, \frac{1}{\sqrt{2^n}}\sum_{x:|x|=n}|x\rangle|1\rangle) = 1$.

Since we do not assume (*) here, our goal is to find a machine $\hat{M}_n$ with $m_{(n)} = \langle \hat{M}_n \rangle$ such that the distance $dis(\frac{1}{\sqrt{2^n}}\sum_{x:|x|=n}|x\rangle|r_{m_{(n)},x}\rangle, \frac{1}{\sqrt{2^n}}\sum_{x:|x|=n}|x\rangle|1\rangle)$ is as large as possible. Letting $\#_{R_n}(\hat{M}_n, n) = |\{x \in \{0,1\}^n \mid r_{m_{(n)},x} = 1\}|$, the distance equals $\frac{1}{2^n} \cdot |\{x \in \{0,1\}^n \mid r_{m_{(n)},x} = 1\}|$, which is expressed as $\frac{1}{2^n} \cdot \#_{R_n}(\hat{M}_n, n)$. It thus suffices to compute $\#_{R_n}(\hat{M}_n, n)$.

Second Quantum Algorithm:

1. From $|0^m\rangle|0^n\rangle$, generate $|\Psi\rangle = \frac{1}{\sqrt{s}}\sum_{m_{(n)}}|m_{(n)}\rangle|0^n\rangle$ similarly to Step 1 of $\mathcal{A}$.

2. For each value $m_{(n)}$ with $m_{(n)} = \langle \hat{M}_n \rangle$, estimate the value $\#_{R_n}(\hat{M}_n, n)$ by running a quantum counting algorithm. Let $\widetilde{\#_{R_n}}(\hat{M}_n, n)$ denote the obtained estimation of $\#_{R_n}(\hat{M}_n, n)$.

3. Assume that the obtained quantum state has the form $\sum_{m_{(n)}} \alpha_{m_{(n)},n} |m_{(n)}\rangle |\phi_{m_{(n)},n}\rangle$ with $\alpha_{m_{(n)},n} \in \mathbb{C}$ and $\| |\phi_{m_{(n)},n}\rangle \| = 1$. We then approximate the maximum value stored in the second register.

4. Perform a measurement on the first register.

We explain each step of the second quantum algorithm.

(Step 2) We utilize a quantum counting algorithm of Brassard, Høyer, and Tapp [7]. Meanwhile, we fix $m_{(n)}$ arbitrarily and start this step with the quantum state $|0^n\rangle$. We then transform $|0^n\rangle$ to $\frac{1}{\sqrt{2^n}} \sum_{x:|x|=n} |x\rangle$ by applying $WH^{\otimes n}$. We change $|m_{(n)}\rangle |x\rangle$ to $-|m_{(n)}\rangle |x\rangle$ if $M_n(x) =_\eta R_n(x)$, and $|m_{(n)}\rangle |x\rangle$ otherwise. For each bit $b \in \{0,1\}$, let $|\Xi_b\rangle$ denote the quantum state $\frac{1}{\sqrt{2^n}} \sum_{x:|x|=n \wedge r_{m_{(n)},x}=b} |x\rangle$. Let $\gamma = \langle \Xi_1 | \Xi_1 \rangle$. Note that $\#_{R_n}(\hat{M}_n, n) = \gamma \cdot 2^n$. To estimate $\#_{R_n}(\hat{M}_n, n)$, it suffices to estimate γ. The estimation of γ can be done by running a quantum phase estimation algorithm (as stated before). Let $\tilde{\gamma}$ denote the obtained estimation of γ. From this $\tilde{\gamma}$, we obtain $\widetilde{\#_{R_n}}(\hat{M}_n, n) = \tilde{\gamma} \cdot 2^n$ with probability at least $\frac{4}{\pi^2}$.

(Step 3) For this purpose, we search for $\hat{M}_n$ with $m_{(n)} = \langle \hat{M}_n \rangle$ such that $|\widetilde{\#_{R_n}}(\hat{M}_n, n)\rangle$ is the nearest value to $|2^n\rangle$. We use a basic idea of quantum algorithm of Dürr and Høyer [13] who demonstrated how to find the minimum value in an unstructured database. This algorithm is founded on a quantum search algorithm of Boyer, Brassard, Høyer, and Tapp [9].

We generate $\frac{1}{\sqrt{2^n}} \sum_{0 \leq t < 2^n} |t\rangle$. Assume that $|\phi_{m_{(n)},n}\rangle = \sum_y \beta_{m_{(n)},y} |r_{m_{(n)},y}\rangle$. Upon a query, we change $|m_{(n)}\rangle |r_{m_{(n)},y}\rangle |t\rangle$ to $-|m_{(n)}\rangle |r_{m_{(n)},y}\rangle |t\rangle$ if $r_{m_{(n)},y} > t$, and $|m_{(n)}\rangle |r_{m_{(n)},y}\rangle |t\rangle$ if $r_{m_{(n)},y} \leq t$. Let $|\Psi\rangle$ denote the obtained quantum state. We then split it into two parts $|\Psi_0\rangle + |\Psi_1\rangle$, where $|\Psi_1\rangle$ has the form $\frac{1}{\sqrt{2^n}} \sum_t \sum_{m_{(n)}} \sum_{y:r_{m_{(n)},y} > t} \gamma_{m_{(n)},y,t} |m_{(n)}\rangle |r_{m_{(n)},y}\rangle |t\rangle$ and $|\Psi_0\rangle$ has the form $\frac{1}{\sqrt{2^n}} \sum_t \sum_{m_{(n)}} \sum_{y:r_{m_{(n)},y} \leq t} \gamma_{m_{(n)},y,t} |m_{(n)}\rangle |r_{m_{(n)},y}\rangle |t\rangle$. After running a quantum amplitude amplification algorithm, we can observe $|m_{(n)}\rangle |\widetilde{\#_{R_n}}(\hat{M}_n, n)\rangle |t\rangle$. We then replace t by $\widetilde{\#_{R_n}}(\hat{M}_n, n)$ and continue the above procedure. Note that the value of t increases to reach the maximum value.

It turns out that the query complexity of the entire algorithm is $O(\sqrt{2^n})$.

5 Summary and Open Problems

Throughout this work, we have engaged in a task of learning a "unary" relation, which is simply a subset of $\{0,1\}^n$ and we have attempted to complete the simple learning task given in Sect. 1.3 using AEQSs, which were originally introduced in [26]. In particular, we have focused on AEQSs whose Hamiltonians are generated by running 1qqaf's. Our core strategy is based on a paradigm shift from learning by AEQSs to approximating 1qqaf's. We have then provided two quantum algorithms to approximate these 1qqaf's.

In this preliminary report, we have discussed only the simple learning task and left more general tasks untreated. It is thus desirable to study more general tasks and more complex machine families, such as 2-way quantum automata families as well as pushdown automata families. Moreover, it may be desirable to simplify the entire quantum algorithms by combining several procedures taken in the first and the second quantum algorithms in Sect. 4.

There remain a number of important questions unsolved in this work. A few of them are listed below.

1. It is desirable to conduct a detailed complexity analysis of the two learning algorithms given in Sect. 4.
2. For each choice of quantum automata family (1qqaf's and other models), which class of relations is learnable by AEQSs?
3. Are more relations learnable efficiently by AEQSs controlled by quantum automata families than by any classical learning algorithms?

References

1. Aimeur, E., Brassard, G., Gambs, S.: Quantum clustering algorithms. In: Proceedings of the 24th International Conference on Machine Learning, pp. 1–8 (2007)
2. Aharonov, D., van Dam, W., Kemp, J., Landau, Z., Lloyd, S., Regev, O.: Adiabatic quantum computation is equivalent to standard quantum computation. SIAM J. Comput. **37**, 166–194 (2007)
3. Ambainis, A., Yakaryilmaz, A.: Automata and quantum computing. arXiv:1507.01988 (2015)
4. Bccr, K., ct al.: Training dccp quantum neural networks. Nat. Commun. **11**, article no. 808 (2020)
5. Benioff, P.: The computer as a physical system: a microscopic quantum mechanical Hamiltonian model of computers represented by Turing machines. J. Stat. Phys. **22**, 563–591 (1980)
6. Bernstein, E., Vazirani, U.: Quantum complexity theory. SIAM J. Comput. **26**, 1411–1473 (1997)
7. Brassard, G., Høyer, P., Tapp, A.: Quantum counting. In: Larsen, K.G., Skyum, S., Winskel, G. (eds.) ICALP 1998. LNCS, vol. 1443, pp. 820–831. Springer, Heidelberg (1998). https://doi.org/10.1007/BFb0055105
8. Brassard, G., Høyer, P., Mosca, M., Tapp, A.: Quantum amplitude amplification and estimation. In: Lomonaco, S.J. (ed.) Quantum Computation and Quantum Information, vol. 305, pp. 53–74. American Mathematical Society (2002)
9. Boyer, M., Brassard, G., Høyer, P., Tapp, A.: Tight bounds on quantum searching. Fortschr. Phys. **46**, 493–505 (1998)
10. Cleve, R., Ekert, A., Macchiavello, C., Mosca, M.: Quantum algorithms revisited. Proc. Roy. Soc. A: Math. Phys. Eng. Sci. **454**, 339–354 (1998)
11. Deutsch, D.: Quantum theory, the church-turing principle, and the universal quantum computer. Proc. Roy. Soc. London Ser. A **400**, 97–117 (1985)

12. Deutsch, D.: Quantum computational networks. Proc. Roy. Soc. London Ser. A **425**, 73–90 (1989)
13. Dürr, C., Høyer, P.: A quantum algorithm for finding the minimum. arXiv:quant-ph/9607014 (1996)
14. Farhi, E., Goldstone, J., Gutmann, S.: A quantum approximate optimization algorithm applied to a bounded occurrence constraint problem. arXiv:1412.6062 (2014)
15. Grover, L.: A fast quantum mechanical algorithm for database search. In: Proceedings of the 28th Annual ACM Symposium on the Theory of Computing (STOC 1996), pp. 212–219 (1996)
16. Grover, L.: Quantum mechanics helps in search for a needle in a haystack. Phys. Rev. Lett. **79**, 325 (1997)
17. Gruska, J.: Quantum Computing. McGraw-Hill (2000)
18. Hopcroft, J.E., Ullman, J.D.: Introduction to Automata Theory, Languages and Computation. Addison-Wesley Publishing Company (1979)
19. Imre, S.: Quantum existence testing and its application for finding extreme values in unsorted databases. IEEE Trans. Comput. **56**(5), 706–710 (2007)
20. Kato, T.: On the adiabatic theorem of quantum mechanics. J. Phys. Soc. Jpn. **5**, 435–439 (1951)
21. Kitaev, A.Y.: Quantum measurements and Thew Abelian stabilizer problem. arXiv:quant-ph/9511026 (1995)
22. Messiah, A.: Quantum Mechanics. Wiley, New York (1958)
23. Miyamoto, K., Iwamura, M., Kise, K.: A quantum algorithm for finding k-minima. arXiv:1907.03315 (2019)
24. Nielsen, M.A., Chuang, I.L.: Quantum Computation and Quantum Information. Cambridge University Press (2000)
25. Yamakami, T.: A schematic definition of quantum polynomial time computability. J. Symb. Log. **85**, 1546–1587 (2020). A preliminary report appeared in the Proc. of NCMA 2017, Osterreichische Computer Gesellschaft 2017, pp. 243–258 (2017)
26. Yamakami, T.: How does adiabatic quantum computation fit into quantum automata theory? Inf. Comput. **284**, article no. 104694 (2022). A preliminary version appeared in the Proc. of DCFS 2019, LNCS vol. 11612, pp. 285–297, Springer (2019)
27. Yamakami, T.: Nonuniform families of polynomial-size quantum finite automata and quantum logarithmic-space computation with polynomial-size advice. In: Martín-Vide, C., Okhotin, A., Shapira, D. (eds.) LATA 2019. LNCS, vol. 11417, pp. 134–145. Springer, Cham (2019). https://doi.org/10.1007/978-3-030-13435-8_10

On Time-Varying Insertion-Deletion Systems

Henning Fernau[1], Lakshmanan Kuppusamy[2],
and Indhumathi Raman[3]($\boxtimes$)

[1] Fachbereich 4 – Abteilung Informatikwissenschaften, Universität Trier,
54286 Trier, Germany
`fernau@uni-trier.de`
[2] School of Computer Science and Engineering, VIT University,
Vellore 632 014, India
`klakshma@vit.ac.in`
[3] Department of Computing Technologies, SRM Institute of Science and Technology,
Kattankulathur, Chennai 603203, India
`indhumar2@srmist.edu.in`

Abstract. Graph-controlled insertion-deletion (GCID) systems are regulated extensions of insertion-deletion systems. In 2022, Alhazov *et al.* introduced *time-varying* systems as a restriction of GCID systems; in these systems, the components are cyclically ordered. A rule is applied to a string in a component and the resultant string moves to the next component as specified by this order. The language of the system is the set of all terminal strings collected in a distinguished component that is both the initial and the final one. With this restriction, we obtain several new computational completeness results for some typical descriptional complexity measures well-studied in the area of GCID systems.

Keywords: Graph-control · Time-varying Insertion-Deletion Systems · Computational Completeness · Descriptional Complexity Measures

1 Introduction

Two basic operations namely *insertion* and *deletion* are frequently used in both linguistics [13] and molecular biology [3,14]. These two fundamental operations were introduced into formal language theory in [12] through a grammatical mechanism called *insertion-deletion system* (shortly termed as ins-del system). The insertion operation refers to inserting a string η between two strings w_1 and w_2, thereby yielding $w_1\eta w_2$ from w_1w_2, whereas the deletion operation is deleting a substring δ from the string $w_1\delta w_2$ to yield the resultant string w_1w_2.

One of the important variants of ins-del systems is graph-controlled ins-del (shortly GCID) system which was introduced in [10]. In this variant, the rules of the ins-del system are distributed among several *components*. A transition is performed by choosing any applicable rule from the set of rules within the

© The Author(s), under exclusive license to Springer Nature Switzerland AG 2026
E. Formenti and L. Manzoni (Eds.): UCNC 2025, LNCS 16364, pp. 245–261, 2026.
https://doi.org/10.1007/978-3-032-15641-9_17

current component and by moving the resultant string to the target component specified in the rule. The transition is initiated with an axiom in the initial component and the terminal strings collected in the final components are listed in the language of the GCID system.

Several types of GCID systems have been studied, e.g., matrix ins-del systems [7,15], path-controlled ins-del systems [6] and star-controlled ins-del systems [8]. Alhazov *et al.* introduced *time-varying* GCID systems. Such a system has a regular control of the form $(C_1 C_2 \cdots C_k)^*$, with C_i being the finite set of rules, also known as the ith component. The rules are applied periodically over the components; the period k is the number of components. This coincides with a GCID system of the following type: the underlying directed graph is a directed cycle (without loops) with the additional requirement that the final component coincides with the initial component. Since no specific notation has been introduced for this class of GCID systems, we will use GCID_C for this variation in the following, in line with previously introduced notations such as GCID_P and GCID_S for path- and star-controlled GCID systems, respectively.

The *descriptional complexity* of a GCID system is measured by its *size* $(k; n, i', i''; m, j', j'')$, where the parameters from left to right denote the following: (i) the number k of components, (ii) the maximal length n of any insertion string, (iii) the maximal lengths i' and i'' of the left context and right context used in insertion rules, respectively, (iv) the maximal length m of any deletion string, and (v) the maximal lengths j' and j'' of the left context and right context used in deletion rules, respectively. The sextuple of the last six parameters is known as *ID size*. In GCID_C systems, the *period* p of the directed cycle seems to emerge as an additional parameter. However, this is exactly the number of components k. If a GCID_C system is able to describe any recursively enumerable language (RE), then it is said to be *computationally complete*. In [1], it is shown that the following sizes are computationally complete:

- $\mathrm{GCID}_C(5; 1, 1, 0; 1, 1, 0)$ (Theorem 10, [1]) and
- $\mathrm{GCID}_C(3; 1, 1, 0; 1, 1, 1)$ (Theorem 13, [1]).

Symmetrically, the families $\mathrm{GCID}_C(5; 1, 0, 1; 1, 0, 1)$ and $\mathrm{GCID}_C(3; 1, 0, 1; 1, 1, 1)$ are also computationally complete as the class RE of recursively enumerable languages is closed under reversal; also cf. Proposition 1. The first fact was mentioned as Corollary 11 in [1], the second fact not. Notice that it is known that ID systems with such sizes are not computationally complete, see Table 3 in [2]. Moreover, one can carry over all that is known for ID systems, interpreting their application as a cycle with period 1. For instance, this yields the computational completeness of $\mathrm{GCID}_C(1; 1, 1, 1; 1, 1, 1)$, or of $\mathrm{GCID}_C(1; 2, 0, 1; 2, 0, 0)$, or of $\mathrm{GCID}_C(1; 3, 0, 0; 2, 0, 0)$. As previous simulation results on star-controlled GCID systems with two components always end their work in a specific component, the corresponding results also immediately translate into results on time-varying insertion-deletion systems (as one could easily adapt these constructions so that the specific component is actually the intial one) and vice versa. However, so far all these results could be immediately improved by our first observation on ID systems as they correspond to a cycle with period 1. More specifically, this applies to ID sizes $(2, 0, 1; 2, 0, 0)$ and $(3, 0, 0; 2, 0, 0)$.

It is clear that a time-varying (ins-del) system with k components can be simulated or even considered as a matrix (ins-del) system with all matrices of length k. However, the reverse is not true. Therefore, it is also interesting to compare the computational completeness results between time-varying ins-del and matrix ins-del systems with the same ID parameters. For instance, we know that for ID size $(1, 1, 0; 1, 1, 0)$, matrices of length 3 suffice to characterize RE, see [15]. From this, an analogous result for ID size $(1, 0, 1; 1, 0, 1)$ was derived in [4]. In [4], it was also shown that for ID size $(1, 1, 1; 1, 1, 0)$, matrices of length 2 suffice. In both cases, the cited results for time-varying ID systems demonstrate a trade-off: the number of components increases to conform to this type of normal form for matrix ID systems. We will make similar observations in this paper concerning other ID sizes.

We should also mention that there is quite a good motivation for considering time-varying grammars in the context of insertion-deletion systems (although time-varying grammars have been originally introduced into the area of regulated rewriting for different reasons, see [16]): this can model different environmental conditions such as some ecological balance or a bio-geo-chemical cycle that happens to occur periodically and may influence the set of applicable rules. More specifically, biological processes that evolve in cyclic environments such as water cycle (where water evaporates, condenses, and precipitates), carbon cycle, seasonal changes, day and night cycle, etc. can be modelled with such a system.

2   Formal Definitions and Some Observations

We assume that the reader has some basic knowledge of formal languages. We will use standard definitions from that field freely. Therefore, in this section we only formally introduce (variants of) the Special Geffert Normal Form as our main tool in the computational completeness proofs, as well as ID, GCID and time-varying GCID as our main objects of study.

We will now define type-0 grammars in Special Geffert Normal Form (SGNF), see [10], followed by ssSGNF, a variant of SGNF [5].

Definition 1 [10]. *A type-0 grammar $G = (N, T, P, S)$ is said to be in* Special Geffert Normal Form, *SGNF for short, if*

- *N decomposes as $N = N' \cup N''$, where $N'' = \{A, B, C, D\}$ and N' contains at least the two nonterminals S and S',*
- *(i) the context-free rules of P are of the following forms:*
  $X \to Yb$ or $X \to bY$ or $S' \to \lambda$, where $X, Y \in N'$, $X \neq Y$, $b \in T \cup N''$;
- *(ii) the two non-context-free erasing rules in P are $AB \to \lambda$ and $CD \to \lambda$;*

The way the normal form is constructed is described in [10] and is based on [11]. We can assume that S' does not appear on the left-hand side of any non-erasing rule. The derivation in G has two phases. In Phase I, only the context-free non-erasing rules are applied repeatedly and the Phase I is completed by applying the context-free erasing rule $S' \to \lambda$ in the derivation. In Phase II, only the non-context-free erasing rules are applied repeatedly and the derivation ends. Note that as long as the N' symbol is present in the derivation, it separates A and C from B and D. This prevents Phase II from starting prematurely. One feature of SGNF derivations is that any string that can be derived can contain at most one substring AB or CD in its so-called *center*. If such a center is present, the derivation is in Phase II; also, then no nonterminal from N' occurs in Phase II. Also, as the context-free rules of P are like linear, having only one nonterminal from N', in Phase I, exactly one such nonterminal is present (in the center). Therefore, we can differentiate two cases within (i) for $X, Y \in N' \setminus \{S'\}$ with $X \neq Y$: either, we have a rule $X \to bY$, with $b \in \{A, C\}$, or we have a rule $X \to Yb$, with $b \in T \cup \{B, D\}$.

Below, we use a *space-separating* variant of SGNF, introduced in [5] and further elaborated in [9]. The advantage of this normal form is that there will be no two consecutive As or Bs or Cs or Ds present in any derivable string as they are separated by E or F as can be seen below.

Definition 2. *A type-0 grammar $G = (N, T, P, S)$ is in ssSGNF [5] if*

- *N decomposes as $N = N_0' \cup N_1' \cup N''$, where $N'' = \{A, B, C, D, E, F\}$ and N_0' contains at least the two nonterminals S and S';*
- *the three non-context-free rules in P are $AB \to \lambda$, $CD \to \lambda$ and $EF \to \lambda$;*
- *the context-free rules in P are either (i) $S' \to \lambda$, or (ii) $X \to Y_1 Y_2$, where $X \in N_0' \setminus \{S'\}$ and $Y_1 Y_2 \in (\{A, C\} N_1') \cup (N_1'(T \cup \{B, D\}))$, or (iii) $Y \to X_1 X_2$, where $Y \in N_1'$ and $X_1 X_2 \in (\{E\} N_0') \cup (N_0'(T \cup \{F\}))$.*

Notice that by construction, a nonterminal $X \in N' = N_0' \cup N_1'$ cannot be both on the left-hand and on the right-hand side of the same rule. The clear structure of (ss)SGNF offers many advantages over other normal forms like those of Penttonen or Kuroda; hence, SGNF has become the predominant normal form for proving computational completeness results for variations of insertion-deletion systems in recent years.

Before introducing time-varying insertion-deletion systems, we first give the basic definition of basic insertion-deletion systems.

Definition 3 [12,14]. *An insertion-deletion system, or ins-del system for short, is a construct $\Gamma = (V, T, A, R)$, where V is an alphabet, $T \subseteq V$ is the terminal alphabet, A is a finite language over V, R is a finite set of triplets of the form $(u, \eta, v)_I$ or $(u, \delta, v)_D$, where $(u, v) \in V^* \times V^*$, $\eta, \delta \in V^+$.*

The pair (u, v) is called the *context*, where we differentiate between the *left context* u and the *right context* v, η is called the *insertion string*, δ is called the *deletion string* and $x \in A$ is called an *axiom*. The language $L(\Gamma)$ generated by Γ can be explained as the possible outputs of the following nondeterministic program. (1) Choose some $w \in A$. (2) While $w \notin T^*$: Update w by applying any rule from R to w. (3) Output w. As usual, a single rule application yields the relation $\Rightarrow$, so that we can also write $L(\Gamma) = \{w \in T^* \mid \exists \omega \in A : \omega \Rightarrow^* w\}$. The *descriptional complexity* of an ins-del system is measured by its *size* $s = (n, i', i''; m, j', j'')$, where the parameters represent resource bounds, see Table 1.

Table 1. Parameters in the size of ins-del system.

$n = \max\{|\eta| : (u, \eta, v)_I \in R\}$	$m = \max\{|\delta| : (u, \delta, v)_D \in R\}$
$i' = \max\{|u| : (u, \eta, v)_I \in R\}$	$j' = \max\{|u| : (u, \delta, v)_D \in R\}$
$i'' = \max\{|v| : (u, \eta, v)_I \in R\}$	$j'' = \max\{|v| : (u, \delta, v)_D \in R\}$

Definition 4. *A time-varying ins-del system (or,* GCID_C *system for short) is a construct* $\Pi = (k, V, T, A, C_1, \ldots, C_k)$, *with* $R = \bigcup_{i=1}^{k} C_i$, *where (i)* k *is the number of components, also known as its* period, *and (ii)* (V, T, A, R) *is the underlying ins-del system.*

The processing of Π system starts by first choosing an axiom $w_0 \in A$, then applying a rule from the first component C_1 to yield a word w_1, then applying a rule from C_2 to get w_2, followed by applying a rule from C_3 to get w_3 and so on. After applying a rule from C_k to get w_k, we can restart again by applying a rule from C_1 to obtain w_{k+1}. This cycle can be repeated arbitrarily often. The language $L(\Pi)$ generated by Π consists of all words $w_n \in T^*$ that can be produced in this way, where $n \equiv 0 \pmod{k}$. Hence, we can interpret GCID_C as a GCID system where the underlying control graph is a directed cycle and C_1 is both the initial and final component. In a different interpretation, the application of rules obeys the regular control $(C_1 C_2 \cdots C_k)^*$. In this way, time-varying insertion-deletion systems have been defined in [1].

As with GCID systems, the *descriptional complexity* of a time-varying ins-del system is measured by its *size* $s = (k; n, i', i''; m, j', j'')$, where k is the period and other parameters $(n, i', i''; m, j', j'')$ are the size of the underlying ins-del system. The language class generated by time-varying ins-del systems of size (at most) s is denoted by $\mathrm{GCID}_C(s)$. Now, we shall discuss a few examples for $GCID_C$ systems which will provide a better understanding to the readers on the working of the system.

Example 1. Consider the language $L_1 = \{a^n b^n c^n \mid n \geq 1\}$. This non-context-free language can be generated by a GCID_C system Π_1 of size $(2; 2, 1, 1; 0, 0, 0)$ with $L(\Pi_1) = L_1$; Π_1 is described as follows.

$$\Pi_1 = (2, \{a, b, c\}, \{a, b, c\}, \{abc\}, C_1, C_2)$$

with $R = C_1 \cup C_2$, where $C_1 = \{r1.1 : (a, a, \lambda)_I\}$, $C_2 = \{r2.1 : (b, bc, c)_I\}$. Starting with the axiom abc in the initial component C_1 (this abc can be listed in the language $L(\Pi_1)$ as C_1 is also the final component), the only rule present in C_1 is $r1.1$ and applying it, we will get the string a^2bc in C_2. In C_2 (which is not the final component), the only available rule is $r2.1$ and applying it, we will get the string $aabbcc$ in C_1 (which is collected in $L(\Pi_1)$). On applying the rules $r1.1$ and $r2.2$ repeatedly, we can see that the language of Π_1 is L_1. A sample derivation for $a^3b^3c^3$ is given now:

$$(abc)_1 \Rightarrow_{r1.1} (aabc)_2 \Rightarrow_{r2.1} (aabbcc)_1 \Rightarrow_{r2.1} (aaabbcc)_2 \Rightarrow_{r2.1} (aaabbbccc)_1.$$

Notice that there is no deletion rule used in the system. $\square$

Example 2. Consider the language $L_2 = \{ww^r \mid w \in \{a,b\}^*\}$ where w^r is the reversal of w. This context-free language can be generated by a $GCID_C$ system Π_2 of size $(2;1,1,2;1,0,0)$ with $L(\Pi_2) = L_2$; Π_2 is described as follows.

$$\Pi_2 = (2, \{\$_\ell, \$_r, a, b\}, \{a, b\}, \{\$_\ell\$_r\}, C_1, C_2)$$

with $R = C_1 \cup C_2$, where the rules of C_i, $i = 1, 2$ are given as follows.

C_1	C_2
$r1.1 : (\$_r, a, \lambda)_I$	$r2.1 : (\lambda, a, \$_r a)_I$
$r1.2 : (\$_r, b, \lambda)_I$	$r2.2 : (\lambda, b, \$_r b)_I$
$r1.3 : (\lambda, \$_r, \lambda)_D$	$r2.3 : (\lambda, \$_\ell, \lambda)_D$

Starting with the axiom $\$_\ell\$_r$ in C_1, any of the three rules are applicable. If $r1.1$ is applied, then in C_2, $r2.1$ and $r2.3$ are only applicable. Likewise, if $r1.2$ is applied in C_1, then in C_2, $r2.2$ or $r2.3$ are only applicable. The cases of (repeatedly) applying $r1.1$ followed by $r2.1$ or $r1.2$ followed by $r2.2$ are part of the intended derivation. The derivation will successfully end when $r1.3$ is applied in C_1 followed by $r2.3$ in C_2. This will produce strings in the language L_2. The interesting case need to be discussed is as follows. During some point of a derivation, when $r1.1$ or $r1.2$ is applied in C_1 and if $r2.3$ is applied next in C_2, then a string of the form $w\$_r cw^r$ with $c \in \{a, b\}$ can be available in C_1. Now, in C_1 if $r1.1$ is applied, then in C_2, the rule $r2.1$ only can be applied. Likewise, if $r1.2$ is applied, then $r2.2$ only is applicable in C_2. At some point, if $r1.3$ is applied, then there will be no rule to be applicable in C_2, thus such strings are stuck in C_2. A sample and intended derivation for $baab$ is given as follows.

$$(\$_\ell\$_r)_1 \Rightarrow_{r1.2} (\$_\ell\$_r b)_2 \Rightarrow_{r2.2} (\$_\ell b\$_r b)_1 \Rightarrow_{r1.1} (\$_\ell b\$_r ab)_2$$
$$\Rightarrow_{r2.2} (\$_\ell ba\$_r ab)_1 \Rightarrow_{r1.3} (\$_\ell baab)_2 \Rightarrow_{r2.3} baab.$$

$\square$

3   Main Results

The first result in this section relies on the fact that the class RE is closed under reversal. We will use this observation without mentioning in the following.

Proposition 1. *Let $i', i'', j', j'', k, n, m$ be natural numbers.*
If RE $=$ $\mathrm{GCID}_C(k; n, i', i''; m, j', j'')$, *then* *also*
RE $= \mathrm{GCID}_C(k; n, i'', i'; m, j'', j')$.

Our first computational completeness result is a modification of [1, Theorem 10].

Theorem 1. $\mathrm{GCID}_C(5; 1, 1, 0; 1, 0, 1) = \mathrm{GCID}_C(5; 1, 0, 1; 1, 1, 0) = \mathrm{RE}$.

Again, it is known that ID systems with such sizes are not computationally complete, see Table 3 in [2]. Also, for matrix ID systems of the same ID size, matrices of length 3 suffice, see [15] and also [4].

Proof. We are simulating a type-0 grammar in Special GNF. For $p : X \to bY$, we take an additional marker p and introduce the following rules into each of the five components, now visualized as a table:

$$\boxed{(\lambda, p, \lambda)_I}\ \boxed{(\lambda, X, p)_D}\ \boxed{(p, Y, \lambda)_I}\ \boxed{(p, b, \lambda)_I}\ \boxed{(\lambda, p, \lambda)_D}$$

Rules $q : X \to Yb$ are treated alike:

$$\boxed{(\lambda, q, \lambda)_I}\ \boxed{(\lambda, X, q)_D}\ \boxed{(q, b, \lambda)_I}\ \boxed{(q, Y, \lambda)_I}\ \boxed{(\lambda, q, \lambda)_D}$$

For the rule $h : S' \to \lambda$, we use the following simulation:

$$\boxed{(\lambda, h, \lambda)_I}\ \boxed{(\lambda, S', h)_D}\ \boxed{(h, h, \lambda)_I}\ \boxed{(\lambda, h, h)_D}\ \boxed{(\lambda, h, \lambda)_D}$$

As a non-context-free erasing rule, we use $r : ABC \to \lambda$ and simulate this with

$$\boxed{(\lambda, r, \lambda)_I}\ \boxed{(\lambda, C, r)_D}\ \boxed{(\lambda, B, r)_D}\ \boxed{(\lambda, A, r)_D}\ \boxed{(\lambda, r, \lambda)_D}$$

Once a marker is introduced in component $C1$, say p, then (only) the respective SGNF p-rule is simulated. After the introduction of the marker p in $C1$, the simulation is moved to second component $C2$, where all the rules are deletion rules with emptyword λ as the left context and a marker as the right context. Since we have introduced marker p in our previous step, we are forced to apply $(\lambda, X, p)_D$ in $C2$. This rule application deletes the nonterminal X only if it is to the left of the marker p. In other words, the marker p replaces X and the simulation moves into component $C3$. The presence of p forces us to apply the rules $(p, Y, \lambda)_I$, $(p, b, \lambda)_I$, $(\lambda, p, \lambda)_D$ in components $C3, C4, C5$ respectively. The above described simulation is summarized as: $X \Rightarrow Xp \Rightarrow p \Rightarrow pY \Rightarrow pbY \Rightarrow bY$. The correctness of this simulation should be obvious. □

Theorem 2. $\mathrm{GCID}_C(5; 2, 0, 0; 1, 0, 1) = \mathrm{GCID}_C(5; 2, 0, 0; 1, 1, 0) = \mathrm{RE}$.

Again, it is known that ID systems with such sizes are not computationally complete, see Table 3 in [2]. Moreover, matrix systems of length two are already known to be computationally complete with ID sizes $(2, 0, 0; 1, 0, 1)$ or $(2, 0, 0; 1, 1, 0)$, see [4]. Since GCID_C and GCID_S nearly coincide when they are restricted to two components only, it is interesting to compare this result with the following sizes of star-controlled systems which are known to be computationally complete:

- $\mathrm{GCID}_S(4; 2, 0, 0; 1, 0, 1) = \mathrm{GCID}_S(4; 2, 0, 0; 1, 1, 0) = \mathrm{RE}$, see [9];
- $\mathrm{GCID}_S(3; 2, 1, 0; 1, 0, 1) = \mathrm{GCID}_S(3; 2, 0, 1; 1, 1, 0) = \mathrm{RE}$, see [9];
- $\mathrm{GCID}_S(3; 2, 1, 0; 1, 1, 0) = \mathrm{GCID}_S(3; 2, 0, 1; 1, 0, 1) = \mathrm{RE}$, see [9].

Observe that for the ID size $(2, 1, 0; 1, 0, 1)$ or $(2, 1, 0; 1, 1, 0)$ (and symmetrical cases), it is unknown whether RE can be reached this way. Therefore, it is also unknown if $\mathrm{GCID}_C(1; 2, 1, 0; 1, 0, 1) = \mathrm{RE}$ or if $\mathrm{GCID}_C(1; 2, 1, 0; 1, 1, 0) = \mathrm{RE}$. It is therefore interesting to also study these sizes in the context of time-varying systems. We will do this in Theorem 7 below. There, we show analogues to the star-controlled case.

Proof. We are simulating a type-0 grammar G in Special GNF. Recall that the nonterminal alphabet N of G splits into N' and N'', with $N'' = \{A, B, C, D\}$, and N' containing at least S and S'. Let L collect all labels of rules from G that will also serve as markers in the simulating time-varying GCID system Π, and let L' collect primed variants thereof. Then, $\hat{N} = N \cup L \cup L'$ is the nonterminal alphabet of Π, sharing the start symbol S with G.

In order to simulate $p : X \to bY$, we take additional markers p, p' and introduce the following rules into each of the five components:

$(\lambda, pY, \lambda)_I$	$(\lambda, X, p)_D$	$(\lambda, bp', \lambda)_I$	$(\lambda, p', p)_D$	$(\lambda, p, \lambda)_D$

Similarly, we simulate $q : X \to Yb$ and $h : S' \to \lambda$:

$(\lambda, qb, \lambda)_I$	$(\lambda, X, q)_D$	$(\lambda, Yq', \lambda)_I$	$(\lambda, q', q)_D$	$(\lambda, q, \lambda)_D$
$(\lambda, h, \lambda)_I$	$(\lambda, S', h)_D$	$(\lambda, h', \lambda)_I$	$(\lambda, h', h)_D$	$(\lambda, h, \lambda)_D$

As non-context-free erasing rule, we consider $f : AB \to \lambda$ and $g : CD \to \lambda$ and simulate these with

$(\lambda, ff', \lambda)_I$	$(\lambda, B, f)_D$	$(\lambda, f, f')_D$	$(\lambda, A, f')_D$	$(\lambda, f', \lambda)_D$
$(\lambda, gg', \lambda)_I$	$(\lambda, D, g)_D$	$(\lambda, g, g')_D$	$(\lambda, C, g')_D$	$(\lambda, g', \lambda)_D$

Towards a proof of correctness, observe that in the third component, markers can be inserted when actually simulating, say, $f : AB \to \lambda$, but if one attempts this, the said marker cannot be erased again. Also, one cannot return to the f-simulation afterwards, as ff' would be still present. $\square$

The next theorem swaps the insertion and deletion parameters compared to the previous one. Surprisingly, we need one component less in our simulation.

Theorem 3. $\mathrm{GCID}_C(4; 1, 1, 0; 2, 0, 0) = \mathrm{GCID}_C(4; 1, 0, 1; 2, 0, 0) = \mathrm{RE}$.

Again, it is known that ID systems with such sizes are not computationally complete, see Table 3 in [2]. For matrix systems of ID size $(1, 1, 0; 2, 0, 0)$ or $(1, 0, 1; 2, 0, 0)$, we know that matrix length two suffices to achieve computational completeness, see [15] (matrix length 3) and [4]. However, these results do not carry over to our case as we cannot afford to nondeterministically choose between all rules that should be executed as first rules in the matrices.

Proof. We simulate a SGNF grammar again to prove this result.

$C1$	$C2$	$C3$	$C4$	simulating
$(X, p, \lambda)_I$	$(p, Y, \lambda)_I$	$(p, b, \lambda)_I$	$(\lambda, Xp, \lambda)_D$	$p : X \to bY$
$(X, q, \lambda)_I$	$(q, b, \lambda)_I$	$(q, Y, \lambda)_I$	$(\lambda, Xq, \lambda)_D$	$q : X \to Yb$
$(S', h, \lambda)_I$	$(h, h', \lambda)_I$	$(\lambda, h, \lambda)_D$	$(\lambda, S'h', \lambda)_D$	$h : S' \to \lambda$
$(\lambda, f, \lambda)_I$	$(f, f', \lambda)_I$	$(\lambda, AB, \lambda)_D$	$(\lambda, ff', \lambda)_D$	$f : AB \to \lambda$
$(\lambda, g, \lambda)_I$	$(g, g', \lambda)_I$	$(\lambda, CD, \lambda)_D$	$(\lambda, gg', \lambda)_D$	$g : CD \to \lambda$

The correctness of this simulation is obvious. $\qquad\square$

We are now turning to a different idea in the simulation, we can be described as "filling up"; we are going to first give a result using matrix insertion-deletion systems, without defining them formally, as this is only for illustration purposes of our technique. It also highlights somewhat the differences between matrix systems with matrices of length two and 2-component time-varying systems. This somewhat intermediate result (that could explain this "filling up" result could be) is based on Theorem 5 of [4]:

Theorem 4. *Every RE language can be described by a matrix system of size* $(2; 1, 1, 0; 1, 1, 1)$ *where each matrix has length two.*

In fact, this also follows directly from Theorem 5 of [4], where the formal definition of matrix ins-del system is explained, but we show this theorem with a different proof which can be interesting in itself.

Proof. We modify the proof of the mentioned theorem as follows.

- We start with two axioms: $S\#\$$ and $S\#\#\$$ (reason for this is given later).
- We replace each matrix $[r]$ of length one by two matrices of length two:
  - $[(\#, \#, \lambda)_I, r]$ and
  - $[(\#, \#, \lambda)_D, r]$.

The remaining simulation remains the same.

To give this in more details, we will arrive at the following construction. Consider a type-0 grammar $G = (N, T, P, S)$ in SGNF, with the rules uniquely labelled with $[1 \dots |P|]$. Define the label subsets P_{rl} and P_{ll} containing the right-linear type rule labels p with $p : X \to bY$ and the left-linear type rule labels q with $q : X \to Yb$, respectively. Recall the decomposition $N = N' \cup N''$ by SGNF. We now construct a matrix ins-del system $\Gamma = (V, T, \{S\#\$, S\#\#\$\}, M)$ with alphabet

$$V = N \cup T \cup \{\#, \$\} \cup \{x, x', x'', x''' \mid x \in P_{rl} \cup P_{ll}\} \cup \{f, f', g, g'\}.$$

The set of matrices M is defined as follows.

The rules $p\colon X \to bY$ and $q\colon X \to Yb$ are simulated by the following set of ins-del matrices shown on the left and right side, respectively:

$$
\begin{aligned}
p1 &= [(X,p,\lambda)_I,\ (\#,p',\lambda)_I] & q1 &= [(X,q,\lambda)_I,\ (\#,q',\lambda)_I] \\
p2 &= [(\lambda,X,p)_D,\ (\#,p'',\lambda)_I] & q2 &= [(\lambda,X,q)_D,\ (\#,q'',\lambda)_I] \\
p3 &= [(p,Y,\lambda)_I,\ (\#,p''',\lambda)_I] & q3 &= [(q,b,\lambda)_I,\ (\#,q''',\lambda)_I] \\
p4 &= [(p,b,\lambda)_I,\ (p''',p'',p')_D] & q4 &= [(q,Y,\lambda)_I,\ (q''',q'',q')_D] \\
p5 &= [(\lambda,p,b)_D,\ (p''',p',\$)_D] & q5 &= [(\lambda,q,Y)_D,\ (q''',q',\$)_D] \\
p6 &= [(\#,\#,\lambda)_I,\ (\#,p''',\$)_D] & q6 &= [(\#,\#,\lambda)_I,\ (\#,q''',\$)_D] \\
p6' &= [(\#,\#,\lambda)_D,\ (\#,p''',\$)_D] & q6' &= [(\#,\#,\lambda)_D,\ (\#,q''',\$)_D]
\end{aligned}
$$

The rules $f\colon AB \to \lambda$ and $g\colon CD \to \lambda$ are simulated by the set of matrices shown on the left and right side, respectively:

$$
\begin{aligned}
f1 &= [(B,f,\lambda)_I,\ (\#,f',\lambda)_I] & g1 &= [(D,g,\lambda)_I,\ (\#,g',\lambda)_I] \\
f2 &= [(\lambda,B,f)_D,\ (\lambda,A,f)_D] & g2 &= [(\lambda,D,g)_D,\ (\lambda,C,g)_D] \\
f3 &= [(\lambda,f,\lambda)_D,\ (\#,f',\$)_D] & g3 &= [(\lambda,g,\lambda)_D,\ (\#,g',\$)_D]
\end{aligned}
$$

A rule $h\colon S' \to \lambda$ is simulated by the matrices $[(\#,\#,\lambda)_I,\ (\lambda,S',\lambda)_D]$ and $[(\#,\#,\lambda)_D,\ (\lambda,S',\lambda)_D]$. The whole simulation is then terminated by applying the matrix $[(\#,\$,\lambda)_D,(\lambda,\#,\lambda)_D]$.

We now explain the simulation (and what has to be taken into account compared to the original one). At the very beginning, by selecting either the axiom $S\#\$$ or the axiom $S\#\#\$$, one decides if an even or odd number of (previously) single-rule matrices will be applied. More precisely, if there is an even number of such applications, then we choose $S\#\$$. Then, whenever the nth single-rule matrix $[r]$ is going to be applied according to the original derivation, we will apply the variant $[(\#,\#,\lambda)_I, r]$ if n is odd and we will apply the variant $[(\#,\#,\lambda)_D, r]$ if n is even. This way, in any sentential form, there are always one or two occurrences of $\#$ and $\#$ serves as an oddity marker. Also, since the last (say, $n_{\text{fin}}^{\text{th}}$) application is if $n_{\text{fin}}^{\text{th}}$ is even, there will be exactly one occurrence in the sentential form at that stage. This means that the matrix $[(\#,\$,\lambda)_D,(\lambda,\#,\lambda)_D]$ can get rid of both the occurrence of $\#$ and the occurrence of $\$$ in the string. In the case when there is an odd number of such single-rule matrix applications of the original simulation, we start with the axiom $S\#\#\$$ and we will simulate the first single-rule matrix application of, say, $[r]$, by $[(\#,\#,\lambda)_D, r]$ and then again alternate between applying first the insertion variant $[(\#,\#,\lambda)_I, r']$ and then the deletion variant $[(\#,\#,\lambda)_D, r']$ of any single-rule matrix $[r']$ that is applied in the original derivation, keeping up the invariant that after an odd number of such matrix applications, only one occurrence of $\#$ will exist in the string. The matrix $[(\#,\$,\lambda)_D,(\lambda,\#,\lambda)_D]$ can eliminate both $\#$ and $\$$ from the string. The remaining details are easy to fill in. $\square$

We have now, however, only matrices of length two, we cannot transfer this result to time-varying systems, because the sequence of rule applications and the rule selection as done by the matrix mechanism are crucial for the correctness of this construction.

We will explain this technique in the context of time-varying systems. The following construction is based on Theorem 13 of [1] but at the same time strengthens this result. It is worth mentioning that this result is optimal with respect to the parameter 'number of components' as it is known that ID system of size $(1, 0, 1; 1, 1, 1)$ are not computationally universal. When interpreted as matrix systems, this result can also be seen as an alternative proof of the previous theorem. However, one can see that the proof is completely different.

Table 2. A time-varying system Π of size $(2; 1, 1, 0; 1, 1, 1)$ simulating an ID system Γ.

$C1$	$C2$	simulated rule	$C1$	$C2$	simulated rule
$(\lambda, b_p, b)_I$	$(\lambda, p, b_p)_I$	$p : (a, x, b)_I$	$(\lambda, b_q, b)_I$	$(a, x, b_q)_D$	$q : (a, x, b)_D$
$(\lambda, x, b_p)_I$	$(x, b_p, b)_D$		$(\lambda, \#, \$_\ell)_I$	$(a, b_q, b)_D$	
$(\lambda, \#, \$_\ell)_I$	$(a, p, x)_D$		$(\lambda, \#, \$_\ell)_D$	$(a, b_q, b)_D$	
$(\lambda, \#, \$_\ell)_D$	$(a, p, x)_D$		$(\lambda, \$_r, \lambda)_D$	$(\lambda, \$_\ell, \lambda)_D$	the last two steps

Theorem 5. $\mathrm{GCID}_C(2; 1, 0, 1; 1, 1, 1) = \mathrm{GCID}_C(2; 1, 1, 0; 1, 1, 1) = \mathrm{RE}$.

Proof. As in the proof of Theorem 13 of [1], we start with some ID system $\Gamma = (V, T, A, R)$, where we assume that all rules are either of the form $(a, x, b)_I$ or $(a, x, b)_D$ with $|a| = |x| = |b| = 1$ and $a \neq x$ except for one rule, which is $(\lambda, \$, \lambda)_D$ (where $|a| = |b| = 0$). This normal form requires that all axioms $\omega \in A$ are of the form $\omega = \$\omega'\$$, where ω' does not contain any occurrences of \$, i.e., \$ marks the boundary and enables context checks to be performed to the left and right of any other rule. This also means that we can assume that the special rule $(\lambda, \$, \lambda)_D$ is only applied at the very end, and it will be applied twice. In our modified simulation, we actually start with a slightly different construction. We differentiate between the left variation of \$ and call the left-end marker as $\$_\ell$, while the right-end marker is called $\$_r$. This means that we have two deletion rules $(\lambda, \$_\ell, \lambda)_D$ and $(\lambda, \$_r, \lambda)_D$ to be applied at the end. Further modifications to the rules themselves are obvious: checking \$-context to the left is substituted by checking $\$_\ell$-context to the left, and checking \$-context to the right is substituted by checking $\$_r$-context to the right. We still call the resulting ID system $\Gamma = (V, T, A, R)$. We will take, in the simulating time-varying system Π, the axiom set $A' = A \cup \{\#\}A$, where $\#$ is another special symbol. The whole rule set is summarized in Table 2. This completes the description of Π. The intended simulation is as follows. First, we nondeterministically guess if an even or if an odd number of applications of rules is going to happen in a terminating derivation; in the first case, we start with an axiom $\omega = \$_\ell\omega'\$_r$, in the second case, with an axiom $\omega = \#\$_\ell\omega'\$_r$. Therefore, we also call $\#$ an *oddity marker*. Let us consider the first case in the following and consider a derivation

$$w_0 = \omega \Rightarrow w_1 \Rightarrow \cdots \Rightarrow w_n \in T^*$$

of a terminal word in Γ, i.e., n is even. We can assume that the last two rules of the ID system that were applied have been $(\lambda, \$, \lambda)_D$, so that $w_{n-2} = \$_\ell w_n \$_r$. We can simulate these last two steps by taking $(\lambda, \$_r, \lambda)_D$ first from $C1$ and then $(\lambda, \$_\ell, \lambda)_D$ from $C2$. Let us consider the jth derivation step $w_{j-1} \Rightarrow w_j$ now.

If this step was due to applying a deletion rule $q : (a, x, b)_D$, with q being a unique label, we can simulate this by applying $(\lambda, b_q, b)_I$ from $C1$, followed by $(a, x, b_q)_D$ from $C2$, then $(\lambda, \#, \$_\ell)_D$ from $C1$ and $(a, b_q, b)_D$ from $C2$ if j is even and by applying $(\lambda, \#, \$_\ell)_I$ from $C1$ and $(a, b_q, b)_D$ from $C2$ if j is odd. Hence, the simulating sentential form should be $w'_j = w_j$ if j is even and $w'_j = \# w_j$ if j is odd, where $w'_{j-1} \Rightarrow^4 w'_j$ in the simulating system.

For an insertion rule $p : (a, x, b)_I$ (again with p being a unique label of the rule), the simulation works as follows. Assume $w_{j-1} = uabv$. Then, $w'_{j-1} = w_{j-1}$ and set $u' = u$ if j is odd and $w'_{j-1} = \$ w_{j-1}$ if j is even, with $u' = \$ u$ we have then $w'_{j-1} = u'abv$ in any case. Also, we are about to apply any rule from $C1$ now. The idea is to use the rules $(\lambda, b_p, b)_I$, $(\lambda, p, b_p)_I$, $(\lambda, x, b_p)_I$, $(x, b_p, b)_D$, $(\lambda, \#, \$_\ell)_D$ (or $(\lambda, \#, \$_\ell)_I$), and $(a, p, x)_D$ in this sequence, yielding (in Π)

$$w'_{j-1} \Rightarrow u'ab_pbv \Rightarrow u'apb_pbv \Rightarrow u'apxb_pbv \Rightarrow u'apxbv \Rightarrow u''apxbv \Rightarrow u''axbv = w'_j \, ,$$

where $u'' = \$_\ell u$ if $u' = \#\$_\ell u$ and $u'' = \#\$_\ell u$ if $u' = \$_\ell u$.

As $w'_{n-2} = w_{n-2}$, the simulation works out properly by a simple induction. For a deletion rule $q : (a, x, b)_D$, the simulation works as follows. $(\lambda, b_q, b)_I$, $(a, x, b_q)_D$, $(\lambda, \$, \$_\ell)_{I/D}$, $(a, b_q, b)_D$, yielding (in Π)

$$w'_{j-1} \Rightarrow u'axb_qbv \Rightarrow u'ab_qbv \Rightarrow u''ab_qbv \Rightarrow u''abv = w'_j \, ,$$

where $u'' = \$_\ell u$ if $u' = \#\$_\ell u$ and $u'' = \#\$_\ell u$ if $u' = \$_\ell u$.

As $w'_{n-2} = w_{n-2}$, the simulation works out properly by a simple induction.

Conversely, consider any string $\$_\ell w \$_r$ or $\#\$_\ell w \$_r$ derivable in Π such that $\$_\ell w \$_r$ is derivable in Γ and assume that we want to apply a rule from $C1$. As no markers are present, apart from the rules that insert or delete occurrences of $\#$ (Cases (A) and (B) in discussions below) or even delete $\$_r$, only a rule of the form (λ, b_p, b) is applicable. This means that w splits as $w = ubv$, so we come to apply rules from $C2$ to the string $\Lambda ub_pbv \$_r$ for $\Lambda \in \{\$_\ell, \#\$_\ell\}$. Apart from the possibility to erase the unique occurrence of $\$_\ell$ right now (a case (C) that we will discuss later), due to the marker b_p that is introduced, we can only apply one rule (in case p is labeling a deletion rule) or two rules (in case p is labeling an insertion rule) now from $C2$. We will discuss these possibilities now in details.

First, we will look at the simpler case of p labeling a deletion rule first (referred as q before). Assume $p : (a, x, b)_D$. Then, we can apply $(a, x, b_p)_D$ in $C2$. This brings us back to $C1$ with the string $\Lambda u'ab_pbv \$_r$ where $u = u'ax$. If we now apply $(\lambda, c_{p'}, c)$, we might start another simulation (of a rule p') that could be disentangled from the simulation of p by sequentialization. If $b = c$, then we might apply this rule to the same occurrence of b as before, but this produces a substring like $b_p b_{p'}$ that can never be deleted again. Hence, we should now apply an insertion or deletion rule for $\#$. Subsequently, we can apply (in $C2$)

the rule (a, b_p, b) that concludes the simulation cycle. Notice that we have either incremented or decremented the number of occurrences of $\#$ in this simulation.

We return to the discussion of the two possibilities to continue with the string $\Lambda u b_p b v \$_r$ for $\Lambda \in \{\$_\ell, \#\$_\ell\}$ for an insertion rule p: either $(\lambda, p, b_p)_I$ (as intended, leading to the string $\Lambda u p b_p b v \$_r$) or $(x, b_p, b)_D$. If we apply the latter rule, we arrive back to the initial situation, so that we can neglect this case. Hence, assume we have applied $(\lambda, p, b_p)_I$. This means that the string $\Lambda u p b_p b v \$_r$ is now waiting for some rules from $C1$ to be applied to. If we decide to insert or delete $\#$ or $\$_r$ now, we are in a case similar to Case (A) and (B) above; we will not discuss similar details here. The presence of markers in $\Lambda u p b_p b v \$_r$ disables many of the rules. If we apply $(\lambda, b_p, b)_I$ now, then the resulting string $\Lambda u p b_p b_p b v \$_r$ would not allow for a termination finally, because the substring $b_p b_p b$ will prevent the deletion of the marker b_p. We would hence have to apply it to a different occurrence of b, which might exist in the string. This might (sort of prematurely) start another simulation of the insertion rule p that we could disentangle by executing them sequentially, or we could undo this attempt by applying $(x, b_p, b)_D$ now in $C2$, as we might have inserted b_p inbetween x and b. We continue discussing $\Lambda u p b_p b v \$_r$ in $C1$. We have to apply $(\lambda, x, b_p)_I$, yielding $\Lambda u p x b_p b v \$_r$, waiting for some rule from $C2$ to be applied.

We can now either apply $(x, b_p, b)_D$ now (as intended), giving $\Lambda u p x b v \$_r$, or delete the unique occurrence of $\$_\ell$ in $\Lambda u p x b_p b v \$_r$. In either case, back in $C1$, apart from possibly adding or deleting an occurrence of $\#$ (only possible if $\$_\ell$ was not deleted), we can apply $(\lambda, b_p, b)_I$ or $(\lambda, x, b_p)_I$ (only if $\$_\ell$ was deleted, but then the derivation will be stuck) or $(a, p, x)_D$ only if $u = u'a$. In the intended derivation, $\Lambda' u p x b v \$_r$ will be passed to $C2$, with $\Lambda' = \#\Lambda$ or $\Lambda = \#\Lambda'$, from which we can get $\Lambda' u x b v \$_r$ as intended (again, applying a deletion rule to $\$_\ell$ is not fruitful). If we apply $(\lambda, b_p, b)_I$ instead, it means a somewhat premature start of the simulation of another simulation cycle that we can again disentangle (sequentialize). Hence, the intended derivations are the only possibly terminating derivations, apart from possibly producing more than one occurrence of $\#$ in the string (that must later be counter-balanced by deletion rules). $\qquad\square$

Table 3. A time-varying system Π of size $(2; 1, 1, 1; 1, 1, 0)$ simulating an ID system Γ.

$C1$	$C2$	simulated rule	$C1$	$C2$	simulated rule
$(a, q, x)_I$	$(x, q', b)_I$	$q : (a, x, b)_D$	$(a, p, b)_I$	$(a, x, p)_I$	$p : (a, x, b)_I$
$(q, x, \lambda)_D$	$(a, q, \lambda)_D$		$(\$_\ell, \#, \lambda)_I$	$(x, p, \lambda)_D$	
$(\$_\ell, \#, \lambda)_I$	$(a, q', \lambda)_D$		$(\$_\ell, \#, \lambda)_D$	$(x, p, \lambda)_D$	
$(\$_\ell, \#, \lambda)_D$	$(a, q', \lambda)_D$		$(\lambda, \$_r, \lambda)_D$	$(\lambda, \$_\ell, \lambda)_D$	the last two steps

Note that we have shown computational completeness of matrix ID systems of ID sizes $(1, 1, 1; 1, 1, 0)$ and $(1, 1, 0; 1, 1, 1)$ in [4]. However, these results were based on simulation of SGNF, while here we are following the idea of [1] and simulated $(1, 1, 1; 1, 1, 1)$ ID systems. These simulations appear to be somewhat simpler, as can be seen by comparing the proofs of Theorems 4 and 5.

Theorem 6. $\mathrm{GCID}_C(2; 1, 1, 1; 1, 0, 1) = \mathrm{GCID}_C(2; 1, 1, 1; 1, 1, 0) = \mathrm{RE}$.

Proof. As the rules in Fig. 2 are very similar to those in the previous simulation, we only provide a sketch here. We start again with some ID system $\Gamma = (V, T, A, R)$ of size $(1, 1, 1; 1, 1, 1)$ in the normal form as in the proof of Theorem 13 of [1]. As in the previous proof, we provide an oddity marker $\#$, this time placed to the right of the left-end marker $\$_\ell$. With the help of Table 3, the reader should be able to construct the details of the time-varying system Π of size $(2; 1, 1, 1; 1, 1, 0)$ simulating Γ.

Let us consider a sentential form $w_\Pi = \Lambda \alpha a x b \beta \$_r$ of Π, with $\Lambda \in \{\$_\ell \#, \$_\ell\}$, $\alpha, \beta \in V^*$, corresponding to a sentential form $w_\Gamma = \$\alpha a x b \beta \$$ of Γ. We could apply $q : (a, x, b)_D$ to w_Γ. In Π, this is simulated by: $w_\Pi \Rightarrow_1 \Lambda \alpha a q x b \beta \$_r \Rightarrow_2 \Lambda \alpha a q x q' b \beta \$_r \Rightarrow_1 \Lambda \alpha a q q' b \beta \$_r \Rightarrow_2 \Lambda \alpha a q' b \beta \$_r \Rightarrow_1 \Lambda' \alpha a q' b \beta \$_r \Rightarrow_2 \Lambda' \alpha a b \beta \$_r$, where $\Rightarrow_i$ refers to applying a rule from Ci, and $\{\Lambda, \Lambda'\} = \{\$_\ell \#, \$_\ell\}$. Similarly (even a bit simpler), a deletion rule can be simulated. The fact that no unintended terminating derivations are possible in Π (apart from possibly having more than one occurrence of $\#$ in the course of a derivation, but taking the number of these occurrences modulo 2 will always give the intended number), is based on the following observations: (a) In the first component, rules are placed that either introduce a certain rule marker, or assume the existence of this rule marker, or increment or decrement the number of $\#$ in the string. Clearly, the introduction of a rule marker must precede its use. (b) In the second component, we always assume the existence of a rule marker in every rule. (c) After introducing a rule marker in $C1$ with $(a, r, x)_I$, the rule $(a, r, x)_D$ of $C2$ should not be applied immediately thereafter, as this would simply undo the previous operation. Therefore, we need to apply a rule $(x', r', b')_I$ of $C2$. (d) After applying some rule from $C1$, apart from introducing another primed s-marker when applying some $(x'', s', b'')_I$ of $C2$, only $(a, q, \lambda)_D$ or $(a, q', \lambda)_D$ (if r corresponds to a deletion rule, i.e., $r = q$) is available, which enforces the application of $(q, x, \lambda)_D$ in $C1$. Then only deletion rules can be applied to the markers. (e) After applying a rule affecting the number of oddity markers in $C1$, apart from again introducing another primed s-marker when applying some $(x'', s', b'')_I$, we have to delete a rule marker as intended. (f) If we have introduced additional primed rule markers, we might be able to delete them again, but as nothing productive has been done with them, an unnecessary detour. $\qquad\square$

Theorem 7. $\mathrm{GCID}_C(3; 2, 1, 0; 1, 0, 1) = \mathrm{GCID}_C(3; 2, 0, 1; 1, 1, 0) = \mathrm{RE}$. $\mathrm{GCID}_C(3; 2, 1, 0; 1, 1, 0) = \mathrm{GCID}_C(3; 2, 0, 1; 1, 0, 1) = \mathrm{RE}$.

Proof (Sketch). We have to start with ssSGNF now. Moreover, we have a special symbol $\$$ attached to the starting symbol to form the only axiom $\$S$. We will have two more special symbols, $\#$ and t. As usual with this type of simulation, we will also employ (primed versions of) rule markers. For reasons of space, we leave the formal details of the construction to the reader, only presenting the ideas. We only show the results for deletion parameters $(1, 0, 1)$, i.e., we prove

$$\mathrm{GCID}_C(3; 2, 1, 0; 1, 0, 1) = \mathrm{GCID}_C(3; 2, 0, 1; 1, 0, 1) = \mathrm{RE}.$$

The other two assertions will follow with Proposition 1.

The f- and g-rules (always together with the e-rule) are simulated as follows:

$C1$	$C2$	$C3$	$C1$	$C2$	$C3$
$(\lambda, f, \lambda)_I$	$(\lambda, F, f)_D$	$(\lambda, B, f)_D$	$(\lambda, g, \lambda)_I$	$(\lambda, F, g)_D$	$(\lambda, D, g)_D$
$(\lambda, A, f)_D$	$(\lambda, E, f)_D$	$(\lambda, f, \lambda)_D$	$(\lambda, C, g)_D$	$(\lambda, E, g)_D$	$(\lambda, g, \lambda)_D$

Notice that if (erroneously) we choose to delete E after introducing f, then ssSGNF will come to our aid, as we will be finally stuck, because our intended derivation will either delete the (expected) substring $EABF$ or $ECDF$ but when E is missing, the simulation cannot be completed. For the context-free rules, we propose the following rules to be added to the three components; now we have to differentiate between insertion parameters $(2, 1, 0)$ and $(2, 0, 1)$.

Insertion parameter $(2, 1, 0)$ **Insertion parameter $(2, 0, 1)$**

$C1$	$C2$	$C3$
$(\lambda, p, \lambda)_I$	$(\lambda, X, p)_D$	$(\$, \#, \lambda)_I$
$(p, bY, \lambda)_I$	$(\lambda, p, \lambda)_D$	$(\lambda, \#, \lambda)_D$
$(\lambda, q, \lambda)_I$	$(\lambda, X, q)_D$	$(\$, \#, \lambda)_I$
$(q, Yb, \lambda)_I$	$(\lambda, q, \lambda)_D$	$(\lambda, \#, \lambda)_D$
$(\lambda, h, \lambda)_I$	$(\lambda, S', h)_D$	$(\lambda, h, \lambda)_D$

$C1$	$C2$	$C3$
$(\lambda, p, \lambda)_I$	$(\lambda, X, p)_D$	$(\lambda, \#, \$)_I$
$(\lambda, bY, p)_I$	$(\lambda, p, \lambda)_D$	$(\lambda, \#, \lambda)_D$
$(\lambda, q, \lambda)_I$	$(\lambda, X, q)_D$	$(\lambda, \#, \$)_I$
$(\lambda, Yb, q)_I$	$(\lambda, q, \lambda)_D$	$(\lambda, \#, \lambda)_D$
$(\lambda, h, \lambda)_I$	$(\lambda, S', h)_D$	$(\lambda, h, \lambda)_D$

The intended derivations obtained with these simulating rules are obvious. As we always work with markers, it should be clear that no unintended derivations are possible. These rules also explain the role of the special symbols $\$, \#$: We need them to "fill up" a simulation cycle that would be (otherwise) too short or too long; we simply need a period of three here. We have to delete $\$$ at the end, which we will do with a special terminal marker t:

$C1$	$C2$	$C3$
$(\$, t, \lambda)_I$	$(\lambda, \$, t)_D$	$(\lambda, t, \lambda)_D$

When we apply these rules before finishing with simulating the context-free rules, the derivation gets stuck. However, we can apply these rules at any time when simulating the non-context-free rules. $\square$

4 Conclusions

We explored the Pareto frontier of the descriptional complexity parameters of time-varying GCID systems, see Table 4. Are any of these bounds sharp? We lack tools to determine when a parameter restriction is insufficient to describe RE. This should be addressed in future work. The idea of oddity markers seems to be crucial for small periods: they also work together with ssSGNF, see Theorem 7.

Table 4. Comparing the results of this paper with those of [1].

Result	Reference	Result	Reference
$\mathrm{GCID}_C(5;1,1,0;1,1,0) = \mathrm{RE}$	[1]	$\mathrm{GCID}_C(5;1,0,1;1,0,1) = \mathrm{RE}$	[1]
$\mathrm{GCID}_C(5;1,1,0;1,0,1) = \mathrm{RE}$	Theorem 1	$\mathrm{GCID}_C(5;1,0,1;1,1,0) = \mathrm{RE}$	Theorem 1
$\mathrm{GCID}_C(5;2,0,0;1,0,1) = \mathrm{RE}$	Theorem 2	$\mathrm{GCID}_C(5;2,0,0;1,1,0) = \mathrm{RE}$	Theorem 2
$\mathrm{GCID}_C(4;1,1,0;2,0,0) = \mathrm{RE}$	Theorem 3	$\mathrm{GCID}_C(4;1,0,1;2,0,0) = \mathrm{RE}$	Theorem 3
$\mathrm{GCID}_C(3;1,1,0;1,1,1) = \mathrm{RE}$	[1]	$\mathrm{GCID}_C(3;1,0,1;1,1,1) = \mathrm{RE}$	Introduction
$\mathrm{GCID}_C(2;1,1,0;1,1,1) = \mathrm{RE}$	Theorem 4	$\mathrm{GCID}_C(2;1,0,1;1,1,1) = \mathrm{RE}$	Theorem 4
$\mathrm{GCID}_C(2;1,1,1;1,0,1) = \mathrm{RE}$	Theorem 6	$\mathrm{GCID}_C(2;1,1,1;1,1,0) = \mathrm{RE}$	Theorem 6
$\mathrm{GCID}_C(3;2,1,0;1,0,1) = \mathrm{RE}$	Theorem 7	$\mathrm{GCID}_C(3;2,0,1;1,1,0) = \mathrm{RE}$	Theorem 7
$\mathrm{GCID}_C(3;2,1,0;1,1,0) = \mathrm{RE}$	Theorem 7	$\mathrm{GCID}_C(3;2,0,1;1,0,1) = \mathrm{RE}$	Theorem 7

Acknowledgement. The first and the second authors are grateful for the support provided by the DST-DAAD project (DST/IND/DAAD P-01-2024) that enabled fruitful discussions and interactions.

References

1. Alhazov, A., Freund, R., Ivanov, S., Verlan, S.: Regulated insertion-deletion systems. J. Autom. Lang. Combin. **27**(1-3), 15–45 (2022)
2. Alhazov, A., Ivanov, S., Verlan, S.: A 15-year retrospective on insertion-deletion systems: progress, evolution, and future directions. Comput. Sci. J. Moldova **32**(3(96)), 332–371 (2024)
3. Benne, R. (ed.): RNA Editing: The Alteration of Protein Coding Sequences of RNA. Series in Molecular Biology. Ellis Horwood, Chichester, UK (1993)
4. Fernau, H., Kuppusamy, L., Raman, I.: Investigations on the power of matrix insertion-deletion systems with small sizes. Nat. Comput. **17**(2), 249–269 (2018)
5. Fernau, H., Kuppusamy, L., Raman, I.: Computational completeness of simple semi-conditional insertion-deletion systems of degree $(2,1)$. Nat. Comput. **18**(3), 563–577 (2019)
6. Fernau, H., Kuppusamy, L., Raman, I.: On path-controlled insertion-deletion systems. Acta Inform. **56**(1), 35–59 (2019)

7. Fernau, H., Kuppusamy, L., Raman, I.: On the generative capacity of matrix insertion-deletion systems of small sum-norm. Nat. Comput. **20**(4), 671–689 (2021). https://doi.org/10.1007/s11047-021-09866-y

8. Fernau, H., Kuppusamy, L., Raman, I.: When stars control a grammar's work. In: Gazdag, Z., Iván, S., Kovásznai, G. (eds.) Proceedings of the 16th International Conference on Automata and Formal Languages, AFL. EPTCS, vol. 386, pp. 96–111. Open Publishing Association (2023)

9. Fernau, H., Kuppusamy, L., Raman, I.: Succinct star-controlled insertion-deletion systems using space separating normal forms. In: Formenti, E., Durand-Lose, J. (eds.) MCU 2024. LNCS, vol. 15270, pp. 17–34. Springer, Cham (2025). https://doi.org/10.1007/978-3-031-81202-6_2

10. Freund, R., Kogler, M., Rogozhin, Y., Verlan, S.: Graph-controlled insertion-deletion systems. In: McQuillan, I., Pighizzini, G. (eds.) Proceedings Twelfth Annual Workshop on Descriptional Complexity of Formal Systems, DCFS. EPTCS, vol. 31, pp. 88–98. Open Publishing Association (2010)

11. Geffert, V.: Normal forms for phrase-structure grammars. RAIRO Theoret. Inform. Appl. **25**, 473–498 (1991)

12. Kari, L., Thierrin, G.: Contextual insertions/deletions and computability. Inf. Comput. **131**(1), 47–61 (1996)

13. Marcus, S.: Contextual grammars. Rev. Roum. Mathém. Pures Appl. **14**, 1525–1534 (1969)

14. Păun, Gh., Rozenberg, G., Salomaa, A.: DNA Computing: New Computing Paradigms. Springer, Cham (1998)

15. Petre, I., Verlan, S.: Matrix insertion-deletion systems. Theoret. Comput. Sci. **456**, 80–88 (2012)

16. Salomaa, A.K.: Periodically time-variant context-free grammars. Inf. Control **17**, 294–311 (1970)

Evaluating ESNs Against Lagged Input Regression Computation

David Griffin[1]([⊠])(iD), James Stovold[2](iD), Simon O'Keefe[1](iD),
and Susan Stepney[1](iD)

[1] Department of Computer Science, University of York, York, UK
`{david.griffin,simon.okeefe,susan.stepney}@york.ac.uk`
[2] Lancaster University Leipzig, Leipzig, Germany
`j.stovold@lancaster.ac.uk`

Abstract. Echo State Networks (ESNs) are stated in literature to use a random structure to project an input sequence into a higher dimensional space where the input becomes linearly separable. However, the linear mathematics used for this projection are incapable of increasing the dimensionality of the input, and the commonly used tanh() activation function tends not to produce much nonlinearity. Therefore, any increase in dimensionality is due to the echoes of the ESN. We introduce Lagged Input Regression Computation to investigate what types of ESN can be replaced with simpler non-randomised structures. We show that tanh()-based ESNs behave as simple linear memory systems, whereas LeakyReLU provides a more effective non-linearity. We also show that the use of certain orthogonal polynomials in defining nonlinear memory capacity benchmarks gives a misleading impression of nonlinearity, due to the relevant high order polynomials nevertheless containing a linear term.

Keywords: Reservoir Computing · Echo State Networks · Memory Systems

1 Introduction

Artificial Neural Networks (ANNs) [4] have become a popular tool: they are widely used in Computer Science, and have wide ranging applications from Pattern Recognition to Large Language Models. Of the various types of ANNs, Recurrent Neural Networks (RNNs) [2] arguably have the most complexity, allowing arbitrary connections between simulated neurons, rather than the more structured approach seen in Feed Forward Neural Networks [4]. RNNs have the primary advantage that they capture data within their recurrent connections, making them suitable for temporal data or pattern recognition [2]. However, this also comes with a disadvantage: they are more complex to train. For example, to apply the backpropagation training techniques, it is necessary to unravel the recurrent connections by applying backpropagation through time [17].

In Sect. 3 we examine the various parts of the intuition behind ESNs and show that, as commonly used in literature, the random recurrence of ESNs may

© The Author(s), under exclusive license to Springer Nature Switzerland AG 2026
E. Formenti and L. Manzoni (Eds.): UCNC 2025, LNCS 16364, pp. 262–276, 2026.
https://doi.org/10.1007/978-3-032-15641-9_18

not be as interesting as previously thought. In particular, we compare common configurations of ESNs using the tanh() activation function with a completely linear lagged input system. The result shows both that many ESNs do not exhibit particularly nonlinear behaviour, as well as highlighting the potential power of linear regression techniques on simple memory.

In Sect. 4, we introduce Lagged Input Regression Computation (LIRC), which uses simple linear memory with a regression layer. We evaluate ESNs against LIRC to show that tanh() ESNs can be approximated with a purely linear system. We also show that the alternative LeakyReLU activation function does not have the same behaviour, and is only approximable by a LIRC system under specific, favourable conditions. Given that other authors [8] have shown that ESNs exhibit nonlinear memory, we revisit this definition of nonlinear memory and show that linear effects unintentionally dominate this measure.

2 Related Work

An ESN [14] is a randomly connected, sparse, recurrent neural network with a fixed activation function. In the literature, sigmoidal functions are typically recommended, with the tanh() function normally being assumed [14]. Although other authors have suggested using a rectified linear unit (ReLU) [12]. Input is supplied to the ESN nodes through a randomly generated input weight matrix. The internal connections of an ESN are given by a randomly generated matrix that specifies the connections and their weights. When generating the weight matrix, typically both the sparsity and spectral radius (the maximum of the absolute values of the eigenvalues of the weight matrix) are specified. It is desirable for ESNs to have the *echo state property*, which can be obtained with a sigmoidal activation function and a spectral radius < 1 [6]. ESNs can also have feedback, which is typically specified by a third randomly generated matrix. For simplicity, here we consider only the case where feedback is not present.

A simple, non-feedback ESN is typically defined by the equation:

$$\mathbf{x}(n+1) = f(\mathbf{W}\mathbf{x}(n) + \mathbf{W}^{in}\mathbf{u}(n)) \tag{1}$$

where $\mathbf{W}$ is the random weight matrix, $\mathbf{W}^{in}$ is the random input weight matrix, $\mathbf{x}(n)$ is the state of the ESN at time n, $\mathbf{u}(n)$ is the input at time n, and f is the activation function. While this equation has no output feedback, the ESN may have internal feedback through the graph described by $\mathbf{W}$.

To extract output, the values stored in the nodes of the ESN are passed through a trained linear output layer.

$$\mathbf{y}(n+1) = \mathbf{W}^{out}\mathbf{x}(n) \tag{2}$$

where $\mathbf{W}^{out}$ is the trained output weight matrix, and $\mathbf{y}(n)$ is the output of the ESN at time n. Typically, the output layer is trained through either linear regression [23] or ridge regression [13] methods, which allow the output layer to be trained in a relatively trivial amount of time.

ESNs have found use in predicting or classifying temporal data [14], where the inherent recurrence of the ESN is most useful. Wringe et al. [22] present a review of benchmarks used in the broader Reservoir Computing literature; the majority of benchmarks used are indeed temporal, such as NARMA (Nonlinear AutoRegressive Moving Average) [21] or Mackey–Glass [16].

Recent work has called into question the way in which ESNs are generated and used. Gauthier et al. [9] found that many of the tasks ESNs excel could also be accomplished by a nonlinear vector autoregression (NVAR) machine [5]. NVAR machines make predictions based on (a) a number of previous timesteps and (b) a number of nonlinear transformations. Those authors were not able to explain which nonlinearites are most useful to which tasks, other than observing that polynomials seem to have broad applicability.

Ma et al. [15] continued that work, introducing a non-random RC that cast inputs into a substantially higher dimensional space, and using a nonlinear read-out instead of an activation function. It resulted in an increase in model accuracy, albeit with a larger state space than an NVAR machine. That work did not identify the cause of the result as being either the changes to the model readout or the increase in dimensionality.

One pattern in prior research is a belief that the projection an ESN's input into a higher dimensional space leads to linear separability. If the dimensionality of the ESN is increased in some way, then this leads to an increase in separability. This appears to be similar to a well known result in the study of ANNs, Cover's Theorem [7]. However, Cover's Theorem specifically utilises *nonlinear* projections to achieve linear separability of nonlinear properties. As we show in this paper, the linearity of the input projection $\mathbf{W}^{in}$, and the subsequent use of the sigmoid activation function in its linear regime render the overall behaviour of many ESNs to be predominantly linear.

3   Dissecting an ESN

Using the non-feedback ESN of Eq. 1, we investigate how they achieve the linear separability property. The ESN equation has three main components: the activation function f, the projection of input into a higher dimensional space given by $\mathbf{W}^{in}$, and the movement of information within the ESN structure given by $\mathbf{W}$. We examine these three components in turn.

3.1   Activation Function

The activation function f is commonly defined to be a sigmoidal function such as $\tanh()$ or the logistic function $1/(1 - e^x)$, see Fig. 1. $\tanh()$ is close to the identity function in the range $[-1, 1]$, and virtually identical to it in the range $[-0.5, 0.5]$. The Maclaurin series for $\tanh(x)$ [1] is $x - x^3/3 + O(x^5)$, and so in the latter range $\frac{|\tanh(x) - x|}{|x|} \approx \frac{0.5^3}{3 \cdot 0.5} = 0.083$, or a less than 10% difference. Shrinking the range further reduces this magnitude further.

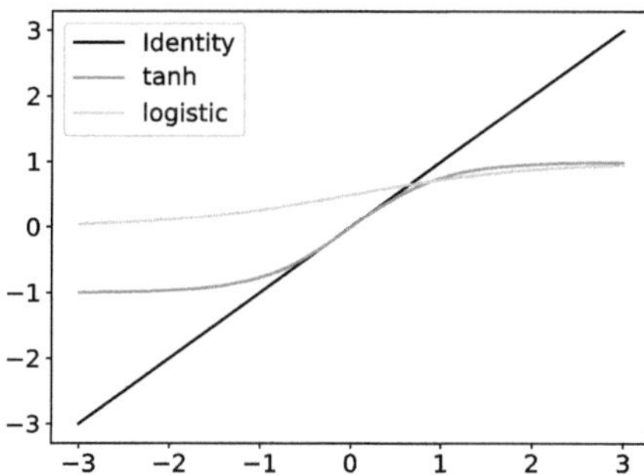

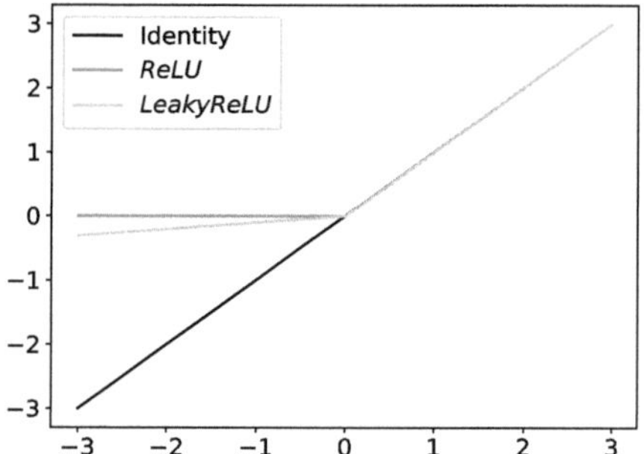

Fig. 1. Sigmoidal activation functions

Fig. 2. Rectifier activation functions

Within an echo state network, tanh() activation provides two properties: a nonlinearity, and the ability to constrain values in the range $[-1, 1]$. The latter is seen as an important reminder of physicality; if an ESN approximates a physical system, then it cannot encode infinity. However, the similarity of tanh() to the identity within the region where input is typically normalised may limit its nonlinearity contribution. This combination of constraining to $[-1, 1]$ and the near-linearity in this region seems a major drawback to the use of tanh() as an activation function in the context of an ESN.

Does the internal dynamics of the ESN counter this, and move the value into the nonlinear region of tanh()? We argue no. Consider the internal state $\mathbf{x}$ of inputs to a given node. Each component is multiplied by a random weight as given by the internal weight matrix $\mathbf{W}$. If these weights have a mean of 0 (such as for uniform weights, or Gaussian weights), then approximately equal numbers of these weights have a positive and negative sign. The internal state values stored in an ESN are all within the image of the activation function, $[-1, 1]$. In literature, the weights of $\mathbf{W}$ are typically drawn from the uniform distribution of $[-1, 1]$ and then scaled by ρ, the spectral radius which is typically < 1, resulting in weights in the interval $[-\rho, \rho]$.

Then the sum of the weighted inputs to a node is also likely to be in the range $[-\rho, \rho]$. This is because the most likely outcome is for roughly an equal number of input weights to be positive and negative, resulting in the expected value of the sum to be close to 0. This can be seen empirically in Fig. 3, which shows the distribution of values over time evolves in an example ESN. The state of the ESN is initialised to be uniform random $[-1, 1]$ at time 0. By time 8, the distribution of values has converged to a normal distribution. This experiment has been repeated with a variety of different parameterisations, and provided that the ESN is sufficiently large, this convergence is always seen.

As the number of input connections increases (increasing number of nodes), we expect the central limit theorem to further compress the range of likely values towards 0. While one could construct parameters to maximise the probability of getting a positive or negative bias in the weights to a node – for example, a small value of N, a higher sparsity in W, or a higher spectral radius – reducing

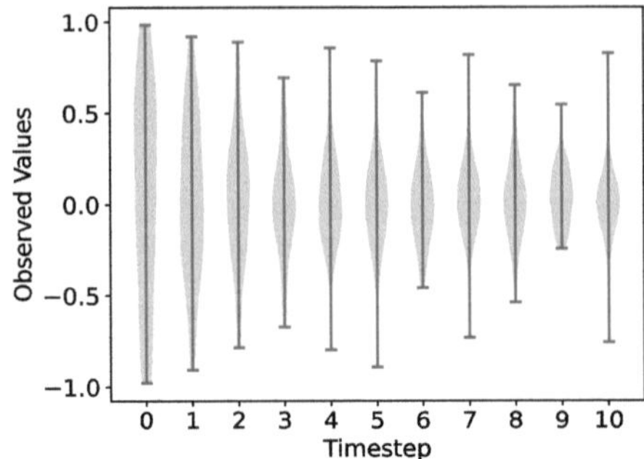

Fig. 3. Distribution of values over time in a 100-node ESN with a spectral radius of 0.8 and connectivity of 0.1 with a tanh() activation function

the number of values summed *also* reduces the probability of getting an extreme value. And most values in the ESN will still be in the linear region, as this remains the most likely outcome.

An effective example of the apparent ineffectiveness of tanh() as a nonlinearity is to apply the power method [11] to an ESN. The power method iterativly calculates the maximum eigenvalue of a matrix using:

$$\lambda_1 = \lim_{k \to \infty} \frac{||\mathbf{A}^k \mathbf{b}||}{||\mathbf{A}^{k-1} \mathbf{b}||} \tag{3}$$

where $\mathbf{b}$ is an arbitrarily chosen initial input vector, $\mathbf{A}$ is a matrix and $||.||$ is a vector norm. Provided that $\mathbf{b}$ is nonzero in the direction of the associated eigenvector and the maximum eigenvalue is strictly greater in magnitude than all other eigenvalues, this sequence converges on the maximum eigenvalue. If not (for example, in the case of complex conjugate eigenvalues), it oscillates between the greatest eigenvalues.

Using a highly restricted case of ESNs where we can set $\mathbf{u}(0) = \mathbf{b}, \mathbf{u}(n+1) = \mathbf{x}(t)$ with $\mathbf{W}^{in} = \mathbf{1}$ the identity, then $\mathbf{x}(n) = (tanh \otimes \mathbf{W})^n(\mathbf{b})$. If $tanh \otimes \mathbf{W} \approx \mathbf{W}$, then $\mathbf{x}(n) \approx \mathbf{W}^n(\mathbf{b})$, and the power iteration is applicable. Conversely, if $tanh \otimes \mathbf{W} \not\approx \mathbf{W}$, then the power iteration would not be applicable. In practice, $tanh \otimes \mathbf{W} \approx \mathbf{W}$ if the spectral radius of $\mathbf{W}$ lies within the mostly linear region of tanh() between $[-1, 1]$. Experimentally, this approach has been applied to $10,000$ ESNs with unique maximum eigenvalues, with the size of the ESN varying between $[10, 100]$, spectral radius between $[0.1, 0.99]$ and connectivity between $[0.05, 1.0]$, and in every case the power method converged.

Similar arguments as presented on tanh() can also be applied to the standard logistic function, Fig. 1. While the logistic function is mostly linear in the region $[-1, 1]$ it is not close to the identity. However, as the image of the standard logistic function is $[0, 1]$, repeated applications once again keep it within a mostly linear section. If a linear transformation is applied to map the standard logistic sigmoids values back to the range $[-1, 1]$, then the previously mentioned suite of power method tests passes. While this transformation may seem to be adding an

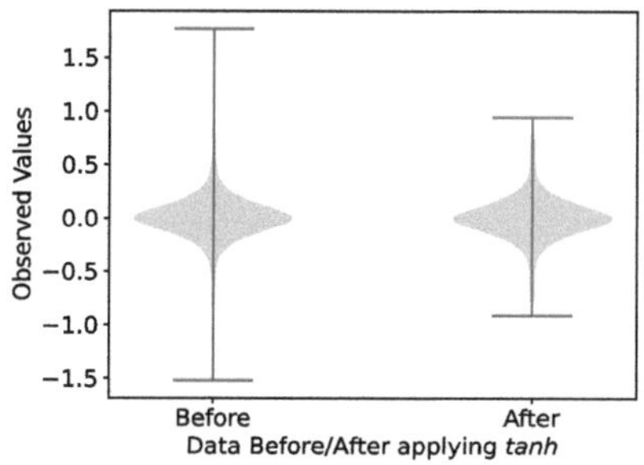

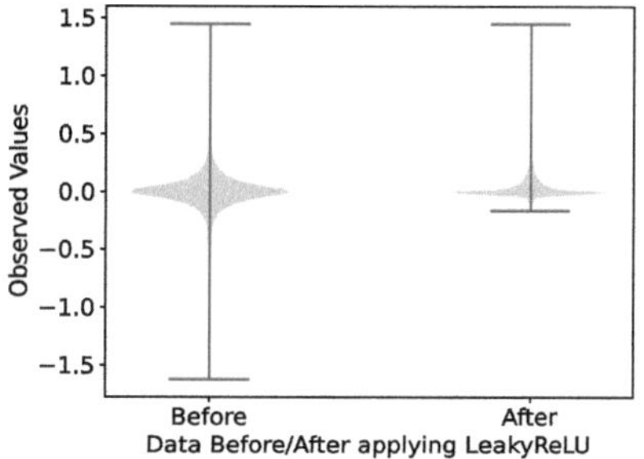

Fig. 4. Effect of tanh() activation function on values in an ESN

Fig. 5. Effect of LeakyReLU activation function on values in an ESN

additional step, an ESN includes a linear readout layer, which can incorporate this transformation.

The Rectified Linear Unit (ReLU) is an alternative activation function (Fig. 2). ReLU maps all negative values to 0, and all positive values to themselves. This mapping of negative values to 0 causes any information represented by those values to be lost. Again, if weights are distributed around 0, one may expect that the sum of values is also distributed around 0. Absent any bias, an ESN with the ReLU activation function may delete around 50% of the information it contains with every timestep, and the remaining 50% is strictly linear. This property makes it very difficult to apply arguments on how it behaves, and typically the power method will not converge if the dominant eigenvector has a negative value. Unlike with the standard logistic sigmoid, due to ReLU deleting information there is no linear transformation that can recover from this. The range of ReLU not being finite is not relevant for the lack of convergence, as the only requirement is that ReLU is mostly linear in its image.

The LeakyReLU activation function is also somewhat common. Rather than mapping values less than 0 to 0, LeakyReLU instead applies a small multiplier, which means that information is no longer being deleted. For LeakyReLU, the image of the function, Im(LeakyReLU), is not confined to a region where the function is strictly linear across the entire range. Instead, LeakyReLU is strictly linear across exactly half of its image. This means that iterative methods are still likely to succeed, as LeakyReLU is a mostly linear invertible function. However, again, there is no linear transformation that can undo LeakyReLU over its entire range. This results in the power method converging for LeakyReLU, provided that the spectral radius is not close to 0.

An experimental evaluation of the activation functions can be performed by plotting the distribution of values in an ESN before and after an activation function is applied. This is shown in Figs. 4 and 5, for a 100 node ESN, although these results are repeatable with many sizes or parameterisations. As in Fig. 3, we see values consistent with a normal distribution. For LeakyReLU the spread of the distribution is lower due to it compressing negative values, thus reducing the average magnitude of values within the ESN. In Fig. 4, we see that values outside of the range $[-1, 1]$ are compressed, but otherwise tanh() does not substantially change the distribution, with the new values still being normally distributed. This is also true of the logistic function, although the distribution of values afterwards is shifted to around 0.5. Figure 5 shows LeakyReLU, and in this case the distribution is substantially altered below 0, although in a manner that can be recovered with a linear transformation of the values below 0.

3.2 Linear Projections into Higher Dimensions

Having made an argument that standard sigmoid activation functions may not be sufficiently nonlinear, we move onto the next portion of the ESN equation: the input projection.

In order to examine the claim that an ESN projects input data into a higher dimensional space where it becomes linearly separable, we examine $\mathbf{W}^{in}$, which implements the initial projection.

Recall some properties of functions. A function $f : X \to Y$ maps all points in X onto some set of points in Y. If there is no $x_1 \neq x_2 \in X$ such that $f(x_1) = f(x_2)$, then f is *injective*. If for all $y \in Y$ there exists $x \in X$ such that $f(x) = y$, then f is *surjective*. If f is both injective and surjective, then f is bijective.

In order to project into a higher dimension, we need a function that maps $\mathbb{R}^m \to \mathbb{R}^n$, where $m < n$. There are bijective functions which do this; for example, consider the function $C : \mathbb{R} \to \mathbb{R}^2$ defined by:

$$C(x) = (x_1, x_2) \text{ where}$$

x_1 comprises the odd digits in the decimal expansion of x

x_2 comprises the even digits in the decimal expansion of x (4)

However, such functions are rather unusual, and tend to have properties that do not fit the spirit of ESNs as a theoretical model for reservoir computing, such as here being discontinuous everywhere. In particular, no such bijective function can be implemented by linear algebra, such as multiplication by a matrix. Indeed, any linear projection from $\mathbb{R} \to \mathbb{R}^n$ results in a 1-dimensional line within $\mathbb{R}^n$.

We can show that linear algebra cannot map onto the whole higher-dimensional space by applying the Rank-Nullity Theorem [20], which is also known as the Fundamental Theorem of Linear Algebra. If $\mathbf{W}^{in}$ is the matrix projecting the m-dimensional input $\mathbf{u}$ into n-dimensional space, then the Rank-Nullity Theorem gives $rank(\mathbf{W}^{in}) + nullity(\mathbf{W}^{in}) = m$, where $rank$ is the matrix rank function, representing the dimension of the image, and $nullity$ is the nullity of the matrix. The nullity represents the dimension of the kernel of the matrix (unrelated to the Kernel Rank measure in Reservoir Computing), i.e. the dimension of the space of vectors where $\mathbf{W}^{in}\mathbf{u}_1 = \mathbf{W}^{in}\mathbf{u}_2$. This trivially implies $dim(Im(\mathbf{W}^{in})) \leq dim(Domain(\mathbf{W}^{in}))$.

Therefore the transformed inputs occupy at most an m-dimensional hyperplane of the n-dimensional echo space. Contrary to the literature, the projection provided by $\mathbf{W}^{in}$ cannot increase the dimensionality of the input, and may even *decrease* it, if not chosen well. If the values of $\mathbf{W}^{in}$ are chosen in a uniform random manner from the range $[-1, 1]$, the probability of a decrease in dimensionality is negligible. However, the decrease in dimensionality is plausible if the values of $\mathbf{W}^{in}$ are chosen from a discrete set, such as $\{-1, 0, 1\}$.

If $nullity(\mathbf{W}^{in}) = 0$, then $dim(Im(\mathbf{W}^{in})) = dim(Domain(\mathbf{W}^{in}))$, and the use of linear algebra guarantees the transformation is surjective. We can therefore define $\mathbf{W}^{in\prime}$ such that $\mathbf{W}^{in\prime}(\mathbf{W}^{in}\mathbf{u}) = \mathbf{u}$. Note that this is *not* a bijection, as there exist points in $\mathbb{R}^n$ which are not in $Im(\mathbf{W}^{in})$.

With $\mathbf{W}^{in\prime}$ defined, we can show that the projection into a higher dimensional space does not contribute to linear separability. Suppose that there exist two sets of inputs A, B such that for all $\mathbf{u}_a \in A$ and $\mathbf{u}_b \in B$, $\mathbf{W}^{in}\mathbf{u}_a$ and $\mathbf{W}^{in}\mathbf{u}_b$ are linearly separable by the hyperplane $H_{sep} \in \mathbb{R}^n$. As $\mathbf{W}^{in}$ is a linear transformation, we know that all points $\mathbf{W}^{in}\mathbf{u}_a, \mathbf{W}^{in}\mathbf{u}_b$ lie on some hyperplane H_{inp} given by $Im(\mathbf{W}^{in})$. As H_{sep} separates points that lie on H_{inp}, by hypothesis, we know these two hyperplanes are not parallel. Therefore we can intersect H_{inp} and H_{sep} to get H_{int}. Finally, as points on H_{int} trivially lie in $Im(\mathbf{W}^{in})$, we can transform an orthognal basis of H_{int} using $\mathbf{W}^{in\prime}$, and find that $\mathbf{W}^{in\prime}(H_{int})$ is a hyperplane that separates A and B.

From the above it is clear that the ability to linearly separate points in echo space is *not* a function of the projection into the higher dimensional space of the ESN. Combined with the apparent insufficiently nonlinear activation functions, we move onto the final part of the ESN equation – the connectivity of the nodes – to attempt to identify how an ESN achieves linear separability.

3.3   Chains and Recurrence as Memory

Having established that the projection into the echo space of an ESN is not responsible for any enhanced linear separability, the next step is to identify where this capability comes from.

An ESN can be treated as a randomly connected graph. Recall that when doing so, the weight matrix of the ESN can be treated as a weighted adjacency matrix, allowing a matrix to be interpreted as a directed graph. Examples of this are shown in Figs. 6 and 7. In these figures, the nodes labelled u_i represent where inputs $\mathbf{u}(t)$ enter the ESN. Other nodes sum the weighted inputs and then apply the activation function.

Assuming that the activation function is not particularly significant, we can identify two clear structures that have the potential for memory. First are terminated chains, as shown in Fig. 6. These are a simple chain of nodes within the ESN with a clear end. From an input node, a terminated chain of length t provides information on the past t inputs. If the spectral radius of the ESN is less than 1, then the values in these chains diminish as they progress along the chain. They diminish by a fixed amount: the product of the weights of the chain to that point in history. Therefore, this diminishment is largely irrelevant, as the linear regression output layer can learn these weights to recover the history.

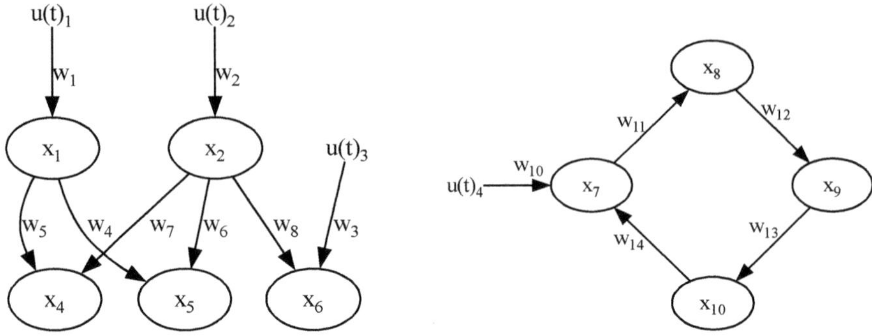

Fig. 6. Multiplexed chains in an ESN   **Fig. 7.** Loop in an ESN

The information provided by a chain is effectively equivalent to a single dimension of the input. This means that multiple-dimensioned inputs require multiple chains to attain the same memory. These chains can be multiplexed, because the linear regression layer can solve linear algebra equations. For example, in Fig. 6, the node values x_4, x_5 and x_6 sum some combination of values from all three inputs, at different stages of time. We can express the values of these nodes at time t as:

$$x_4 = w_1 w_5 u(t-2)_1 + w_2 w_7 u(t-2)_2 \tag{5}$$

$$x_5 = w_1 w_4 u(t-2)_1 + w_2 w_6 u(t-2)_2 \tag{6}$$

$$x_6 = w_2 w_8 u(t-2)_2 + w_3 u(t-1)_3 \tag{7}$$

Given that the weights are typically randomly generated floating point values, they are unlikely to be identical, and so these equations are likely to be independent. Therefore, as we have three equations and the three values $u(t-2)_1$, $u(t-2)_2$ and $u(t-1)_3$ to recover, there is a linear solution, and therefore linear regression can recover all three values. While this example used a sparse input matrix for simplicity, meaning that not all nodes received all inputs, this also holds for ESNs with a densely connected $\mathbf{W}^{in}$.

The second structure is a cycle of nodes, such as in Fig. 7. A cycle has similar properties to a chain, especially if the number of nodes in the loop is large. In this case, the diminishing of value due to the spectral radius ensures that even a large initial input will be mostly insignificant once it has made its way around the cycle. However, if the size of the loop is small, or the spectral radius is close to 1, then a cycle is noticeably different from a chain. In Fig. 7, the cycle exhibits memory modulo 4: the input is added to the residual value from 4 timesteps ago. In practical terms, this means that a cycle can be sensitive to periodic inputs, provided that the periodic input is of great enough magnitude to overcome the diminishing of being passed through the cycle. However, a similar effect can be accomplished with linear regression, provided that the stored history is long enough.

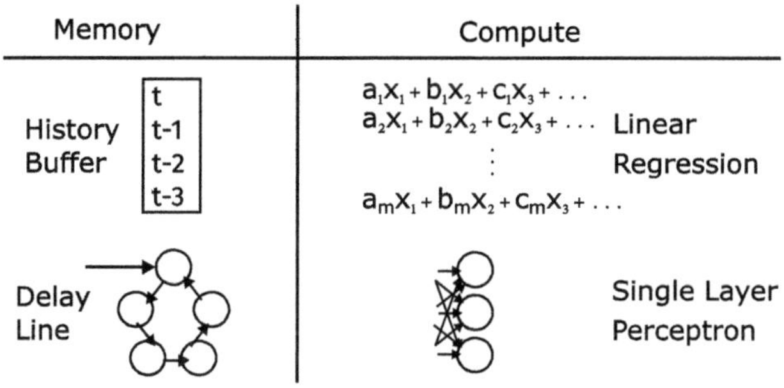

Memory | Compute

History Buffer: t, t-1, t-2, t-3

$a_1x_1 + b_1x_2 + c_1x_3 + \ldots$
$a_2x_1 + b_2x_2 + c_2x_3 + \ldots$ Linear Regression
$a_mx_1 + b_mx_2 + c_mx_3 + \ldots$

Delay Line

Single Layer Perceptron

Fig. 8. LIRC Method

As an ESN is randomly connected, these chain and loop structures occur at random. They may also be combined with other structures, for example, a chain-like structure with extra inputs. These combinations are not necessarily significant, as they all constitute linear algebra. A combination of values may change the way in which information is recovered, but provided that the history of time t is comprised of weights that form a basis for the input, a linear regression layer can extract that history.

Unlike the cases of input projection and the activation function, the chains of nodes within an ESN do increase the dimensionality of the state of the ESN. By encoding the values of previous inputs, by storing history, the amount of data available increases.

This raises a question: can we increase the dimensionality by simpler means?

4   LIRC: Lagged Input Regression Computation

To recap: we have shown that the tanh() activation function and linear projection into a higher dimensional space do not yield an increase in dimensionality. In this situation, the only increase in dimensionality is given by the memory of the ESN storing past values. One question that arises is can the setup be simplified and memory provided by simply time-lagging inputs?

Time-lagged data is not a new idea in reservoir computing: delay-lines [19] are commonly used to implement feedback in physical systems. An alternative is the so-called "single dynamical node" approach of Appeltant et al. [3], which stores the last n outputs. In both cases, either the virtual nodes of the delay line or output buffer of the single dynamical node are then utilised by a linear regression layer which can be trained to a task. Both of these approaches can be adapted to store information on past inputs.

While delay-lines and the single dynamical node approach use linear regression, more recent work has identified that the capability for nonlinear readout [9,15] can be beneficial. Therefore, in addition to using Linear Regression, we also propose the use of a trained nonlinear readout, in the form of a single layer

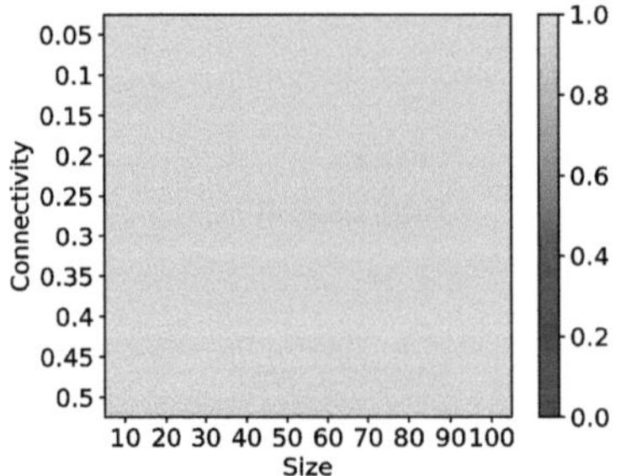
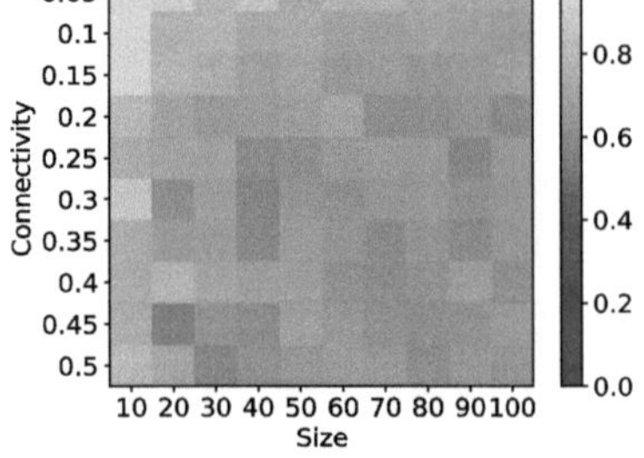

Fig. 9. R^2 of tanh() ESNs approximated by Linear Regression on a Single Dynamical Node of length 10

Fig. 10. R^2 of LeakyReLU ESNs approximated by Linear Regression on a Single Dynamical Node of length 10

perceptron (SLP). Use of an SLP sidesteps the need to define a precise nonlinearity, whilst still being easy to train.

The combination of time lagged inputs and regression results is here named Lagged Input Regression Computation (LIRC). LIRC separates the memory and computational aspects of a system (Fig. 8). This enhances the understandability of the system, and in some configurations may be simpler to implement.

4.1   Comparison to ESNs

The first evaluation of LIRC as an approach is to compare against ESNs. In this case, randomly generated ESNs are used as a benchmark that LIRC methods attempt to predict. Depending on the LIRC methods with the greatest accuracy, we can make statements about the complexity of the ESN. To ensure a wide variety of configurations were tested, we conducted a parameter sweep on the spectral radius, connectivity, size and activation function of the ESN, with all tests being carried out against all combinations of LIRC approximations. All experiments were repeated 5 times with different random seeds to verify the results. Accuracy is measured using the coefficient of determination R^2 [10]; for the presented heatmaps, values below 0 are deemed sufficiently inaccurate to be unusable and clamped to 0. More parameters were explored than can be represented on a 2D heatmap; each cell represents the average of all experiments with those parameters (100 experiments per cell). The effects of parameters not mentioned in the graphs, such as spectral radius, were observed to be minimal.

Figure 9 shows that ESNs with a tanh() activation function are trivially approximable via linear regression on time lagged inputs. Even in cases where the ESN is small or the connectivity is low, this holds true, with a minimum accuracy of 98% being observed. However, linear regression is unable to handle the LeakyReLU activation function (Fig. 10), due to the nonlinearity of LeakyReLU occurring at 0, thus meaning the linear regression layer can train to fit only half of the output data.

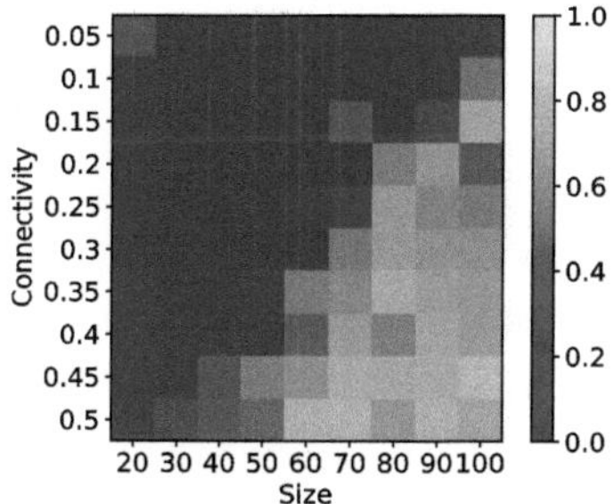

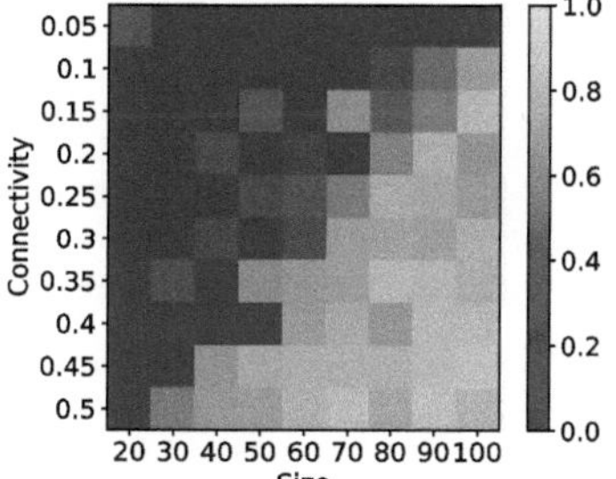

Fig. 11. R^2 of LeakyReLU ESNs approximated by an SLP on a Single Dynamical Node of length 20

Fig. 12. R^2 of LeakyReLU ESNs approximated by an SLP on a Delay Line of length 20

In order to achieve more accurate results for LeakyReLU, it is necessary to use the SLP (Figs. 11 and 12); however, even this requires favourable conditions to achieve a high level of accuracy, with a large size, unreasonably high connectivity and increased available memory required to achieve greater than 80% accuracy. This illustrates that when choosing a nonlinearity for an ESN, the probability of the nonlinearity having a significant impact must be taken into account.

There is a noticeable improvement in the accuracy of the approximation when using a Delay Line for memory with LeakyReLU, as shown in Fig. 12, where the greatest accuracy is 0.92, as opposed to 0.79 in Fig. 11. This suggests that the unbounded nature of the LeakyReLU activation may be capable of causing data to persist in cycles more than is the case with tanh. However, an increase in accuracy of the approximation indicates that the ESN is not providing more computational ability than a LIRC-SLP with the same number of nodes. As SLPs are simpler to compute and better studied, this raises questions around the efficacy of large ESNs. Smaller ESNs are less approximable, and therefore give a different computation to an equivalently sized LIRC-SLP.

4.2 Nonlinear Memory Capacity

Next we turn to evaluating the linear regression based LIRC models for nonlinear memory capacity. This may seem unusual; a linear system should not be able to exhibit nonlinear behaviour. However, we have shown that we can approximate an ESN with the tanh() activation function to a high degree of accuracy with a linear system. This is apparently at odds with the results of Dambre et al. [8], who show that ESNs exhibit nonlinear memory of odd degree, measured using the orthogonal set of Legendre Polynomials (used for uniform random input).

This same nonlinear memory capacity exists in our linear system for odd degrees of Legendre Polynomials (e.g. Figure 13 for P_3). This is because all odd degrees of Legendre Polynomials include a linear term. For example, $P_3(x) = \frac{1}{2}(5x^3 - 3x)$. As $x \to 0$, the linear term dominates. Therefore, there is a region of the odd degree Legendre Polynomials that is approximable by a linear system,

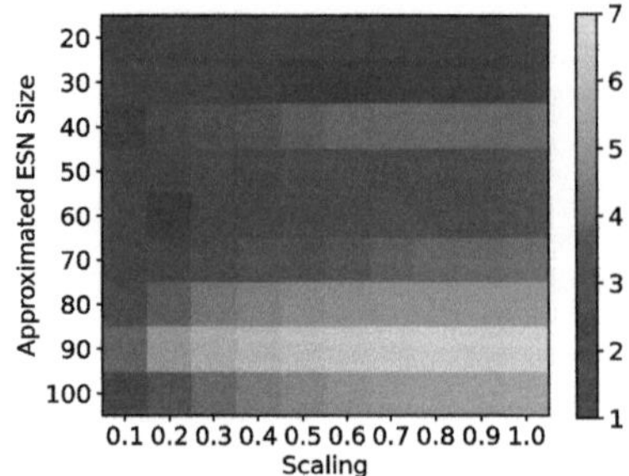

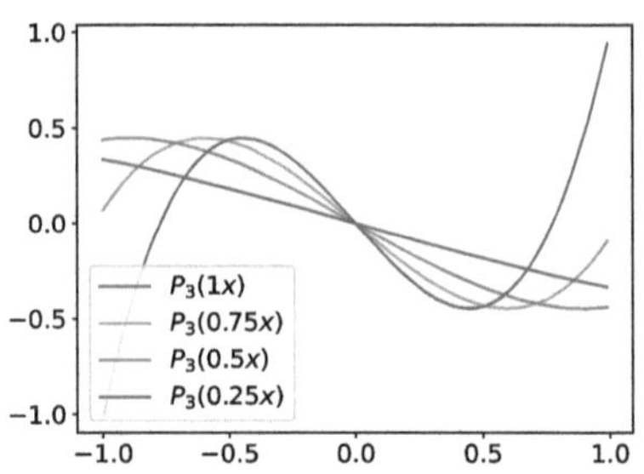

Fig. 13. Memory Capacity of Lagged Input Linear Regression Approximating a tanh() ESN

Fig. 14. Legendre Polynomial P_3 with different input scalings

with the size of this region decreasing for higher degrees of Legendre Polynomial, explaining the apparent nonlinear memory capacity. This observation also applies to other orthogonal polynomials, such as Hermite Polynomials used with Gaussian inputs [8].

Even degrees of the Legendre Polynomial lack this linear term, and so are not approximable in this manner, but are zero for the reason Dambre et al. [8] give: tanh() is an odd function, which matches the odd degrees.

Dambre et al. [8] also observe that by changing the input scaling, the system appears to become more nonlinear, as the contribution of higher degrees becomes greater. This is correct, but somewhat misleading. Changing the input scaling stetches the polynomial (Fig. 14), which increases the range of the linearly approximable region. For $P_3(0.25x)$, the function is almost entirely linear across $[-1, 1]$. If this range exceeds $[-1, 1]$, then the tanh() activation clamps the values, resulting in the linear section becoming nonlinear. The effect of this is that as the scaling changes, so does the Legendre Polynomial that is most approximable by a linear system. While this scaling is not reflected exactly in an ESN or linear approximation due to the summation of values in the ESN, in Fig. 13 we do see that the optimal P_3 memory is at a scaling of 0.8.

5   Conclusion

Several of the claims in the literature about how ESNs perform their computations, and what computations they perform, may be overstated.

ESNs with the commonly-used tanh() activation function can be accurately approximated by performing linear regression on time lagged inputs. This is due to the nonlinearity of tanh() not being substantial for the majority of values likely to occur within such an ESN. tanh() ESNs tend to behave as linear memory systems, with the linear regression layer unscrambling the random weighted connections within an ESN to recover data. Linear regression over simple time-lagged memory structures is sufficient to solve many problems, and should be used as a control case when evaluating ESN performance.

The LeakyReLU activation function is *not* approximable by linear regression to the same degree. Even when using the more powerful SLP-based regression method, LeakyReLU ESNs require favourable conditions to be well approximated. Despite being effectively a choice between two linear functions, LeakyReLU produces more complex dynamics in an ESN by virtue of the non-linearity being more likely to be exploited.

Other authors have demonstrated nonlinear memory capacity in tanh() ESNs. These results can be explained by the odd order Legendre Polynomials containing a linear term, which dominates the nonlinear terms for sufficiently small input values. When this linear term dominates in the range $[-1, 1]$, the Legendre Polynomial is approximable by linear regression. By varying the scaling factor, it is possible to vary which of the Legendre Polynomials contributes most nonlinear memory capacity. This gives the illusion of nonlinear memory capacity, when in fact only linear memory capacity is being demonstrated.

Further work is needed to evaluate other activation functions, and to determine the desirable properties for an activation function of an ESN, or indeed if an ESN is still a useful tool given recent work that utilised simplified structures to achieve similar results [9,18]. In addition, it may be necessary to revisit the use of orthogonal polynomials for determining nonlinear memory capacity.

Statements and Declarations. The authors acknowledge funding from the MARCH project, EPSRC grant numbers EP/V006029/1. The authors have no relevant financial or non-financial interests to disclose.

References

1. Abramowitz, M., Stegun, I.A.: Handbook of Mathematical Functions: With Formulas, Graphs, and Mathematical Tables, vol. 55. Courier Corporation (1965)
2. Amari, S.I.: Learning patterns and pattern sequences by self-organizing nets of threshold elements. IEEE Trans. Comput. **C-21**(11), 1197–1206 (1972). https:// doi.org/10.1109/T-C.1972.223477
3. Appeltant, L., et al.: Information processing using a single dynamical node as complex system. Nat. Commun. **2**(1), 468 (2011)
4. Bishop, C.M.: Pattern Recognition and Machine Learning (Information Science and Statistics). Springer, Heidelberg (2006)
5. Bollt, E.: On explaining the surprising success of reservoir computing forecaster of chaos? the universal machine learning dynamical system with contrast to VAR and DMD. Chaos: Interdisc. J. Nonlinear Sci. **31**(1) (2021)
6. Buehner, M., Young, P.: A tighter bound for the echo state property. IEEE Trans. Neural Netw. **17**(3), 820–824 (2006). https://doi.org/10.1109/TNN.2006.872357
7. Cover, T.M.: Geometrical and statistical properties of systems of linear inequalities with applications in pattern recognition. IEEE Trans. Electron. Comput. **EC-14**(3), 326–334 (1965). https://doi.org/10.1109/PGEC.1965.264137
8. Dambre, J., Verstraeten, D., Schrauwen, B., Massar, S.: Information processing capacity of dynamical systems. Sci. Rep. **2**(1), 514 (2012)
9. Gauthier, D.J., Bollt, E., Griffith, A., Barbosa, W.A.S.: Next generation reservoir computing. Nat. Commun. **12**(1) (2021). https://doi.org/10.1038/s41467-021-25801-2

10. Glantz, S.A., Slinker, B.K., Neilands, T.B.: Primer of applied regression & analysis of variance. (No Title) (1990)
11. Gu, B.: Lecture notes: the power method for spectral radius (2023). https://users.wpi.edu/~bgu/sp23/ma3257/lecture_notes/MA3257_L22.pdf
12. Hart, A.G., Olding, K.R., Cox, A.M.G., Isupova, O., Dawes, J.H.P.: Using echo state networks to approximate value functions for control (2021). https://arxiv.org/abs/2102.06258
13. Hoerl, A.E., Kennard, R.W.: Ridge regression: biased estimation for nonorthogonal problems. Technometrics **12**(1), 55–67 (1970). https://doi.org/10.1080/00401706.1970.10488634
14. Jaeger, H., Haas, H.: Harnessing nonlinearity: predicting chaotic systems and saving energy in wireless communication. Science **304**(5667), 78–80 (2004)
15. Ma, H., Prosperino, D., Räth, C.: A novel approach to minimal reservoir computing. Sci. Rep. **13**(1), 12970 (2023)
16. Mackey, M.C., Glass, L.: Oscillation and chaos in physiological control systems. Science **197**(4300), 287–289 (1977)
17. Mozer, M.C.: A focused backpropagation algorithm for temporal pattern recognition. In: Backpropagation, pp. 137–169. Psychology Press (2013). https://doi.org/10.4324/9780203763247
18. Prosperino, D., Ma, H., Räth, C.: Tailored minimal reservoir computing: on the bidirectional connection between nonlinearities in the reservoir and in data (2025). https://arxiv.org/abs/2504.17503
19. Soriano, M.C., et al.: Delay-based reservoir computing: noise effects in a combined analog and digital implementation. IEEE Trans. Neural Netw. Learn. Syst. **26**(2), 388–393 (2014)
20. Strang, G.: The fundamental theorem of linear algebra. Am. Math. Monthly **100**(9), 848–855 (1993). http://www.jstor.org/stable/2324660
21. Waheeb, W., Ghazali, R., Shah, H.: Nonlinear autoregressive moving-average time series forecasting using neural networks. In: 2019 International Conference on Computer and Information Sciences, pp. 1–5 (2019). https://doi.org/10.1109/ICCISci.2019.8716417
22. Wringe, C., Trefzer, M., Stepney, S.: Reservoir computing benchmarks: a tutorial review and critique. Int. J. Parallel Emerg. Distrib. Syst. 1–39 (2025)
23. Yan, X., Su, X.: Linear Regression Analysis: Theory and Computing. World Scientific (2009)

Determining Isomorphic Crystal Structures

Nataša Jonoska[(✉)], Milé Krajčevski[(✉)], and Greg McColm

University of South Florida, Tampa, FL 33620, USA
{jonoska,mile,mccolm}@usf.edu

Abstract. A common representation of crystal structures is by periodic graphs, i.e., graphs whose automorphism groups have subgroups of translational symmetries. Such graphs may be represented as (finite) quotient graphs (called voltage graphs) whose edges are labeled by corresponding elements of their translational subgroup. We outline a polynomial time algorithm for determining whether two finite bi-deterministically edge-labeled voltage graphs with voltages from corresponding translational subgroups generate isomorphic periodic graphs. This algorithm aids in the classification of crystal structures and provides a method for determining whether two sets of building blocks can assemble isomorphic crystals.

Keywords: Periodic Graphs · Isomorphic crystal structures · voltage graphs · graph coverings · bi-deterministic edge labeling

1   Introduction

In crystallography, periodic graphs - i.e., graphs with virtually free abelian automorphism groups - are often used to represent covalent crystals, with the vertices representing atoms or molecular building blocks, and the edges representing bonds or ligands, respectively [7]. Databases such as the EPINET [25], RCSR [24], Systre [6], or TOPOS [2], aim to catalog isomorphism classes of "crystal nets" [6] where two "crystal structures" are equivalent if their periodic graphs are isomorphic. Recent advances in DNA nanotechnology have experimentally obtained a variety of DNA based crystal structures with the tensegrety triangle (Fig. 1) and its variations as building blocks [15, 31].

Determining whether two crystal nets are isomorphic by considering appropriate representations of their unit cells is at least as hard as the finite Graph Isomorphism Problem [9, §7.1], which is widely suspected to be neither polynomial time computable nor NPTIME-complete [32]. However, in some cases for crystal structure design (e.g. [8,23]), it may suffice to compare appropriately labeled periodic graphs. For example, in deriving a periodic graph, a vertex may have several edges in a common orbit, the computer program will necessarily uniquely index these edges, and that indexing may be important in the design process, a prospect explored in [16,22] (also similar to [20]). In predicting the formation of a crystal, some analyses (e.g. for rigidity [4]) may require that

© The Author(s), under exclusive license to Springer Nature Switzerland AG 2026
E. Formenti and L. Manzoni (Eds.): UCNC 2025, LNCS 16364, pp. 277–293, 2026.
https://doi.org/10.1007/978-3-032-15641-9_19

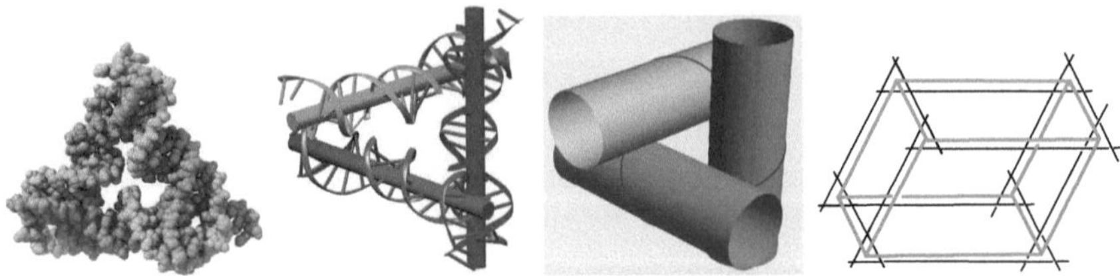

Fig. 1. Experimental assembly of a DNA tensegrity triangle from [33] with schematics of the DNA and cylindrical representation. The building block forms one corner of a rhombohedral unit block of a crystal as shematically represented to the right.(Color figure online)

all edges be uniquely identifiable. In the manufacturing process, the individual molecular building blocks might not have a symmetry reflected in the crystal net, and that asymmetry could be represented in the labeling. Finally, recalling proposals for three-dimensional periodic or crystalline computer architecture (e.g., [21,28,29]), edge labeling could facilitate addressing and information flow.

Here we use "bi-deterministic" labeling (aka "bi-resolving"), meaning that at each vertex, all incoming edges carry distinct labels and all outgoing edges do as well. To address periodic graphs - fundamentally infinite structures - we follow [5] by representing them using voltage graphs [11, §2] (see also [10]), whose edges are labeled with elements from their translational symmetry subgroup. In [17], we developed a method for determining when two voltage graphs yield isomorphic derived graphs, provided the groups generated by the voltages had certain properties. With bi-deterministic labeling and voltages taken from free abelian groups (which we take to be powers of $\mathbb{Z}$), the necessary properties for the derived graphs isomorphism can be checked. Notice that an edge labeling on a graph of order n and size m can be determined to be bi-deterministic within time $O(nm \log m)$. Building on this foundation, we present a polynomial time algorithm that determines whether two voltage graphs have isomorphic derived periodic graphs. Our primary result is (for a more precise statement see Theorem 4):

Theorem 1. *Given two bi-deterministically labeled voltage graphs with periodic derived graphs, there exists a polynomial time algorithm for determining whether the two derived graphs are isomorphic.*

In this article, following a section on preliminaries, we outline algorithms that prove Theorem 1. We assume that periodic graphs are given with their quotient graphs, labeled by elements of respective free abelian groups, that is, as voltage graphs [11] (in this note, groups are abelian). We outline a procedure for constructing a common underlying graph that simultaneously covers the given voltage graphs, provided such a common graph exists and is covered by both corresponding derived periodic graphs. Next, we show how to 'lift' and condense voltage assignments into the common graph. Finally, we determine whether such obtained voltage assignments of the common graph generate a group isomorphism, this occurs if and only if the supposed label-preserving isomorphism of the periodic graphs exists.

2   Preliminaries

We review some graph and group theoretic notions we use, mostly found in standard texts [1,3,12,30]. In addition, we use some material on voltage graphs [10] and topological crystallography [27]. Following voltage graph theory, some combinatorial and geometric group theory, and theoretical crystallography literature, our graphs have oriented edges[1].

A *graph* is a quadruple $\Gamma = (V, E, \iota, \tau)$ where V is a set of *vertices*, E is a set of *edges*, and $\iota, \tau : E \to V$ are *initial endpoint* and *terminal endpoint* functions, respectively. To ease notation, we write $\Gamma = (V, E)$ and assume ι, τ as being specified. A *pointed graph* is a pair (Γ, v) where Γ is a graph and v a vertex in Γ, which we call the *base vertex* of (Γ, v). The *order* of a graph is the number of vertices, while the *size of the graph* is the number of edges.

Given a graph $\Gamma = (V, E)$ and a subset $E' \subseteq E$, the *subgraph of Γ induced by E'* (written $\Gamma' \leq \Gamma$) is the graph $\Gamma = (V', E')$ where $V' = \{\iota(e), \tau(e) : e \in E'\}$.

The *orientation* of e is the ordered pair $(\iota(e), \tau(e))$. Given a graph Γ and a vertex $v \in V$, we define the *inset of v*, $\mathrm{In}(v)$, and *outset* of v, $\mathrm{Out}(v)$, to be

$$\mathrm{In}(v) = \{e \in E : \tau(e) = v\} \text{ and } \mathrm{Out}(v) = \{e \in E : \iota(e) = v\}.$$

The *neighborhood* of a vertex v is defined as $\mathrm{Nbd}(v) = \mathrm{In}(v) \cup \mathrm{Out}(v)$.

Given a graph Γ with vertices $v, v' \in V$, a *walk* from v to v' is a finite sequence of vertices and edges $v = v_0, e_1, v_1, e_2, v_2, e_3, \ldots, e_n, v_n = v'$, $n \geq 0$, such that for each i, $1 \leq i \leq n$, $\{v_{i-1}, v_i\} = \{\iota(e_i), \tau(e_i)\}$. If $v_{i-1} = \iota(e_i)$ and $v_i = \tau(e_i)$, the walk traverses the edge e_i *along* e_i's orientation, while if $v_{i-1} = \tau(e_i)$ and $v_i = \iota(e_i)$, the walk traverses the edge e_i *against* e_i's orientation. The *length* of the above walk is n, the number of edges traversed (counting repetitions). The *distance* between any two vertices in Γ is the minimum length of all walks from one of the vertices to the other. A graph is *connected* if, for any two vertices in the graph, there is a walk from one to the other. If a walk starts and ends at the same vertex we say it is *cyclic*; a walk in which no vertex occurs more than once is a *path*.

An alphabet is a finite set Σ whose elements are symbols. We add the set $\bar{\Sigma} = \{\bar{s} : s \in \Sigma\}$ of "dual symbols" such that $\Sigma \cap \bar{\Sigma} = \varnothing$. The one-to-one mapping $s \mapsto \bar{s}$, $s \in \Sigma$ is an involution, i.e., $\bar{\bar{s}} = s$.

Definition 1. *A labeled graph is a pair (Γ, λ) where $\Gamma - (V, E)$ is a graph, and $\lambda : E \to \Sigma$ is a labeling map. A labeled graph (Γ, λ) is bi-deterministic if, for each vertex v, λ is one-to-one on $\mathrm{In}(v)$ and on $\mathrm{Out}(v)$.*

In this article, every labeled graph is bi-deterministic. Notice that the number of elements in Σ is bounded below by $\max\{|\mathrm{In}(v)|, |\mathrm{Out}(v)| : v \in V\}$.

---

[1] In combinatorics, graphs with oriented edges are called "digraphs".

We say that two vertices v_1 and v_2 in labeled graphs have congruent neighborhoods, denoted with $\mathrm{Nbd}(v_1) \cong \mathrm{Nbd}(v_2)$, if there are label preserving bijections $\mathrm{In}(v_1) \to \mathrm{In}(v_2)$ and $\mathrm{Out}(v_1) \to \mathrm{Out}(v_2)$.

A *word* is a sequence of symbols from $\Sigma \cup \bar{\Sigma}$, and the set of all words is denoted $(\Sigma \cup \bar{\Sigma})^*$; the empty word is denoted ϵ and we define $\bar{\epsilon} = \epsilon$. For a word $x = s_1 s_2 \cdots s_{n-1} s_n \in (\Sigma \cup \bar{\Sigma})^*$, $n \geq 0$, the *reversal* of x is the word $\bar{x} = \bar{s}_n \bar{s}_{n-1} \cdots \bar{s}_2 \bar{s}_1$. The length of x is $|x| = n$; in particular, $|\epsilon| = 0$. Given a labeled graph (Γ, λ) and a walk $p = v_0, e_1, v_1, \ldots, e_n, v_n$, the *label* of p is the word $\lambda(p) = x = s_1 s_2 \cdots s_n$ in $(\Sigma \cup \bar{\Sigma})^*$ such that for each i, $s_i = \lambda(e_i)$ if $\iota(e_i) = v_{i-1}$ and $\tau(e_i) = v_i$ (p is along e_i's orientation) or $s_i = \overline{\lambda(e_i)}$ if $\tau(e_i) = v_{i-1}$ and $\iota(e_i) = v_i$ (p is against e_i's orientation). Since (Γ, λ) is bi-deterministic, for a given word x and a vertex v there is at most one walk $p_{v,x}$ that starts at v and has label x. We use the notation vx to denote the end vertex of the walk $p_{v,x}$. If there is no walk from v with label x, we say that vx is *undefined*, and write $vx = \emptyset$; see Fig. 2. It follows that $v\epsilon = v$. Notice that for a bi-deterministic labeling, if $vx = u$, then starting at u and reversing x, we return to v, that is, $u\bar{x} = v$. If vx is defined then $vx\bar{x}$ is also defined and $vx\bar{x} = v$.

Given a labeled graph Γ, we say that the words x_1 and x_2 are *equivalent from v* if $vx_1 = vx_2$. Two words might be equivalent from one vertex of a graph but not from another in the same graph. Notice that two words x_1 and x_2 are equivalent from v if and only if $vx_1\overline{x_2} = v$. Given a word x and a vertex v, denote the set of words equivalent to x from v by $[x]_v = \{x' \in (\Sigma \cup \bar{\Sigma})^* : vx = vx'\}$. In particular, $[\epsilon]_v$ is the set of words x such that $vx = v$, which is the set of all cyclic walks from v to v.

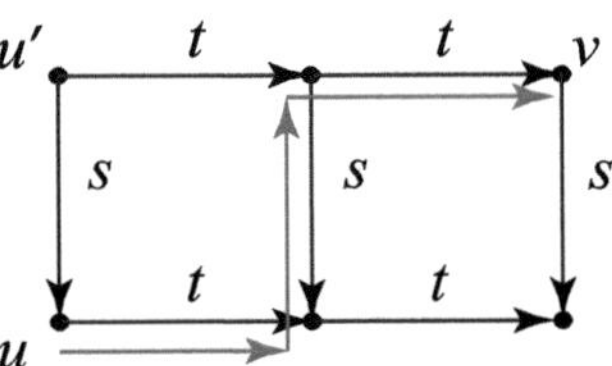

Fig. 2. In the above graph, the word $x = t\bar{s}t$ determines a path $p_{u,x}$ from the vertex u with terminal vertex v, so $ux = ut\bar{s}t = v$. On the other hand, $u'x = u't\bar{s}t = \emptyset$.

As is standard, for two graphs $\Gamma_1 = (V_1, E_1)$ and $\Gamma_2 = (V_2, E_2)$, a function $\varphi : \Gamma_1 \to \Gamma_2$ is a *homomorphism* if, for every $e \in E_1$, $\iota(\varphi(e)) = \varphi(\iota(e))$, and $\tau(\varphi(e)) = \varphi(\tau(e))$. A bijective homomorphism is an *isomorphism*. An isomorphism from a graph Γ to itself, is called an *automorphism*. If Γ_1, Γ_2 are pointed, then a homomorphism maps the base vertex of Γ_1 to the base vertex of Γ_2.

Definition 2. *Given labeled graphs (Γ_1, λ_1) and (Γ_2, λ_2), a homomorphism $\varphi : \Gamma_1 \to \Gamma_2$ is* label-preserving *if, for each edge e in Γ_1, we have $\lambda_2(\varphi(e)) = \lambda_1(e)$.*

A label-preserving isomorphism is called a λ-*isomorphism*. If there is a label-preserving isomorphism from Γ_1 onto Γ_2, we say that they are λ-*isomorphic* and write $\Gamma_1 \cong_\lambda \Gamma_2$. For a labeled digraph (Γ, λ), the group of its λ-automorphisms is denoted $\mathrm{Aut}_\lambda(\Gamma)$. Notice that if φ is an isomorphism from Γ_1 onto Γ_2, and if $v \in V$, and $x \in (\Sigma \cup \bar{\Sigma})^*$, then vx exists if and only if $\varphi(v)x$ exists; further, if they exist then $\varphi(vx) = \varphi(v)x$. Thus, if $vx = g(v)$ for some $g \in \mathrm{Aut}_\lambda(\Gamma)$, and $h \in \mathrm{Aut}_\lambda(\Gamma)$, then $h(v)x = h(vx) = h(g(v))$.

Definition 3. *Let Γ and Δ be graphs. An onto homomorphism $\phi : \Gamma \to \Delta$ is a* covering *if, for each $v \in V_\Gamma$, the restrictions $\phi|_{\mathrm{Out}(v)}$ and $\phi|_{\mathrm{In}(v)}$ are bijections from $\mathrm{Out}(v)$ onto $\mathrm{Out}(\phi(v))$ and from $\mathrm{In}(v)$ onto $\mathrm{In}(\phi(v))$, respectively. If (Γ, v) and (Δ, v') are pointed graphs, then ϕ is a* pointed covering *if it is a pointed homomorphism. If (Γ, v) and (Δ, v') are labeled graphs, then ϕ is a* label-preserving covering *if it is a label-preserving homomorphism. We say that Γ is a (pointed, label-preserving, resp.)* cover *of Δ.*

If Γ and Δ are both pointed, label-preserving, or both, we take the cover $\Gamma \to \Delta$ to be pointed, label-preserving, or both, per the context.

The group $G_\phi = \{\alpha \in \mathrm{Aut}(\Gamma) : \phi\alpha = \phi\} \le \mathrm{Aut}(\Gamma)$ is the *cover group* of the covering ϕ. Figure 3 shows a covering $\phi : (\Gamma, v) \to (\Delta, v')$ where Γ is pointed with base v being the square vertex and Δ the pointed quotient to the right with corresponding base v'. As $\phi(v) = v'$, ϕ is a pointed covering with $G_\phi = \mathbb{Z}$. One can observe $\mathbb{Z}$ 'copies' of Δ in Γ (the 0'the copy indicated).

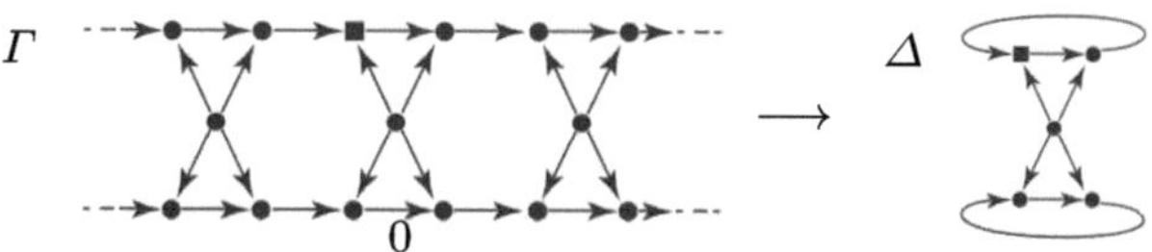

Fig. 3. An infinite cyclic group as the cover group generated by the rightwards shift. ■ is the pointed vertex.

Definition 4. *Let Δ be connected and $\phi : \Gamma \to \Delta$ be a covering. For any edge or vertex of Δ the* fiber *of x is the set $\phi^{-1}(x)$, and an element of $\phi^{-1}(x)$ is a* lift *of x. A* lift *of Δ is a connected subgraph $\Lambda \le \Gamma$ of Γ such that ϕ restricted to the edges of Λ is bijective, inducing a homomorphism $\phi|_\Lambda : \Lambda \to \Delta$.*

Let Δ be a graph, and let G be a group. A *voltage graph* is a pair (Δ, γ), where $\gamma : E \to G$ is a labeling on the edges of Δ with the elements of G. We call γ a *voltage assignment* on the edges of Δ. Although the definition below holds for any group, in this paper we consider only abelian groups and we use '+' for the operation notation.

Definition 5. *Given the voltage graph (Δ, γ), the* derived graph *of Δ by γ is the graph $\Gamma = (V, E)$ defined by $V = V_\Delta \times G$, $E = E_\Delta \times G$, and*

$$\iota(e, g) = (\iota(e), g), \quad \tau(e, g) = (\tau(e), g + \gamma(e)).$$

We denote the derived graph $\Gamma = \Delta \times_\gamma G$.

Observe that the derived graph $\Delta \times_\gamma G$ is a *natural* cover of Δ by the canonical projection map $\phi : \Delta \times_\gamma G \to \Delta$ defined by $\phi(v,g) = v$ and $\phi(e,g) = e$.

Given a graph Δ, a *spanning tree* is a subgraph of Δ that is a tree, i.e., it is acyclic and connected, and contains all vertices of Δ. A voltage assignment γ in a voltage graph (Δ, γ) is *condensed* if there exists a spanning tree S of Δ such that for each $e \in E_S$, $\gamma(e) = 0_G$, the identity of the voltage group (recall that in this paper, our groups are abelian, written additively). Given a voltage assignment γ of a voltage graph (Δ, γ), the *condensation* of γ with respect to a spanning tree S of Δ is a voltage assignment γ' such that for each $e \in E_S$, $\gamma'(e) = 0_G$, and such that for any cyclic walk $v_0, e_1, v_1, \ldots, v_n = v_0$ in Δ that has exactly one edge e_k not in S, $\sum_{j=1}^{n} \epsilon_j \gamma(e_j) = \epsilon_k \gamma'(e_k)$, where $\epsilon_j = +$ if e_j is traversed along its orientation, while $\epsilon_j = -$ otherwise. From [11, Theorem 2.5.4] and [17]:

Theorem 2. *Let (Δ, γ) be a voltage graph such that Δ is connected and $\gamma(E_\Delta)$ generates G. Let S be any spanning tree of Δ and γ' is a condensation of γ with respect to S. If $\Delta \times_\gamma G$ is connected then $\Delta \times_\gamma G \cong \Delta \times_{\gamma'} G'$, where G' is the subgroup of G generated by $\gamma'(E_\Delta)$. If Δ is bi-deterministically labeled, then the isomorphism is a λ-isomorphism.*

Let Δ be a connected and bi-deterministically labeled graph, γ_1 and γ_2 are voltage assignments to edges in Δ that generate groups G_1 and G_2. Suppose further that γ_1 and γ_2 are condensed with respect to a spanning tree S. Let $\phi_i : \Delta \times_{\gamma_i} G_i \to \Delta$ be the canonical cover for $i = 1, 2$ defined with $(v, g) \mapsto v$. Then we have the following Theorem 3, which is a generalization of a similar result in [19] and [26] and is adapted in [17].

Theorem 3. *There exists a λ-isomorphism $\varphi : \Delta \times_{\gamma_1} G_1 \to \Delta \times_{\gamma_2} G_2$ such that $\phi_2 \varphi = \phi_1$ if and only if the assignment $\gamma_1(e) \mapsto \gamma_2(e)$, $e \in E_\Delta$, induces a group isomorphism of G_1 onto G_2.*

Definition 6. *Let (Δ, γ) be a voltage graph and let $\mu : \Delta_\mathsf{T} \to \Delta$ be a covering. The* lift *of γ via μ is the voltage assignment γ' such that for each edge e of Δ_T, $\gamma'(e) = \gamma(\mu(e))$.*

The following is from [17, Theorem 4.15] with assistance from [27, Theorems 5.2 & 5.5]. See Fig. 4.

Recall that a group acts *freely* on a set X if no elements of the group, save the identity, fixes any elements of X. The lattice group of a periodic graph acts freely on that graph.

Proposition 1. *Let Γ be a bi-deterministically labeled connected graph and let $\nu : \Gamma \to \Delta_\mathsf{T}$ and $\mu : \Delta_\mathsf{T} \to \Delta$ be label-preserving coverings. Suppose that $\phi = \mu\nu$ is such that G_ϕ is abelian and acts freely on Γ, and let $\gamma : E_\Delta \to G_\phi$ be a condensed voltage assignment whose voltages generate G_ϕ. Let γ_T be the condensation of the lift of γ with respect to a spanning tree of Δ_T.*

Then $\Gamma \cong_\lambda \Delta_\mathsf{T} \times_{\gamma_\mathsf{T}} G'$ where G' is the subgroup of G_ϕ generated by the voltages $\gamma_\mathsf{T}(E_{\Delta_\mathsf{T}})$.

If $\Delta_\mathsf{T} = \Delta$ and μ is the identity so that γ'_S is the condensation of γ to S, this becomes [11, Theorem 2.5.4].

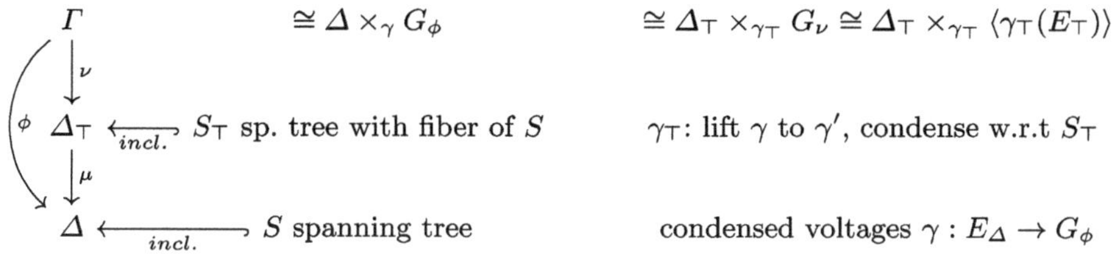

$$\Gamma \qquad\qquad \cong \Delta \times_\gamma G_\phi \qquad\qquad \cong \Delta_\mathsf{T} \times_{\gamma_\mathsf{T}} G_\nu \cong \Delta_\mathsf{T} \times_{\gamma_\mathsf{T}} \langle \gamma_\mathsf{T}(E_\mathsf{T})\rangle$$

$$\phi \quad \Delta_\mathsf{T} \xleftarrow[\text{incl.}]{} S_\mathsf{T} \text{ sp. tree with fiber of } S \qquad \gamma_\mathsf{T}\colon \text{lift } \gamma \text{ to } \gamma', \text{ condense w.r.t } S_\mathsf{T}$$

$$\Delta \xleftarrow[\text{incl.}]{} S \text{ spanning tree} \qquad\qquad \text{condensed voltages } \gamma : E_\Delta \to G_\phi$$

(with maps ν and μ from Γ to Δ_T and Δ_T to Δ.)

Fig. 4. Situation of Proposition 1.

3   Complexity of Determining Isomorphism

This section outlines a proof of Theorem 1. Consider two bi-deterministically labeled voltage graphs Δ_1, Δ_2 of order at most n and size at most m, with voltages from $\mathbb{Z}^d$, so that the voltage groups are subgroups of $\mathbb{Z}^d$. The coordinates of the voltages are of absolute value at most B (which can be represented by strings of length at most $\log B$). We outline a polynomial time algorithm for determining whether the d-periodic derived graphs from the voltage graphs Δ_1 and Δ_2 are λ-isomorphic. The algorithm functions in time $O(n^5 m^3 d^4 \omega)$, where ω is polylogarithmic in n, m, d and B. Notice that as a bi-deterministic labeling of voltage graphs of order n and size m requires at most mn labels, which could be represented by strings of length $O(\log(mn))$, so the labels are subsumed by the polylogarithmic terms as well. In the rest of this section we assume parameters m, n, B, d given with Δ_1, Δ_2.

Throughout this section we consider each Δ_i as bi-deterministically labeled voltage graphs whose voltages generate a group G_i ($i = 1, 2$). The derived graphs of Δ_i are denoted with Γ_i, i.e., $\Gamma_i = \Delta_i \times_{\gamma_i} G_i$. If $\Gamma_1 \cong_\lambda \Gamma_2$, then each Γ_i covers both Δ_1 and Δ_2. In that case, by [17, Prop. 3.11], there exists a least labeled graph (modulo λ-isomorphism) Δ_T covered by each Γ_i and covering each Δ_i. By Proposition 1, $\Delta_\mathsf{T} \times_{\gamma_S} G_{\nu_i} \cong \Gamma_i$ so Theorem 3 can be applied, i.e., the pair of labeled periodic graphs Γ_i admits a λ-isomorphism if and only if they have a common quotient Δ_T in which the correspondence between the voltages of their quotients induces a group isomorphism. We start by attempting to construct Δ_T; if we are successful, we can then check for λ-isomorphism using Theorem 3; if we are unsuccessful, there is no common cover of Δ_1 and Δ_2 that is covered by both Γ_i, and they are not λ-isomorphic.

The algorithm takes two voltage graphs (Δ_1, γ_1) and (Δ_2, γ_2) and attempts to construct a graph Δ_T (which plays the role of Δ in Theorem 3) that covers both Δ_1 and Δ_2 while at the same time is covered by both derived graphs Γ_1 and Γ_2. If construction of Δ_T is successful, we then lift the voltage assignments γ_1 and γ_2 to Δ_T and condense them with respect to the same spanning tree to get voltage assignments γ_1' and γ_2' (Fig. 5).

By Proposition 1, each graph $\Gamma_i = \Delta_i \times_{\gamma_i} G_i$ is λ-isomorphic to the ith derived graph of Δ_T with respect to its voltage group, i.e., $\Gamma_i \cong_\lambda \Delta_\mathsf{T} \times_{\gamma_i'} \langle \gamma_i'(E_\mathsf{T})\rangle$ for $i = 1, 2$. Then we check whether $\gamma_1'(e) \mapsto \gamma_2'(e)$ in Δ_T generates a group isomorphism, which by Theorem 3, implies $\Delta_T \times_{\gamma_1'} \langle \gamma_1'(E_T) \rangle \cong_\lambda \Delta_T \times_{\gamma_2'} \langle \gamma_2'(E_T) \rangle$, i.e., and thus whether $\Gamma_1 = \Delta_1 \times_{\gamma_1} G_1 \cong_\lambda \Delta_2 \times_{\gamma_2} G_2 = \Gamma_2$.

But if the construction of Δ_T fails, or Δ_T is constructed, but the map $\gamma_1'(e) \mapsto \gamma_2'(e)$ fails to generate a group isomorphism, then the two derived graphs Γ_1 and Γ_2 are not λ-isomorphic.

Incidentally, by [17, Prop. 3.11], this algorithm constructs the minimal such common cover Δ_T, but by Theorem 3, any such common cover would decide the issue. By [17, Cor. 3.12] Δ_T is isomorphic to a subgraph of the product $\Delta_1 \times \Delta_2$, which we use in our construction.

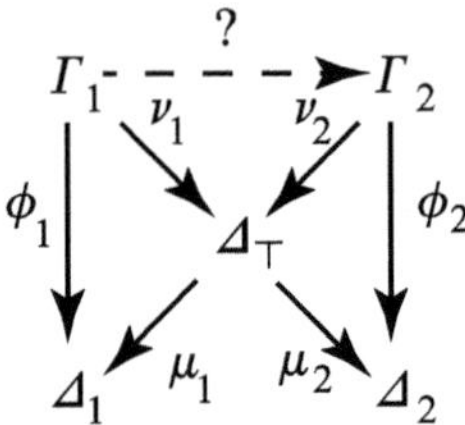

Fig. 5. Δ_T is the common cover for ϕ_1 and ϕ_2. We seek an isomorphism $\varphi : \Gamma_1 \to \Gamma_2$ such that the diagram commutes.

Definition 7. *Given coverings $\phi_1 : \Gamma_1 \to \Delta_1$ and $\phi_2 : \Gamma_2 \to \Delta_2$, a* common cover *for ϕ_1 and ϕ_2 is a graph Δ_T such that there exist coverings $\nu_1 : \Gamma_1 \to \Delta_T$, $\nu_2 : \Gamma_2 \to \Delta_T$, $\mu_1 : \Delta_T \to \Delta_1$, $\mu_2 : \Delta_T \to \Delta_2$ such that $\mu_1 \circ \nu_1 = \phi_1$ and $\mu_2 \circ \nu_2 = \phi_2$.*

In Definition 7, the common cover Δ_T is pointed, that is, for base vertices v_1^* in Γ_1 and v_2^* in Γ_2 if $\nu_1(v_1^*) = \nu_2(v_2^*)$, then $\phi_1(v_1^*) = \mu_1(\nu_1(v_1^*))$ and $\phi_2(v_2^*) = \mu_2(\nu_2(v_2^*))$. The algorithm below in fact attempts to construct a pointed common cover Δ_T, not just a graph that covers both Δ_1 and Δ_2, but one that makes the diagram in Fig. 5 commute. Exploiting the bi-deterministic labeling we ensure that the diagram commutes.

We fix a vertex v_1 in Δ_1, and for each vertex v_2 in Δ_2, the algorithm attempts to construct a common cover Δ_T with a vertex $v_T = (v_1, v_2)$ such that there are two coverings $\mu_i : \Delta_T \to \Delta_i$, $i = 1, 2$, with $\mu_1(v_T) = v_1$ and $\mu_2(v_T) = v_2$. If the construction fails for all v_2 in Δ_2, then there is no λ-isomorphism $\varphi : \Gamma_1 \to \Gamma_2$ such that $\phi_2\varphi(v_1, 0) = v_2$ for all v_2, and therefore $\Gamma_1 \not\cong_\lambda \Gamma_2$.

Let V_2' be the set of vertices in V_2 that have not been checked yet, and start with $V_2' = V_2$.

3.1   Constructing the Common Cover

As outlined above, there are two steps in the process of determining whether two derived graphs Γ_1, Γ_2, where $\Gamma_i = \Delta_i \times_{\gamma_i} G_i$, are isomorphic. The first step is to determine whether Δ_1 and Δ_2 can be covered by a finite (minimal) bi-deterministically labeled, graph Δ_T. If so, the second step is to lift the two voltage assignments, condense them with respect to a common spanning tree on Δ_T, and per Theorem 3 check whether the correspondence between the voltage assignments (induced by the edge assignments) induce a group isomorphism.

For the first step we construct the graph Δ_T as a subgraph of the cross product of Δ_1 and Δ_2. The idea is that if $\Delta_1 \times_{\gamma_1} G_1 \cong_\lambda \Delta_2 \times_{\gamma_2} G_2$, then G_1 and G_2 can be regarded as groups acting on the same graph isomorphic to both derived graphs, which we can call Γ. Then there is a covering $\nu : \Gamma \to \Delta_T$ such that for each G_1-orbit O_1 and each G_2-orbit O_2 in Γ, $O_1 \cap O_2 \neq \varnothing$ if and only if there is exactly one vertex or edge in Δ_T that ν maps all of $O_1 \cap O_2$ to that vertex or edge in Δ_T. The cross product construction below ensures this intersection of orbits.

First, we construct an auxiliary graph Λ whose vertices and edges are 5-tuples $(u_1, g_1; u_2, g_2; x)$ where u_i are vertices (edges) in Δ_i, $g_i \in G_i$ and $x \in (\Sigma \cup \bar{\Sigma})^*$. Such a vertex can be identified with the vertex (u_1, g_1) in Γ_1 and (u_2, g_2) in Γ_2. We add in Λ one tuple at a time, starting with $(v_1, 0; v_2, 0, \epsilon)$. At each step, a new vertex $(u_1, g_1; u_2, g_2; x)$ is added by carefully examining its neighborhood and checking whether traversing an edge in Δ_1 from u_1 and an edge in Δ_2 from u_2 is compatible.

We check whether $\mathrm{Nbd}(u_1) \cong \mathrm{Nbd}(u_2)$, that is, whether incoming and outgoing degrees are the same and whether the incoming/outgoing edges at u_1 and u_2 have the same labels. Once this is accomplished, we check whether traversing an edge e_i in Δ_i for $i = 1, 2$ simultaneously, starting at u_i with label s leads to a new pair (u_1', u_2'); if so, we add a new vertex in Λ, $(u_1', g_1'; u_2', g_2'; xs)$ where g_i' are appropriate updates of voltages $g_i' = g_i + \gamma_i(e_i)$. The addition of the new vertex is done cautiously so that there is no case that $u_1' = w_1$, $g_1' = h_1$ but either $u_2' \neq w_2$ or $g_2' \neq h_2$ for some already existent vertex $(w_1, h_1; w_2, h_2, z)$, avoiding associating a vertex (u_1', g_1') in Γ_1 with two potential vertices (u_2', g_2') and (w_2, h_2) in Γ_2. In a sense, the second and the fourth coordinates follow updates on voltages, while the fifth coordinate is a word label of the path that is traversed from the initial vertices v_1, v_2, and we do not generate non-compatible new vertices.

If traversing an outgoing edge e from (u_1, u_2) along s does not lead to a new pair (u_1', u_2'), i.e., $(u_1', h_1; u_2', h_2; z)$ already exists in Λ, we check whether the voltages are compatible, i.e., $g_1' = h_1$ iff $g_2' = h_2$. If so, we label the vertex as 'boundary' and we don't use it for extension and generation of new vertices in Λ.

If Λ is completed, the sought graph Δ_T is obtained as a projection on the first and the third coordinates.

Proposition 2. *Given two connected bi-deterministically labeled voltage graphs Δ_1 and Δ_2, with a base vertex v_1 in Δ_1 and an ordered set $V_2' \subseteq V_2$, each graph of order at most n, and size at most m, and given voltages of rank at most d and coordinates of absolute value at most B, there is a polynomial time algorithm in time $O(n^4 m^4 d\omega)$ that determines whether a bi-deterministically labeled pointed graph Δ_T covering both Δ_1 and Δ_2 exists.*

If $\Delta_\top$ exists, the algorithm selects the first plausible base vertex $v_2 \in V_2'$ and constructs $\Delta_\top$ with base $v_\top = (v_1, v_2)$.

Proof. (Sketch) Let $(\Delta_i, \gamma_i, \lambda)$ for $i = 1, 2$ be bi-deterministic labeled voltage graphs where $\Delta_i = (V_i, E_i)$ and the group generated by $\gamma_i(E_i)$ is G_i. We denote the derived graphs $\Delta_i \times_{\gamma_i} G_i$ with Γ_i for $i = 1, 2$. Algorithm 1 constructs a connected graph $\Delta_\top = (V_\top, E_\top)$ such that $V_\top \subseteq V_1 \times V_2$ and $E_\top \subseteq E_1 \times E_2$. The algorithm goes through a construction of an auxiliary graph Λ with vertices of form $(u_1, g_1; v_2, g_2; x) \in V_1 \times G_1 \times V_2 \times G_2 \times (\Sigma \cup \bar{\Sigma})^*$ keeping track of voltage accumulation with g_1, g_2 while traversing a path with label x.

There are two checks in the algorithm to assure a construction of $\Delta_\top$. First, starting at v_1 and v_2 and following labeled paths, if the algorithm attempts to associate a vertex u_1 with a vertex u_2 whose incident edges have different labels, the algorithm fails (in Step 1 or Step 2.1). Second, if the algorithm attempts to add $(u_1, g_1; u_2, g_2; x)$ when either $(u_1, g_1; w_2, h_2; x)$ or $(w_1, h_1; u_2, h_2; x)$ with $(u_i, g_i) \neq (w_i, h_i)$ is already in Λ, the algorithm fails in Step 2.2. This check avoids the situation when a vertex of Γ_1 (or Γ_2) is associated with two vertices of Γ_2 (i.e., Γ_1).

We estimate the time complexity of Algorithm 1 as follows. In Step 1, we check at most n vertices, and for each one, we check its incident edges for their labels; as there are at most m edges incident to each vertex; that is a total of $O(nm)$ edges to check. At each vertex, the labels of incoming edges are sorted (within time $O(m \log m)$ and compared with the labels of incoming edges to v_1 (within time $O(m \log m)$). Altogether Step 1 takes time $O(nm^2\omega)$ where ω is logarithmic in m.

Having found v_2 (if it exists), at most n^2 putative vertices for Λ (and hence for $\Delta_\top$) will be generated in the loop of Step 2. For each vertex popped from the queue, Step 2.1 takes time $O(m\omega)$, and once checked, we compare that vertex with each of the at most n^2 vertices already generated (Step 2.2); each comparison (Step 2.2) takes time $O(md\omega)$, for a total time $O(n^4m^2d\omega)$. Step 2.3 entails adding, checking, and computing voltages for at most $2m$ edges, taking time at most $O(m^2d\omega)$; alternatively, in Step 2.4 adding a vertex (and edges) to the boundary takes time at most $O(md\omega)$. Thus Steps 2.* take time $O(n^4m^4d\omega)$.

So running Algorithm 1 takes time at most $O(n^4m^4d\omega)$.

Observe that by construction of $\Delta_\top$, Algorithm 1 ensures that x is a label of a walk from v_1 to u_1 in Δ_1 iff x is a label of walk from v_2 to u_2 in Δ_2 iff x is a label of walk in $\Delta_\top$ from $v_\top = (v_1, v_2)$ to (u_1, u_2).

Consider the derived graphs $\Gamma_i = \Delta_i \times_{\gamma_i} G_i$ for $i = 1, 2$ and $\phi_i : \Gamma_i \to \Delta_i$ be the canonical covers. We define maps $\nu_i : \Gamma_i \to \Delta_\top$ in the following way: for $(u, g) \in \Gamma_i$ let x be the label of a path from $(v_i, 0)$ to (u, g), then $\nu_i(u, g) = (w, w') \in \Delta_\top$ if there is a path (a walk of non-repeating edges) with label x from (v_1, v_2) to (w, w'). Similarly for the edges. Since the labeling is bi-deterministic, the maps are well defined, and because the algorithm follows the accumulation of the voltages, vertices $(u_1, g_1; v_2, g_2; x)$ in Λ are associated with vertices (u_1, u_2) in $\Delta_\top$, these maps are covers.

Algorithm 1. An algorithm computing $\Delta_\top$ covering Δ_1 and Δ_2

Input: $(\Delta_i, \gamma_i, \lambda)$; $i = 1, 2$, v_1, $V_2' \subseteq V_2$

Initialize sets $V_\Lambda, E_\Lambda, V_\top, E_\top = \emptyset$

(*Step 1*) Find next $v_2 \in V_2'$ such that $\mathrm{Nbd}(v_1) \cong \mathrm{Nbd}(v_2)$

    if $V_2' = \emptyset$ or no other such $v_2 \in V_2'$ exists, Output: $\Delta_\top$ does not exist.

    otherwise $(v_1, 0; v_2, 0, \epsilon) \in V_\Lambda$, $v_\top = (v_1, v_2)$ and set Queue $= \{(v_1, 0; v_2, 0, \epsilon)\}$.

(*Step 2*) While Queue $\neq \emptyset$

    Remove the first element $\mathbf{u} := (u_1, g_1; u_2, g_2; x)$ from Queue

        (*Step 2.1*) If $\mathrm{Nbd}(u_1) \not\cong \mathrm{Nbd}(u_2)$, stop; $\Delta_\top$ does not exist with base (v_1, v_2).

            Initialize sets $V_\Lambda, E_\Lambda, V_\top, E_\top = \emptyset$, remove v_2 from V_2', empty Queue

            go to Step 1.

        (*Step 2.2*) For all $\mathbf{w} = (w_1, h_1; w_2, h_2; y)$ in V_Λ

            If $w_1 = u_1$ and $h_1 = g_1$ but $h_2 \neq g_2$ or $w_2 \neq u_2$, stop;

                $\Delta_\top$ does not exist with base (v_1, v_2).

                Initialize sets $V_\Lambda, E_\Lambda, V_\top, E_\top = \emptyset$, remove v_2 from V_2', empty Queue

                go to Step 1.

            If $w_2 = u_2$ and $h_2 = g_2$ but $h_1 \neq g_1$ or $w_1 \neq u_1$, stop;

                $\Delta_\top$ does not exist with base (v_1, v_2).

                Initialize sets $V_\Lambda, E_\Lambda, V_\top, E_\top = \emptyset$, remove v_2 from V_2', empty Queue

                go to Step 1.

        (*Step 2.3*) If $(u_1, u_2) \notin V_\top$ then

            Add $\mathbf{u}$ to V_Λ and (u_1, u_2) to $V_\top$

            for each $e_1 \in \mathrm{Out}(u_1)$ and $e_2 \in \mathrm{Out}(u_2)$ with $\lambda(e_1) = \lambda(e_2) = s$

                add $\mathbf{e} = (e_1, g_1 + \gamma_1(e_1); e_2, g_2 + \gamma_2(e_2); xs)$ to E_Λ

                add (e_1, e_2) to $E_\top$ with label $\lambda_\top(e_1, e_2) = s$.

                add the endpoint $(\tau(e_1), g_1 + \gamma_1(e_1); \tau(e_2), g_2 + \gamma_2(e_2); xs)$ to Queue,

                      unless it is already in Queue or V_Λ

        (*Step 2.4*) else ($\mathbf{u}$ is a 'boundary' vertex, with $(u_1, h_1; u_2, h_2; y) \in V_\Lambda$)

            if $g_1 = h_1$ but $g_2 \neq h_2$ or $g_2 = h_2$ but $g_1 \neq h_1$ then stop

            Output: $\Delta_\top$ does not exist with base (v_1, v_2).

            Initialize sets $V_\Lambda, E_\Lambda, V_\top, E_\top = \emptyset$, remove v_2 from V_2', empty Queue

            go to Step 1.

Output $\Delta_\top = (V_\top, E_\top)$ with base $v_\top = (v_1, v_2)$, and labels $\lambda_\top$

From the construction in Algorithm 1 we observe that $\Delta_\top$, if constructed, is a common cover of Δ_1 and Δ_2.

Lemma 1. *The maps $\nu_i : \Gamma_i \to \Delta_\top$ are covers. Similarly, $\nu_i : \Delta_\top \to \Delta_i$ defined with $\mu_i(u_1, u_2) = u_i$ for $i = 1, 2$ are covers and $\phi_i = \mu_i \nu_i$.*

We conclude this subsection by observing that the converse of the above lemma does not hold. Therefore, in order to determine the isomorphism of the derived graphs, it is necessary that once $\Delta_\top$ is constructed, the voltage assignments of the edges (after condensation) must also generate a group isomorphism. For example, consider the two connected bi-deterministically labeled voltage graphs Δ_1 and Δ_2 at the bottom of Fig. 6. They have a minimal common cover $\Delta_\top$ obtained by Algorithm 1. Yet their derived graphs are not isomorphic, much less λ-isomorphic. There is no voltage assignment on $\Delta_\top$ that would match both assignments in Δ_1 and Δ_2. All blue edges in Δ_2 have label $(0,1)$, and they need to coincide with blue edges in Δ_1, one having a voltage $(0,2)$ and the other $(0,1)$.

In comparing two voltage graphs, we need to take the voltages into account.

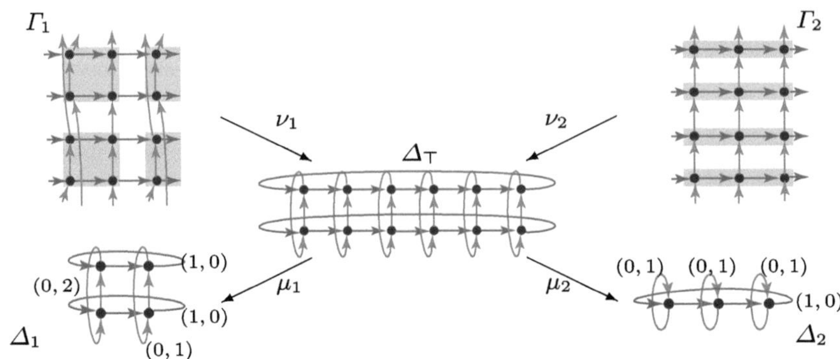

Fig. 6. Two voltage graphs Δ_1 and Δ_2 at the bottom with their labels indicated with red and blue and their voltages. All non-indicated voltages are $(0,0)$. The derived periodic Γ_1 and Γ_2 are left and right at the top with respective labels red and blue. The lifts of Δ_1 in Γ_1 are on yellow squares while those of Δ_2 in Γ_2 in on yellow rectangles; in each, the voltage to shift from a lift to the lift at the right is $(1,0)$ while the voltage to shift from a lift to the lift above is $(0,1)$. Thus each has the voltage group $\mathbb{Z} \times \mathbb{Z}$. Δ_1 and Δ_2 have a common cover Δ_T, but their derived graphs are not isomorphic. Compare with Fig. 5. (Color figure online)

3.2 Voltages of the Common Cover

The second step in determining whether the two derived graphs Γ_1 and Γ_2 are isomorphic is to check whether the voltage assignments to Δ_T reflecting voltages in Δ_i $(i = 1, 2)$ generate a group isomorphism. We assume the output of Algorithm 1. We need the bound of the size of Δ_T.

Lemma 2. *If the order and size of Δ_i, $(i = 1, 2)$, are less than n and m, respectively, then the size of Δ_T is at most nm.*

Proof. From the proof of Proposition 2, the edges of the auxiliary graph Λ are tuples $(e_1, g_1; e_2, g_2; z)$, projected to edges $(e_1, e_2) \in E_T$, are those whose initial vertices $(u_1, g_2; u_2, g_2; x)$ are projected to $(u_1, u_2) \in V_T$.

For each vertex $u_1 \in V_1$ and edge $e_2 \in E_2$, there is at most one edge $e_1 \in E_1$ such that $\iota(e_1) = u_1$ and $\lambda_1(e_1) = \lambda_2(e_2)$. Thus Δ_T has at most one edge of the form $(e_1, g_1; e_2, g_1; *)$ of initial vertex $(u_1, *; \iota(e_2), *; *)$. Hence, for each pair $u_1 \in V_1$, $e_2 \in E_2$, there exists at most two edges in Δ_T.

Proposition 3. *Given a common cover Δ_T of voltage graphs (Δ_i, γ_i), $i = 1, 2$, where Δ_T is covered by both derived graphs, there is a polynomial time algorithm that determines whether the condensation of the lifts of γ_i, assignments $\gamma_1''(e) \mapsto \gamma_2''(e)$, generate a group isomorphism, within time $O(n^4 m^4 d\omega)$.*

Proof. (Sketch) We sketch an algorithm that provides the result of the Proposition. We assume that we have the output of Algorithm 1: Δ_T with $v_\mathsf{T} = (v_1, v_2)$ as well as the two voltage graphs Δ_1 and Δ_2.

Algorithm 2. An algorithm that determines whether voltage assignments in Δ_T generate a group isomorphism

---

Input: $(\Delta_i, \gamma_i, \lambda)$, Δ_T, $v_\mathsf{T} = (v_1, v_2)$
(*Step 1*) For each $e = (e_1, e_2) \in E_\mathsf{T}$ set $\gamma_1'(e_1) = \gamma_1(e_1)$ and $\gamma_2'(e_2) = \gamma_2(e_2)$
(*Step 2*) Use depth-first algorithm to construct a spanning tree S_T in Δ_T
(*Step 3*) Condense γ_1' and γ_2' with respect to S_T
  For each $e \in S_\mathsf{T}$ set $\gamma_i'' = \mathbf{0}$ for $i = 1, 2$
   For $e \in E_\mathsf{T}$ and $e \notin S_\mathsf{T}$ determine a cycle C_e
    such that C_e passes through v_T and all edges in C_e are in S_T except e.
   set $\gamma_i''(e)$ such that $\sum_{e \in C_e} \gamma_i'(e) = \sum_{e \in C_e} \gamma_i''(e)$ for $i = 1, 2$
(*Step 4*) Form matrices M_i such that for each $e \notin S_\mathsf{T}$ add a row $\gamma_i''(e)$ to M_i ($i = 1, 2$).
(*Step 5*) Perform simultaneous row operations over $\mathbb{Z}$ on M_1 and M_2,
  until M_1 is in Hermite Normal Form.
(*Step 6*) check whether M_1 and M_2 are reduced to matrices whose rows form bases of the same size.
  If yes, output: isomorphism found.
  Otherwise output: isomorphism not found.

In Algorithm 2, as there are at most nm edges in Δ_T, Step 1 takes time $O(nmd\omega)$ and Step 2 takes time $O(nm\omega)$. Condensation in Step 3 entails checking each of the at most nm edges e not in S_T, and for each of these, finding the unique path from $\tau(e)$ to $\iota(e)$ through S_T, then adding the voltages on that path, and adding that to the voltage on e itself; finding the path takes time $O(nm\omega)$ and adding the voltages takes time $O(mnd\omega)$, so Step 3 takes time $O(n^2 m^2 d\omega)$. Step 4 takes time $O(nmd\omega)$ as there are d columns and at most nm rows. For Step 5, first observe that $m \geq d$, for otherwise, the voltages of Δ_i would not generate G_i, $i = 1, 2$. Then by [18] and [14], computing the Hermite Normal Form (HNM) of a matrix of d columns and $O(nm)$ rows, whose entries have absolute value at most B, can be computed in time $O((nm)^4 \omega)$ entry operations; by [13], multiplication of integers of length $\ell \approx \log B$ can be done in time $O(\ell \log \ell)$, so Step 5 can be done in time $O(n^4 m^4 d\omega)$. For Step 6, checking that both matrices have the same number of nonzero rows can be done in time $O(nmd)$, and as the first is already in HNM, it suffices to check that the latter is also a basis, which can be done by reducing it to HNM too. Altogether, Algorithm 2 will answer within time $O(n^4 m^4 d\omega)$.

Theorem 4. *Given 2 finite bi-deterministically labeled connected voltage graphs Δ_1 and Δ_2 of order at most n and size at most m, whose voltage groups are lattice groups of dimension d and whose voltages are integer vectors whose coordinates are all of absolute value at most B, the question of whether their derived graphs are λ-isomorphic can be resolved in time $O(n^5 m^4 d^4 \omega)$, where ω is polylogarithmic in n, m, d, and B.*

Proof. The result in the theorem follows by executing the following algorithm. We recall that $v_1 \in V_1$ is fixed.

Algorithm 3. An algorithm that determines whether the derived graphs of Δ_1 and Δ_2 are isomorphic

---

Input: $(\Delta_i, \gamma_i, \lambda)$ $i = 1, 2$
Fix v_1 in Δ_1 set $V_2' = V_2$
While $V_2' \neq \emptyset$
(*Step 1*) If $V_2' \neq \emptyset$ perform Algorithm 1 on $(\Delta_i, \gamma_i, \lambda)$ and V_2', v_1
    If Δ_T is constructed then
        (*Step 2*) Perform Algorithm 2 on $(\Delta_i, \gamma_i, \lambda)$, Δ_T, $v_T = (v_1, v_2)$
        If isomorphism found, Output: Derived graphs are isomorphic. Halt.
        else remove v_2 form V_2'
            go to Step 1.
Output: Derived graphs are not isomorphic.

Recall that using Theorem 3, any common cover of Δ_1 and Δ_2 (covered by their derived graphs) will suffice. Turning to the complexity, there are at most n vertices v_2 in Δ_2, and for each of these, Step 1 takes time $O(n^4 m^4 d\omega)$ while Step 2 takes time $O(n^4 m^4 d\omega)$. Thus Algorithm 3 takes time $O(n^5 m^4 d\omega)$.

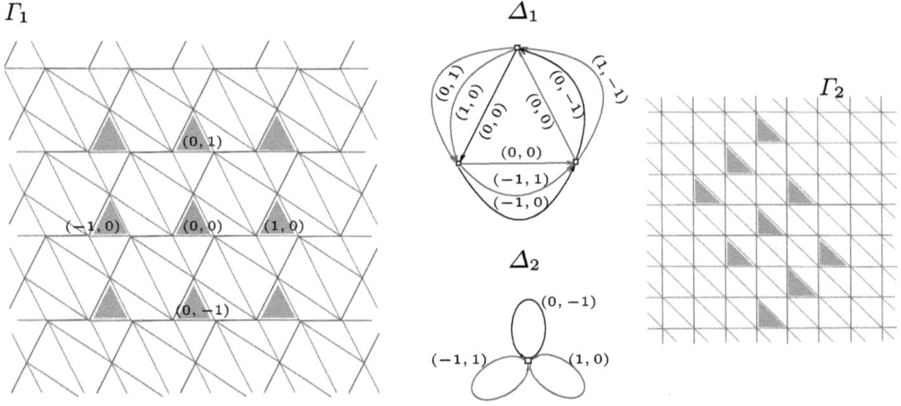

Fig. 7. Two voltage graphs Δ_1 and Δ_2 with their respectrive derived graphs Γ_1 and Γ_2. The derived graph Γ_1 has a copy of the three vertex graph Δ_1 at each point of the $\mathbb{Z}^2$ lattice. The coordinates of the origin and the unit vectors are indicated in the figure. Arrows not depicted. Each vertex of the integer lattice $\mathbb{Z}^2$ in Γ_2 corresponds to a copy of the single vertex graph Δ_2. The two derived graphs are isomorphic. Although the green triangle in Γ_2 has vertices lattice points $(0, 0)$, $(1, 0)$ and $(0, 1)$, it is an image of the green triangle in Γ_1 positioned at lattice point $(0, 0)$ under the constructed isomorphism. The isomorphism between Γ_1 and Γ_2 maps the neighboring colored triangles respectively. (Color figure online)

We show an example of the construction in Figs. 7 and 8. The graphs Δ_1 and Δ_2 are shown with their respective voltages and the labeling is indicated with the colors of the edges. The derived graphs Γ_1 and Γ_2 are obtained by $\Delta_i \times_{\gamma_i} \mathbb{Z}^2$ for $i = 1, 2$. Since Δ_1 is a three-vertex graph, at each vertex of the integer lattice of Γ_1 there is a three-vertex copy of Δ_1. These are indicated with coordinate labels of the origin and the unit vectors in Γ_1. We do not depict the arrows.

In Γ_2, each vertex is a copy of Δ_2 as Δ_2 is a single vertex graph. Algorithm 1 produces a common cover Δ_T with three vertices, essentially isomorphic to Δ_1 (since Δ_2 is a singleton and the cross product does not build a larger graph). Moreover, Δ_T inherits the voltages of both graphs such that $\gamma_1' = \gamma_1$ while γ_2' copies the values of the voltages of Δ_2, i.e., with γ_2' all edges of the same color have the same voltage inherited from γ_2.

We choose a spanning tree in Δ_T to be the $(0,0)$ labeled red and blue edges in Δ_1. With this spanning tree, $\gamma_1' = \gamma_1$ is already condensed. However, γ_2' becomes condensed γ_2'' giving new voltage assignments as depicted in Fig. 8 to the right. Then the corresponding voltage assignments $\gamma_1(e) \mapsto \gamma_2''(e)$ for all edges in Δ_T provide a group isomorphism $(0,-1) \mapsto (1,-2)$ and $(1,0) \mapsto (1,1)$.

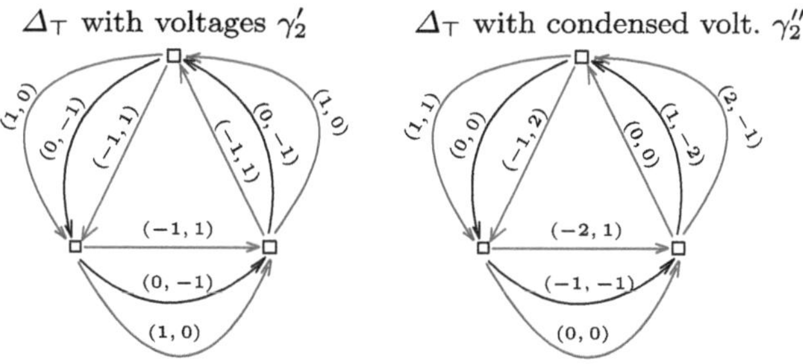

Fig. 8. (left) Lift of voltages γ_2 of Δ_2 from Fig. 7 to γ_2' of Δ_T. (right) Condensation γ_2'' of γ_2' with respect to the spanning tree consisting of the $(0,0)$-labeled red and blue edges of Δ_1.

A deterministic edge labeling would not be sufficient to facilitate an equivalent PTIME algorithm. For example, any finite simple graph (i.e., with unoriented edges) may be represented by a deterministically labeled graph (with oriented edges[2]) as follows. Let Γ be a simple graph with vertex set V and edge set E, and suppose that E consists of 2-element sets $\{u, v\}$, $u, v \in V$. From this we can define a graph Γ' of vertex set $V \cup V'$, where $V' = \{(u, v) : \{u, v\} \in E\}$, and edge set $\{((u, v), u), ((u, v), v) : \{u, v\} \in E\}$; note that each oriented edge runs from one of the new vertices to an old vertex of Γ. Label the edges $((u, v), u)$ with the color red and the edges $((u, v), v)$ with the color blue (Fig. 9). Then Γ' is deterministically labeled. Any two finite graphs Γ_1 and Γ_2 are isomorphic if and only if there is a label preserving isomorphism from Γ_1' onto Γ_2'. But the existence of such a label preserving isomorphism between Γ_1' and Γ_2' is no easier than the existence of any isomorphism between Γ_1 and Γ_2, so deterministic labeling was no help here.

---

[2] As in a digraph.

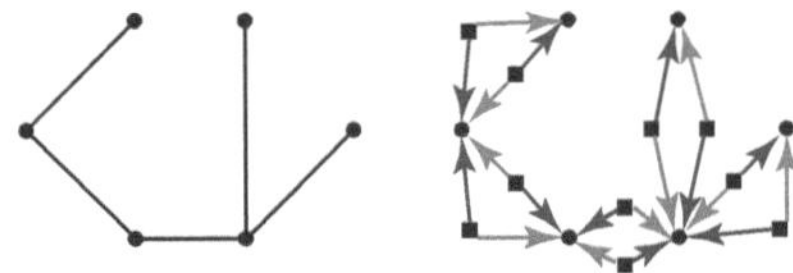

Fig. 9. Given the simple graph at left, we can associate a deterministically labeled graph such that each edge is replaced by a subgraph with two new vertices, each with one outgoing red edge and one outgoing blue edge. (Color figure online)

References

1. Behzad, M., Chartrand, G., Lesniak-Foster, L.: Graphs & Digraphs. Wadsworth (1986)
2. Blatov, V.A., Shevchenko, A.P., Proserpio, D.M.: Applied topological analysis of crystal structures with the program package ToposPro. Cryst. Growth Des. **14**(7), 3576–3586 (2014)
3. Bondy, A., Murphy, U.S.R.: Graph Theory and Its Applications. Elsevier (1976)
4. Borcea, C.S., Streinu, I.: Frameworks with crystallographic symmetry. Phil. Trans. R. Soc. A **372**, 20120143 (2014)
5. Chung, S.J., Hahn, T., Klee, W.E.: Nomenclature and generation of three-periodic nets: the vector method. Acta Crys. **A40**, 42–50 (1984)
6. Delgado-Friedrichs, O., O'Keeffe, M.: Identification and symmetry computation for crystal nets. Acta Cryst. A **59**, 351–360 (2003)
7. Delgado-Friedrichs, O., O'Keeffe, M.: Crystal nets as graphs: terminology and definitions. J. Solid State Chem. **178**(8), 2480–2485 (2005)
8. Desiraju, G.R., Vittal, J.J., Ramanan, A.: Crystal Engineering: A Textbook. World Scientific (2011)
9. Garey, M.R., Johnson, D.S.: Computers and Intractibility: A Guide to the Theory of NP-Completeness. W. H. Freeman (1979)
10. Gross, J.L.: Voltage graphs. In: Gross, J.L., Yellen, J., Zhang, P. (eds.) Handbook of Graph Theory, pp. 783–802. CRC Press, Boca Raton (2014)
11. Gross, J.L., Tucker, T.W.: Topological Graph Theory. Wiley (1987)
12. Harary, F.: Graph Theory. Basic Books (1972)
13. Harvey, D., van der Hoeven, J.: Integer multiplication in time $o(n \log n)$. Ann. Math. **193**(2), 563–617 (2021)
14. Havas, G., Majewski, B.S., Matthews, K.R.: Extended GCD and Hermite normal form algorithms via lattice basis reduction. Exp. Math. **7**, 125–136 (1998)
15. Janowski, J., et al.: Engineering tertiary chirality in helical biopolymers. Proc. Natl. Acad. Sci. **121**(19), e2321992121 (2024). https://doi.org/10.1073/pnas.2321992121, https://www.pnas.org/doi/abs/10.1073/pnas.2321992121
16. Jonoska, N., Krajcevski, M., McColm, G.: Counter machines and crystallographic structures. Nat. Comput. **15**(1), 97–113 (2016)
17. Jonoska, N., Krajcevski, M., McColm, G.: Lifting voltages in graph covers. arXiv:2501.17135 [math.CO] (2025)
18. van der Kallen, W.: Complexity of the Havas, Majewski, Matthews LLL Hermite normal form algorithm. J. Symb. Comput. **30**, 329–337 (2000)
19. Kwak, J.H., Lee, J.: Isomorphism classes of graph bundles. Can. J. Math. **XLII**, 747 – 761 (1990)

20. Li, F., Shang, H., Woo, P.Y.: Determination of isomorphism and its applications for arbitrary graphs based on circuit simulation. Circuits Syst. Signal Process. **27**, 749–761 (2008)
21. Li, H., et al.: Memristive crossbar arrays for storage and computing applications. Adv. Intell. Syst. **3**, 2100017 (2021)
22. McColm, G.L.: Realizations of crystal nets. I. (Generalized) derived graphs. Acta Crystallographica 18–32 (2024)
23. Oganov, A.R.: Modern Methods of Crystal Structure Prediction. Wiley (2011)
24. O'Keeffe, M., Peskov, M.A., Ramsden, S.J., Yaghi, O.M.: The reticular chemistry structure resource (RCSR) database of, and symbols for, crystal nets. Acc. Chem. Res. **41**(12), 1782–1789 (2008)
25. Ramsden, S., Robins, V., Hyde, S.: Three-dimensional Euclidean nets from two-dimensional hyperbolic tilings: kaleidoscopic examples. Acta Cryst. A **65**, 81–108 (2009)
26. Skoviera, M.: A contribution to the theory of voltage graphs. Discret. Math. **61**, 281–292 (1986)
27. Sunada, T.: Topological Crystallography: With a View Towards Discrete Geometric Analysis. Springer Surveys and Tutorials in the Applied Mathematical Sciences, vol. 6 (2013)
28. Toffoli, T., Margolus, N.: Programmable matter: concepts and realization. Phys. D **47**(1–2), 263–272 (1991)
29. Wang, Y., Zhang, X., Corcovilos, T.A., Kumar, A., Weiss, D.S.: Coherent addressing of individual neutral atoms in a 3D optical lattice. Phys. Rev. Lett. **115**(4), 043003 (2015)
30. West, D.B.: Graph Theory. Pearson (1995)
31. Woloszyn, K., et al.: Augmented DNA nanoarchitectures: a structural library of 3D self-assembling tensegrity triangle variants. Adv. Mater. **34**(49) (2022). https://doi.org/10.1002/adma.202206876, https://www.osti.gov/biblio/1902569
32. Zemlyachenko, V.N., Tyshkevich, N.M.K.R.I.: Graph isomorphism problem. J. Sov. Math. **29**(4), 1426–1481 (1985)
33. Zheng, J., et al.: From molecular to macroscopic via the rational design of a self-assembled 3D DNA crystal. Nature **461**(7260), 74–77 (2009)

Enhancing MFCC Feature Extraction Through Reservoir Computing

Rinku Sebastian[1]([envelope]) [iD], Simon O' Keefe[2] [iD], and Martin A Trefzer[1] [iD]

[1] School of PET, University of York, York, UK
{rinku.sebastian,martin.trefzer}@york.ac.uk
[2] Department of Computer Science, University of York, York, UK
simon.okeefe@york.ac.uk

Abstract. The extraction of features from speech is the most critical process in speech signal processing. Mel Frequency Cepstral Coefficients (MFCC) are the most widely used features in the majority of the speaker and speech recognition applications, as the filtering in this feature is similar to the filtering taking place in the human ear. But the current MFCC extraction is complex and requires time-frequency translations. Through our investigation, we were able to model a reservoir as a feature extractor capable of extracting the Mel Frequency Cepstral Coefficient (MFCC) without time-frequency domain translations. We have developed a real-time audio signal processing system by simplifying audio signal processing through the utilization of reservoir computers, which are significantly easier to train. We have established an experimental framework for end-to-end audio processing utilizing the reservoir and have investigated its capability to perform end-to-end audio signal processing.

Keywords: Reservoir computing · Audio signal processing · MFCC

1 Introduction

Many current technologies, such as telecommunication, automatic speech transcription and translation, speaker verification, hearing aids, etc., depend heavily on efficient speech processing. [3] However, modern audio processing technologies have not yet surpassed the efficiency of the human hearing system. We are proposing to simplify and further improve the efficiency of audio signal processing by using reservoir computing (RC). Reservoirs are promising because they can operate exclusively in the time domain, and training reservoirs can be achieved by a simple linear regression on its readout weights.

The application of neural networks in the domain of audio signal processing has been extensively investigated for an extended period, given the inherent complexity of the task. State-of-the-art audio signal processing methods still require complex feature extraction techniques before the neural network stage. A majority of the literature employs this feature extraction in the frequency domain and hence, significant computation is performed when transferring the signal to

© The Author(s), under exclusive license to Springer Nature Switzerland AG 2026
E. Formenti and L. Manzoni (Eds.): UCNC 2025, LNCS 16364, pp. 294–306, 2026.
https://doi.org/10.1007/978-3-032-15641-9_20

the frequency domain, extracting the required information, and transferring the result back to the time domain. For example, in MFCC extraction, we first need to convert the signal to the frequency domain using FFT after pre-emphasis, windowing, and framing. This frequency-transformed signal is then modified by a predefined filter-bank and the resulting spectrum is converted to cepstrum (by applying logarithm to the spectrum), and discrete cosine transform is finally applied to obtain the MFCC coefficients. This is a computationally complex and expensive process. We are investigating methods to extract features like MFCC directly extracted using a reservoir computing approach. To achieve end-to-end processing with a reservoir, we propose modeling reservoirs to both create the MFCC and perform classification.

A strong temporal component is present in many difficult computing problems [19]. Recurrent neural networks are well suited to handle temporal pattern classification and regression applications. However, the gradual convergence of the majority of present learning rules makes their implementation difficult. The use of reservoirs provides a convenient solution to this issue. Reservoir Computing (RC) is increasingly being used as a conceptually simple yet powerful method for temporal processing [20], including audio applications like acoustic modeling, automatic speech recognition, etc. [18].

Reservoir computing is based on a recurrent neural network (RNN) [16]. That is, it uses feedback to acquire state representation. Recurrent neural networks appear to be a highly promising tool for nonlinear time series processing applications because they exhibit recurrent connection pathways similar to human brain. Despite its widely accepted potential, the application of RNNs remained limited for a long time due to the burden of training. The intrinsic properties of a reservoir make them suitable for processing time-varying inputs [11]. In this work, we are analyzing the usefulness and efficiency of reservoir computers for real time audio signal processing. Our aim is for the system to work directly on audio samples in the time domain, thereby reducing the burden of feature extraction and computational complexity.

In this paper we are investigating on methods to improve the performance of MFCC coefficient extraction using a reservoir.

2   Reservoir Computing

Reservoir computing is an established bio-inspired paradigm in machine learning [9]. It is a framework for computation that was developed from the notion of recurrent neural networks that maps input signals into higher-dimensional computational spaces via the dynamics of a fixed, nonlinear system known as a reservoir. After the input signal is fed into the reservoir, which is treated as a 'black box', a straightforward readout mechanism is trained to read the state of the reservoir and map it to the desired output [8]. An RNN is created at random and it is just the readout which trained in reservoir computing, typically using some regression based on least squares. Figure 1 shows a classical reservoir computer. An input layer that is randomly connected to each of the N

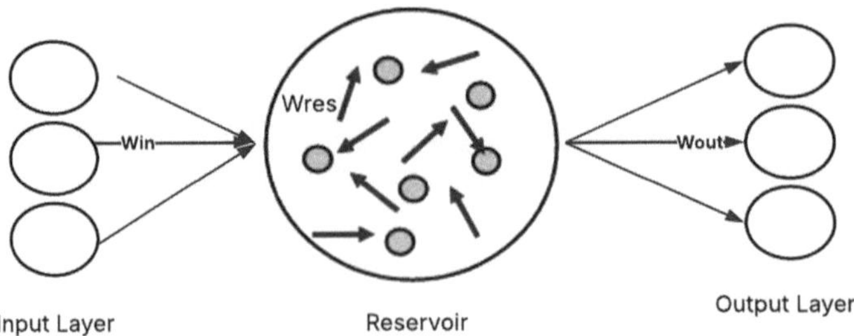

Fig. 1. Topology of Reservoir computer.

reservoir nodes receives the input. The reservoir itself is left untrained since the connections and weights between its nodes are fixed and selected at random. An output layer reads out the transient dynamical response of the reservoir using linear weighted summing of the node states. [9] The drawbacks of gradient-descent RNN training are avoided by the RC paradigm. This made it much easier to use RNNs in real-world applications and outperformed traditional fully trained RNNs in many tasks [11].

In the reservoir framework, since the training is limited to the readout part, the burden of training is reduced. Interference between tasks is also minimized when performing multiple tasks by training multiple readouts on the same reservoir. Hence, it is possible to solve several tasks with based on the same input by adding multiple readouts to a single reservoir. The echo state property of a reservoir gives the system memory so that it can process time series [11]. The fading memory property of reservoir allows the system not to saturate. Furthermore, the reservoir has the ability to perform nonlinear transformations. All these qualities of a reservoir show that it is a suitable fit for temporal signal processing [4].

3   Audio Signal Processing

Extracting features from speech is the most critical process in Speech signal processing. Feature extraction is a method of extracting the dominant and distinctive qualities of a signal. The process of feature extraction involves converting an audio waveform into a parametric representation at a data rate that is relatively low for further processing and analysis. The goal of feature extraction is to represent an audio signal using a fixed number of components. This is due to the fact that processing all of the information in the acoustic signal would be intractable, and some of it is not relevant for the purpose of recognition or classification tasks [1]. An appropriate feature mimics a signal's characteristics in a more condensed manner. Mel Frequency Cepstral Coefficients (MFCC) are the most widely used features in the majority of the speaker and speech recognition applications as the filtering in this feature is similar to the filtering taking place in human ear. The following section describes Mel Frequency Cepstral Coefficient in detail.

3.1 MFCC

Mel-frequency Cepstral Coefficients are referred to as MFCC. The Mel-scale used is to map between linear frequency scale of speech signals to logarithmic scale for frequencies higher than 1 kHz. This makes the spectral frequency characteristics of a signal closely corresponding to human auditory perception and hence, MFCCs are a feature that is frequently used in automatic speech and speaker recognition. The mel-frequency cepstrum (MFC), which is based on a linear cosine transform of a log power spectrum on a nonlinear Mel scale of frequency, is a representation of the short-term power spectrum of a sound. An MFC is made up of a number of coefficients known as Mel-frequency cepstral coefficients (MFCCs). The frequency bands of the MFC are evenly spaced on the Mel scale. This frequency warping may make it possible to depict sound more accurately.

MFCCs are commonly derived as follows:

- Step 1: Take the Fourier transform of (a windowed excerpt of) a signal.
- Step 2: Map the powers of the spectrum obtained above onto the Mel scale, using triangular overlapping windows or alternatively, cosine overlapping windows.
- Step 3: Take the logs of the powers at each of the Mel frequencies.
- Step 4: Take discrete cosine transform of the list of Mel log powers.
- The MFCCs are the amplitudes of the resulting spectrum

(See Fig. 2).

Framing and Windowing: The MFCC algorithm needs to be transformed from the time domain to the frequency domain because it is based on spectral analysis. The goal of the overlapping analysis is to ensure that each speech sound in the input sequence is centered within a specific frame. The signal is tapered towards the frame borders on each frame by applying a window. Hanning or Hamming windows are typically used. While applying the discrete Fourier transform (DFT) to the signal, this is done to improve the harmonics, soften the edges, and to reduce edge effects like aliasing.

DFT Spectrum: Each windowed frame is converted into frequency spectrum by applying DFT.

$$X(k) = \sum_{n=0}^{N-1} x(n) * e^{(-j2\pi nk/N)} \tag{1}$$

Mel Spectrum: Mel spectrum is computed by passing the Fourier transformed signal through a set of band-pass filters known as Mel-filter bank. A Mel is a unit of measurement of how loudness is perceived by the human ear. Since the human auditory system reportedly does not detect pitch linearly, it does not correspond linearly to the tonal frequency physically present in the sound. The frequency spacing for the Mel scale is roughly linear below 1 kHz and logarithmic above 1 kHz. Mel can be approximated by physical frequency using the formula

$$f_{Mel} = 2595 log_{10}(1 + f/700) \tag{2}$$

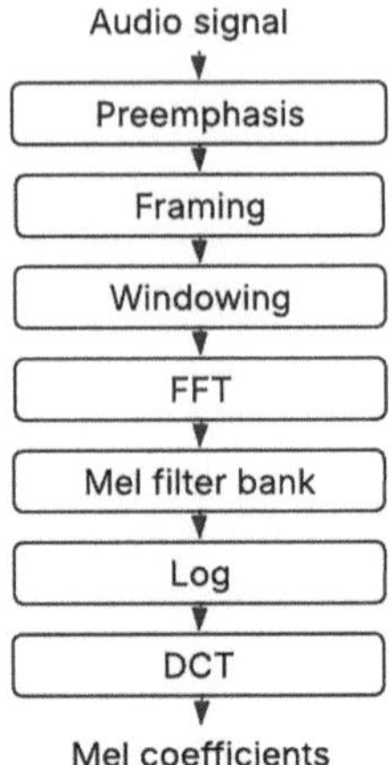

Fig. 2. MFCC extraction.

where f denotes the physical frequency in Hz, and f_{Mel} denotes the perceived frequency Both the frequency domain and the time domain are capable of representing filter banks. Filter banks are typically built in the frequency domain for MFCC calculations. On the frequency axis, the center frequencies of the filters are typically uniformly spaced. However, the warped axis, in accordance with the nonlinear function provided in Eq. (2), is implemented in order to match the human ear's perception [10]. The filter bank typically consists of overlapping triangular filters [12]. Figure 3 shows the generated Mel filter bank. The frequency spectrum of the signal (i.e., X(k) from Eq. (2) is multiplied with the filter bank to obtain mel frequency spectrum. Thus mapping the power-spectrum of the signal on to the Mel scale.

Discrete Cosine Transform (DCT): The vocal tract is smooth and hence there is a tendency for adjacent bands' energy levels to correlate. The DCT is used to create a set of cepstral coefficients from the transformed Mel frequency coefficients. The Mel spectrum is typically displayed on a log scale before being subjected to DCT. In the cepstral domain, this produces a signal with a quefrency peak that corresponds to the signal's pitch and a number of formants that represent low quefrency peaks. Since the first few MFCC coefficients constitute the majority of the signal information, the system can be made robust by extracting only those coefficients while ignoring or truncating higher-order DCT components.

Finally, MFCC is calculated as

$$c(n) = \sum_{m=0}^{M-1} log_{10}(s(m))cos(\pi n(m - 0.5)/M) \tag{3}$$

n = 0, 1,2....C-1. where c(n) are the cepstral coefficients, and C is the number of MFCCs. MFCC systems use only 8–13 cepstral coefficients. [10]

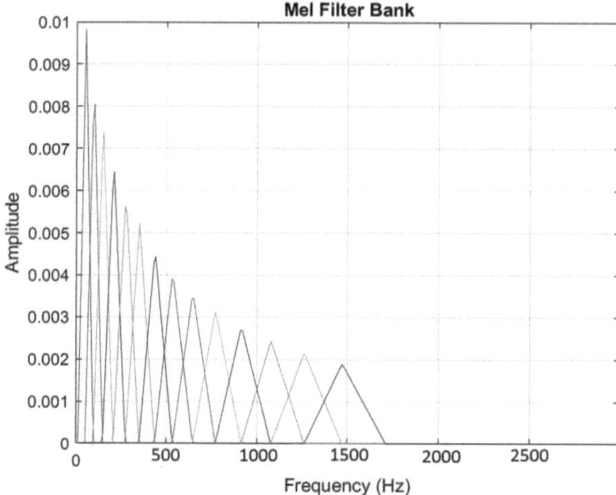

Fig. 3. Mel filter bank.

The log Mel spectrum is converted back to the time domain in this last phase, resulting in the MFCCs. The Mel Frequency Cepstrum Coefficients are the outcome (MFCC). The Mel coefficients are transformed back into the time domain using the discrete cosine transform [17].

4 Methodology

In this work, we are analyzing the usefulness and efficiency of a reservoir computer for real time audio signal processing. Our aim is to work on audio samples directly in the time domain. We are investigating whether reservoirs can be modeled to implement the functions performed by every stage of audio signal processing as shown in Fig. 4. In addition to using RC as a classifier, we aim to model a reservoir as a feature extractor that operates to mimic the MFCC Matlab function. Thus, simplifying this stage of audio signal processing.

For pre-processing, we have used the Mel frequency Cepstral Coefficient (MFCC) to extract information from the speech signals. We extracted the first 14 MFCC coefficients from the speech signal, which represent the short-term spectral features of the audio. These coefficients capture the shape of the vocal tract and are commonly used for speech signal processing. Figure 4(a) shows the reservoir as a classifier. We have used the TI-46 dataset, which consists of eight female speakers uttering digits 0 to 9 10 times each. Additionally, we used the Audio-Mnist data set, which consists of 60 speakers uttering numbers 0–9 50 times each. To confirm the functionality of the system, we conducted experiments that involved both speaker and digit recognition.

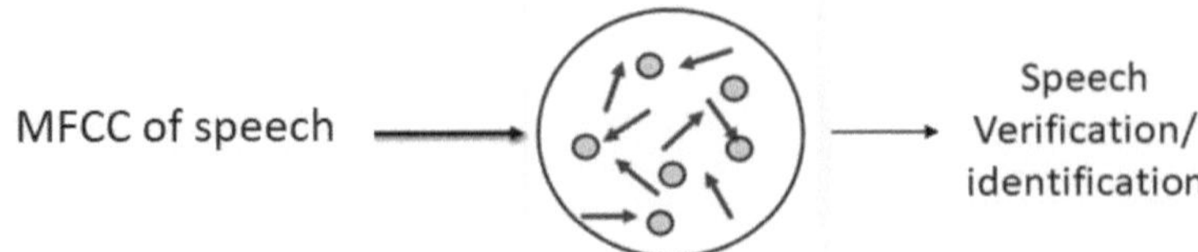

(a) Reservoir as classifier

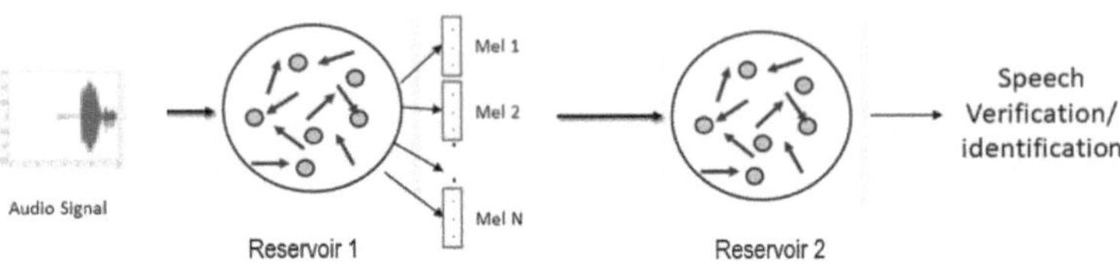

(b) Reservoir as feature extractor and classifier

Fig. 4. A comparison between the conventional approach (a) where the reservoir is used as a classifier after complex pre-processing by different means and the approach proposed here (b) where RC is used as an end-to-end audio processing concept.

4.1 RC-Based MFCC Feature Extraction

We are modeling a reservoir to perform feature extraction. Here, we have considered the MFCC function. We aim to model a reservoir that takes raw audio as input and provides MFCC coefficients as the output. First, we train the reservoir to perform MFCC extraction. Our objective is to replicate the MFCC algorithm by utilizing the values, as calculated by Matlab's MFCC function, as the training target for the reservoir.

We have calculated the normalized mean square error (NRMSE) of the reservoir output and the target to measure the reservoir performance. To begin with, we used the TI-46 data of female speakers uttering digits. The target matrix is the MFCC coefficient for the corresponding audio sample.

Figure 4b shows the reservoir-based end-to-end audio processing. The MFCC coefficients of the speech signal are extracted by reservoir 1 (RC-1), followed by a second reservoir performing audio identification/verification.

The audio data (input into reservoir 1) is usually of size $[N \times 1]$. The MFCC of these data (Output of reservoir 1 as shown in Fig. 4) is of size $[M \times 14]$, where $N \gg M$. To effectively utilize a reservoir, we need to apply some windowing technique. We window the audio signal in such a way that each window results in one data-point of the MFCC coefficient. To find the window size, we use the equation. $Windowsize = S/N$, where S is the size of the audio data, N is the number of non-overlapping windows. To calculate the number of non-overlapping windows, we use the equation $N = (M + 2)/2$, where M is the number of MFCC samples.

Audio signal corresponding to one window length was given as input to the reservoir. So we have created a reservoir whose input size is equal to the window length of the signal. The target for the reservoir is the corresponding MFCC coefficient. We evaluated the Normalized mean square error of the output of the reservoir with the target signal to evaluate its performance.

We also used the MFCC that we obtained from the reservoir as input to the classifier reservoir and evaluated its performance. We have performed speaker and digit recognition using this method. We have also verified its performance using the larger Audio Mnist dataset.

We have found that this method works fine. We are also investigating around different factors that can impact and hence improve the performance of the method.

4.2 Methods Improving Performance of RC-Based MFCC Feature Extraction

We have found that the window length we consider has a significant impact on the performance of the system. Based on our findings, we are analyzing three scenarios in this paper.

As explained earlier, to calculate the number of non-overlapping windows, we use the equation $N = (M) + 2)/2$, where M is the number of MFCC samples. In the first method, we are using the window length obtained based on this equation.

In the second method, we increase the number of overlapping windows by decreasing the window size. This method performs better than the first method.

We have found that in the mel spectrum obtained by the RC based method, higher frequencies (higher mel coefficients) have a greater impact in speaker identification and lower frequencies (lower mel coefficients) have a greater impact on digit identification. So to verify this, we developed a third method in which, we increased the number of overlapping windows by decreasing the window size for higher frequencies(Higher mel coefficients) for speaker identification and increased the number of overlapping windows by decreasing the window size for lower frequencies(Lower mel coefficients) for digit identification.

5 Results and Discussion

In our investigation, we are utilizing a reservoir to mimic the MFCC extraction. The normalized mean square error of reservoir 1, which is trained to mimic the MFCC as calculated by Matlab's function, is displayed in Table 1. The reservoir effectively mimics MFCC extraction, with low NMSE values for most coefficients. This demonstrates the reservoir's capability to perform this task. The NRMSE value clearly emphasizes the ability of a reservoir to mimic MFCC extraction.

We utilize the output of this first reservoir(feature extraction reservoir-RC-1) as the input to the second reservoir(Classification reservoir RC-2).

The training and testing performance of the MFCC obtained from Matlab function are shown as boxplot. The plot labeled *Experiment 1*, is the output of method 1. Whereas the one labeled *Experiment 2*, is the output of method 2. And the plot labeled *Experiment 3*, is the output of method 3. The Tables 2 and 3 shows a comparison of the performance of different audio signal processing methods using Ti-46 and Audio-Mnist datasets respectively for digit Recognition.

Table 1. NRMSE of Reservoir-1 extracting MFCC, based on 14 mel coefficients

MEL	Mel 1	Mel 2	Mel 3	Mel 4	Mel 5	Mel 6	Mel 7
NRMSE-Ti	0.129	0.175	0.085	0.032	0.038	0.024	0.013
NRMSE-AMnist	0.130	0.176	0.083	0.038	0.044	0.026	0.016
MEL	Mel 8	Mel 9	Mel 10	Mel 11	Mel 12	Mel 13	Mel14
NRMSE-Ti	0.018	0.015	0.013	0.015	0.013	0.014	0.009
NRMSE-AMnist	0.018	0.017	0.015	0.019	0.014	0.015	0.011

Table 2. Comparison of Performance of models with Ti-46 dataset for digit recognition

Models	Accuracy (%)
LSM [19]	94.0
Liquid-SNN [15]	77.7
Reservoir Computing (MEMS) [7]	78
Reservoir as classifier(Using Matlab MFCC)	92.9
Reservoir-based(Using Reservoir MFCC)	76.45

(See Figs. 5, 6 and 7).

Table 4 shows a comparison of the number of neurons used by different audio signal processing methods. The number of neurons in a network is calculated based on architecture implementation and hyper parameters such as the number of hidden layers, the number of units (neurons) in each layer, and the input/output dimensions. As can be seen from the table, Our method is the most lightweight and effective model, achieving high accuracy with smaller number of neurons. By utilizing significantly less parameters while maintaining the MFCC extraction in the time domain, our method demonstrates good performance.

This method's primary benefit is that it reduces computing complexity, particularly when executing the Fourier transform, by doing away with the necessity for intricate time-frequency conversion. It also facilitates the implementation of

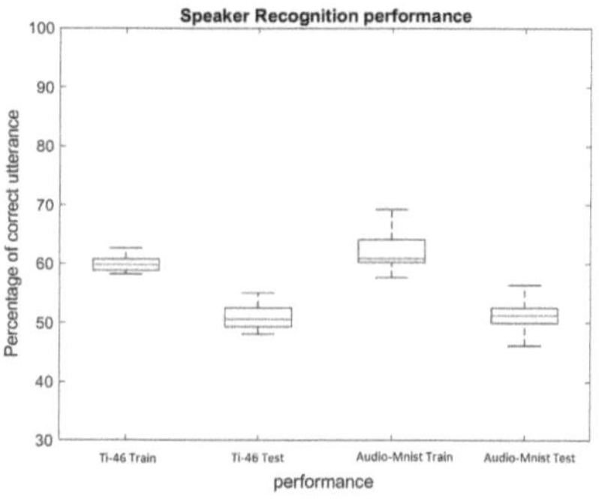
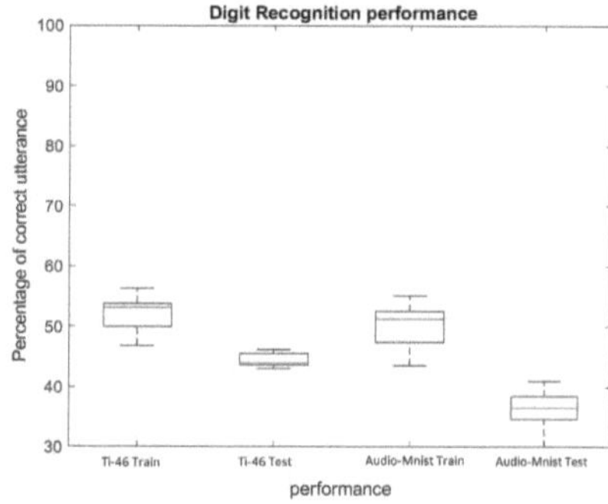

(a) Speaker Recognition Performance of ex-
periment1

(b) Digit Recognition Performance of experiment 1

Fig. 5. Speaker and Digit Recognition performance of Experiment 1.

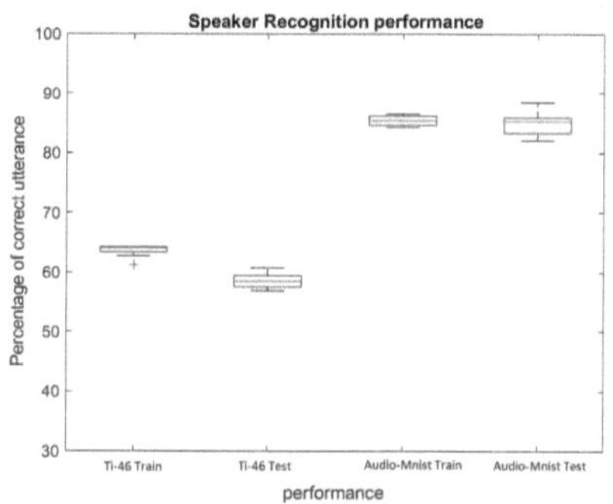
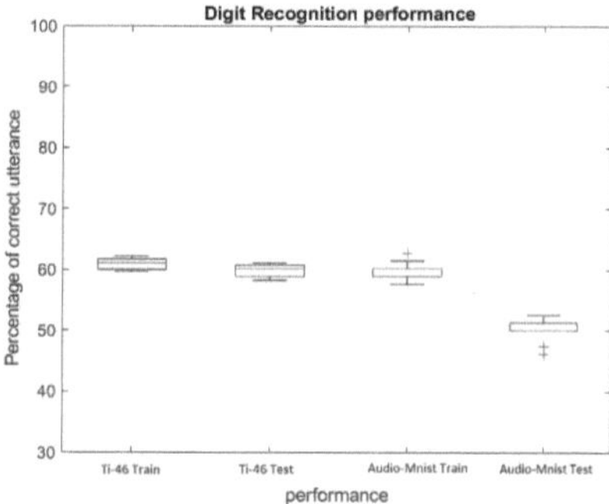

(a) Speaker Recognition Performance of ex-
periment2

(b) Digit Recognition Performance of experiment 2

Fig. 6. Speaker and Digit Recognition performance of Experiment 2.

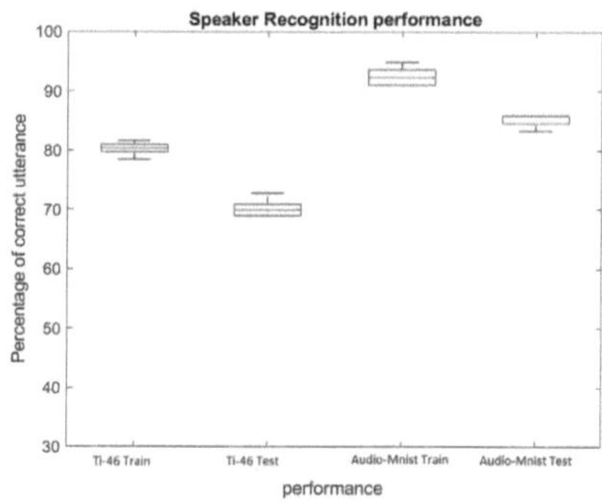
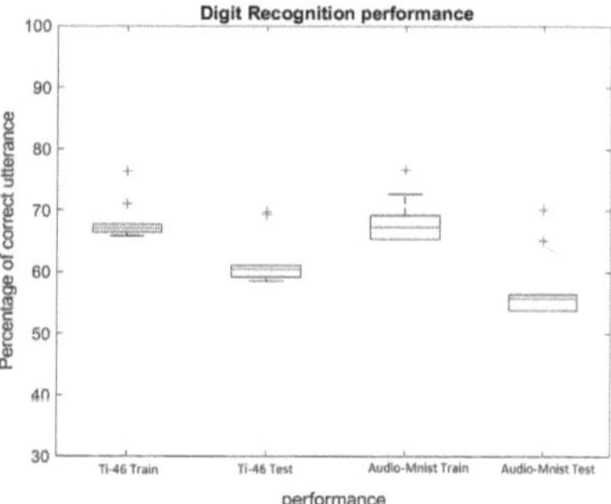

(a) Speaker Recognition Performance of ex-
periment3

(b) Digit Recognition Performance of experiment 3

Fig. 7. Speaker and Digit Recognition performance of Experiment 3.

Table 3. Comparison of Performance of models with Audio-Mnist dataset for digit recognition.

Models	Accuracy (%)
CNN [14]	96.4
LSTM [14]	95.23
AudioNet(Deep-NN) [5]	92.53
Liquid-SNN [15]	82.65
Reservoir as classifier(Using Matlab MFCC)	93.08
Reservoir-based(Using Reservoir MFCC)	76.66

Table 4. Measures of performance acquired for various models [2,6,13]

Models	Train accuracy%.	Test accuracy%	Average number of neurons
CNN	100	98.63	2M-10M parameters
Word embedding	95.50	92.20	1M-5M parameters
Logistic regression	64.32	61.95	- (Depends on dataset size)
Naive Bayes	50.25	49.75	100K parameters
SVM	82.88	83.32	(Depends on support vectors)
Random forest classifier VGG16	72.42	71.90	10K–100K trees
ResNet50	91.30	80.20	5M parameters
CapsNet	91.80	88.76	10M–20M parameters
2D ConvNet bidirectional GRU	68.85	65.23	10M–20M parameters
Acoustic model	75.69	73.23	(Depends on dataset size)
CNN LSTM	83.25	80.52	5M-15M parameters
Logistic regression 1-vector [26]	84.30	80.23	1M parameters
LSTM-CNN [29]	70.21	68.33	5M-15M parameters
RC based(Using Matlab MFCC)	98.05	96.3	400 Neurons
RC based(Using Reservoir MFCC)	94.87	85.89	RC-1 = 950 neurons, RC-2 = 400 neurons

a real-time audio signal processor, which is another significant benefit. A third benefit is that, by avoiding sophisticated computations, the system may be implemented with a basic hardware reservoir, leading to a processing system that can easily duplicate the MFCC extraction. This reduces the need for more complex systems.

In terms of speaker recognition, the outcomes of our method is are comparable to the results of Matlab based MFCC. In experiment 2 and 3 we are able to improve the performance of speaker recognition compared to experiment 1. Digit recognition is more challenging than speaker recognition. In experiment 2 and 3 we are able to improve the performance of Digit recognition. This shows that the performance is significantly impacted by the window size. In short, our approach to MFCC extraction is effective and seems promising. Our next objective will be to optimize the reservoir configuration. Additionally, we want to create a different technique that uses a reservoir to extract MFCC or any other superior speech feature.

6   Conclusion

In this work, we presented an efficient end-to-end audio processing system that operates directly on time-domain samples, eliminating the need for complex time-frequency transformations. By mimicking MFCC extraction using a reservoir computing approach, our method simplifies the traditional pipeline while maintaining performance. The proposed system processes audio signals through a reservoir (or cascaded reservoirs), directly yielding the desired output without intermediate domain conversions.

Our experimental results demonstrate the viability of this approach, where the reservoir-based system achieves competitive performance while significantly reducing computational overhead. This success underscores the potential of time-domain reservoir computing as a low-complexity alternative to conventional frequency-domain methods. Future work will focus on optimizing reservoir architectures for broader audio processing applications and further improving accuracy.

References

1. Alim, S.A., Rashid, N.K.A.: Some Commonly Used Speech Feature Extraction Algorithms. IntechOpen, 12 December 2018
2. Sandesara, A., Parikh, S., Sapovadiya, P., Rahevar, M.: A comparative study on speech emotion recognition, November 2020
3. Anusuya, M.A., Katti, S.K.: Speech recognition by machine: a review, 2009
4. Ghani, A.: Neuro-inspired speech recognition based on reservoir computing. In: Shabtai, N. (ed.), Advances in Speech Recognition, chapter 2. IntechOpen, Rijeka, 2010. Advances in Speech Recognition chapter 2 IntechOpen
5. Becker, S., et al.: AudioMNIST: exploring explainable artificial intelligence for audio analysis on a simple benchmark. arXiv [cs.SD], July 2018
6. Deng, T.: Effect of the number of hidden layer neurons on the accuracy of the back propagation neural network. Highlights Sci. Eng. Technol. **74**, 462–468 (2023)
7. Dion, G., Mejaouri, S., Sylvestre, J.: Reservoir computing with a single delay-coupled non-linear mechanical oscillator. J. Appl. Phys. **124**(15), 152132 (2018)

8. Fernando, C., Sojakka, S.: Pattern recognition in a bucket. In: Banzhaf, W., Ziegler, J., Christaller, T., Dittrich, P., Kim, J.T. (eds.) Advances in Artificial Life. ECAL 2003. LNCS, vol. 2801, pp. 588–597. Springer, Berlin, Heidelberg (2003). https://doi.org/10.1007/978-3-540-39432-7_63

9. Gan, T., Stepney, S., Trefzer, M.A.: Combining multiple inputs to a delay-line reservoir computer: control of a forced van der pol oscillator system. In: 2023 International Joint Conference on Neural Networks (IJCNN), pp. 1–7, 2023

10. Rao, K.S., Manjunath, K.E.: Speech Recognition Using Articulatory and Excitation Source Features. Springer, Cham (2017). https://doi.org/10.1007/978-3-319-49220-9 part of book series: SpringerBriefs in Speech Technology (BRIEF-SSPEECHTECH)

11. Lukovsevivcius, M., Jaeger, H.: Reservoir computing approaches to recurrent neural network training. Comput. Sci. Rev. 3(3), 127–149 (2009)

12. Molau, S., Pitz, M., Schluter, R., Ney, H.: Computing mel frequency cepstral coefficient on the power spectrum. In: 2001 IEEE International Conference on Acoustics, Speech, and Signal Processing. Proceedings (Cat. No.01CH37221), 2001. 2001 IEEE International Conference on Acoustics, Speech, and Signal Processing. Proceedings (Cat. No.01CH37221) pages=73-76 vol.1

13. Singh, G., Sharma, S., Kumar, V., Kaur, M., Baz, M., Masud, M.: Spoken language identification using deep learning. Comput. Intell. Neurosci. 2021(1) (2021)

14. Sridhar, C., Kanhe, A.: Performance comparison of various neural networks for speech recognition. J. Phys.: Conf. Ser. 2466(1), 012008 (2023)

15. Srinivasan, G., Panda, P., Roy, K.: SpiLinC: spiking liquid-ensemble computing for unsupervised speech and image recognition. Front. Neurosci. 12, 524 (2018)

16. Stepney, S.: Physical reservoir computing: a tutorial. Nat. Comput. 23(4), 665–685 (2024)

17. Tiwari, V.: MFCC and its applications in speaker recognition. Int. J. Emerg. Technol. 1(1), 19–22 (2010). ISSN : 0975-8364e, 2010

18. Triefenbach, F., Jalalvand, A., Schrauwen, B., Martens, J.P.: Phoneme recognition with large hierarchical reservoirs. Adv. Neural Inf. Process. Syst. 23 (2010)

19. Verstraeten, D., Schrauwen, B., Stroobandt, D., Van Campenhout, J.: Isolated word recognition with the liquid state machine: a case study. Inf. Process. Lett. 95(6), 521–528 (2005)

20. Verstraeten, D., Dambre, J., Dutoit, X., Schrauwen, B.: Memory versus nonlinearity in reservoirs. In: The 2010 International Joint Conference on Neural Networks (IJCNN), pp. 1–8, July 2010

Efficient Algorithms for Quantum Hashing

Ilnar Zinnatullin[1,2]([✉]) and Kamil Khadiev[1,2]

[1] Institute of Computational Mathematics and Information Technologies,
Kazan Federal University, Kazan, Tatarstan, Russia
[2] Zavoisky Physical-Technical Institute, FRC Kazan Scientific Center of RAS,
Kazan, Tatarstan, Russia
`IlnGZinnatullin@kpfu.ru`

Abstract. Quantum hashing is a useful technique that allows us to construct memory-efficient algorithms and secure quantum protocols. First, we present a circuit that implements the phase form of quantum hashing using 2^{n-1} $CNOT$ gates, where n is the number of control qubits. Our method outperforms existing approaches and reduces the circuit depth. Second, we propose an algorithm that provides a trade-off between the number of $CNOT$ gates (and consequently, the circuit depth) and the precision of rotation angles. This is particularly important in the context of NISQ (Noisy Intermediate-Scale Quantum) devices, where hardware-imposed angle precision limit remains a critical constraint.

Keywords: Quantum hashing · Quantum circuit decomposition · Uniformly controlled rotation

1 Introduction

Nowadays, quantum hardware still belongs to the so-called Noisy Intermediate-Scale Quantum (NISQ) [28] era, a term introduced by Preskill in 2018. This term reflects the fact that quantum computations are noisy and that the number of available qubits is limited. One of the major challenges is decoherence which disrupts the fragile quantum states necessary for computation and constrain the capabilities of the current quantum computers. Thus, the algorithm's execution time must be short enough before quantum states are broken.

Quantum computers provide a set of elementary gates. Typically, such a set includes one-qubit and two-qubit gates, and it is universal in the sense that any unitary transformation can be decomposed into a circuit consisting only of elementary gates [11,13]. In most cases, the $CNOT$ gate is used as the two-qubit gate. It is known that quantum gates are prone to errors. In particular, two-qubit gates have higher error rates compared to their one-qubit counterparts [16]. So, minimizing the total number of $CNOT$ gates is essential for improving reliability. Hardware-imposed angle precision limit [23] is another problem that we need to take into consideration. Because of limited angle precision, some small-angle rotations cannot be performed accurately on the current hardware.

© The Author(s), under exclusive license to Springer Nature Switzerland AG 2026
E. Formenti and L. Manzoni (Eds.): UCNC 2025, LNCS 16364, pp. 307–322, 2026.
https://doi.org/10.1007/978-3-032-15641-9_21

As a result, we face numerous challenges when implementing quantum algorithms on NISQ devices. Therefore, it is crucial to design quantum algorithms optimized according to various metrics, such as the circuit depth, the number of $CNOT$ gates and the rotation angle precision.

Hashing is a well-established technique used in a variety of computational and cryptographic scenarios. In classical computing, a hash function maps an input of arbitrary length to a fixed-length output (a hash), enabling fast search, data integrity check, etc. In cryptography, hash functions must be one-way and collision-resistant, making them essential for digital signatures, message authentication codes (MACs), secure password storage, and blockchain security. A quantum hash function [6] is a classical-quantum one-way function that maps classical inputs to quantum states in such a way that states corresponding to different inputs are "nearly" orthogonal. Because quantum hashes are exponentially smaller than the original inputs, quantum hash functions are non-invertible by virtue of fundamental quantum information theory. Moreover, collision resistance allows us to distinguish different quantum hashes with high probability.

Ambainis and Freivalds [9] proposed a technique called quantum fingerprinting to construct quantum automata recognizing the unary $MOD_p = \{a^i : i \bmod p = 0\}$ language, where p is prime. Then, Buhrman et al. [12] used quantum fingerprinting that is based on binary error correcting codes to construct communication protocol for equality problem. Ablayev and Vasiliev [8] proposed a non-binary classical-quantum one-way function as a generalization of quantum fingerprinting and introduced the notion of quantum hashing. Later, Vasiliev proposed a phase form approach for quantum hashing in [30]. Efficient construction of branching programs for quantum hashing in terms of both the circuit depth and width was investigated in [1]. In [1], it is stated that when ε-biased sets are used to hash elements of $\mathbb{Z}_q$, the algorithm has the running time of $\Omega(\log q)$ and requires $\Omega(\log \log q)$ qubits. The quantum hashing (fingerprinting) approach has been widely used in various areas such as stream processing algorithms [24], query model algorithms [4,7], online algorithms [17–19,21], branching programs [5,20], development of quantum devices [29], automata [9,10,14,15], etc.

In this paper, we focus on designing efficient quantum circuit (a program for quantum computer) for quantum hashing. By "efficient", we mean a circuit with the minimal number of $CNOT$ gates. The base element of quantum circuit for the method is the uniformly controlled rotation gate, for which the most efficient representation was suggested by Möttönen et al. [26]. The technique allows us to represent the uniformly controlled rotation (and, as a result, a circuit for quantum hashing) with n control qubits using 2^n $CNOT$ gates.

We consider the phase-form quantum hashing algorithm (the uniformly controlled rotation with a specific initial state), and present a circuit with 2^{n-1} $CNOT$ gates. Reducing the number of $CNOT$ gates by half is important as two-qubit gates are more costly to implement on the current quantum devices.

Another issue with representing uniformly controlled rotation by our technique and technique of Möttönen et al. [26] is rotation angle precision. These techniques require much higher precision for the modified angles compared to the original ones. We provide an algorithm that demonstrates a trade-off between the number of $CNOT$ gates and the rotation angle precision. Similar trade-off was demonstrated by Khadieva et al. [22]. Asymptotically, our algorithm is equivalent to the existing one, yet it requires fewer $CNOT$ gates in practice.

Optimization of the circuit for quantum hashing algorithm is considered from different points of view. In [34], the authors adapt quantum circuits implementing quantum hashing for specific quantum processor architectures and optimize them with respect to the number of $CNOT$ gates. Researchers consider shallow circuits for approximate versions of quantum hashing in [31–33].

The paper is organized as follows. Section 2 provides the necessary preliminaries. In Sect. 3, we demonstrate how to construct an efficient quantum circuit that implements the phase form of quantum hashing. Section 4 is devoted to the algorithm that offers a trade-off between the number of $CNOT$ gates and the precision of the rotation angles. Finally, we summarize our results in Sect. 5.

2   Preliminaries

Quantum Computation. We use Dirac notation. In general case, a qubit is represented as a column vector (ket vector) $|\psi\rangle = \alpha_0|0\rangle + \alpha_1|1\rangle \in \mathcal{H}^2$ with complex amplitudes satisfying $|\alpha_0|^2 + |\alpha_1|^2 = 1$. A state of n qubits is described by a ket vector from $(\mathcal{H}^2)^{\otimes n}$ and has the form $\sum_{j=0}^{2^n-1} \alpha_j|j\rangle$, where $\alpha_j \in \mathbb{C}$ and $\sum_{j=0}^{2^n-1} |\alpha_j|^2 = 1$.

Quantum circuit model is a model for quantum computation, similar to classical circuits, which consists of qubits (represented as horizontal lines), quantum gates corresponding to unitary transformations and measurements to extract classical information from the qubits. A quantum circuit is a quantum algorithm that is characterized by two parameters: depth and width. The circuit depth describes the time complexity of the algorithm, while the width corresponds to its space complexity (i.e. the number of qubits required). More details about quantum circuits can be found in [27].

Next, we present matrix representations of the gates used in our work: negation gate $X = \begin{pmatrix} 0 & 1 \\ 1 & 0 \end{pmatrix}$, Hadamard gate $H = \begin{pmatrix} \frac{1}{\sqrt{2}} & \frac{1}{\sqrt{2}} \\ \frac{1}{\sqrt{2}} & -\frac{1}{\sqrt{2}} \end{pmatrix}$, rotation about the z-axis $R_z(\theta) = \begin{pmatrix} e^{-\frac{i\theta}{2}} & 0 \\ 0 & e^{\frac{i\theta}{2}} \end{pmatrix}$, rotation about the y-axis $R_y(\theta) = \begin{pmatrix} \cos(\frac{\theta}{2}) & -\sin(\frac{\theta}{2}) \\ \sin(\frac{\theta}{2}) & \cos(\frac{\theta}{2}) \end{pmatrix}$, relative phase shift $P(\theta) = \begin{pmatrix} 1 & 0 \\ 0 & e^{\frac{i\theta}{2}} \end{pmatrix}$, two-qubit gate $CNOT = \begin{pmatrix} 1 & 0 & 0 & 0 \\ 0 & 1 & 0 & 0 \\ 0 & 0 & 0 & 1 \\ 0 & 0 & 1 & 0 \end{pmatrix}$.

By $C^{n-1}(R)$, we denote an n-qubit controlled R gate with $n-1$ control qubits that applies R to the qubit with index n if and only if all control qubits are in the $|1\rangle$ state, where $R = \begin{pmatrix} r_{11} & r_{12} \\ r_{21} & r_{22} \end{pmatrix}$. The $2^n \times 2^n$ matrices corresponding to the $C^{n-1}(R_z(\theta))$ and $C^{n-1}(R_y(\theta))$ gates are

$$C^{n-1}(R_z(\theta)) = \begin{pmatrix} 1 & & & \\ & \ddots & & \\ & & 1 & \\ & & & R_z(\theta) \end{pmatrix}, \quad C^{n-1}(R_y(\theta)) = \begin{pmatrix} 1 & & & \\ & \ddots & & \\ & & 1 & \\ & & & R_y(\theta) \end{pmatrix}.$$

By UCR_a^{n-1}, we denote an n-qubit uniformly controlled rotation about the a-axis [26] that employs $n-1$ control qubits and uses all possible control states to rotate the qubit with index n.

Quantum hashing. Let $S = \{s_0, s_1, \ldots, s_{d-1}\} \subseteq \mathbb{Z}_q$ be an ε-biased set, i.e., set of parameters satisfying $\frac{1}{d}\left|\sum_{j=0}^{d-1} e^{i2\pi s_j x/q}\right| \le \varepsilon$, for every $x \in \mathbb{Z}_q\backslash\{0\}$.

For $x \in \mathbb{Z}_q$, we define its quantum hash in amplitude and phase forms. An $(n-1)$-qubit quantum hash in the phase form [30] is given by

$$|\psi(x)\rangle = \tfrac{1}{\sqrt{d}} \sum_{j=0}^{d-1} e^{i2\pi s_j x/q}|j\rangle. \tag{1}$$

An n-qubit quantum hash in the amplitude form [8] is defined as

$$|\psi(x)\rangle = \tfrac{1}{\sqrt{d}} \sum_{j=0}^{d-1}|j\rangle \left(\cos\left(\tfrac{2\pi s_j x}{q}\right)|0\rangle + \sin\left(\tfrac{2\pi s_j x}{q}\right)|1\rangle\right). \tag{2}$$

Note that $n-1 = \log d$ and $d = O\left(\tfrac{\log q}{\varepsilon^2}\right)$ [8,30]. We give the following formula for quantum hash that is a generalization of Eqs. (1) and (2):

$$|\psi(x)\rangle = \tfrac{1}{\sqrt{d}} \sum_{j=0}^{d-1}|j\rangle \left(R_a\left(\theta_j\right)|q_n\rangle\right), \tag{3}$$

where R_a is a rotation about the a-axis on the Bloch sphere, $\theta_j = \tfrac{4\pi s_j x}{q}$. For the phase form, we use R_z gates and the target qubit $|q_n\rangle = |1\rangle$. For the amplitude form, we use R_y gates and the target qubit $|q_n\rangle = |0\rangle$. Note that in Eq. (3) an ancilla (namely, the target qubit $|q_n\rangle$) is used to create a quantum hash while Eq. (1) lacks it. An algorithm for quantum hashing with parameter $\tilde{\theta} = (\theta_0, \theta_1, \ldots, \theta_{d-1})$ in the quantum circuit model is presented in Fig. 1. For further details on quantum hashing, please refer to [2,3].

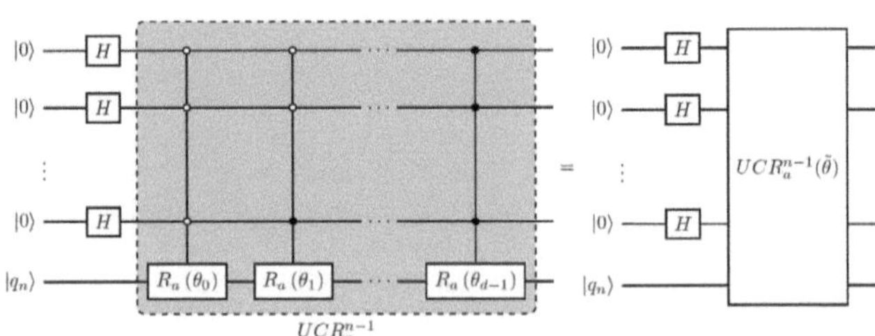

Fig. 1. Algorithm for quantum hashing.

2.1   Efficient Decomposition of UCR Gates

Möttönen et al. presented a method [26] for decomposing a UCR gate into a circuit consisting of an alternating sequence of single-qubit rotations and $CNOT$ gates. This method requires a modification of the original rotation angles. The decomposition is constructed recursively by applying the decomposition step from [25] (see Fig. 2) to each UCR gate until we get a circuit consisting only of one-qubit rotations and $CNOT$ gates. The parameters of the UCR gates shown in Fig. 2 are defined as follows: $\tilde{\theta}_1^0 = (\theta_0^0 = \theta_0, \theta_1^0 = \theta_1, \ldots, \theta_{2r-1}^0 = \theta_{2r-1})$, $\tilde{\theta}_1^1 = (\theta_0^1, \theta_1^1, \ldots, \theta_{r-1}^1), \tilde{\theta}_2^1 = (\theta_0^2, \theta_1^2, \ldots, \theta_{r-1}^2)$, where $r = 2^{n-2}$.

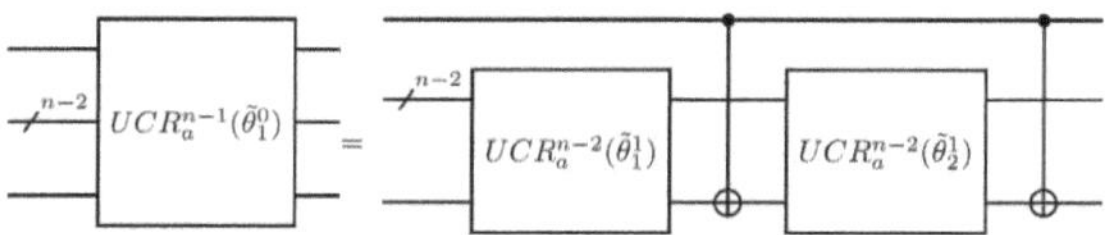

Fig. 2. Decomposition step for Möttönen et al.'s method.

The connection between the original and the modified angles for the decomposition step is the following: $\theta_0^0 = \theta_0^1 + \theta_0^2, \ldots, \theta_{r-1}^0 = \theta_{r-1}^1 + \theta_{r-1}^2, \theta_r^0 = \theta_0^1 - \theta_0^2, \ldots, \theta_{2r-1}^0 = \theta_{r-1}^1 - \theta_{r-1}^2$. So, we get $\theta_i^1 = \frac{\theta_i^0 + \theta_{r+i}^0}{2} = \frac{\theta_i + \theta_{r+i}}{2}$, $\theta_i^2 = \frac{\theta_i^0 - \theta_{r+i}^0}{2} = \frac{\theta_i - \theta_{r+i}}{2}$, where $i \in [0, r-1]$.

Finally, we obtain a circuit containing $2^{n-1} = d$ one-qubit R_a gates and $2^{n-1} = d\ CNOT$ gates. Let $\tilde{\theta}' = (\theta_0', \theta_1', \ldots, \theta_{d-1}')$ denote the vector of modified angles in the resulting circuit. The relationship between the angles is given by

$$\tilde{\theta}'^T = \tfrac{1}{d} M^T \tilde{\theta}^T, \tag{4}$$

where M is a $d \times d$ matrix with entries $M_{ij} = (-1)^{(b_{i-1} \cdot g_{j-1})}, 1 \le i, j \le d$, and $(b_{i-1} \cdot g_{j-1})$ denotes the dot product of the $(i-1)$th codeword of the standard binary code and the $(j-1)$th codeword of Gray code. Figure 3 shows an example for $d = 8$.

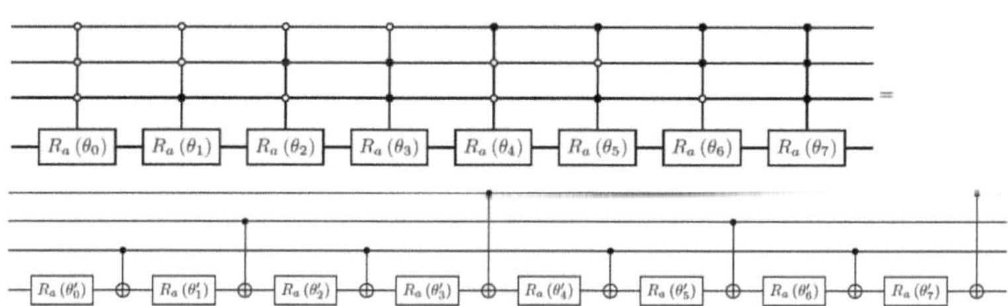

Fig. 3. Decomposition for $d = 8$.

3   Circuit Optimization

In this section, we propose an algorithm that reduces the number of $CNOT$ gates in the circuit that implements the phase form of quantum hashing by half compared to the best existing method.

Firstly, we present lemmas that show several equivalences of quantum circuits. These results will be used in the circuit optimization process.

Lemma 1. *The circuit equivalence presented in Fig. 4 holds.*

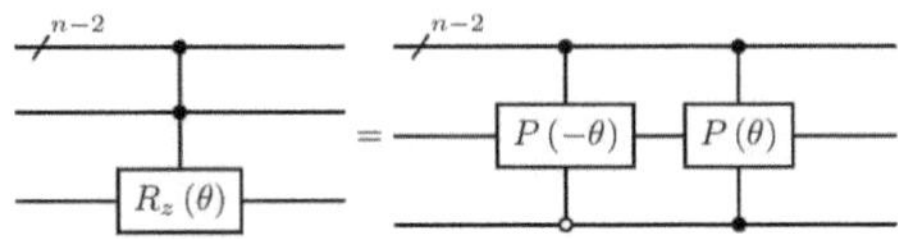

Fig. 4. Circuit equivalence for the n-qubit controlled $R_z(\theta)$ rotation.

Proof. See [35].

Lemma 2. *Relative phase shift P is equivalent, up to a global phase factor, to the half-argument rotation R_z, that is, $P(\theta) = e^{i\theta/4} R_z(\theta/2)$.*

Proof. Obviously, $P(\theta) = \begin{pmatrix} 1 & 0 \\ 0 & e^{i\theta/2} \end{pmatrix} = e^{i\theta/4} \begin{pmatrix} e^{-i\theta/4} & 0 \\ 0 & e^{i\theta/4} \end{pmatrix} = e^{i\theta/4} R_z(\theta/2)$.

Lemma 3. *The circuit equivalence shown in Fig. 5 holds. In mathematical form, this equivalence is expressed as $(I^{\otimes(n-1)} \otimes X)C^{n-1}(R_z(\theta)) = C^{n-1}(R_z(-\theta))(I^{\otimes(n-1)} \otimes X)$.*

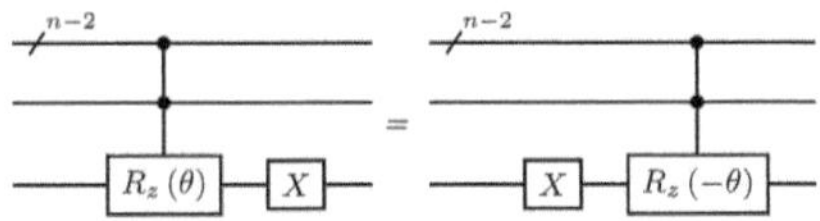

Fig. 5. Circuit equivalence for n-qubit controlled $R_z(\theta)$ rotation and single X gate.

Proof. See [35].

Our circuit optimization algorithm that reduces the number of $CNOT$ gates is based on the next theorem.

Theorem 1. *The phase version of the quantum hashing algorithm can be represented as a quantum circuit with 2^{n-1} $CNOT$ gates.*

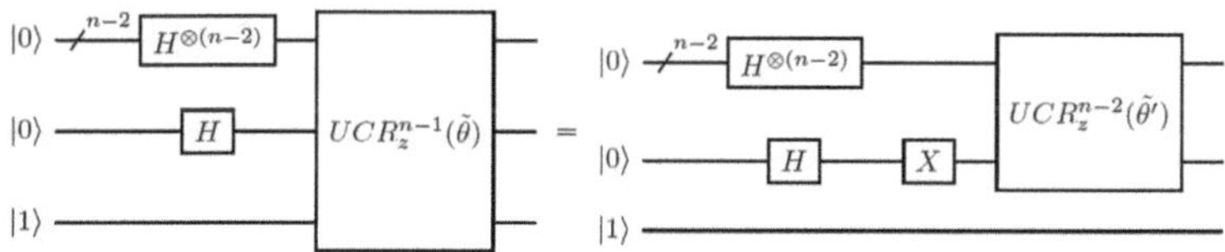

Fig. 6. Elimination of the ancilla in the circuit implementing the phase form of quantum hashing.

Proof. For proving the claim of the theorem, we should show the equivalence that presented in Fig. 6. So, in that case, we can use technique from Sect. 2.1 for representation of UCR gate and obtain a circuit with 2^{n-1} $CNOT$ gates.

We start with the original circuit shown in Fig. 7. The objective is to eliminate the ancilla qubit (i.e. the qubit with index n), thereby reducing the number of control qubits in the UCR_z^{n-1} gate by one. Subsequently, we apply the method described in Sect. 2.1 to the resulting UCR_z^{n-2} gate and get a circuit consisting of 2^{n-2} $CNOT$ gates.

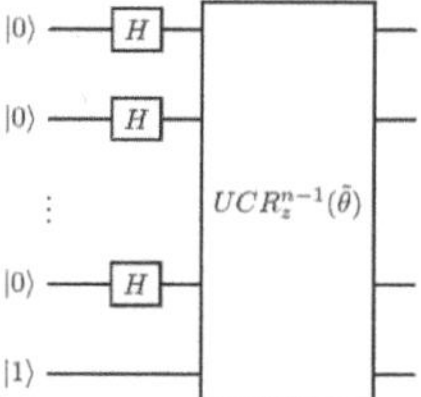

Fig. 7. Algorithm for the phase form of quantum hashing.

Let us reorder n-qubit controlled rotations in the UCR_z^{n-1} gate using Gray code instead of the standard binary code (see Fig. 8). This reordered structure is then applied to the circuit presented in Fig. 7 (see Fig. 9). We use Gray code because adjacent codewords differ in exactly one position. This allows us to switch between states of control qubits by applying a single X gate. Note that the circuit shown in Fig. 9 consists solely of n-qubit controlled rotations of the form $C^{n-1}(R_z)$, which require all control qubits to be in the $|1\rangle$ state.

For simplicity, we fix $n = 4$ (see Fig. 10) and follow the optimization process step by step. Our focus is on optimizing the red border-boxed gate depicted in Fig. 10. First, we apply Lemma 1 to each blue border-boxed gate in Fig. 10 and get a circuit shown in Fig. 11. Since the last qubit is in the $|1\rangle$ state, multi-qubit controlled relative phase shifts with negative arguments are not applied. Consequently, the circuit is further simplified to the one shown in Fig. 12. Next, we replace relative phase shifts with rotations about the z-axis using Lemma 13 (see Fig. 14). Subsequently, we apply Lemma 3 to the blue border-boxed segments of the circuit shown in Fig. 13. This step reduces the number of multi-qubit controlled rotations by half and eliminates some of the X gates, as each blue border-boxed pair of multi-qubit controlled rotations in Fig. 14 can be merged into one multi-qubit controlled rotation. Consecutive X gates (that are in red border-boxes) applied to the third qubit cancel each other out. So, we obtain the circuit shown in Fig. 16. Careful analysis reveals that Gray code is used to iterate over all multi-qubit controlled rotations in Fig. 16. Alternatively, we can use the standard binary code for this purpose, resulting in the circuit presented in Fig. 16. As a result, we obtain the circuit identity shown in Fig. 17.

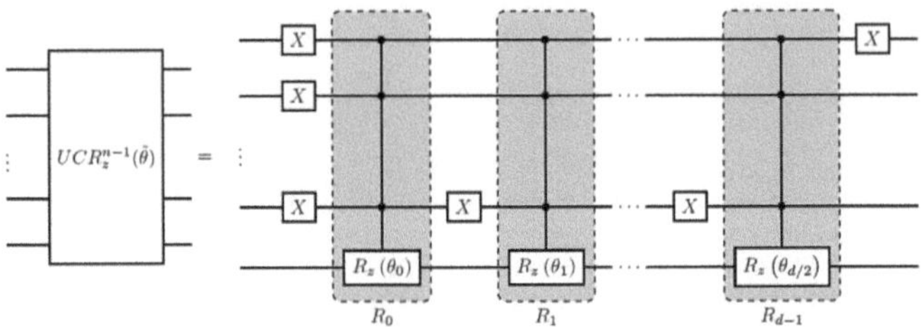

Fig. 8. Decomposition of UCR_z^{n-1} gate using Gray code.

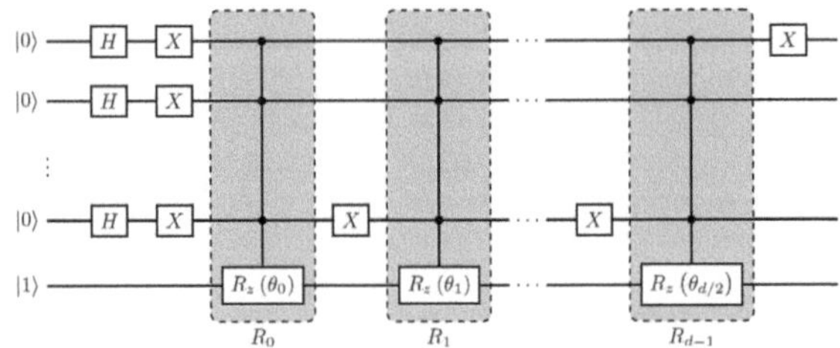

Fig. 9. Intermediate decomposition for the circuit implementing the phase form of quantum hashing.

In general case, we get the circuit equivalence shown in Fig. 18. Using the technique from Sect. 2.1, we obtain the decomposition consisting of 2^{n-2} $CNOT$ gates, whereas the original circuit on the left in Fig. 17 requires 2^{n-1} $CNOT$ gates to decompose it. Finally, we get the circuit shown in Fig. 6 with modified angles $\theta_i' = (\theta_{2i} - \theta_{2i+1})/2$. Thus, the target qubit in the $|1\rangle$ state is eliminated.

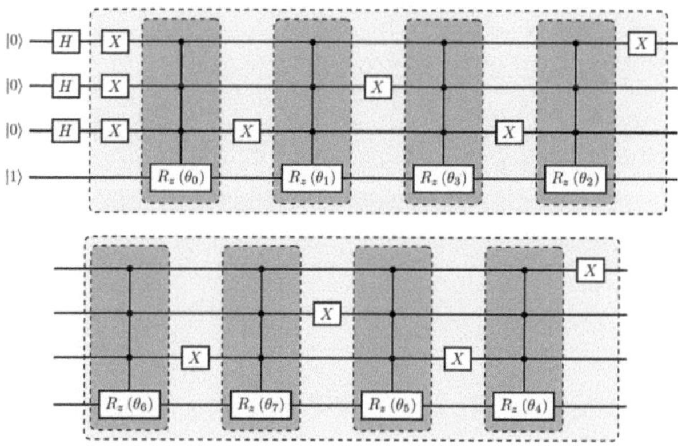

Fig. 10. Initial circuit for $n = 4$.

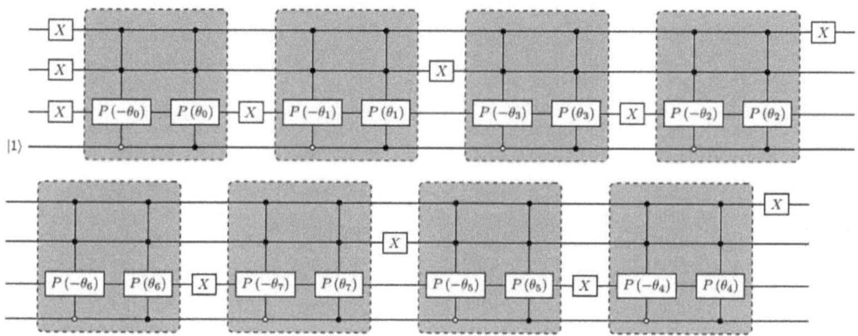

Fig. 11. The circuit after applying Lemma 1.

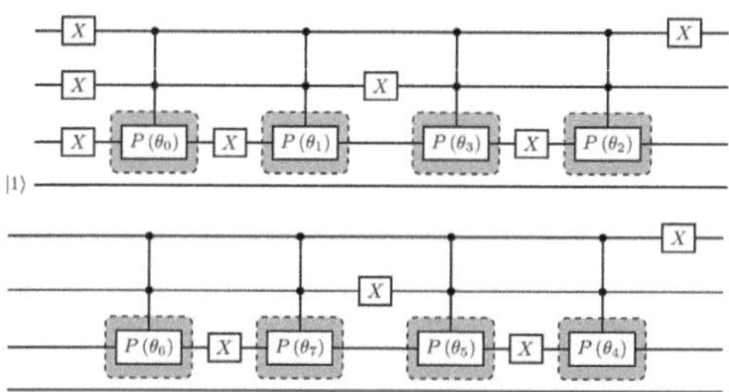

Fig. 12. The simplified circuit.

4 Efficient Algorithm for Quantum Hashing: Circuit Depth vs. Angle Precision

In this section, we propose an algorithm that offers a trade-off between the number of $CNOT$ gates (and, consequently, the circuit depth) and the precision

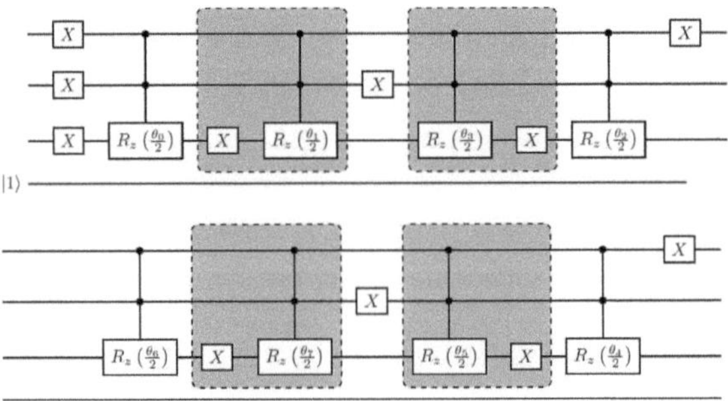

Fig. 13. The circuit after applying Lemma 2

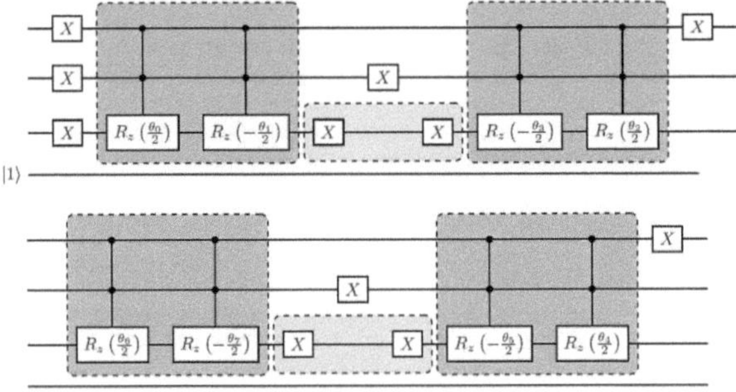

Fig. 14. The circuit after applying Lemma 3.

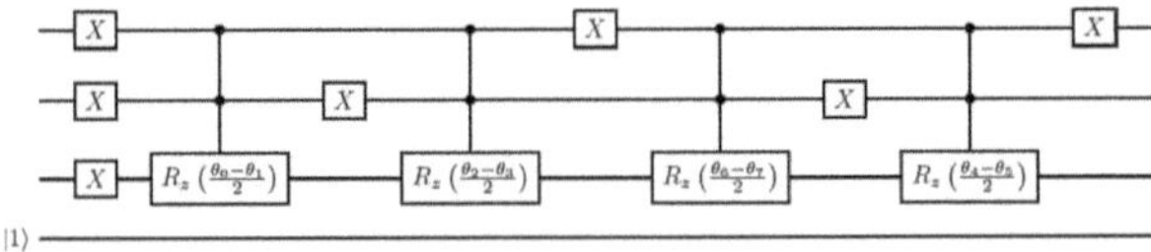

Fig. 15. The decomposition for a UCR_z^{n-2} gate using Gray code with additional X gate.

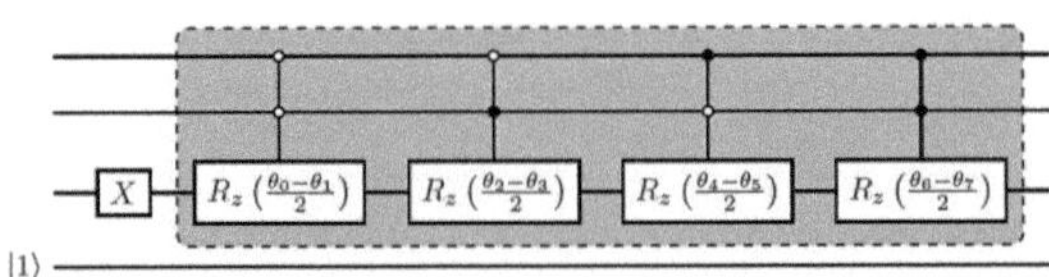

Fig. 16. The decomposition for the UCR_z^{n-2} gate using the standard binary code with additional X gate.

of angles. This is particularly important in the context of NISQ devices, as hardware-imposed angle precision limit remains a critical aspect.

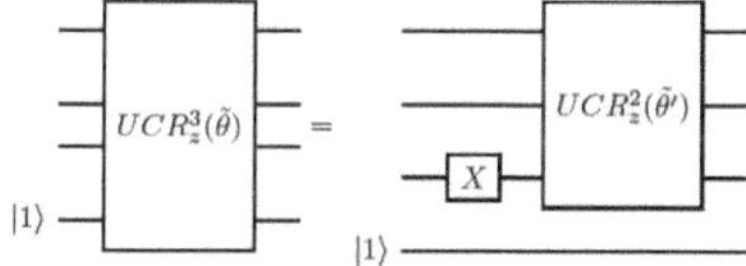

Fig. 17. Transformation of a UCR_z^3 gate into a UCR_z^2 gate.

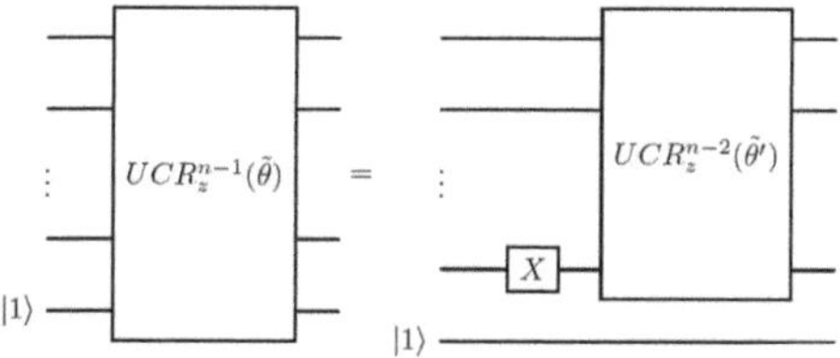

Fig. 18. Transformation of a UCR_z^{n-1} gate into a UCR_z^{n-2} gate.

Let us recall the generalization formula (Eq. (3)) for quantum hashing: $|\psi(x)\rangle = \frac{1}{\sqrt{d}} \sum_{j=0}^{d-1} |j\rangle \left(R_a \left(\theta_j \right) |q_n\rangle \right)$, where $\theta_j = \frac{4\pi s_j x}{q}$. Let us also recall that to implement quantum hashing, the circuit shown in Fig. 19 is used.

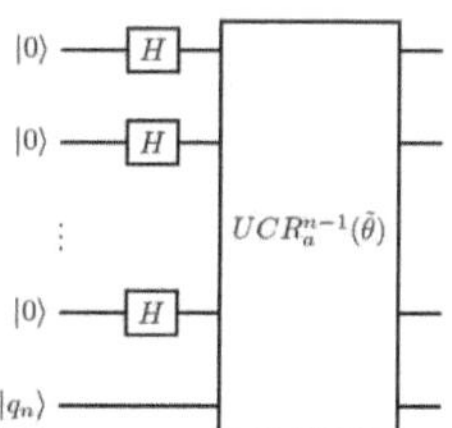

Fig. 19. General algorithm for quantum hashing.

Our goal is to efficiently decompose the UCR_a^{n-1} gate. To achieve this, we employ the technique described in Sect. 2.1. In this case, the resulting circuit consists of an alternating sequence of 2^{n-1} one-qubit rotations R_a and 2^{n-1} $CNOT$ gates. In the resulting circuit, the original angles $\tilde{\theta}$ are modified, and the relationship between the original angles $\tilde{\theta}$ and the modified angles $\tilde{\theta}'$ is given by Eq. (4): $\tilde{\theta'}^T = \frac{1}{d} M^T \tilde{\theta}$.

The precision of the original angles is $O\left(\frac{1}{q}\right)$. Using Eq. (4), we derive that the precision of the modified angles becomes $O\left(\frac{1}{dq}\right) = O\left(\frac{1}{q \log q}\right)$. Consequently, the modified angles are more sensitive. Implementing a circuit with such parameters can be challenging for current NISQ devices [23].

Let *Decomposition* 1 and *Decomposition* 2 be some procedures that will be described later. We propose the following algorithm: (i) Recursively apply

Decomposition 1 k times to construct a circuit consisting of 2^k UCR_a^{n-k-1} gates and 2^k $CNOT$ gates. (ii) Apply *Decomposition* 2 to each of the 2^k UCR_a^{n-k-1} gates.

Decomposition 1 involves applying the decomposition step shown in Fig. 2 to each UCR gate.

Below, we describe *Decomposition* 2 procedure. First, we reorder multi-qubit controlled rotations in a UCR_a^{n-k-1} gate using Gray code. Next, we highlight 2^{n-k-3} circuit segments shown in Fig. 20.

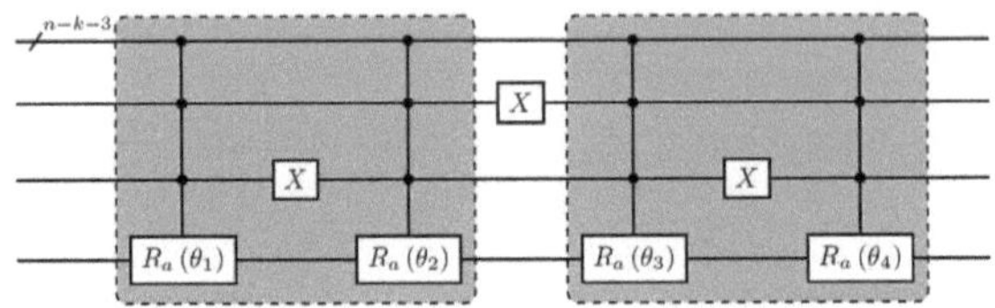

Fig. 20. The circuit segment.

Let us propose a decomposition for the border-boxed circuit segments in Fig. 20. Thus, we focus on decomposing the smaller segment presented in Fig. 21.

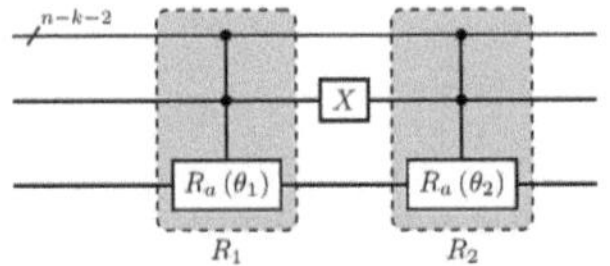

Fig. 21. A pair of multi-qubit controlled rotations.

In [11], a decomposition of a multi-qubit controlled rotation is presented (for more details see Sect. 4.1 in [35]). We apply this decomposition to the R_1 gate and its mirrored version to the R_2 gate. As a result, it produces the circuit depicted in Fig. 22. It can be seen that border-boxed $C^{n-k-2}(X)$ gates in Fig. 22 cancel each other out, and we obtain the circuit shown in Fig. 23. We decompose the blue border-boxed fragment in Fig. 23 using the circuit shown in Fig. 24. Next, we apply the decomposition shown in Fig. 23 to border-boxed circuit segments depicted in Fig. 20. See Fig. 25 and note that border-boxed controlled rotations are merged into one two-qubit controlled rotation. This trick is done for all pairs of "outside" two-qubit controlled rotations.

Now, let us count the total number of $CNOT$ gates in the resulting circuit. The decomposition of $C^{n-k-2}(X)$ [11] requires $24 \cdot (n - k - 2) - 52 = 24 \cdot (n - k) - 100$ $CNOT$ gates. Note that each of the 2^{n-k-2} circuit segments includes two $C^{n-k-2}(X)$ gates, one $U_a(\theta_1, \theta_2)$ gate, and one $C^1(R_a)$ gate. In addition, the first $C^1(R_a)$ gate must be taken into account. Thus, the decomposition of

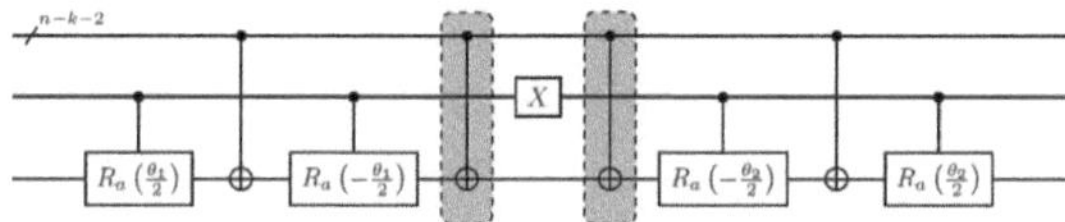

Fig. 22. Decomposition for a pair of multi-qubit controlled rotations.

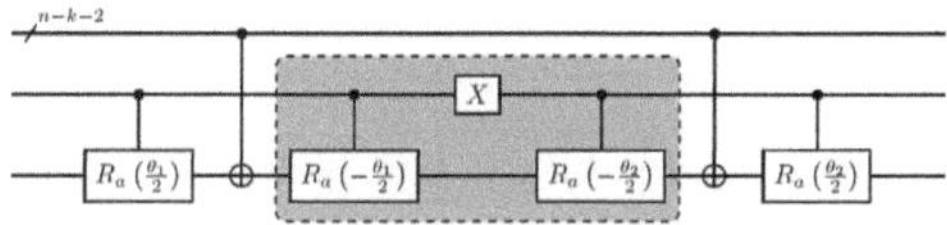

Fig. 23. Simplified decomposition for a pair of multi-qubit controlled rotations.

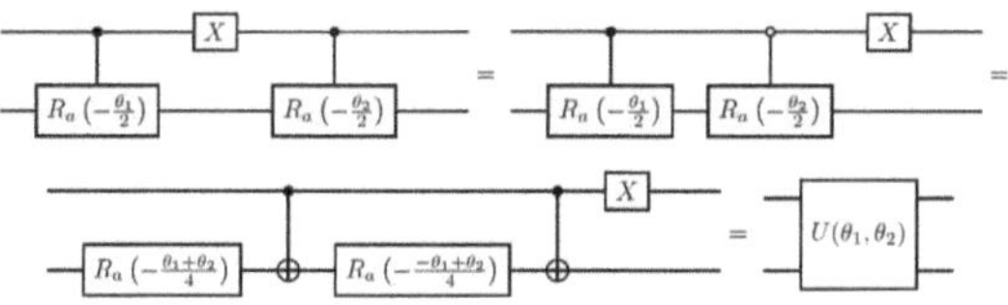

Fig. 24. Decomposition for the circuit fragment.

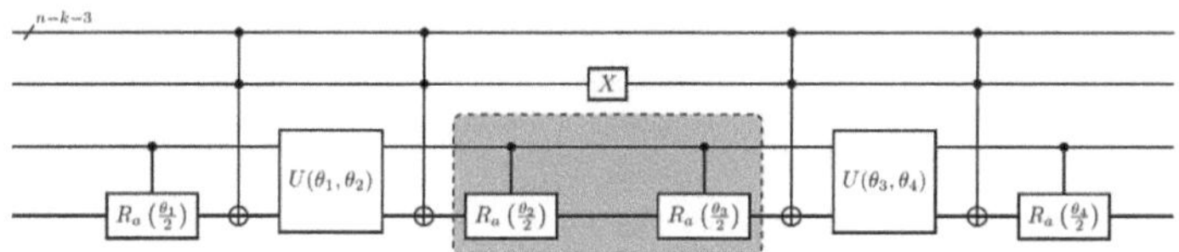

Fig. 25. Intermediate decomposition for the circuit segment.

each UCR_a^{n-k-1} gate requires $2^{n-k-2} \cdot (2 \cdot (24 \cdot (n - k) - 100) + 2 + 2) + 2 = 2^{n-k} \cdot (12 \cdot (n - k) - 49) + 2$ $CNOT$ gates.

So, the total number of $CNOT$ gates after k iterations is the following: $3 \cdot 2^k + 2^n \cdot (12 \cdot (n - k) - 49)$ or $3 \cdot 2^k + d \cdot (24 \cdot (\log d - k) - 74)$, where $k \leq n - 5 = \log d - 4$.

Note that the circuit depicted in Fig. 23 can be further simplified if we focus on the phase form of quantum hashing (for more details, see Sect. 4.3 in [35]).

Theorem 2. *Application of the proposed decomposition algorithm to the original circuit that implements quantum hashing yields a circuit with $3 \cdot 2^k + 2^n \cdot (12 \cdot (n - k) - 49)$ or $3 \cdot 2^k + d \cdot (24 \cdot (\log d - k) - 74)$ $CNOT$ gates, for $k \leq n - 5 = \log d - 4$, where k is the number of applications of Decomposition 1.*

We can see that as k increases, the circuit depth decreases from $O(\log q \log \log q)$ to $O(\log q)$, while the angle precision changes from $O\left(1/q\right)$ to $O\left(1/(q \log q)\right)$.

Authors of [22] propose an algorithm that balances between the number of $CNOT$ gates and the precision of rotation angles that yields a circuit with

$2^k + d(96(\log d - k) - 384))$ $CNOT$ gates for $k \leq \log d - 5$, matching the asymptotic complexity of our algorithm. Thus, our approach reduces the number of $CNOT$ gates by $d(72(\log d - k) - 310) - 2^{k+1} \geq \frac{799}{16}d$ for $k \leq \log d - 5$, achieving asymptotic savings of $O(d)$ $CNOT$ gates.

5   Conclusion

Implementation of algorithms on current quantum computers faces significant challenges. Therefore, it is essential to design efficient algorithms optimized with respect to multiple metrics including the circuit depth, the number of two-qubit gates, the precision of rotation angles.

We propose algorithms for quantum hashing that are inspired by NISQ quantum devices. First, we demonstrated how to eliminate an ancilla qubit used in constructing the phase-form quantum hash. Our technique reduces the number of qubits used by one and reduces the number of $CNOT$ gates and the circuit depth by half, thereby saving both time and space resources. Our technique also outperforms all existing methods. Second, we present an algorithm that provides a trade-off between the number of $CNOT$ gates (the circuit depth) and the rotation angle precision. Since there are limitations on the precision of rotation angles, our algorithm allows the circuit to avoid impractically small rotations. Although asymptotically our algorithm is equivalent to the one from [22], it is more efficient in terms of the number of $CNOT$ gates.

Acknowledgments. The research (Sects. 1, 2, 3) has been supported by Russian Science Foundation Grant 24-21-00406, https://rscf.ru/en/project/24-21-00406/.

The study in Sect. 4 was funded by the subsidy allocated to Kazan Federal University for the state assignment in the sphere of scientific activities (Project No. FZSM-2024-0013).

References

1. Ablayev, M.F.: Efficient branching programs for quantum hash functions generated by small-biased sets. Lobachevskii J. Math. **39**(7), 961–966 (2018). https://doi.org/10.1134/S199508021807003X
2. Ablayev, F.M., Ablayev, M.F., Vasiliev, A.V.: Quantum Hashing: Effective Constructions. Lambert Academic Publishing, Saarbrücken, Germany (2023). in Russian
3. Ablayev, F.M., Vasiliev, A.V.: Quantum Hashing for Quantum Communications. LAMBERT Academic Publishing, Saarbrücken, Germany (2015). in Russian
4. Ablayev, F., Ablayev, M., Khadiev, K., Salihova, N., Vasiliev, A.: Quantum algorithms for string processing. In: Badriev, I.B., Banderov, V., Lapin, S.A. (eds.) Proceedings of the 13th International Conference on Mesh Methods for Boundary-Value Problems and Applications. LNCSE, vol. 141, pp. 1–14. Springer, Cham, Switzerland (2022). https://doi.org/10.1007/978-3-030-87809-2_1

5. Ablayev, F., Ablayev, M., Khadiev, K., Vasiliev, A.: Classical and quantum computations with restricted memory. In: Böckenhauer, H.J., Komm, D., Unger, W. (eds.) Adventures Between Lower Bounds and Higher Altitudes: Essays Dedicated to Juraj Hromkovič on the Occasion of His 60th Birthday, LNCS, vol. 11011, pp. 129–155. Springer, Cham, Switzerland (2018). https://doi.org/10.1007/978-3-319-98355-4_9

6. Ablayev, F., Khadiev, K., Vasiliev, A., Ziiatdinov, M.: Theory and applications of quantum hashing. Quantum Rep. **7**(2), 24 (2025). https://doi.org/10.3390/quantum7020024

7. Ablayev, F., Salikhova, N., Ablayev, M.: Hybrid classical–quantum text search based on hashing. Mathematics **12**(12), 1858 (2024). https://doi.org/10.3390/math12121858

8. Ablayev, F.M., Vasiliev, A.V.: Cryptographic quantum hashing. Laser Phys. Lett. **11**(2), 025202 (2013). https://doi.org/10.1088/1612-2011/11/2/025202

9. Ambainis, A., Freivalds, R.: 1-way quantum finite automata: Strengths, weaknesses and generalizations. In: Proceedings of the 39th Annual Symposium on Foundations of Computer Science (FOCS'98), pp. 332–341. IEEE Computer Society (1998). https://doi.org/10.1109/SFCS.1998.743469

10. Ambainis, A., Nahimovs, N.: Improved constructions of quantum automata. Theor. Comput. Sci. **410**(20), 1916–1922 (2009). https://doi.org/10.1016/j.tcs.2009.01.027

11. Barenco, A., et al.: Elementary gates for quantum computation. Phys. Rev. A **52**(5), 3457–3467 (1995). https://doi.org/10.1103/PhysRevA.52.3457

12. Buhrman, H., Cleve, R., Watrous, J., de Wolf, R.: Quantum fingerprinting. Phys. Rev. Lett. **87**(16), 167902 (2001). https://doi.org/10.1103/PhysRevLett.87.167902

13. DiVincenzo, D.P.: Two-bit gates are universal for quantum computation. Phys. Rev. A **51**(2), 1015–1022 (1995). https://doi.org/10.1103/PhysRevA.51.1015

14. Gainutdinova, A., Yakaryılmaz, A.: Unary probabilistic and quantum automata on promise problems. Quantum Inf. Process. **17**(2), 28 (2018)

15. Gainutdinova, A., Yakaryılmaz, A.: Nondeterministic unitary obdds. In: Weil, P. (ed.) Computer Science - Theory and Applications - 12th International Computer Science Symposium in Russia, CSR 2017, Kazan, Russia, 8–12 June 2017, Proceedings. LNCS, vol. 10304, pp. 126–140. Springer, Cham (2017). https://doi.org/10.1007/978-3-319-58747-9_13

16. Hyyppä, E.: Reducing leakage in single-qubit gates on superconducting quantum processors using analytical control pulse envelopes, 23 September 2024. https://www.meetiqm.com/newsroom/blog/reducing-leakage-in-single-qubit-gates. Accessed 1 Oct 2024

17. Khadiev, K., Khadieva, A.: Two-way quantum and classical machines with small memory for online minimization problems. In: International Conference on Micro- and Nano-Electronics 2018. Proc. SPIE, vol. 11022, p. 110222T (2019). https://doi.org/10.1117/12.2522462

18. Khadiev, K., Khadieva, A.: Quantum online streaming algorithms with logarithmic memory. Int. J. Theor. Phys. **60**(2), 608–616 (2019). https://doi.org/10.1007/s10773-019-04209-1

19. Khadiev, K., Khadieva, A.: Quantum and classical log-bounded automata for the online disjointness problem. Mathematics **10**(1), 143 (2022). https://doi.org/10.3390/math10010143

20. Khadiev, K., Khadieva, A., Knop, A.: Exponential separation between quantum and classical ordered binary decision diagrams, reordering method and hierarchies.

Nat. Comput. **22**(4), 723–736 (2023). https://doi.org/10.1007/s11047-022-09904-3

21. Khadiev, K., et al.: Two-way and one-way quantum and classical automata with advice for online minimization problems. Theor. Comput. Sci. (2022). https://doi.org/10.1016/j.tcs.2022.02.026

22. Khadieva, A., Salehi, Ö., Yakaryılmaz, A.: A representative framework for implementing quantum finite automata on real devices. In: Cho, D.J., Kim, J. (eds.) Proceedings of the 21st International Conference on Unconventional Computation and Natural Computation (UCNC 2024), Pohang, South Korea, 17–21 June 2024. LNCS, vol. 14776, pp. 163–177. Springer, Cham (2024). https://doi.org/10.1007/978-3-031-63742-1_12

23. Koczor, B., Morton, J.J.L., Benjamin, S.C.: Probabilistic interpolation of quantum rotation angles. Phys. Rev. Lett. **132**(13), 130602 (2024). https://doi.org/10.1103/PhysRevLett.132.130602

24. Le Gall, F.: Exponential separation of quantum and classical online space complexity. Theory Comput. Syst. **45**(2), 188–202 (2009). https://doi.org/10.1007/s00224-007-9097-3

25. Möttönen, M., Vartiainen, J.J.: Decompositions of general quantum gates. In: Shannon, S. (ed.) Trends in Quantum Computing Research, pp. 149–172. Nova Science Publishers, New York, USA (2006)

26. Möttönen, M., Vartiainen, J.J., Bergholm, V., Salomaa, M.M.: Quantum circuits for general multiqubit gates. Phys. Rev. Lett. **93**(13), 130502 (2004). https://doi.org/10.1103/PhysRevLett.93.130502

27. Nielsen, M.A., Chuang, I.L.: Quantum Computation and Quantum Information: 10th Anniversary Edition. Cambridge University Press, Cambridge, UK (2010)

28. Preskill, J.: Quantum computing in the NISQ era and beyond. Quantum **2**, 79 (2018). https://doi.org/10.22331/q-2018-08-06-79

29. Vasiliev, A.: A model of quantum communication device for quantum hashing. J. Phys: Conf. Ser. **681**(1), 012020 (2016). https://doi.org/10.1088/1742-6596/681/1/012020

30. Vasiliev, A.: Quantum hashing for finite abelian groups. Lobachevskii J. Math. **37**(6), 753–757 (2016). https://doi.org/10.1134/S1995080216060184

31. Vasiliev, A.: Constant-depth algorithm for quantum hashing. Russ. Microlectron. **52**(Suppl. 1), S399–S402 (2023). https://doi.org/10.1134/S106373972360067X

32. Ziiatdinov, M., Khadieva, A., Khadiev, K.: Shallow implementation of quantum fingerprinting with application to quantum finite automata. Front. Comput. Sci. **7**, 1519212 (2025). https://doi.org/10.3389/fcomp.2025.1519212

33. Ziiatdinov, M., Khadieva, A., Yakaryılmaz, A.: Gaps for shallow implementation of quantum finite automata. In: Proceedings of the 16th International Conference on Automata and Formal Languages (AFL 2023), pp. 269–280. Electronic Proceedings in Theoretical Computer Science (2023). https://doi.org/10.4204/EPTCS.386.21

34. Zinnatullin, I., Khadiev, K., Khadieva, A.: Efficient implementation of amplitude form of quantum hashing using state-of-the-art quantum processors. Russ. Microlectron. **52**(1), S390–S394 (2023). https://doi.org/10.1134/S1063739723600620

35. Zinnatullin, I., Khadiev, K.: Efficient algorithms for quantum hashing. https://arxiv.org/abs/2507.07002 (2025). arXiv:2507.07002 [quant-ph]

Reachability in Interactive Chemical Reaction Networks

Aberto Avila-Jimenez[(✉)], Bin Fu, Elise Grizzell, Robert Schweller,
and Tim Wylie

University of Texas Rio Grande Valley, Edinburg, USA
`alberto.avilajimenez01@utrgv.edu`

Abstract. This paper studies the effects of interactivity on molecular
computation, specifically in the Step Chemical Reaction Networks model
(Step CRNs), by adding the ability for a user to interact with the system
by selecting which species to add at each step, or by having some control
over which reactions execute. The two proposed variants are Interactive
CRNs and Randomized Interactive CRNs. We show that in Interactive
CRNs, even when restricted to *void* (deletion-only) rules of relatively
small size, if a user can decide which species to add at each step based on
the configuration, reachability is PSPACE-complete when bounded and
EXPTIME-hard when unbounded. In Randomized Interactive CRNs, we
prove that reachability with void rules is PSPACE-complete.

1 Introduction

Molecular computing is an important topic that has generated a lot of interest
and research within the past few decades. This has led to a host of theoretical
models based on different laboratory techniques such as DNA Tiles and Origami.
A popular model based solely on the interaction of species is Chemical Reaction
Networks [4], which is fundamental in describing distributed processes and is
equivalent to both Vector Addition Systems [12] and Petri-nets [14].

In general, many molecular theoretical models, including Chemical Reaction
Networks (CRNs), are not designed to allow for the addition of more inputs
later. Instead, most work under the operating assumption of designing a computation for the model and letting it run until complete with no intermediate
interaction. Obviously, this is limiting for many computational tasks and would
require additional systems to run experiments under different parameters.

The Step Chemical Reaction Network (Step CRN) model was introduced
to address this limitation by allowing staged addition of chemicals [2,3]. It was
shown that even with void rules, threshold circuits can be simulated. The model,
however, still requires all added species for every step to be fixed as input, and
thus does not allow flexible interaction with the computation.

This research was supported in part by National Science Foundation Grant CCF-
2329918.

© The Author(s), under exclusive license to Springer Nature Switzerland AG 2026
E. Formenti and L. Manzoni (Eds.): UCNC 2025, LNCS 16364, pp. 323–343, 2026.
https://doi.org/10.1007/978-3-032-15641-9_22

In this paper, we investigate interaction with molecular computation in two generalizations of the Step CRN. First, we explore reachability when a user is allowed to choose which sets of species to add at each step. These Interactive Chemical Reaction Networks, even with void rules, are extremely powerful. Then, we include randomization to the interaction process. The Randomized Interactive CRNs introduce randomized elements by allowing random selection of species additions, a significant step toward realistic biochemical computation scenarios. We demonstrate that reachability remains PSPACE-hard within this randomized context.

1.1 Previous Work

CRNs and Step CRNs. Chemical Reaction Networks (CRNs) are a well-studied field of theoretical chemistry. CRNs abstract chemical interactions through chemical *species* and a set of reaction *rules* that dictate how the species interact. Models that are similar to CRNs include Vector Addition Systems (VASs) [12] and Petri-nets [14]. The traditional CRN model studies an environment in which all chemicals that will be used in the system are added to a single container simultaneously. Step Chemical Reaction Networks [2] are a natural extension of CRNs, aligning more closely with actual laboratory procedures. Step CRNs with void rules have been used to simulate threshold circuits [2] and study reachability questions [9], proving NP-completeness for traditional Step CRNs.

Void Rules. Reachability with void rules in CRNs was first studied in [1]. Previous studies have included void rules as a part of their systems; it is even possible to program such rules in CRN++, but they were never studied exclusively. Void rules can also be considered a special case of reaction extinction [18].

Mixing Systems. Another generalization of CRNs related to the step model is I/O CRNs [8], where additional inputs can be added at timed intervals. Still, those inputs are read-only in the system (used exclusively as catalysts). Step CRNs generalize I/O CRNs as the inputs are not read-only and are rate-independent, unlike I/O CRNs. Staged systems have been explored in many self-assembly models [5–7,13].

1.2 Our Contributions

This paper introduces two laboratory-motivated generalizations of the Step CRN: the Interactive CRN model and the Randomized Interactive CRN model. We start in Sect. 2 by defining these two models and the reachability problem. In Sect. 3, we first show PSPACE membership of reachability in Interactive CRNs. Then, in Sect. 4, we present two different reductions from 2-Player Constraint Logic to prove hardness for the reachability problem. Later in Sect. 5, we show the hardness of reachability in the Randomized Interactive CRN by simulating Interactive Proofs. Table 1 gives an overview of the main results.

Table 1. Summary of results: reachability hardness in two different Step CRN generalization models using only void rules.

Interactive CRNs		
Void Rules	Complexity	Ref.
(10,8)	PSPACE-Complete	Thm. 2
(13,10)	EXPTIME-hard	Thm. 3
Randomized Interactive CRNs		
Void Rules	Complexity	Ref.
(2,0)	PSPACE-hard	Thm. 4
(3,0)	PSPACE-hard	Thm. 5

2 Preliminaries

2.1 Chemical Reaction Networks

Let $\Lambda = \{\lambda_1, \lambda_2, \ldots, \lambda_{|\Lambda|}\}$ denote some ordered alphabet of *species*. A configuration over Λ is a length-$|\Lambda|$ vector of non-negative integers that denotes the number of copies of each present species. A *rule* or *reaction* has two multisets, the first containing one or more *reactant* (species), used for creating the resulting *product* (species) contained in the second multiset. Each rule is represented as an ordered pair of configuration vectors $R = (\overrightarrow{R_r}, \overrightarrow{R_p})$. $\overrightarrow{R_r}$ contains the minimum counts of each reactant species necessary for reaction R to occur, where reactant species are either *consumed* by the rule in some count or leveraged as *catalysts* (not consumed); in some cases a combination of the two. The product vector $\overrightarrow{R_p}$ has the count of each species *produced* by the *application* of rule R, effectively replacing vector $\overrightarrow{R_r}$. The species corresponding to the non-zero elements of $\overrightarrow{R_r}$ and $\overrightarrow{R_p}$ are termed *reactants* and *products* of R, respectively.

The *application* vector of R is $\overrightarrow{R_a} = \overrightarrow{R_p} - \overrightarrow{R_r}$, which shows the net change in species counts after applying rule R once. For a configuration $\overrightarrow{C}$ and rule R, we say R is applicable to $\overrightarrow{C}$ if the count of species i in configuration $\overrightarrow{C}$, $\overrightarrow{C}[i] \geq \overrightarrow{R_r}[i]$ for all $1 \leq i \leq |\Lambda|$, and we define the *application* of R to $\overrightarrow{C}$ as the configuration $\overrightarrow{C}' = \overrightarrow{C} + \overrightarrow{R_a}$. For a set of rules Γ, a configuration $\overrightarrow{C}$, and rule $R \in \Gamma$ applicable to $\overrightarrow{C}$ that produces $\overrightarrow{C} = \overrightarrow{C} + \overrightarrow{R_a}$, we say $\overrightarrow{C} \rightarrow^1_\Gamma \overrightarrow{C}'$, a relation denoting that $\overrightarrow{C}$ can transition to $\overrightarrow{C}'$ by way of a single rule application from Γ. We further use the notation $\overrightarrow{C} \rightarrow^*_\Gamma \overrightarrow{C}'$ to signify the transitive closure of $\rightarrow^1_\Gamma$ and say $\overrightarrow{C}'$ is *reachable* from $\overrightarrow{C}$ under Γ, i.e., $\overrightarrow{C}'$ can be reached by applying a sequence of applicable rules from Γ to the initial configuration $\overrightarrow{C}$. A configuration is *terminal* if no applicable rules exist. Here, we use the following notation to depict a rule $R = (\overrightarrow{R_r}, \overrightarrow{R_p})$: $\sum_{i=1}^{|\Lambda|} \overrightarrow{R_r}[i]s_i \rightarrow \sum_{i=1}^{|\Lambda|} \overrightarrow{R_p}[i]s_i$.

Using this notation, a rule turning two copies of species H and one copy of species O into one copy of species W would be written as $2H + O \rightarrow W$.

Definition 1 (Discrete Chemical Reaction Network). *A discrete chemical reaction network (CRN) is an ordered pair (Λ, Γ) where Λ is an ordered alphabet of species, and Γ is a set of rules over Λ.*

Definition 2 (Void rules). *A rule $R = (\overrightarrow{R_r}, \overrightarrow{R_p})$ is a void rule if $\overrightarrow{R_a} = \overrightarrow{R_p} - \overrightarrow{R_r}$ has no positive entries and at least one negative entry. A void rule is a rule whose product multiset is a strict sub-multiset of the reactant. In catalytic void rules, such as $(2,1)$ rules, one or more reactants remain, and one or more is deleted after the rule is applied.*

Definition 3. *The size/volume of a configuration vector $\overrightarrow{C}$ is $\mathtt{volume}(\overrightarrow{C}) = \sum \overrightarrow{C}[i]$.*

Definition 4 (size-(i,j) rules). *A rule $R = (\overrightarrow{R_r}, \overrightarrow{R_p})$ is said to be a size-(i,j) rule if $(i,j) = (\mathtt{volume}(\overrightarrow{R_r}), \mathtt{volume}(\overrightarrow{R_p}))$.*

2.2   Interactive CRNs

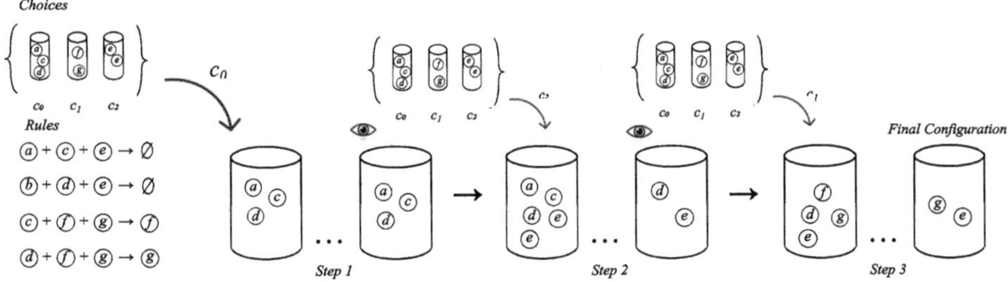

Fig. 1. An example Interactive CRN system. c_j is added once all the possible rules have been applied (i.e., the CRN has reached a terminal configuration).

An Interactive CRN is a generalization of a Step CRN. Step CRNs define a number of copies of each species to add at each step. This can be viewed as a single "choice". Interactive CRNs offer a fixed set of k choices $\{\overrightarrow{c_0}, \overrightarrow{c_1} \ldots, \overrightarrow{c_{k-1}}\}$, where, like Step CRNs, each $\overrightarrow{c_j}$ is a sequence of length-$|\Lambda|$ vectors of non-negative integers denoting how many copies of each species type are present within that choice. In this model, whenever a reaction from Γ applies to the current configuration $\overrightarrow{A}$, the system executes that reaction in the usual fashion, producing $\overrightarrow{B} = \overrightarrow{A} - \overrightarrow{R_r} + \overrightarrow{R_p}$. When the configuration is terminal (no reaction can fire), the experimenter observes the current state of the system and selects one of the choices to add. The process of adding $\overrightarrow{c_j}$ and waiting for the CRN to reach a terminal configuration is denoted as a step. See Fig. 1.

Definition 5 (Interactive CRN). *An Interactive CRN is a tuple* $C_{\text{int}} = ((\Lambda, \Gamma), \{\vec{c_0}, \vec{c_1} \ldots, \vec{c}_{k-1}\})$, *where* Λ *is a finite species set,* $\Gamma \subseteq \mathbb{N}^\Lambda \times \mathbb{N}^\Lambda$ *is the reaction set, and* $\{\vec{c_0}, \vec{c_1} \ldots, \vec{c}_{k-1}\}$ *is a finite set of vectors* $\vec{c_j} \in \mathbb{N}^\Lambda$. *The one-step transition* $\xrightarrow{C_{\text{int}}}$ *on configurations* $\vec{A}, \vec{B} \in \mathbb{N}^\Lambda$ *is defined by* $\vec{A} \xrightarrow{C_{\text{int}}} \vec{B}$ *if there exists* $(\vec{R_r}, \vec{R_p}) \in \Gamma$ *with* $\vec{R_r} \leq \vec{A}$ *and* $\vec{B} = \vec{A} - \vec{R_r} + \vec{R_p}$, *or, when no reaction in* Γ *applies to* $\vec{A}$, *if there is some* $\vec{c_j} \in \{\vec{c_0}, \vec{c_1} \ldots, \vec{c}_{k-1}\}$ *with* $\vec{B} = \vec{A} + \vec{c_j}$.

Note that a *step* is counted only when the terminal configuration is different from the initial. If a choice vector is added and subsequent the reactions yield the same configuration, it is a *null* step and does not increase step count.

Given an Interactive CRN, we define all configurations that are terminal as $TERM$. A *strategy* is a polynomial time computable function $\sigma(\vec{X}) \in \{\vec{c_0}, \vec{c_1} \ldots, \vec{c}_{k-1}\}, \forall \vec{X} \in TERM$ that computes which $\vec{c_j}$ to add given a terminal configuration $\vec{X}$. Let $REACH_1[\sigma]$ be the set of reachable configurations of C_{int} with initial configuration $\vec{A}$ at step 1 plus $\sigma(\vec{A})$, and let $TERM_1[\sigma]$ be the subset of reachable configurations that are terminal. Define $REACH_2[\sigma]$ to be the union of all reachable configurations from each possible starting configuration $\vec{C} \in TERM_1[\sigma]$ plus $\sigma(\vec{C})$. Let $TERM_2[\sigma]$ be the subset of the configurations in $REACH_2[\sigma]$ that are terminal. Similarly, define $REACH_i[\sigma]$ to be the union of all reachable sets attained by using some terminal configuration $\vec{X} \in TERM_{i-1}[\sigma]$ plus $\sigma(\vec{X})$, and let $TERM_i[\sigma]$ denote the subset of these configurations that are terminal.

Definition 6 (Interactive CRN Forced Reachability problem). *Let* m *be an integer. Given start configuration* $\vec{A}$, *target configuration* $\vec{B}$, *and* $C_{\text{int}} = ((\Lambda, \Gamma), \{\vec{c_0}, \vec{c_1} \ldots, \vec{c}_{k-1}\})$, *determine whether there exists a strategy* σ *that ensures* $TERM_m[\sigma] = \{\vec{B}\}$. *In short, is there a strategy* σ *that picks which choice* c_j *to add at every step and guarantees* $\vec{B}$ *is the only terminal configuration reachable after following* σ *for* m *steps?*

3   PSPACE Membership of Interactive CRNs

Theorem 1. *Let* $C_{\text{int}} = ((\Lambda, \Gamma), \{\vec{c_0}, \vec{c_1} \ldots, \vec{c}_{k-1}\})$ *be an Interactive CRN. Then the Forced Interactive reachability problem for* $TERM_m[\sigma]$ *is in space* $(m + v_0 + \max_i\{|c_i|\})^{O(1)}$, *where initial volume is* v_0 *and* $|c_i|$ *is the sum of elements in the vector* c_i.

Proof. We describe a polynomial-space algorithm based on a depth-first search traversal of the configuration tree. Consider a start configuration $\vec{A}$, a target $\vec{B}$ and some integer m. From $\vec{A}$, the system evolves by applying reactions or by adding some $\vec{c_j}$ chosen by σ. Each step adds a bounded number of species, The volume is bounded by $v_0 + m \times \max_i\{|c_i|\}$ where v_0 is the initial volume, and

c_i is the sum of elements in $\vec{c_j}$. Since we only use void rules, all sequences of rule applications can happen in polynomial time. We construct a configuration tree with polynomial depth bounded by the step count m and the volume v. Each node represents a configuration, and every node has a child node for a $\vec{c_j}$ addition or a rule application. We now run depth-first search on this tree, we store only the current configuration using polynomial space. The algorithm checks if $\vec{B}$ is the only terminal configuration after exactly m steps. Since depth and volume are polynomially bounded, we have a polynomial space algorithm.

4   Hardness for Interactive CRNs

In this section, we prove hardness for the Interactive CRN Forced Reachability problem using void rules by reductions from bounded and unbounded two-player constraint logic (2CL). We interpret the IA-CRN dynamics as a game between the experimenter and the CRN. Bounded and unbounded 2CL using AND, OR, CHOICE, FANOUT, and VARIABLE vertices were shown to be PSPACE-complete (bounded) and EXPTIME-hard (unbounded) by Hearn in 2006 [10].

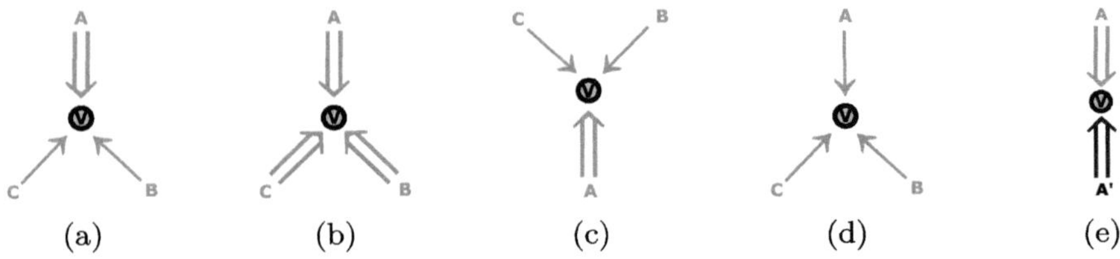

Fig. 2. Bounded 2 player constraint logic vertex variations. Left to right: (a) AND vertex (b) OR vertex (c) FANOUT vertex (d) CHOICE vertex (e) VARIABLE vertex.

4.1   Bounded Systems

Bounded 2CL. An instance of Bounded 2CL, defined by [10], consists of a directed graph $G(V, E)$ with edge weights $\in \{1, 2\}$, vertices of degree 3, and a minimum inflow constraint of 2 for each vertex. Graphs are constructed using the vertex gadgets detailed below and shown in Fig. 2. Players alternate flipping edges on their respective vertices while maintaining vertex inflow constraints. The only vertices where both players have edges that they may flip are the variable vertices. Every edge can be flipped at most once. The decision problem is: given a target edge $e \in E$, does a sequence of legal edge flips exist allowing player 1 to flip edge e?

Vertex gadgets. Bounded 2CL uses five degree-3 vertex types (Fig. 2). *AND*: Fig. 2a. One weight-2 edge and two weight-1 edges; the weight-2 edge can flip outward only when both weight-1 edges point inward. *OR*: Fig. 2b. Three weight-2 edges; any inward-pointing edge satisfies the inflow constraint, allowing either of the other two to flip. *FANOUT*: Fig. 2c. One weight-2 "input" edge and two weight-1 "output" edges; with the input edge directed inward, both outputs may point outward simultaneously. *CHOICE*: Fig. 2d. Three weight-1 edges; at most one may point outward at any time. *VARIABLE*: Fig. 2e. The only vertex where both players influence the inflow of the. It consists of two weight-2 edges, one controlled by each player; whichever edge flips first locks the other in place, since flipping both would violate the inflow constraint.

We transform an instance of 2CL into an instance of Interactive CRN Forced reachability. An edge reversal, or flip, for both players is represented by a single rule where the experimenter (white) adds the species representing the desired flip. If it is a valid move, the rule flips that edge, and the system (black) nondeterministically selects a valid response. The experimenter's objective is to reach a configuration where the species corresponding to the winning edge indicates it has been reversed.

Theorem 2. *The Interactive CRN Forced Reachability Problem is PSPACE-complete using only void rules of at most size (10,8) with a polynomially bounded number of steps.*

Proof. Given a constraint graph $G = (V, E)$ with edges partitioned into black (B) and white (W), we encode the instance into an IA-CRN as follows. Each edge is represented as a species named by its endpoints: an edge from vertex A to vertex B becomes species AB, while its flipped version is $\overline{AB}$. Weight-two edges include a superscript and each edge species has a subscript indicating which player it belongs to (see Fig. 3a). At every step, the white player adds exactly one choice $\overrightarrow{c_j} \in \{\overrightarrow{c_0}, \ldots, \overrightarrow{c_{k-1}}\}$, where $k = |W|$. Each $\overrightarrow{c_j}$ consists of exactly one flipped white edge (the edge the white player attempts to reverse), one copy of every flipped black edge, and $|B| - 1$ copies of a cleanup species DEL_b. The idea is that when a choice is added, the flipped white edge species will consume the unflipped version. Similarly, one out of all of the flipped black edge species will consume the unflipped version. The rest of flipped black edge species that were added will be cleaned up by the cleanup species DEL_b after the reaction representing the edge reversals executes.

We construct the rules Γ as follows. Consider a white edge $e_w \in W$ directed into vertex A. To determine if e_w can be flipped (reversed), we first identify all other edges surrounding A. The different combinations of flipped and unflipped versions of these surrounding edges form local configurations $\overrightarrow{C_{e_w}}$. However, we only retain local configurations that satisfy the constraint when e_w is flipped. For example, in Fig. 2a, if edge AV^2 is flipping away from vertex V, and the other edges directed a V are CV and BV, then the local configurations are $\{(CV, BV), (\overline{CV}, BV), (CV, \overline{BV}), (\overline{CV}, \overline{BV})\}$. Since these edges are weight-1, only the configuration $\overrightarrow{C_{AV}} = \{(CV, BV)\}$ would be retained, since this is the

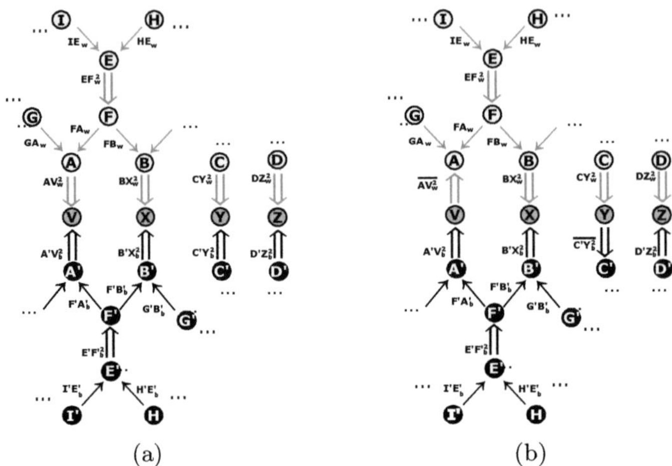

Fig. 3. (a) A fragment of a constraint graph and the species used to represent its initial state. (b) The same graph after one rule application, where player 1 selects vertex V and player 2 selects vertex Y. Edge EF_w^2 cannot be reversed unless white player selects vertices V and X first. Ellipsis ($\dots$) imply omitted portions of the full graph.

only scenario where reversing BA mantains the constraint. Similarly, for each black edge $e_b \in B$, we construct local configurations $\overrightarrow{C_{e_b}}$, retaining only valid configurations that satisfy vertex constraints. Every local configuration $\overrightarrow{w} \in \overrightarrow{C_{e_w}}$ and $\overrightarrow{b} \in \overrightarrow{C_{e_b}}$ will also include the species of the edge that is being reversed and its reversed version as a prefix. Finally, we form rules by combining each valid local configuration $\overrightarrow{w}$ for white edges with each valid configuration $\overrightarrow{b}$ for black edges, provided the moves remain valid when the white player's flip occurs first. Each rule γ_i thus has reactants $\overrightarrow{w} + \overrightarrow{b}$. The unflipped edge species are consumed, so the rest of the reactants are used as catalysts and remain as the product. An example reaction is shown in Fig. 4.

We show our rule construction explicitly using the VARIABLE and AND vertices. The OR, CHOICE, and FANOUT vertices follow similarly by modifying inflow conditions. Full details for these vertices are straightforward and can be found in the appendix. Vertex labels and edge directions correspond to those in Fig. 3.

VARIABLE vertex. See VARIABLE vertex V in Fig. 3a. Initially, AV_w^2 and $A'V_b^2$ are present. For the rule flip AV_w^2, $\overrightarrow{w}$ will have AV_w^2, $\overline{AV_w^2}$ and $A'V_b^2$. These represent the edge that will be flipped, the flipped version of that edge, and the inflow check to vertex V. The second half $\overrightarrow{b}$ will include the same for another variable edge. In a valid flip, the only element consumed will be the species representing the original edges that were flipped by white and black. The complete $(6, 4)$ void reaction for a turn on a VARIABLE vertex is shown in Fig. 4

$$\overbrace{AV_w^2 + \overline{AV_w^2} + A'V_b^2 + C'Y_b^2 + \overline{C'Y_b^2} + CY_w^2}^{\vec{w}} \rightarrow \overline{AV_w^2} + A'V_b^2 + \overline{C'Y_b^2} + CY_w^2$$

with $\underbrace{C'Y_b^2 + \overline{C'Y_b^2} + CY_w^2}_{\vec{b}}$

Fig. 4. $(6,4)$ void rule to play a turn on a VARIABLE vertex. Player 1 selects vertex V and reverses edge AV_w^2. Player 2 selects Vertex Y and reverses edge $C'Y_b^2$. The graph after this turn is shown in Fig. 3b.

AND vertex. Let F be an AND vertex with single edges FA_w, FB_W and double edge EF_w^2. To reverse EF_w^2, we have local configurations $\overrightarrow{C_{EF_w^2}} = \{(\overline{FA_w}, \overline{FB_w})\}$ that make up $\vec{w}$, the first half of the rule. The second half $\vec{b}$ consists of a valid move for Player 2. For simplicity in this example, we show a mirrored reversal of the player 1 move in a different vertex F'. Two copies of any reversed edge of the black player are needed in the reactantsto avoid the black player's reversed edge species added in $\vec{c_j}$ from triggering this reaction in an invalid configuration. It is important to note that after the VARIABLE vertex selections, black moves are irrelevant. We still account for them to preserve turn order. The completed $(10,8)$ void reaction for a turn on an AND vertex is as follows: Player 1 reverses edge EF_w^2. Player 2 reverses edge $E'F_b'^2$,

$$EF_w^2 + \overline{EF_w^2} + \overline{FA_w} + \overline{FB_w} + E'F_b'^2 + \overline{E'F_b'^2} + 2\overline{F'A_b'} + 2\overline{F'B_b'} \rightarrow$$
$$\overline{EF_w^2} + \overline{FA_w} + \overline{FB_w} + \overline{E'F_b'^2} + 2\overline{F'A_b'} + 2\overline{F'B_b'}$$

After a reaction representing a turn executes, the reversed black edge species added in $\vec{c_j}$ will still be in the system. The species DEL_b then cleans up those leftover black edge species, leaving the system with only the species representing the current state of the graph. These rules are:

$$(1) \ 2\overline{A'V_b^2} + DEL_b \rightarrow \overline{A'V_b^2}$$
$$(2) \ A'V_b^2 + \overline{A'V_b^2} + DEL_b \rightarrow A'V_b^2$$
$$(3) \ BX_w^2 + \overline{BX_w^2} \rightarrow BX_w^2, (4) \ 2\overline{BX_w^2} \rightarrow \overline{BX_w^2}$$

First, a $(3,1)$ void rule to clean up unused black player flipped edges. This rule consumes one of two reversed black edge species for the case when the current state of the graph has $A'V_b^2$ in reversed state. Second is a $(2,1)$ void rule for the case where the current state of the graph has $A'V_b^2$ in its initial state. Rules (3) and (4) will delete a white move and return the system to its original state before the choice. (3) takes the case for an invalid white flip. (4) is for the case of trying to flip an already flipped edge.

Using this construction, the IA-CRN reaches a terminal configuration containing the species corresponding to the reversed winning edge if and only if the white player has a winning strategy in the constraint logic game. This establishes the Interactive CRN Forced Reachability problem is PSPACE-hard. By Theorem 1, it follows that the problem is PSPACE-complete.

4.2 Unbounded Systems

In unbounded two-player constraint logic, edges may be flipped and unflipped indefinitely. The graph representation remains similar to the bounded case, but each black player edge is tracked using two species to represent its current direction. To allow reversibility, both flipped and unflipped versions of black edges are included in each choice vector. As a result, rule size increases, with valid rules deleting both versions of a black edge, reaching a maximum size of $(17, 6)$.

Unbounded 2CL. An instance of unbounded 2-player Constraint Logic [10] is similar to that in Sect. 4.1, consisting of a directed graph with edge weights in $1, 2$ and degree-3 vertices enforcing minimum inflow constraints. Unlike the bounded case, edges may be flipped multiple times. While the reduction is less direct, it reuses the vertex gadgets from Sect. 4.1, with modifications, particularly to the AND vertex, which now accommodates mixed white and black edges, as shown in Fig. 5.

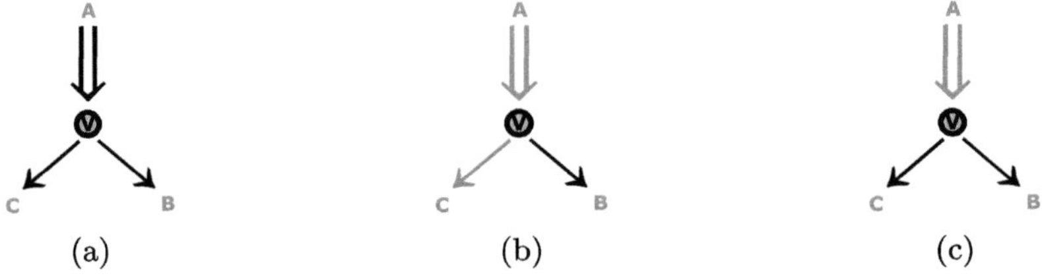

Fig. 5. Unbounded 2-player constraint logic vertex variations. Left to right: (a) Black player AND vertex (b) Mixed AND vertex case 1 (c) Mixed AND vertex case 2.

Theorem 3. *The Interactive CRN reachability problem is EXPTIME-hard, even when restricted to void rules of at most size (13,10) and allowing an unbounded number of steps.*

Proof. Given a constraint graph $G = (V, E)$ with E partitioned into B and W, we encode the instance into the IA-CRN the same as Theorem 2. After every step, the white player adds exactly one choice $\vec{c_j} \in \{\vec{c_0}, \ldots, \vec{c_{k-1}}\}$, where $k = 2|W|$. one for every e_w and one for its reverse $\overline{e_w}$. Each $\vec{c_j}$ consists of exactly one white edge, one copy of every flipped black edge, one copy of every reversed black edge, and $|B| - 1$ copies of a cleanup species DEL_b. The rules are constructed the similar to Theorem 2. They keep a similar reactant structure of $\vec{w} + \vec{b}$. Note that now, they can have mixed edges, so $\vec{w}$ wont be exclusively white edges and the same for $\vec{b}$. For every black edge used as a validation catalyst in $\vec{w}$ and $\vec{b}$ will need 2 copies in its reactants. Having two of every species for every black edge ensures the rule matches the current state (one copy already in the configuration and one from the choice), and prevents the extra black edges in $\vec{c_j}$ from serving as catalysts. The rule keeps the flipped orientations and the existing DEL_b cleanup reactions then remove these duplicates, restoring a single valid edge species.

We explicitly show the unbounded construction with two representative gadgets. First, we revisit the AND vertex used in Theorem 2 and show how a weight-2 white edge can be flipped forward and back under our new "two-copy" convention. Then, we present the mixed-edge AND vertex in Fig. 5c. This case produces the largest rule in the reduction. All other mixed-edge variants from Fig. 5 can be found in the appendix.

Black AND Vertex. The rules for a black AND vertex follow almost the same as the AND vertex in Theorem 2. The only change is that the current state of the black edge that is being flipped requires 2 copies in the reactants. In the following rule, white is flipping edge EF_w^2 and black player is flipping $E'F_b'^2$. Note that now, two copies of $E'F_b'^2$ are required to verify the current state of the edge. See vertex F and F' in Fig. 3 for reference.

$$EF_w^2 + \overline{EF_w^2} + \overline{FA_w} + \overline{FB_w} + 2E'F_b'^2 + \overline{E'F_b'^2} + 2\overline{F'A_b'} + 2\overline{F'B_b'} \rightarrow$$
$$\overline{EF_w^2} + \overline{FA_w} + \overline{FB_w} + \overline{E'F_b'^2} + 2\overline{F'A_b'} + 2\overline{F'B_b'}$$

Rules to reverse an edge again are symmetrical. To reverse edge $\overline{EF_w^2}$ and $\overline{E'F_b'^2}$ again, we must now ensure the valid local configurations of vertices E and E' are present. These now include edges $IE_w, HE_w, I'E_b', H'E_b'$. The $(11,8)$ reaction is the following:

$$\overline{EF_w^2} + EF_w^2 + IE_w + HE_w + 2\overline{E'F_b'^2} + E'F_b'^2 + 2I'E_b' + 2H'E_b' \rightarrow$$
$$EF_w^2 + IE_w + HE_w + E'F_b'^2 + 2I'E_b' + 2H'E_b'$$

Mixed AND Vertex. The largest rule in the reduction is triggered in the scenario where both players attempt to flip edges from an AND vertex with two black single-weight edges and one white weight-2 edge. In this case, all of the validation catalysts require to have a count of two since they are all black edges. Let V be an AND vertex with edges AV_w^2, VC_b, VB_b Let X be a black AND vertex with edges $A'X_b^2, VC_b', VB_b'$. A $(13,10)$ rule to flip both $AV_w^2, A'X_b^2$ in a single turn is as follows: Player 1 reverses edge AV_w^2. Player 2 reverses edge $A'X_b^2$.

$$AV_w^2 + \overline{AV_w^2} + 2\overline{VC_b} + 2\overline{VB_b} + 2A'X_b^2 + \overline{A'X_b^2} + 2\overline{VC_b'} + 2\overline{VB_b'} \rightarrow$$
$$\overline{AV_w^2} + 2\overline{VC_b} + 2\overline{VB_b} + \overline{A'X_b^2} + 2\overline{VC_b'} + 2\overline{VB_b'}+$$

Similar to Theorem 2, after each turn rule executes the system enters a cleanup phase that uses DEL_b species to delete every unused edge left by the choice vector:

$$(1)\ 2\,\overline{A'V_b^2} + A'V_b^2 + DEL_b \ \rightarrow\ \overline{A'V_b^2} \quad (2)\ 2\,A'V_b^2 + \overline{A'V_b^2} + DEL_b \ \rightarrow\ A'V_b^2$$
$$(3)\ BX_w^2 + \overline{BX_w^2} + \overrightarrow{w} \ \rightarrow\ BX_w^2 + \overrightarrow{w} \quad (4)\ BX_w^2 + \overline{BX_w^2} + \overrightarrow{\overline{w}} \ \rightarrow\ \overline{BX_w^2} + \overrightarrow{\overline{w}}$$
$$(5)\ 2\,BX_w^2 \ \rightarrow\ BX_w^2, \ \ 2\overline{BX_w^2} \ \rightarrow\ \overline{BX_w^2}$$

(1) and (2) are $(3,1)$ void reactions that discard the unused black edges introduced by the choice vector. Each consumes one DEL_b and preserves the edge's current state. (3) and (4) discard an invalid white edge flip. Recall $\vec{w}$ (respectively $\overrightarrow{\overline{w}}$) represents a valid local configuration, so the stray copy is deleted while the valid orientation is retained. (5) handles the redundant case in which the choice vector adds a duplicate of the already-present white orientation.

With this construction, the IA-CRN reaches the target terminal configuration iff White has a winning strategy in the unbounded 2CL. Hence, Interactive CRN forced reachability with no bound in m is EXPTIME-hard.

5 Hardness for Randomized Interactive CRNs

In this section, we prove hardness for the reachability problem for Randomized Interactive CRNs defined in Definition 7 that allows some random number of copies of species to be added each step. Using this approach, we can transform the classical result IP=PSPACE to the PSPACE-hardness of the reachability problem of Randomized Interactive CRN with only $(2,0)$ rules.

Definition 7 (Randomized Interactive Step CRN). *A Randomized Interactive CRN is a tuple $C_{\mathrm{rand}} = \big((\Lambda, \Gamma), h, m, n\big)$, where Λ is a finite species set, $\Gamma \subseteq \mathbb{N}^\Lambda \times \mathbb{N}^\Lambda$ is the reaction set, m is the length of random bits in one step, n is the input-length for a problem (to solve with randomized interactive CRN). A series of interaction steps $S_0, R_0, S_1, R_1, \ldots, S_{k-1}, R_{k-1}, S_k$ with $k \leq h$ is allowed in CRN computation, where (1) each S_i is a species vector in $\mathbb{N}^\Lambda$ and depends on $S_0, R_0, \ldots, S_{i-1}, R_{i-1}$, and (2) R_i in $\{0,1\}^\Lambda$ encodes a sequence of m random bits. If A is the configuration right before a step R_i, it enters the configuration $B = A + R_i$, and then enters a terminal configuration A' after applying reactions.*

In Randomized Interactive CRNs, instead of adding a defined number of species at each step, the model allows the addition of species to encode some random bits. In this model, reactions fire in the usual fashion whenever they are enabled. The experimenter adds the next step whenever the system reaches a terminal configuration.

We define the reachability of Randomized Interactive CRNs below. A terminal configuration is a "Yes" configuration if it contains at least one copy of species "y".

Definition 8. *Let $C_{\mathrm{rand}} = \big((\Lambda, \Gamma), h, m, n\big)$ be a randomized interactive CRN. We define: A terminal configuration is "Yes" if it contains at least one copy of species "y".*

A configuration "Yes" is fully reachable by C_{rand} if for any $R_0, R_1, \ldots, R_{k-1}$, with $k \leq h$ there is a series of steps $S_0, R_0, S_1, R_1, \ldots, S_{k-1}, R_{k-1}, S_k$ and it enters "Yes" configuration.

A configuration "Yes" is highly unreachable by C_{rand} if a sequence of steps $S_0, R_0, S_1, R_1, \ldots$ does not enter Yes configuration with probability at least $3/4$.

The reachability problem for C_{rand} is to determine if Yes configuration is fully reachable.

The randomized interactive CRN C_{rand} is of a polynomial size if $|\Lambda| + |\Gamma| + h + m \leq p(n)$ for some polynomial $p(.)$.

5.1   Interactive Proof Systems

An IP consists of a Prover **P** (unbounded computational power) and a Verifier **V** (probabilistic polynomial-time machine) that exchange messages. The dialogue protocol proceeds as follows. First, P asserts a statement S. Then P and V engage in k rounds of interaction: in round i, P sends proof π_i and V responds with challenge c_i. Finally, V either accepts or rejects S based on the transcript $(\pi_1, c_1, \ldots, \pi_k, c_k)$. Shamir [16](later simplified by [17]) showed that IP = PSPACE by constructing an IP for the TQBF problem via arithmetization. The quantified boolean formula is converted to a polynomial P over some prime field $\mathbb{F}_p$, and in each round the prover sends a one-variable polynomial that the verifier checks with a random number in the field. The protocol runs for n rounds, one per quantified variable. In the final round, the verifier plugs in the last random value, evaluates the degree-1 polynomial directly, and halts with accept or reject.

Lemma 1. *In the TQBF interactive-proof protocol, after reducing each quantified polynomial modulo $x^2 - x$, the verifier's check in any round consists of evaluating a one-variable polynomial of degree at most 1 over $\mathbb{F}_p$. Furthermore, the verifier uses a random number in the range $[0, 2^i - 1]$, where i is the integer with $2^i \leq p < 2^{i+1}$.*

Proof. It is easy to see that the polynomial P the Verifier has to evaluate could be of a large degree as the size of the formula increases. In [17], the authors add the operation RxP after the arithmetization of each of the quantifiers. This operation reduces the degree of the polynomial $P \mod x^2 - x$. This means that any x^l where $l > 1$ is changed to x. This leaves P as a polynomial of degree ≤ 1 while keeping the correct boolean values. For two different polynomials $p_1(x)$ and $p_2(x)$ of degrees at most k over F_p, with probability at most $\frac{2k}{p}$, $p_1(x) = p_2(x)$ for a random number x in $[0, 2^i - 1]$. This is because the number of integers in the interval $[0, 2^i - 1]$ is at least $\frac{p}{2}$. Therefore, if p is selected to be at least double the size of the old choice, we can satisfy the same probability upper bound for a failure case in the proof system.

Using a random number in $[0, 2^i - 1]$ instead of $[0, p - 1]$ is easy to implement in CRNs. For an m bits random 0, 1-sequence $b_1 b_2 \ldots b_m$, we may use a copy of species $x_{i,0}$ to represent the case $b_i = 0$, and a copy of species $x_{i,1}$ to represent the case $b_i = 1$.

5.2   Circuits and Circuit Simulation

TC^0 is a class of circuits with constant depth and polynomial size using only AND, OR, NOT, and MAJORITY gates. It is known that TC circuits can do arithmetic. See Lemma 3 in the Appendix for a brief explanation. Reif and Tate (1992) [15] show multiplication and addition, and Hesse, Allender, and Barrington (2002) [11] show division. We use this class of circuits as a verifier in the protocol for TQBF. Step CRNs are known to simulate threshold circuits.

See Lemma 4 in the Appendix for a brief explanation. Anderson et al. (2024) [2] show Step CRNs with $(3,0)$ rules can simulate threshold circuits. We directly extend this to the Randomized Interactive CRN.

We show that $(3,0)$ rules are PSPACE-hard in Theorem 4. We extend that proof and PSPACE-hardness to $(2,0)$ rules in Theorem 5. These proofs rely on Lemma 2 from Anderson et al. [3].

Lemma 2. *A Boolean formula with G gates and D depth can be computed by a $(2,0)$ Step CRN with $\mathcal{O}(G)$ species, $\mathcal{O}(D)$ steps, and $\mathcal{O}(G)$ volume. Anderson et al. [3].*

Theorem 4. *The reachability problem in polynomial size steps of Randomized Interactive CRNs with $(3,0)$ rules is PSPACE-hard.*

Proof. We show that the PSPACE-hard problem TQBF can be transformed into the reachability problem of Randomized Interactive CRNs with $(3,0)$ rules. We let each S_i encode the message from the prover in round i, and R_i encode the random elements in $[0, 2^i - 1] \subseteq [0, p - 1]$ from the verifier in round i. For an instance F for TQBF, by Lemmas 1, 3, and 4, a polynomial size randomized interactive step CRN C_{rand}, which only has $(3,0)$ rules, can be generated in polynomial time. We have $F \in TQBF$ if and only if the yes configuration of C_{rand} is fully reachable.

Theorem 5. *The reachability problem in polynomial size steps of Randomized Interactive CRNs with $(2,0)$ rules is PSPACE-hard.*

Proof. We show that the PSPACE-hard problem TQBF can be transformed into the reachability problem of Interactive Randomized CRNs with $(2,0)$ rules. Each threshold gate can be simulated by an NC_1 circuit of polynomial size and $O(\log n)$ depth. A NC_1 circuit can be transformed into a polynomial-size boolean formula by the classical result $NC_1 = BF$, where BF represents the class of decision problems with polynomial-size boolean formulas. We let each S_i encode the message from the prover in round i, and R_i encode the random elements in $[0, 2^i - 1] \subseteq [0, p - 1]$ from the verifier in round i. For an instance F for TQBF, by Lemmas 1, 3, and 2, a polynomial size randomized interactive CRN C_{rand}, which only has $(2,0)$ rules, can be generated in polynomial time. We have $F \in TQBF$ if and only if the yes configuration of C_{rand} is fully reachable.

6   Discussion and Future Work

In this paper, we introduce Interactive CRN computation in both deterministic and randomized models. The hardness of reachability problems is studied for IA-CRNs with bounded-size void rules. Some interesting directions remain open. *Rule size*: our reductions use void rules of size $(12, 6)$ and $(17, 6)$, narrowing these parameters, or pinpointing the exact threshold where hardness emerges, would sharpen the model. *Plain reachability*: We analyzed the forced variant (requiring a unique terminal state). Determining the complexity of ordinary reachability for IA-CRNs, which we conjecture is still in PSPACE, is a natural next target.

A   Hardness for Interactive CRNs

Theorem 2. *The Interactive CRN Forced Reachability Problem is PSPACE-complete using only void rules of at most size (10,8) with a polynomially bounded number of steps.*

Proof. Given a constraint graph $G = (V, E)$ with edges partitioned into black (B) and white (W), we encode the instance into an IA-CRN as follows. Each edge is represented as a species named by its endpoints: an edge from vertex A to vertex B becomes species AB, while its flipped version is $\overline{AB}$. Weight-two edges include a superscript and each edge species has a subscript indicating which player it belongs to (see Fig. 3a). At every step, the white player adds exactly one choice $\overrightarrow{c_j} \in \{\overrightarrow{c_0}, \ldots, \overrightarrow{c_{k-1}}\}$, where $k = |W|$. Each $\overrightarrow{c_j}$ consists of exactly one flipped white edge (the edge the white player attempts to reverse), one copy of every flipped black edge, and $|B| - 1$ copies of a cleanup species DEL_b. The idea is that when a choice is added, the flipped white edge species will consume the unflipped version. Similarly, one out of all of the flipped black edge species will consume the unflipped version. The rest of flipped black edge species that were added will be cleaned up by the cleanup species DEL_b after the reaction representing the edge reversals executes.

We construct the rules Γ as follows. Consider a white edge $e_w \in W$ directed into vertex A. To determine if e_w can be flipped (reversed), we first identify all other edges surrounding A. The different combinations of flipped and unflipped versions of these surrounding edges form local configurations $\overrightarrow{C_{e_w}}$. However, we only retain local configurations that satisfy the constraint when e_w is flipped. For example, in Fig. 2a, if edge AV^2 is flipping away from vertex V, and the other edges directed a V are CV and BV, then the local configurations are $\{(CV, BV), (\overline{CV}, BV), (CV, \overline{BV}), (\overline{CV}, \overline{BV})\}$. Since these edges are weight-1, only the configuration $\overrightarrow{C_{AV}} = \{(CV, BV)\}$ would be retained, since this is the only scenario where reversing BA mantains the constraint. Similarly, for each black edge $e_b \in B$, we construct local configurations $\overrightarrow{C_{e_b}}$, retaining only valid configurations that satisfy vertex constraints. Every local configuration $\overrightarrow{w} \in \overrightarrow{C_{e_w}}$ and $\overrightarrow{b} \in \overrightarrow{C_{e_b}}$ will also include the species of the edge that is being reversed and its reversed version as a prefix. Finally, we form rules by combining each valid local configuration $\overrightarrow{w}$ for white edges with each valid configuration $\overrightarrow{b}$ for black edges, provided the moves remain valid when the white player's flip occurs first. Each rule γ_i thus has reactants $\overrightarrow{w} + \overrightarrow{b}$. The unflipped edge species are consumed, so the rest of the reactants are used as catalysts and remain as the product. An example reaction is shown in Fig. 4. Vertex labels and edge directions correspond to those in Fig. 3.

VARIABLE Vertex. See VARIABLE vertex V in Fig. 3a . Initially, AV_w^2 and $A'V_b^2$ are present. For the rule flip AV_w^2, $\overrightarrow{w}$ will have AV_w^2, $\overline{AV_w^2}$ and $A'V_b^2$. These represent the edge that will be flipped, the flipped version of that edge, and the inflow check to vertex V. The second half $\overrightarrow{b}$ will include the same for another variable edge. In a valid flip, the only element consumed will be the species representing the original edges that were flipped by white and black. The complete $(6,4)$ void reaction for a turn on a VARIABLE vertex is shown in Fig. 4 (Fig. 6).

$$\overbrace{AV_w^2 + \overline{AV_w^2} + A'V_b^2 + \underbrace{C'Y_b^2 + \overline{C'Y_b^2} + CY_w^2}_{\overrightarrow{b}}}^{\overrightarrow{w}} \rightarrow \overline{AV_w^2} + A'V_b^2 + \overline{C'Y_b^2} + CY_w^2$$

Fig. 6. $(6,4)$ void rule to play a turn on a VARIABLE vertex. Player 1 selects vertex V and reverses edge AV_w^2. Player 2 selects Vertex Y and reverses edge $C'Y_b^2$. The graph after this turn is shown in Fig. 3b.

AND Vertex. Let F be an AND vertex with single edges FA_w, FB_W and double edge EF_w^2. To reverse EF_w^2, we have local configurations $\overrightarrow{C_{EF_w^2}} = \{(\overline{FA_w}, \overline{FB_w})\}$ that make up $\overrightarrow{w}$, the first half of the rule. The second half $\overrightarrow{b}$ consists of a valid move for Player 2. For simplicity in this example, we show a mirrored reversal of the player 1 move in a different vertex F'. Two copies of any reversed edge of the black player are needed in the reactants to avoid the black player's reversed edge species added in $\overrightarrow{c_j}$ from triggering this reaction in an invalid configuration. It is important to note that after the VARIABLE vertex selections, black moves are irrelevant. We still account for them to preserve turn order. The completed $(10,8)$ void reaction for a turn on an AND vertex is as follows: Player 1 reverses edge EF_w^2. Player 2 reverses edge $E'F_b'^2$,

$$EF_w^2 + \overline{EF_w^2} + \overline{FA_w} + \overline{FB_w} + E'F_b'^2 + \overline{E'F_b'^2} + 2\overline{F'A_b'} + 2\overline{F'B_b'} \rightarrow$$
$$\overline{EF_w^2} + \overline{FA_w} + \overline{FB_w} + \overline{E'F_b'^2} + 2\overline{F'A_b'} + 2\overline{F'B_b'}$$

OR Vertex. Use the system in Fig. 3 for reference. Now, let F be an OR vertex with double edges FA_w^2, FB_W^2, and EF_w^2. Vector $\overrightarrow{w}$ will contain $EF_W^2, \overline{EF_w^2}$, and one of $\overline{FA_W^2}, \overline{FB_w^2}$. Note that only one is needed since one double edge satisfies the inflow constraint. There will exist two reactions to reverse the output edge of an OR vertex. Vector $\overrightarrow{b}$ is formed the same as was for the AND vertex. The two $(7,5)$ void reaction rules for a turn on an OR vertex are as follows: Player 1 reverses edge EF_w^2. Player 2 reverses edge $E'F_b'^2$. The first rule uses edges FA_w^2 and $F'A_b'^2$ to ensure the inflow constraint is met. The second one uses FB_w^2 and $F'B_b'^2$,

$$EF_w^2 + \overline{EF_w^2} + \overline{FA_w^2} + E'F_b'^2 + \overline{E'F_b'^2} + 2\overline{F'A_b'^2} \rightarrow \overline{EF_w^2} + \overline{FA_w^2} + \overline{E'F_b'^2} + 2\overline{F'A_b'^2}$$
$$EF_w^2 + \overline{EF_w^2} + \overline{FB_w^2} + E'F_b'^2 + \overline{E'F_b'^2} + 2\overline{F'B_b'^2} \rightarrow \overline{EF_w^2} + \overline{FB_w^2} + \overline{E'F_b'^2} + 2\overline{F'B_b'^2}$$

CHOICE Vertex. In Fig. 2d. Let V be a CHOICE vertex with single edges AV_w, CV_w, BV_w. Vector $\vec{w}$ will consist of $AV_w, \overline{AV_w}$, and the two edges that ensure the inflow constraint is met, CV_w, BV_w. Vector $\vec{b}$ is constructed the same way as the OR and AND vertices. The completed $(8,6)$ reaction for a turn on a CHOICE vertex is as follows: Player 1 reverses edge AV_w. Player 2 reverses edge $A'X_b$. Note that a choice vertex will also have more variations, as different edges can satisfy the constraint.

$$AV_w + \overline{AV_w} + CV_w + BV_w + A'X_b + \overline{A'X_b} + C'X_b' + B'X_b' \rightarrow$$
$$\overline{AV_w} + CV_w + BV_w + \overline{A'X_b} + C'X_b' + B'X_b'$$

FANOUT Vertex. Using Fig. 3. The two rules for each output single edge will need the double edge to ensure the constraint is met. Let A be a FANOUT vertex with single edges GA_w, FA_w and double edge AV_w^2. Vector $\vec{w}$ consists of one of GA_w, FA_w, whichever one is being reversed along with its reversed version, and $\overline{AV_w^2}$. Let B' be a FANOUT vertex with edges $G'B_b', F'B_b', B'X_b^2$. Vector $\vec{b}$ is constructed the same way as in the previous vertices. The two $(7,5)$ void reactions for a turn on a FANOUT vertex are as follows: Player 1 reverses edge GA_w, FA_w. Player 2 reverses edge $G'B_b', F'B_b'$,

$$GA_w + \overline{GA_w} + \overline{AV_w^2} + G'B_b' + \overline{G'B_b'} + 2\overline{B'X_b^2} \rightarrow \overline{GA_w} + \overline{AV_w^2} + \overline{G'B_b'} + \overline{B'X_b^2}$$
$$FA_w + \overline{FA_w} + \overline{AV_w^2} + F'B_b' + 2\overline{F'B_b'} + 2\overline{B'X_b^2} \rightarrow \overline{FA_w} + \overline{AV_w^2} + \overline{F'B_b'} + 2\overline{B'X_b^2}$$

After a reaction representing a turn executes, the reversed black edge species added in $\vec{c_j}$ will still be in the system. The species DEL_b then cleans up those leftover black edge species, leaving the system with only the species representing the current state of the graph. These rules are:

$$(1)\ 2\overline{A'V_b^2} + DEL_b \rightarrow \overline{A'V_b^2}$$
$$(2)\ A'V_b^2 + \overline{A'V_h^2} + DEL_b \rightarrow A'V_b^2$$
$$(3)\ BX_w^2 + \overline{BX_w^2} \rightarrow BX_w^2, (4)\ 2\overline{BX_w^2} \rightarrow \overline{BX_w^2}$$

First, a $(3,1)$ void rule to clean up unused black player flipped edges. This rule consumes one of two reversed black edge species for the case when the current state of the graph has $A'V_b^2$ in reversed state. Second is a $(2,1)$ void rule for the case where the current state of the graph has $A'V_b^2$ in its initial state. Rules (3) and (4) will delete a white move and return the system to its original state before the choice. (3) takes the case for an invalid white flip. (4) is for the case of trying to flip an already flipped edge.

Using this construction, the IA-CRN reaches a terminal configuration containing the species corresponding to the reversed winning edge if and only if the white player has a winning strategy in the constraint logic game. This establishes the Interactive CRN Forced Reachability problem is PSPACE-hard. By Theorem 1, it follows that the problem is PSPACE-complete.

Theorem 3. *The Interactive CRN reachability problem is EXPTIME-hard, even when restricted to void rules of at most size (13,10) and allowing an unbounded number of steps.*

Proof. Given a constraint graph $G = (V, E)$ with E partitioned into B and W, we encode the instance into the IA-CRN the same as Theorem 2. After every step, the white player adds exactly one choice $\vec{c_j} \in \{\vec{c_0}, \ldots, \overrightarrow{c_{k-1}}\}$, where $k = 2|W|$. one for every e_w and one for its reverse $\overline{e_w}$. Each $\vec{c_j}$ consists of exactly one white edge, one copy of every flipped black edge, one copy of every reversed black edge, and $|B| - 1$ copies of a cleanup species DEL_b. The rules are constructed the similar to Theorem 2. They keep a similar reactant structure of $\vec{w} + \vec{b}$. Note that now, they can have mixed edges, so $\vec{w}$ wont be exclusively white edges and the same for $\vec{b}$. For every black edge used as a validation catalyst in $\vec{w}$ and $\vec{b}$ will need 2 copies in its reactants. Having two of every species for every black edge ensures the rule matches the current state (one copy already in the configuration and one from the choice), and prevents the extra black edges in $\vec{c_j}$ from serving as catalysts. The rule keeps the flipped orientations and the existing DEL_b cleanup reactions then remove these duplicates, restoring a single valid edge species.

We explicitly show the unbounded construction with two representative gadgets. First, we revisit the AND vertex used in Theorem 2 and show how a weight-2 white edge can be flipped forward and back under our new "two-copy" convention. Then, we present the mixed-edge AND vertex in Fig. 5c, where there are two single edges for black, and one double for white. This case produces the largest rule in the reduction. All other mixed-edge variants (single/weight-2 combinations and color permutations) can be found in the appendix.

Black AND Vertex. The rules for a black AND vertex follow almost the same as the AND vertex in Theorem 2. The only change is that the current state of the black edge that is being flipped requires 2 copies in the reactants. In the following rule, white is flipping edge EF_w^2 and black player is flipping $E'F_b'^2$. Note that now, two copies of $E'F_b'^2$ are required to verify the current state of the edge. See vertex F and F' in Fig. 3 for reference.

$$EF_w^2 + \overline{EF_w^2} + \overline{FA_w} + \overline{FB_w} + 2E'F_b'^2 + \overline{E'F_b'^2} + 2\overline{F'A_b'} + 2\overline{F'B_b'} \rightarrow$$
$$\overline{EF_w^2} + \overline{FA_w} + \overline{FB_w} + \overline{E'F_b'^2} + 2\overline{F'A_b'} + 2\overline{F'B_b'}$$

Rules to reverse an edge again are symmetrical. To reverse edge $\overline{EF_w^2}$ and $\overline{E'F_b'^2}$ again, we must now ensure the valid local configurations of vertices E and E' are present. These now include edges $IE_w, HE_w, I'E_b', H'E_b'$. The $(11, 8)$ reaction is the following:

$$\overline{EF_w^2} + EF_w^2 + IE_w + HE_w + 2\overline{E'F_b'^2} + E'F_b'^2 + 2I'E_b' + 2H'E_b' \rightarrow$$
$$EF_w^2 + IE_w + HE_w + E'F_b'^2 + 2I'E_b' + 2H'E_b'$$

Mixed AND Vertex. The largest rule in the reduction is triggered in the scenario where both players attempt to flip edges from an AND vertex with two black single-weight edges and one white weight-2 edge. In this case, all of the validation catalysts require to have a count of two since they are all black edges. Let V be an AND vertex with edges AV_w^2, VC_b, VB_b Let X be a black AND vertex with edges $A'X_b^2, XC_b', XB_b'$. A $(13, 10)$ rule to flip both $AV_w^2, A'X_b^2$ in a single turn is as follows: Player 1 reverses edge AV_w^2. Player 2 reverses edge $A'X_b^2$.

$$AV_w^2 + \overline{AV_w^2} + 2\overline{VC_b} + 2\overline{VB_b} + 2A'X_b^2 + \overline{A'X_b^2} + 2\overline{XC_b'} + 2\overline{XB_b'} \rightarrow$$
$$\overline{AV_w^2} + 2\overline{VC_b} + 2\overline{VB_b} + \overline{A'X_b^2} + 2\overline{XC_b'} + 2\overline{XB_b'} +$$

B/W AND Vertex Case 1. There will be instances where the AND vertex will be half white edges and half black edges. Both players must make valid moves when they act on the same vertex. The white player's move happens first, so a rule where a single black edge is flipped into the vertex and then a white double edge is flipped away from the same vertex would not be valid within the same turn. This means we should only account for the white double edge pointing out once both black single edges are pointing in. Let V be an AND vertex with both single black edges VC_b' and VB_b', and a white double edge AV_w^2. Let X be an AND vertex with both single white edges XC_w and XB_w, and a white double edge $A'X_b^2$. A $(13, 6)$ reaction for a turn to flip AV_w^2 and $A'X_b^2$ is as follows:

$$AV_w^2 + \overline{AV_w^2} + 2\overline{VC_b'} + 2\overline{VB_b'} + 2A'X_b^2 + \overline{A'X_b^2} + \overline{XC_w} + \overline{XB_w} \rightarrow$$
$$\overline{AV_w^2} + 2\overline{VC_b'} + 2\overline{VB_b'} + \overline{A'X_b^2} + \overline{XC_w} + \overline{XB_w}$$

Player 1 reverses edge AV_w^2. Player 2 reverses edge $A'X_b^2$. Note that $\overrightarrow{w}$ uses 2 black edges as catalysts and $\overrightarrow{b}$ uses 2 white edges as catalysts.

B/W AND Vertex Case 2. Another case is when each single edge in the AND vertex belongs to a different player. One white and one black single edge means both players can affect the same vertex in a single turn. This will only happen when both single edges are being flipped away from their common vertex. Let V be an AND vertex with one white single edge VC_w, one black single edge VB_b, and one white double edge AV_w^2. A $(6, 3)$ reaction for a turn to flip $\overline{VC_w}, \overline{VB_b}$ on a mixed AND vertex is as follows:
$$\overline{VC_w} + VC_w + AV_w^2 + 2\overline{VB_b} + VB_b \rightarrow VC_w + AV_w^2 + VB_b$$

Here, Player 1 reverses edge $\overline{VC_w}$. Player 2 reverses edge $\overline{VB_b}$.

Similar to Theorem 2, after each turn rule executes the system enters a cleanup phase that uses DEL_b species to delete every unused edge left by the choice vector:

$$(1)\ 2\overline{A'V_b^2} + A'V_b^2 + DEL_b\ \rightarrow\ \overline{A'V_b^2} \quad (2)\ 2A'V_b^2 + \overline{A'V_b^2} + DEL_b\ \rightarrow\ A'V_b^2$$
$$(3)\ BX_w^2 + \overline{BX_w^2} + \overrightarrow{w}\ \rightarrow\ BX_w^2 + \overrightarrow{w} \quad (4)\ BX_w^2 + \overline{BX_w^2} + \overrightarrow{\overline{w}}\ \rightarrow\ \overline{BX_w^2} + \overrightarrow{\overline{w}}$$
$$(5)\ 2BX_w^2\ \rightarrow\ BX_w^2,\ \ 2\overline{BX_w^2}\ \rightarrow\ \overline{BX_w^2}$$

(1) and (2) are $(3, 1)$ void reactions that discard the unused black edges introduced by the choice vector. Each consumes one DEL_b and preserves the edge's current state. (3) and (4) discard an invalid white edge flip. Recall $\overrightarrow{w}$ (respectively $\overrightarrow{\overline{w}}$) represents a valid local configuration, so the stray copy is deleted while the valid orientation is retained. (5) handles the redundant case in which the choice vector adds a duplicate of the already-present white orientation.

With this construction, the IA-CRN reaches the target terminal configuration iff White has a winning strategy in the unbounded 2CL. Hence, Interactive CRN forced reachability with no bound in m is EXPTIME-hard.

B   Randomized Interactive CRNs

Lemma 3. *Polynomial operations, which are based on $+, -, *$ in the field modulo a prime p (F_p), can be implemented using Threshold circuits with constant depth and polynomial size [11, 15].*

Proof. Let p be a fixed prime number. Polynomial operations over the finite field $\mathbb{Z}_p$, such as addition, multiplication, and division, can be executed using constant-depth, polynomial-size threshold circuits. Addition and multiplication of polynomials involve coefficient-wise operations and can be performed in TC^0 circuits [15]. Evaluating a polynomial at a specific point is to solve additions and multiplications in $\mathbb{Z}_p$, all of which are computable in TC^0 [15]. In [11], the authors then show division is also achievable in TC^0. Therefore, all fundamental polynomial operations modulo p are within TC^0 of constant depth and polynomial size.

Lemma 4. *Threshold circuits with G gates and D depth can be computed by a $(3,0)$ Step CRN with $\mathcal{O}(G)$ species, $\mathcal{O}(D)$ steps, and $\mathcal{O}(G)$ volume. [2].*

Proof. In [2], the authors show that a step-CRN with only (2,0) void can simulate threshold formulas of size G and depth D under linear resource bounds. In particular, each formula gate is represented by a constant number of species, all gates at the same depth fire in one CRN step, and only $O(G)$ total volume is ever used. Then [3] refines the construction to handle threshold circuits (allowing unbounded fan-out) by suitably adjusting volume to account for copying across gates. They prove that the same $(2,0)$-step CRN and rules suffice to compute a circuit of size G and depth D with only $O(G)$ species, in $O(D)$ steps, and using $O(G)$ volume when fan-out is bounded, or $O(G\,F_{\mathrm{out}}^D)$ in the general case.

References

1. Alaniz, R.M., Fu, B., Gomez, T., Grizzell, E., Rodriguez, A., Schweller, R., Wylie, T.: Reachability in restricted chemical reaction networks (2022)
2. Anderson, R., Avila, A., Fu, B., et al.: Computing threshold circuits with void reactions in step chemical reaction networks. In: Formenti, E., Durand-Lose, J. (eds.) Machines, Computations, and Universality. MCU 2024. LNCS, vol. 15270, pp. 52–71. Springer, Cham (2025). https://doi.org/10.1007/978-3-031-81202-6_4
3. Anderson, R., Fu, B., Massie, A., et al.: Computing threshold circuits with bimolecular void reactions in step chemical reaction networks. In: Cho, DJ., Kim, J. (eds.) Unconventional Computation and Natural Computation. UCNC 2024. LNCS, vol. 14776, pp. 253–268. Springer, Cham (2024). https://doi.org/10.1007/978-3-031-63742-1_18

4. Aris, R.: Prolegomena to the rational analysis of systems of chemical reactions. Arch. Ration. Mech. Anal. **19**(2), 81–99 (1965). https://doi.org/10.1007/BF00282276

5. Chalk, C., Martinez, E., Schweller, R., Vega, L., Winslow, A., Wylie, T.: Optimal staged self-assembly of general shapes. Algorithmica **80**, 1383–1409 (2018)

6. Cirlos, S.C., Gomez, T., Grizzell, E., Rodriguez, A., Schweller, R., Wylie, T.: Simulation of multiple stages in single bin active tile self-assembly. In: Genova, D., Kari, J. (eds.) Unconventional Computation and Natural Computation. UCNC 2023. LNCS, vol. 14003, pp. 155–170. Springer, Cham (2023). https://doi.org/10.1007/978-3-031-34034-5_11

7. Demaine, E.D., Eisenstat, S., Ishaque, M., Winslow, A.: One-dimensional staged self-assembly. Nat. Comput. **12**(2), 247–258 (2013)

8. Ellis, S.J., Klinge, T.H., Lathrop, J.I.: Robust chemical circuits. Biosystems **186**, 103983 (2019)

9. Fu, B., Gomez, T., Knobel, R., et al.: Brief announcement: reachability in deletion-only chemical reaction networks. In: Proceedings of the Symposium on Algorithmic Foundations of Dynamic Networks. SAND (2025)

10. Hearn, R.A.: Games, Puzzles, and Computation. Ph.D. thesis, Massachusetts Institute of Technology (May 2006)

11. Hesse, W., Allender, E., Mix Barrington, D.A.: Uniform constant-depth threshold circuits for division and iterated multiplication. J. Comput. Syst. Sci. **65**(4), 695–716 (2002). https://doi.org/10.1016/S0022-0000(02)00025-9, special Issue on Complexity 2001

12. Karp, R.M., Miller, R.E.: Parallel program schemata. J. Comput. Syst. Sci. **3**(2), 147–195 (1969). https://doi.org/10.1016/S0022-0000(69)80011-5

13. Mo, D., Stefanovic, D.: Iterative Self-assembly with dynamic strength transformation and temperature control. In: Soloveichik, D., Yurke, B. (eds.) DNA 2013. LNCS, vol. 8141, pp. 147–159. Springer, Cham (2013). https://doi.org/10.1007/978-3-319-01928-4_11

14. Petri, C.A.: Communication with automata (1966)

15. Reif, J.H., Tate, S.R.: On threshold circuits and polynomial computation. SIAM J. Comput. **21**(5), 896–908 (1992). https://doi.org/10.1137/0221053

16. Shamir, A.: IP = PSPACE. J. ACM **39**(4), 869–877 (1992). https://doi.org/10.1145/146585.146609

17. Shen, A.: IP = PSPACE: simplified proof. J. ACM **39**(4), 878–880 (1992). https://doi.org/10.1145/146585.146613

18. Winfree, E.: Chemical reaction networks and stochastic local search. In: Thachuk, C., Liu, Y. (eds.) DNA 2019. LNCS, vol. 11648, pp. 1–20. Springer, Cham (2019). https://doi.org/10.1007/978-3-030-26807-7_1

Gakmoro: An Application of Physical Secure Computation to Card Game

Takaaki Mizuki[1]([✉]) [iD], Tomoki Kuzuma[2], Tomoya Hirano[3], Ririn Oshima[4], and Momofuku Yasuda[4] [iD]

[1] Cyberscience Center, Tohoku University, Sendai, Japan
mizuki+lncs@tohoku.ac.jp
[2] Graduate School of Information Sciences, Tohoku University, Sendai, Japan
[3] School of Engineering, Tohoku University, Sendai, Japan
[4] Faculty of Agriculture, Tohoku University, Sendai, Japan

Abstract. We have created a new card game, Gakmoro, where two players each have seven cards numbered from 1 to 7. In each round, they secretly choose one to three cards to compete based on their total value; the first player to win two rounds wins the game. The unique feature of Gakmoro is that the cards chosen by the players must remain secret, and only the winner of each round is revealed. This secrecy is expected to add depth to the game by emphasizing the strategy of reading an opponent's hand. As designed, the game requires a dealer to secretly calculate the sum of the cards and announce only the winner. In this paper, instead of relying on a human dealer, we use card-based cryptography to virtually fulfill the dealer's role. In other words, we design an 'unconventional' computation protocol that allows players to securely determine the winner without a dealer using a physical deck of cards.

Keywords: Card-based cryptography · Secure computation · Playing cards · Card games

1 Introduction

We introduce a new card game, *Gakmoro*, designed by students (the third, fourth, and fifth authors) and a teaching assistant (the second author) during an "Introduction to Academic Learning" class at Tohoku University. Its name, Gakmoro, is derived from the Japanese name of the class, "Gakumonron Enshu."

1.1 Gakmoro's Rules

The game, Gakmoro, involves two players, Alice and Bob, together with a dealer. First, each of the two players has seven cards with numbers from 1 to 7. For example, Alice holds seven cards 1♠ 2♠ 3♠ 4♠ 5♠ 6♠ 7♠ in her hand, and Bob holds seven cards 1♦ 2♦ 3♦ 4♦ 5♦ 6♦ 7♦ in his hand. The game consists of up to three rounds. The first player to win two rounds wins the game. Each round proceeds as follows.

© The Author(s), under exclusive license to Springer Nature Switzerland AG 2026

E. Formenti and L. Manzoni (Eds.): UCNC 2025, LNCS 16364, pp. 344–360, 2026.
https://doi.org/10.1007/978-3-032-15641-9_23

1. Each player selects one to three cards from his or her hand and submits them secretly to the dealer. (Neither the number of cards nor the card values are revealed to the opponent.)
2. The dealer calculates the sum of the card values submitted by each player. The player with the higher total wins the round. The dealer announces only the winner's name or "a tie" (while keeping the sum, the number of cards, and the card values secret).

Of course, once a card is submitted to the dealer, it cannot be used in subsequent rounds. As mentioned, the first player to achieve two wins is the winner of the game.

1.2 Example of Game Progression

Here is an example of how a game of Gakmoro might unfold.

In the first round, Alice secretly plays $6\spadesuit$ and $7\spadesuit$ (totaling 13), while Bob secretly plays $1\diamondsuit$ and $2\diamondsuit$ (totaling 3). Alice's total is higher, so she wins the round. The dealer only announces, "Alice wins the round."

In the second round, Alice plays $1\spadesuit$ and $2\spadesuit$ (totaling 3), aiming to save her higher cards for a potential third round. Bob, anticipating this, plays his strong cards: $5\diamondsuit$, $6\diamondsuit$, and $7\diamondsuit$ (totaling 18). Bob's total is higher, so he wins the round. Only "Bob wins the round" is announced.

In the third round, they play their remaining cards. Alice plays $3\spadesuit$, $4\spadesuit$, and $5\spadesuit$ (totaling 12). Bob plays $3\diamondsuit$ and $4\diamondsuit$ (totaling 7). Alice's total is higher, so she wins the round and, consequently, she is the winner of the game.

1.3 Playing Without a Dealer

The dealer's role in Gakmoro is crucial for maintaining secrecy and ensuring fair play. The main functions are:

1. Ensuring that each player submits between one and three cards.
2. Correctly adding the values of the submitted cards.
3. Announcing only the winner, not the sums or the number of cards played.

Without a dealer, the game loses its core element of hidden information, which is central to strategic play. The excitement and strategic depth come from the uncertainty of what the opponent has played.

Thus, the role of a dealer is indispensable. What if there are only two players, making it difficult to find a dealer? Are there any solutions for playing Gakmoro with just two people?

1.4 Application of Secure Computations

To the questions raised above, one might think of using *secure computation* [51] to fulfill the dealer's role in Gakmoro. Secure computation is a cryptographic technique that produces the output of a predetermined function while keeping the input information secret. Typically, secure computation protocols are implemented on computers over networks.

However, Gakmoro is envisioned as a casual card game played with physical cards, so relying on computers or networks is not ideal, and we solicit 'unconventional' computing that relies only on daily objects. Therefore, we will use *card-based cryptography* [15,27], which enables secure computation using a physical deck of cards.

1.5   Contribution of This Paper

In this paper, we aim to eliminate the need for a human dealer in Gakmoro by combining existing techniques from card-based cryptography and designing a new protocol.

Card-based cryptography typically uses two types of cards, such as ♣ ♣ ♣ ⋯ and ♥ ♥ ♥ ⋯ , whose backs are identical ? ? ? ⋯ (cf. [3,19,46]). A pair of cards is often used to represent a bit value, i.e., ♣ ♥ = 0, ♥ ♣ = 1. Under such encoding, it is theoretically possible to perform secure computation of addition by representing Alice and Bob's cards in binary numbers, as protocols for secure computations of the half-adder and the full-adder are known [25]. However, it is not user-friendly to handle binary arithmetic in a game intended for both children and adults.

Thus, instead, we will use an encoding that expresses integers by the position of ♥ , as in the following example (details are given in Sect. 2.2):

Under such integer encoding, Ruangwises and Itoh [38] have proposed a protocol to realize secure computation of addition (details are given in Sect. 2.4).

In this paper, we use their protocol to sum the cards submitted by each player. Recalling Gakmoro's rules, besides addition, the other necessary operation is comparison between two integers (to determine the winner). In card-based cryptography, several methods for secure computation of comparison (on integers) are known [21,31], but in this paper, we propose a new comparison protocol (Sect. 3). By combining the above, we show that the dealer's role in Gakmoro can be achieved by a card-based protocol (Sect. 4).

1.6   Related Studies

Card-based cryptography has made remarkable progress in recent years. Most previous studies on card-based cryptography have focused on protocols for secure computations (e.g., [4,39,49,52]) and zero-knowledge proofs (e.g., [17,37,48]), with few studies applying card-based cryptography to card games. To the best of the authors' knowledge, the first such attempt was to create virtual players in a famous card game, Old Maid [42], followed by the work for UNO games [40].

Other examples of applications to games (not necessalily card games) include: a protocol that can secretly form multiple player groups for games like Werewolf [8]; a protocol that can generate only solvable problems for games like the 15-puzzle and Rubik's Cube [41]; a protocol for two-player games that allows players to secretly determine who goes first based on their private preferences [44]; and protocols for generating a random derangement to enjoy secret Santa [3,11,29]. In addition, as implied above, numerous card-based zero-knowledge proof protocols exist, primarily conceived for puzzles and games, such as Topswops [18], the 15-puzzle [47], Moon-or-Sun [7], Usowan [23], Sumplete [9], Dosun-Fuwari [14], and Nurimisaki [36].

1.7   Organization of This Paper

The rest of this paper is organized as follows. Section 2 describes the cards to be used, the integer encoding, and the existing protocol for addition. Section 3 provides the subtraction protocol and the method of secure computation of comparison using it. In Sect. 4, we describe the proposed protocol for playing Gakmoro without a dealer. The implementation method is discussed in Sect. 5. The conclusion is given in Sect. 6.

2   Preliminaries

This section provides the necessary preparation for describing and explaining the proposed method. Specifically, we first describe the cards to be used, the encoding of integers, and the shuffling operation, followed by an introduction to the existing protocol for integer addition.

2.1   Two-Color Deck of Cards

As we saw in Sect. 1, Gakmoro itself is a game played with playing cards. However, since it is not easy to use playing cards to construct secure computation protocols for addition or comparison of the numbers written on the playing cards, we use a *two-color deck* of cards, as is typically used in card-based cryptography.

As briefly introduced in Sect. 1.5, we use two types of cards, namely, black cards ♣ ♣ ♣ ⋯ and red cards ♥ ♥ ♥ ⋯ , whose backs ? ? ? ⋯ are all indistinguishable. In the field of card-based cryptography, many research papers have been published using such a two-color deck of cards, and some practical representative protocols for secure computation of logical products exist, such as the five-card trick [1] and MS-AND protocol [28].

In fact, although it is possible to construct secure computation protocols with playing cards (e.g., [6,16,24,32]), it is generally believed to be more difficult to construct a simple protocol compared to the two-color-deck case (since all playing cards are 'unique,' it is generally more challenging to keep the input secret). It is a future prospect to develop an efficient method that can be implemented on playing cards,[1] which are more readily available.

2.2 Encoding of Integers

Gakmoro requires integers from 1 to 7 as inputs, and their addition and comparison. This paper employs the following encoding of integers using a two-color deck of cards. Suppose that we fix a positive integer m and wish to deal with integers from 0 to m.

Using m ♣ s and one ♥ , setting the $(i+1)$-th card to ♥ denotes the integer $i \in \{0, 1, \ldots, m\}$:

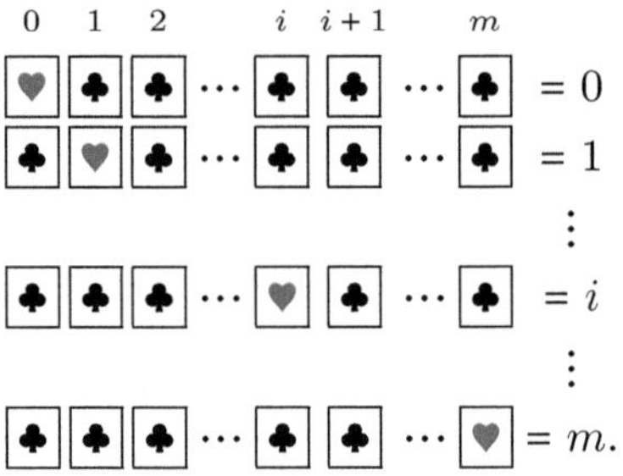

$$
\begin{array}{cccccccc}
0 & 1 & 2 & & i & i+1 & & m \\
\boxed{♥}\,\boxed{♣}\,\boxed{♣}\, \cdots \,\boxed{♣}\,\boxed{♣}\, \cdots \,\boxed{♣} & & & & & & & = 0 \\
\boxed{♣}\,\boxed{♥}\,\boxed{♣}\, \cdots \,\boxed{♣}\,\boxed{♣}\, \cdots \,\boxed{♣} & & & & & & & = 1 \\
\vdots \\
\boxed{♣}\,\boxed{♣}\,\boxed{♣}\, \cdots \,\boxed{♥}\,\boxed{♣}\, \cdots \,\boxed{♣} & & & & & & & = i \\
\vdots \\
\boxed{♣}\,\boxed{♣}\,\boxed{♣}\, \cdots \,\boxed{♣}\,\boxed{♣}\, \cdots \,\boxed{♥} & & & & & & & = m.
\end{array}
$$

In the following, we write $E^{♥}_{m+1}(i)$ as a bundle of cards representing an integer i placed face down under this encoding.

This encoding was first introduced in Crépeau and Kilian's secure derangement generation [3], and then has been used in Bultel et al.'s zero-knowledge proof protocol [2], Ruangwises and Itoh's protocols [38], and so on.

2.3 Pile-Shifting Shuffle

The shuffling action used in our proposed method is the *pile-shifting shuffle* [33,43][2]. This is an operation that considers a bundle of cards of the same size as a *pile* and shuffles the piles in a cyclic manner. The number of shifts is random, and no one knows the number of shifts. This shuffling can be implemented securely by putting each pile of cards in a sleeve (or a card case [13]) and using a technique called Hindu cut [50].

---

[1] The "partial opening" action [10,22] on playing cards is considered a promising technique.

[2] Note that this is different from the pile-scramble shuffle [11].

As an example, applying a pile-shifting shuffle to the following five (two-card) piles yields one of the following five sequences each with a probability of $1/5$:

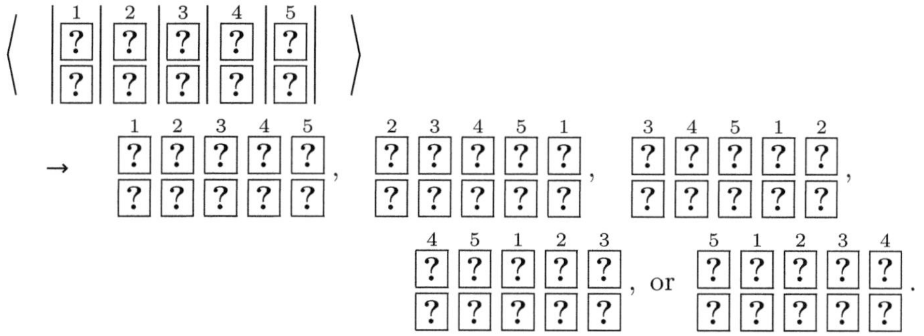

$$\left\langle \begin{array}{ccccc} 1 & 2 & 3 & 4 & 5 \end{array} \right\rangle$$

$$\rightarrow$$

2.4   Addition Protocol

Ruangwises and Itoh [38] constructed a protocol that, given two integers based on the encoding given in Sect. 2.2, securely computes their sum. Since this existing protocol is used in this paper, the protocol procedure is described below.

Let m be fixed, $a, b \in \{0, 1, \ldots, m\}$, and suppose that bundles of cards corresponding to integers a and b are given face down, that is,

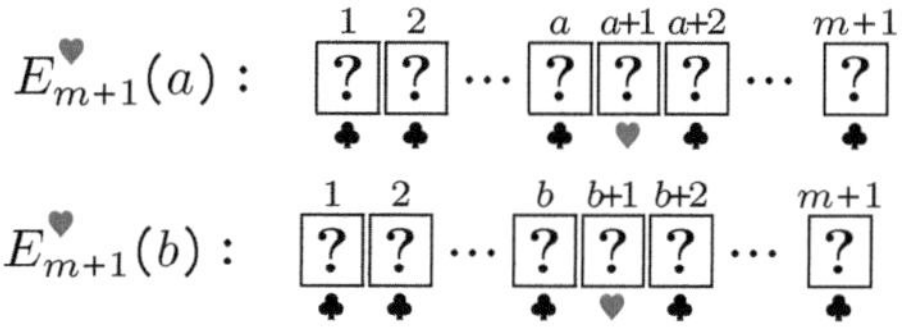

$$E_{m+1}^{\heartsuit}(a):$$

$$E_{m+1}^{\heartsuit}(b):$$

is the input (a total of $2m + 2$ cards). Recall that the positions of $\boxed{\heartsuit}$ determine the values of a and b.

The addition protocol [38] takes $E_{m+1}^{\heartsuit}(a)$ and $E_{m+1}^{\heartsuit}(b)$ as input and works as follows.

1. Rearrange the cards in $E_{m+1}^{\heartsuit}(b)$ such that the left and right sides are reversed. This yields $E_{m+1}^{\heartsuit}(m - b)$:

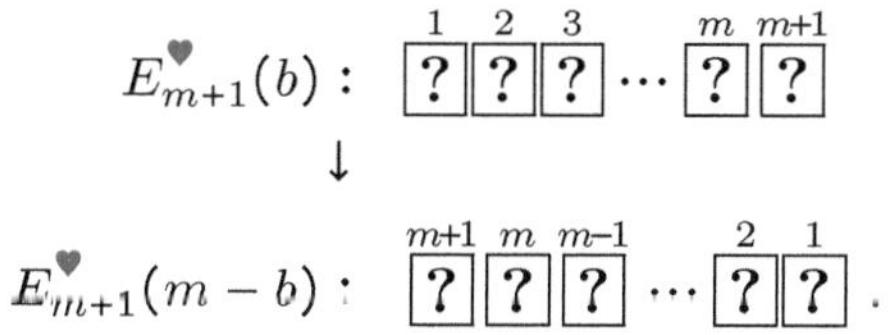

$$E_{m+1}^{\heartsuit}(b):$$

$$\downarrow$$

$$E_{m+1}^{\heartsuit}(m - b):$$

2. $E_{m+1}^{\heartsuit}(a)$ and $E_{m+1}^{\heartsuit}(m - b)$ are arranged in two rows, with one card on top and one card on the bottom, and a pile-shifting shuffle is applied to the piles:

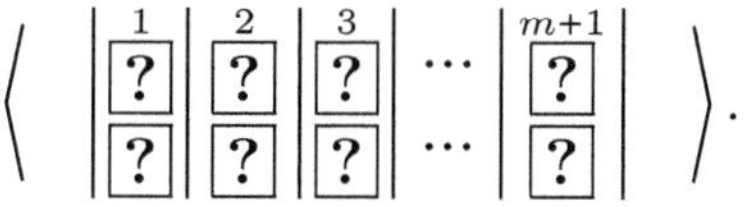

$$\left\langle \begin{array}{ccccc} 1 & 2 & 3 & \cdots & m+1 \end{array} \right\rangle.$$

In this case, a random shift $r \in \{0, 1, \ldots, m\}$ arises, so that we obtain $E_{m+1}^{\heartsuit}(a + r \bmod m + 1)$ and $E_{m+1}^{\heartsuit}(m - b + r \bmod m + 1)$.

3. Turn over all the cards in the bottom row, i.e., $E_{m+1}^{\heartsuit}(m - b + r \bmod m + 1)$, and shift the top and bottom cards in piles until $\boxed{\heartsuit}$ is at the right edge:

$$
\begin{array}{cccccccc}
\boxed{?} & \boxed{?} & \cdots & \boxed{?} & \boxed{?} & \boxed{?} & \cdots & \boxed{?} & \boxed{?} \\
\boxed{\clubsuit} & \boxed{\clubsuit} & \cdots & \boxed{\clubsuit} & \boxed{\heartsuit} & \boxed{\clubsuit} & \cdots & \boxed{\clubsuit} & \boxed{\clubsuit}
\end{array}
$$

$$\downarrow$$

$$
\begin{array}{cccccccc}
\boxed{?} & \boxed{?} & \cdots & \boxed{?} & \boxed{?} & \boxed{?} & \cdots & \boxed{?} & \boxed{?} \\
\boxed{\clubsuit} & \boxed{\clubsuit} & \cdots & \boxed{\clubsuit} & \boxed{\clubsuit} & \boxed{\clubsuit} & \cdots & \boxed{\clubsuit} & \boxed{\heartsuit}
\end{array}
$$

This results in the upper row being shifted by

$$
m - (m - b + r \bmod m + 1) = b - r \bmod m + 1
$$

positions to the right. Therefore, the resulting bundle from $E_{m+1}^{\heartsuit}(a + r \bmod m + 1)$ after this shift will be $E_{m+1}^{\heartsuit}((a + r) + (b - r) \bmod m + 1)$. Thus, $E_{m+1}^{\heartsuit}(a + b \bmod m + 1)$ is obtained.

The above is the existing addition protocol, which can generate the addition result $E_{m+1}^{\heartsuit}(a + b \bmod m + 1)$ from $E_{m+1}^{\heartsuit}(a)$ and $E_{m+1}^{\heartsuit}(b)$ without leaking their values.

3   Subtraction Protocol and Comparison Protocol

In this section, we construct a card-based protocol for secure comparison between integers in order to perform the role of Gakmoro's dealer.

The protocol design policy is that, when integers a and b are to be compared, the subtraction $a - b$ is obtained, and the sign (positive, negative, or zero) of this value is used to determine whether $a > b$, $a < b$, or $a = b$. For this reason, we first construct a subtraction protocol in Sect. 3.1, referring to the addition protocol introduced in Sect. 2.4. Next, in Sect. 3.2, we propose a comparison protocol by making use of the subtraction protocol.

3.1   Subtraction Protocol

The subtraction protocol proposed here takes $E_{m+1}^{\heartsuit}(a)$ and $E_{m+1}^{\heartsuit}(b)$ as input and aims to output the difference $a - b$ (note that this is not $a - b \bmod m + 1$). Therefore, since we need to deal with negative integers, we extend the encoding a little. Fix a positive integer m and represent the integers from $-m$ to m using

$2m + 1$ cards as follows:

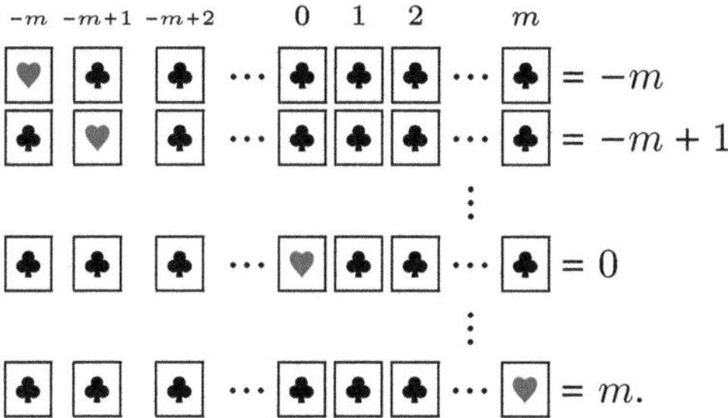

Under such an encoding, when an integer i such that $-m \le i \le m$ is represented by $2m + 1$ cards placed face down, we write $E^{\heartsuit}_{[-m,m]}(i)$ for the bundle of cards. (A similar encoding is used in the literature [9].)

In the following, given $E^{\heartsuit}_{m+1}(a)$ and $E^{\heartsuit}_{m+1}(b)$ along with $2m$ additional black cards, we show a protocol for generating $E^{\heartsuit}_{[-m,m]}(a - b)$.

1. Place $2m$ $\clubsuit$ s face down as follows; this yields $E^{\heartsuit}_{[-m,m]}(a)$ and $E^{\heartsuit}_{[-m,m]}(b - m)$:

$$E^{\heartsuit}_{[-m,m]}(a) : \quad \boxed{?}\,\boxed{?} \cdots \boxed{?}\,\boxed{?}\,\boxed{?}\,\boxed{?} \cdots \boxed{?}\,\boxed{?}$$
$$E^{\heartsuit}_{[-m,m]}(b - m) : \quad \boxed{?}\,\boxed{?} \cdots \boxed{?}\,\boxed{?}\,\boxed{?}\,\boxed{?} \cdots \boxed{?}\,\boxed{?}.$$

Note that if we shift the lower row together with the upper row so that $\boxed{\heartsuit}$ in the lower row is at the left end, the whole is shifted b positions to the left and the value of the upper row becomes $a - b$, yielding $E^{\heartsuit}_{[-m,m]}(a - b)$. Since we cannot just reveal the lower row, we will apply a shuffle in the next step.

2. Apply a pile-shifting shuffle to the two rows:

3. Turn over all the cards in the bottom row and shift the cards in the upper and lower cards in piles until $\boxed{\heartsuit}$ is at the left edge:

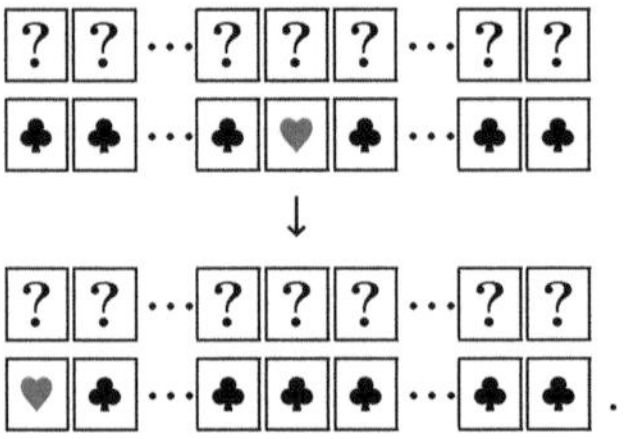

$$\begin{array}{cccccccc} \boxed{?} & \boxed{?} & \cdots & \boxed{?} & \boxed{?} & \boxed{?} & \cdots & \boxed{?} & \boxed{?} \\ \boxed{\clubsuit} & \boxed{\clubsuit} & \cdots & \boxed{\clubsuit} & \boxed{\heartsuit} & \boxed{\clubsuit} & \cdots & \boxed{\clubsuit} & \boxed{\clubsuit} \end{array}$$

$$\downarrow$$

$$\begin{array}{cccccccc} \boxed{?} & \boxed{?} & \cdots & \boxed{?} & \boxed{?} & \boxed{?} & \cdots & \boxed{?} & \boxed{?} \\ \boxed{\heartsuit} & \boxed{\clubsuit} & \cdots & \boxed{\clubsuit} & \boxed{\clubsuit} & \boxed{\clubsuit} & \cdots & \boxed{\clubsuit} & \boxed{\clubsuit} \end{array}.$$

In this case, the upper row becomes $E_{[-m,m]}^{\heartsuit}(a - b)$.

3.2   Secure Computation of Comparison

Suppose that we want to compare $E_{m+1}^{\heartsuit}(a)$ and $E_{m+1}^{\heartsuit}(b)$. With the subtraction protocol constructed in Sect. 3.1, we can obtain

$$E_{[-m,m]}^{\heartsuit}(a - b) : \quad \boxed{?}\,\boxed{?}\cdots\boxed{?}\,\boxed{?}\,\boxed{?}\cdots\boxed{?}\,\boxed{?},$$

and hence, if we reveal this bundle of cards to disclose its value, we can of course determine whether $a > b$, $a < b$, or $a = b$. However, this should be prohibited for Gakmoro, because if the value of $a - b$ is disclosed, the opponent's value can be calculated.

Therefore, shuffling is performed for the following two locations in order to obtain only the result of a comparison:

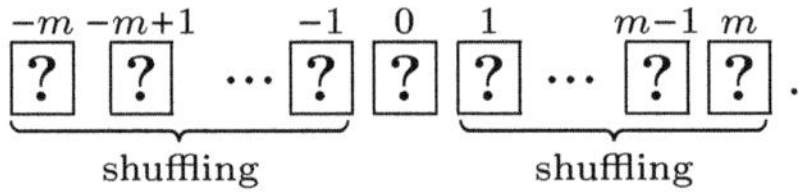

$$\underset{-m}{\boxed{?}}\ \underset{-m+1}{\boxed{?}}\ \cdots\ \underset{-1}{\boxed{?}}\ \underset{0}{\boxed{?}}\ \underset{1}{\boxed{?}}\ \cdots\ \underset{m-1}{\boxed{?}}\ \underset{m}{\boxed{?}}.$$

$$\underbrace{\qquad\qquad\qquad}_{\text{shuffling}} \qquad \underbrace{\qquad\qquad\qquad}_{\text{shuffling}}$$

Then, all cards are turned over, and the position of $\boxed{\heartsuit}$, which is present only once, yields the result of the comparison:

- If $\boxed{\heartsuit}$ appears in the center, then $a = b$;
- If $\boxed{\heartsuit}$ appears on the left, then $a < b$; and
- If $\boxed{\heartsuit}$ appears to the right, then $a > b$.

This is our comparison protocol using the subtraction protocol.

The proposed method is slightly similar to Nuida's "Tug-of-War" technique [34] and also slightly similar to the additive protocol using counters and dummies by Hatsugai et al. [9]. Card-based arithmetic operations were discussed in [35], and those for other kinds of cards appear in [12,45].

4   Gakmoro with Secure Computation

In this section, we describe a card-based protocol for playing Gakmoro with only two players (Alice and Bob) by combining the protocols described and constructed thus far.

Remember that in Gakmoro, each player holds seven cards and submits one to three cards at every round. Let us add two cards of value 0 to each player's hand. Then, we can assume that each player, holding nine cards (namely, those of 0, 0, 1, 2, 3, 4, 5, 6, and 7) at the beginning, submits exactly three cards at every round (because a card of value 0 can serve as a dummy card). Thus, in the sequel, we assume this version of Gakmoro.

Our protocol for playing Gakmoro with secure computations proceeds as follows.

1. For each of Alice and Bob, create nine bundles of cards corresponding to 0, 0, 1, 2, 3, 4, 5, 6, and 7:

$$E_8^{\heartsuit}(0), E_8^{\heartsuit}(0), E_8^{\heartsuit}(1), E_8^{\heartsuit}(2), E_8^{\heartsuit}(3), E_8^{\heartsuit}(4), E_8^{\heartsuit}(5), E_8^{\heartsuit}(6), E_8^{\heartsuit}(7).$$

    These nine bundles are regarded as his or her hand.
2. Each player chooses three bundles from their hand and places them face down on the table.
3. Let $E_8^{\heartsuit}(a_1), E_8^{\heartsuit}(a_2), E_8^{\heartsuit}(a_3)$ be the three bundles placed by Alice. The addition protocol described in Sect. 2.4 is applied to these to compute the total value. In this case, the modulus for addition needs to be expanded. Specifically, when $E_8^{\heartsuit}(a_1)$ and $E_8^{\heartsuit}(a_2)$ are first added, the maximum possible sum is $7+6 = 13$. Therefore, additional cards are prepared, and $E_{14}^{\heartsuit}(a_1)$ and $E_{14}^{\heartsuit}(a_2)$ are used as input to obtain the result of addition $E_{14}^{\heartsuit}(a_1 + a_2)$. Next, we want to add $E_{14}^{\heartsuit}(a_1 + a_2)$ and $E_8^{\heartsuit}(a_3)$. Considering that the maximum possible total value is $7 + 6 + 5 = 18$, we can use $E_{19}^{\heartsuit}(a_1 + a_2)$ and $E_{19}^{\heartsuit}(a_3)$ as input to obtain the result of addition $E_{19}^{\heartsuit}(a_1 + a_2 + a_3)$. In the same way, we obtain Bob's sum $E_{19}^{\heartsuit}(b_1 + b_2 + b_3)$.
4. Perform a comparison of the players' sums, $E_{19}^{\heartsuit}(a_1 + a_2 + a_3)$ and $E_{19}^{\heartsuit}(b_1 + b_2 + b_3)$, using the comparison protocol described in Sect. 3.2, to obtain only the winner's name (or a draw). After obtaining the result of a win or loss, the face-up cards are used as additional cards for the next and subsequent rounds.
5. Step 2 to Step 4 are repeated for up to three rounds, with the first player to win two rounds being declared the winner.

5   Implementation Considerations

In this section, we discuss points to note, implementation methods, and physical space requirements when actually using the methods presented in Sect. 4 in Gakmoro.

5.1　Notes on Input and How to Implement

We proposed a method to enjoy Gakmoro with only two players in Sect. 4. The protocol for outputting the win/loss at each round conforms to the standard computational model for card-based protocols [26],[3] where both players input $E_8^\heartsuit(a_1), E_8^\heartsuit(a_2), E_8^\heartsuit(a_3)$ and $E_8^\heartsuit(b_1), E_8^\heartsuit(b_2), E_8^\heartsuit(b_3)$. After inputting, the protocol can be executed by anyone, and the correct winner is always output.

However, when actually applying and implementing this in Gakmoro, care must be taken as to how each player prepares an input. For example, if we simply give each player 63 $\boxed{\clubsuit}$ s and 9 $\boxed{\heartsuit}$ s to make two $E_8^\heartsuit(0)$ bundles and $E_8^\heartsuit(1)$ through $E_8^\heartsuit(7)$ bundles, a malicious player might create all nine bundles as $E_8^\heartsuit(7)$.[4]

Fig. 1. Preparation for making $E_8^\heartsuit(3)$.

Fig. 2. The cards are placed in a sleeve.

---

[3] There is another computational model (cf. [20,30]).

[4] This is unlikely to happen in a normal secure computation where players input their own secrets and try to obtain a meaningful output.

The following implementation method is possible. First, Alice and Bob cooperatively and publicly make two $E_8^\heartsuit(0)$ and $E_8^\heartsuit(1)$ through $E_8^\heartsuit(7)$. Each bundle of cards is then placed in a sleeve, which is often used in trading card games. Figure 1 shows a photograph of the making of $E_8^\heartsuit(3)$ and the sleeve. Here, a sleeve with a transparent front and an opaque backside is used. Figure 2 shows how eight cards are placed in a sleeve.

After placing two $E_8^\heartsuit(0)$ and one each of $E_8^\heartsuit(1)$ through $E_8^\heartsuit(7)$ in their respective sleeves, place a sticky note on the transparent surface of each sleeve so that the integer values inside can be easily identified. Such a picture is shown in Fig. 3.

Fig. 3. Sleeves with numbers.

Two sets of sleeves made in this way are prepared, one for Alice and one for Bob. The back of the sleeve is opaque, so each player holds the sleeve in his or her hand with the back facing the other player and the sticky note side facing him or her, and make up the hand in units of sleeves, similar to how one plays Gakmoro with regular playing cards. In this case, each player only needs to choose three sleeves for each round (while possibly shuffling the sleeves).

When each player submits a sleeve from his or her hand, he or she peels off the sticky note before placing it on the table (of course, the peeled-off sticky note should not be visible to the opponent). Then, the player simply removes the cards face down from the sleeve, and the protocol is executed.

5.2   Space Requirements in Implementation

This subsection addresses practical (physical) space requirements encountered during implementation.

Recall that when each player performs the addition protocol on the three sleeves, they first apply the addition protocol to two bundles of cards, then apply it again to the result and the remaining bundle. This process is performed twice. As described in Sect. 2.4, the first addition uses $14 \times 2 = 28$ cards and the second addition uses $19 \times 2 = 38$ cards. In the comparison protocol, $(19 + 18) \times 2 = 74$ cards will be used.

To perform these operations, we require enough space to place 74 cards total, arranged in a 2-row by 37-column grid. This space requirement could be a challenge in implementation. Furthermore, for the game to run smoothly, it is desirable to have space for six cards vertically, allowing players to arrange the bundles of cards they submitted vertically.

6 Conclusion

In this paper, we proposed a new card game called "Gakmoro" that typically requires a dealer. We then presented a method that utilizes secure computation to allow two players to enjoy this game without a dealer.

We have demonstrated that leveraging secure computations can increase the flexibility of game play. If secure computation can be applied not only to Gakmoro but also to various other card games, it is expected that a wider range of people will be exposed to the fascinating world of cryptography. Furthermore, such applications could be useful for education in terms of gamification (cf. [5]).

Acknowledgements. We thank the anonymous reviewers, whose comments have helped us improve the presentation of the paper. This work was supported in part by JSPS KAKENHI Grant Numbers JP24K02938 and JP23H00479.

References

1. Boer, B.: More efficient match-making and satisfiability the five card trick. In: Quisquater, J.-J., Vandewalle, J. (eds.) EUROCRYPT 1989. LNCS, vol. 434, pp. 208–217. Springer, Heidelberg (1990). https://doi.org/10.1007/3-540-46885-4_23
2. Bultel, X., et al.: Physical zero-knowledge proof for Makaro. In: Izumi, T., Kuznetsov, P. (eds.) SSS 2018. LNCS, vol. 11201, pp. 111–125. Springer, Cham (2018). https://doi.org/10.1007/978-3-030-03232-6_8
3. Crépeau, C., Kilian, J.: Discreet solitary games. In: Stinson, D.R. (ed.) CRYPTO 1993. LNCS, vol. 773, pp. 319–330. Springer, Heidelberg (1994). https://doi.org/10.1007/3-540-48329-2_27
4. Doi, A., et al.: Card-based protocols for private set intersection and union. New Gener. Comput. **42**, 359–380 (2024). https://doi.org/10.1007/s00354-024-00268-z
5. Gong, X., Xu, W., Yu, S., Ma, J., Qiao, A.: Enhancing computational thinking and spatial reasoning skills in gamification programming learning: a comparative study of tangible, block and paper-and-pencil tools. Br. J. Edu. Technol. **56**(1), 80–102 (2025). https://doi.org/10.1111/bjet.13482
6. Haga, R., Hayashi, Y., Miyahara, D., Mizuki, T.: Card-minimal protocols for three-input functions with standard playing cards. In: Batina, L., Daemen, J. (eds.) AFRICACRYPT 2022. LNCS, vol. 13503, pp. 448–468. Springer, Cham (2022). https://doi.org/10.1007/978-3-031-17433-9_19
7. Hand, S., Koch, A., Lafourcade, P., Miyahara, D., Robert, L.: Efficient card-based ZKP for single loop condition and its application to Moon-or-Sun. New Gener. Comput. **42**, 449–477 (2024). https://doi.org/10.1007/s00354-024-00274-1

8. Hashimoto, Y., Shinagawa, K., Nuida, K., Inamura, M., Hanaoka, G.: Secure grouping protocol using a deck of cards. IEICE Trans. Fundam. **E101.A**(9), 1512–1524 (2018). https://doi.org/10.1587/transfun.E101.A.1512

9. Hatsugai, K., Ruangwises, S., Asano, K., Abe, Y.: NP-completeness and physical zero-knowledge proofs for Sumplete, a puzzle generated by ChatGPT. New Gener. Comput. **42**, 429–448 (2024). https://doi.org/10.1007/s00354-024-00267-0

10. Honda, Y., Shinagawa, K.: Efficient card-based protocols with a standard deck of playing cards using partial opening. In: Minematsu, K., Mimura, M. (eds.) Advances in Information and Computer Security. LNCS, vol. 14977, pp. 85–100. Springer, Singapore (2024). https://doi.org/10.1007/978-981-97-7737-2_5

11. Ishikawa, R., Chida, E., Mizuki, T.: Efficient card-based protocols for generating a hidden random permutation without fixed points. In: Calude, C.S., Dinneen, M.J. (eds.) UCNC 2015. LNCS, vol. 9252, pp. 215–226. Springer, Cham (2015). https://doi.org/10.1007/978-3-319-21819-9_16

12. Isuzugawa, R., Miyahara, D., Mizuki, T.: Zero-knowledge proof protocol for Cryptarithmetic using dihedral cards. In: Kostitsyna, I., Orponen, P. (eds.) Unconventional Computation and Natural Computation. LNCS, vol. 12984, pp. 51–67. Springer, Cham (2021). https://doi.org/10.1007/978-3-030-87993-8_4

13. Ito, Y., Shikata, H., Suganuma, T., Mizuki, T.: Card-based cryptography meets 3D printer. In: Cho, D.J., Kim, J. (eds.) Unconventional Computation and Natural Computation. LNCS, vol. 14776, pp. 74–88. Springer, Cham (2024). https://doi.org/10.1007/978-3-031-63742-1_6

14. Iwamoto, C., Ohara, K.: Card-based zero-knowledge proof for Dosun-Fuwari. IEICE Trans. Fundam. (2025). https://doi.org/10.1587/transfun.2024DML0001

15. Koch, A.: The landscape of security from physical assumptions. In: IEEE Information Theory Workshop, pp. 1–6. IEEE, NY (2021). https://doi.org/10.1109/ITW48936.2021.9611501

16. Koch, A., Schrempp, M., Kirsten, M.: Card-based cryptography meets formal verification. In: Galbraith, S.D., Moriai, S. (eds.) ASIACRYPT 2019. LNCS, vol. 11921, pp. 488–517. Springer, Cham (2019). https://doi.org/10.1007/978-3-030-34578-5_18

17. Komano, Y., Mizuki, T.: Card-based zero-knowledge proof protocol for pancake sorting. In: Bella, G., Doinea, M., Janicke, H. (eds.) Innovative Security Solutions for Information Technology and Communications. LNCS, vol. 13809, pp. 222–239. Springer, Cham (2023). https://doi.org/10.1007/978-3-031-32636-3_13

18. Komano, Y., Mizuki, T.: Physical zero-knowledge proof protocols for Topswops and Botdrops. New Gener. Comput. **42**, 399–428 (2024). https://doi.org/10.1007/s00354-024-00272-3

19. Koyama, H., Miyahara, D., Mizuki, T., Sone, H.: A secure three-input AND protocol with a standard deck of minimal cards. In: Santhanam, R., Musatov, D. (eds.) Computer Science – Theory and Applications. LNCS, vol. 12730, pp. 242–256. Springer, Cham (2021). https://doi.org/10.1007/978-3-030-79416-3_14

20. Manabe, Y., Ono, H.: Card-based cryptographic protocols with a standard deck of cards using private operations. New Gener. Comput. **42**, 305–329 (2024). https://doi.org/10.1007/s00354-024-00257-2

21. Miyahara, D., Hayashi, Y., Mizuki, T., Sone, H.: Practical card-based implementations of Yao's millionaire protocol. Theor. Comput. Sci. **803**, 207–221 (2020). https://doi.org/10.1016/j.tcs.2019.11.005

22. Miyahara, D., Mizuki, T.: Secure computations through checking suits of playing cards. In: Li, M., Sun, X. (eds.) Frontiers in Algorithmics. LNCS, vol. 13461, pp. 110–128. Springer, Cham (2023). https://doi.org/10.1007/978-3-031-20796-9_9

23. Miyahara, D., Robert, L., Lafourcade, P., Mizuki, T.: ZKP protocols for Usowan, Herugolf, and Five Cells. Tsinghua Sci. Technol. **29**(6), 1651–1666 (2024). https://doi.org/10.26599/TST.2023.9010153

24. Mizuki, T.: Efficient and secure multiparty computations using a standard deck of playing cards. In: Foresti, S., Persiano, G. (eds.) CANS 2016. LNCS, vol. 10052, pp. 484–499. Springer, Cham (2016). https://doi.org/10.1007/978-3-319-48965-0_29

25. Mizuki, T., Asiedu, I.K., Sone, H.: Voting with a logarithmic number of cards. In: Mauri, G., Dennunzio, A., Manzoni, L., Porreca, A.E. (eds.) UCNC 2013. LNCS, vol. 7956, pp. 162–173. Springer, Heidelberg (2013). https://doi.org/10.1007/978-3-642-39074-6_16

26. Mizuki, T., Shizuya, H.: A formalization of card-based cryptographic protocols via abstract machine. Int. J. Inf. Secur. **13**(1), 15–23 (2014). https://doi.org/10.1007/s10207-013-0219-4

27. Mizuki, T., Shizuya, H.: Computational model of card-based cryptographic protocols and its applications. IEICE Trans. Fundam. **E100.A**(1), 3–11 (2017). https://doi.org/10.1587/transfun.E100.A.3

28. Mizuki, T., Sone, H.: Six-card secure AND and four-card secure XOR. In: Deng, X., Hopcroft, J.E., Xue, J. (eds.) FAW 2009. LNCS, vol. 5598, pp. 358–369. Springer, Heidelberg (2009). https://doi.org/10.1007/978-3-642-02270-8_36

29. Murata, S., Miyahara, D., Mizuki, T., Sone, H.: Efficient generation of a card-based uniformly distributed random derangement. In: Uehara, R., Hong, S.-H., Nandy, S.C. (eds.) WALCOM 2021. LNCS, vol. 12635, pp. 78–89. Springer, Cham (2021). https://doi.org/10.1007/978-3-030-68211-8_7

30. Nakai, T., Iwanari, K., Ono, T., Abe, Y., Watanabe, Y., Iwamoto, M.: Card-based cryptography with a standard deck of cards, revisited: Efficient protocols in the private model. New Gener. Comput. **42**, 345–358 (2024). https://doi.org/10.1007/s00354-024-00269-y

31. Nakai, T., Tokushige, Y., Misawa, Y., Iwamoto, M., Ohta, K.: Efficient card-based cryptographic protocols for millionaires' problem utilizing private permutations. In: Foresti, S., Persiano, G. (eds.) CANS 2016. LNCS, vol. 10052, pp. 500–517. Springer, Cham (2016). https://doi.org/10.1007/978-3-319-48965-0_30

32. Niemi, V., Renvall, A.: Solitaire zero-knowledge. Fundam. Inf. **38**(1,2), 181–188 (1999). https://doi.org/10.3233/FI-1999-381214

33. Nishimura, A., Hayashi, Y., Mizuki, T., Sone, H.: Pile-shifting scramble for card-based protocols. IEICE Trans. Fundam. **101**(9), 1494–1502 (2018). https://doi.org/10.1587/transfun.E101.A.1494

34. Nuida, K.: Efficient card-based Millionaires' protocols via non-binary input encoding. In: Shikata, J., Kuzuno, H. (eds.) Advances in Information and Computer Security. LNCS, vol. 14128, pp. 237–254. Springer, Cham (2023). https://doi.org/10.1007/978-3-031-41326-1_13

35. Odaka, S., Komano, Y.: Card-based arithmetic operations and application to statistical data aggregation. In: Morogan, L., Roenne, P., Bica, I. (eds.) Innovative Security Solutions for Information Technology and Communications. SecITC 2024. LNCS, vol. 15595, pp. 118–134. Springer, Cham (2025). https://doi.org/10.1007/978-3-031-87760-5_10

36. Robert, L., Miyahara, D., Lafourcade, P., Mizuki, T.: Physical ZKP protocols for Nurimisaki and Kurodoko. Theor. Comput. Sci. **972**, 114071 (2023). https://doi.org/10.1016/j.tcs.2023.114071

37. Ruangwises, S., Itoh, T.: Physical ZKP for connected spanning subgraph: Applications to Bridges puzzle and other problems. In: Kostitsyna, I., Orponen, P. (eds.)

Unconventional Computation and Natural Computation, LNCS, vol. 12984, pp. 149–163. Springer, Cham (2021). https://doi.org/10.1007/978-3-030-87993-8_10

38. Ruangwises, S., Itoh, T.: Securely computing the n-variable equality function with $2n$ cards. Theor. Comput. Sci. **887**, 99–110 (2021). https://doi.org/10.1016/j.tcs.2021.07.007

39. Ruangwises, S., Ono, T., Abe, Y., Hatsugai, K., Iwamoto, M.: Card-based overwriting protocol for equality function and applications. In: Cho, D.J., Kim, J. (eds.) Unconventional Computation and Natural Computation. LNCS, vol. 14776, pp. 18–27. Springer, Cham (2024). https://doi.org/10.1007/978-3-031-63742-1_2

40. Ruangwises, S., Shinagawa, K.: Simulating virtual players for UNO without computers. arXiv preprint (2025). https://doi.org/10.48550/arXiv.2502.05987, https://arxiv.org/abs/2502.05987

41. Shinagawa, K., Kanai, K., Miyamoto, K., Nuida, K.: How to covertly and uniformly scramble the 15 puzzle and Rubik's cube. In: Broder, A.Z., Tamir, T. (eds.) Fun with Algorithms. LIPIcs, vol. 291, pp. 30:1–30:15. Schloss Dagstuhl, Dagstuhl, Germany (2024). https://doi.org/10.4230/LIPIcs.FUN.2024.30

42. Shinagawa, K., Miyahara, D., Mizuki, T.: How to play old maid with virtual players. Theory Comput. Syst. **69**(1) (2025). https://doi.org/10.1007/s00224-024-10203-w

43. Shinagawa, K., et al.: Card-based protocols using regular polygon cards. IEICE Trans. Fundam. **E100.A**(9), 1900–1909 (2017). https://doi.org/10.1587/transfun.E100.A.1900

44. Shinoda, Y., Miyahara, D., Shinagawa, K., Mizuki, T., Sone, H.: Card-based covert lottery. In: Maimut, D., Oprina, A.-G., Sauveron, D. (eds.) SecITC 2020. LNCS, vol. 12596, pp. 257–270. Springer, Cham (2021). https://doi.org/10.1007/978-3-030-69255-1_17

45. Takahashi, Y., Shinagawa, K.: Extended addition protocol and efficient voting protocols using regular polygon cards. New Gener. Comput. **42**, 479–496 (2024). https://doi.org/10.1007/s00354-024-00275-0

46. Takahashi, Y., Shinagawa, K., Shikata, H., Mizuki, T.: Efficient card-based protocols for symmetric functions using four-colored decks. In: ACM ASIA Public-Key Cryptography Workshop, pp. 1–10. ACM, New York (2024). https://doi.org/10.1145/3659467.3659902

47. Tamura, Y., Suzuki, A., Mizuki, T.: Card-based zero-knowledge proof protocols for the 15-puzzle and the token swapping problem. In: ACM ASIA Public-Key Cryptography Workshop, pp. 11–22. ACM, New York (2024). https://doi.org/10.1145/3659467.3659905

48. Tanaka, K., Sasaki, S., Shinagawa, K., Mizuki, T.: Only two shuffles perform card-based zero-knowledge proof for Sudoku of any size. In: 2025 Symposium on Simplicity in Algorithms (SOSA), pp. 94–107. SIAM (2025). https://doi.org/10.1137/1.9781611978315.7

49. Tozawa, K., Morita, H., Mizuki, T.: Single-shuffle card-based protocol with eight cards per gate. In: Genova, D., Kari, J. (eds.) Unconventional Computation and Natural Computation. LNCS, vol. 14003, pp. 171–185. Springer, Cham (2023). https://doi.org/10.1007/978-3-031-34034-5_12

50. Ueda, I., Nishimura, A., Hayashi, Y., Mizuki, T., Sone, H.: How to implement a random bisection cut. In: Martín-Vide, C., Mizuki, T., Vega-Rodríguez, M.A. (eds.) TPNC 2016. LNCS, vol. 10071, pp. 58–69. Springer, Cham (2016). https://doi.org/10.1007/978-3-319-49001-4_5

51. Yao, A.C.: Protocols for secure computations. In: Foundations of Computer Science, pp. 160–164. IEEE Computer Society, Washington, DC, USA (1982). https://doi.org/10.1109/SFCS.1982.88
52. Yoshida, T., Tanaka, K., Nakabayashi, K., Chida, E., Mizuki, T.: Upper bounds on the number of shuffles for two-helping-card multi-input AND protocols. In: Deng, J., Kolesnikov, V., Schwarzmann, A.A. (eds.) Cryptology and Network Security. LNCS, vol. 14342, pp. 211–231. Springer, Singapore (2023). https://doi.org/10.1007/978-981-99-7563-1_10

Pattern Graphs of Cellular Automata
and Reaction Systems

Kyle Ambrose, Daniela Genova[(✉)][iD], and Troy Kidd[iD]

Department of Mathematics and Statistics, University of North Florida, Jacksonville,
FL 32224, USA
d.genova@unf.edu

Abstract. We introduce a framework that connects two discrete models of natural computing, cellular automata (CA) and reaction systems (RS). We define pattern graphs of a CA as one-out digraphs where vertices correspond to totally periodic configurations and edges reflect CA evolution. Using pattern graphs based on periodic configurations we show that every one-dimensional binary CA can be transformed into an RS via its zero-context graph and provide counterexamples for the converse. Modified techniques, such as increasing the number of states and subgraphs of pattern graphs, are used to transform arbitrary RS into CA.

Keywords: Elementary CA · Periodic configurations · RS Graph · Zero-context graph

1   Introduction

Natural computing, a term coined by G. Rozenberg in the 1970s, has grown to encompass vast and rapidly growing areas of study, two of which are cellular automata and reaction systems. In this work, we introduce a framework to connect the two.

Modeling parallel computation with cellular automata allows for a relatively simple, efficient, and easily scalable physical implementation, mainly due to locality of computational dependencies. A more traditional approach is to have groups of complex computational units following a dynamically traversed, often lengthy set of instructions, and interacting with a shared block of memory. To get a meaningful result, the instructions must be carefully designed such that one computational unit never modifies memory in a way that another was not designed to ignore or expect.

In contrast, cellular automata have computational units that only control their own state (with influence from a fixed neighborhood), and each perform a single, often simple computation at every time step. To create a system capable of complex computation, rather than mirroring the traditional approach by encoding a complex and brittle set of instructions for cells to follow, a more natural approach would be to design/grow multicellular structures to perform the

© The Author(s), under exclusive license to Springer Nature Switzerland AG 2026
E. Formenti and L. Manzoni (Eds.): UCNC 2025, LNCS 16364, pp. 361–377, 2026.
https://doi.org/10.1007/978-3-032-15641-9_24

computation. This much more closely parallels the approach biological systems take to perform complex tasks. Developed by J. von Neumann and S. Ulam [21], cellular automata continues to be a very active area of research [7,15,18,22,23].

Reaction systems were introduced in [8] as a formal model of biochemical interactions capturing the two main mechanisms: facilitation and inhibition. Instead of tracking concentrations, as chemical reaction networks, they describe the presence and absence of substances and evolve through stepwise applications of reactions. Each reaction consists of a set of required reactants, a set of inhibitors that must be absent, and a set of products that are produced when the reaction is enabled. Their finite-state nature makes them well-suited for modeling systems where timing and interaction patterns are relevant. The simplicity of the RS paradigm allows for studying reaction systems with many structures such as sets and posets, graphs, and logical formulas, [2,6,8–13,17] as well as to explore biological and computing applications [1–3,5,14].

While cellular automata are typically formalized as acting over an infinite grid of cells, practical implementation must be finitely describable. With binary cell states, a configuration can be translated to presence/absence of entities and the CA dynamics simulated with an RS. Abstracting from totally periodic configurations to build finite graphs that we call *pattern graphs*, we introduce a framework to transform cellular automata to reaction systems. Specifically, we provide a bridge between pattern graphs of CA and zero-context graphs of RS.

The paper is organized as follows. In Sect. 2, we recall basic definitions and facts about cellular automata and periodic configurations. In particular, we discuss and provide examples for extracting patterns from totally periodic configurations and building configurations from patterns to set the stage for defining pattern graphs in Sect. 3. In Sect. 4, we recall definitions of reaction systems and zero-context graphs, and discuss a bijection with power set functions. Using this, we define injective mappings from CA pattern graphs to RS graphs in Sect. 5. Finally, in Sect. 6, we discuss encoding methods that allow multi-state cellular automata to represent arbitrary reaction systems.

2 Cellular Automata and Periodic Configurations

Cellular automata, CA, developed by John von Neumann and Stanisław Ulam [21], model the dynamics of a regular network of cells that simultaneously change their states as a function of neighboring cell states. As with reaction systems, RS, there is typically a simplifying assumption that computation is globally synchronous, discrete, and deterministic. Additionally, cells are assumed to exist at each integer coordinate in an infinite space.

Despite some physical implausibility of these assumptions, the parallelism, spatial homogeneity of the update rule, and local interaction are useful for modeling physical/biological phenomena. Even very simple rules can result in automatic construction of complex spatiotemporal patterns from a random initial configuration, or support computational universality within a specified structure. We follow basic CA definitions and notation from Kari [15], with some changes to standard symbols where needed to distinguish from RS notation.

A *cellular automaton* is a tuple $\mathcal{C} = (d, \Sigma, N, f)$. We denote by $d \in \mathbb{Z}^+$ the *dimension*, with cells being located at each vector $\vec{n} \in \mathbb{Z}^d$. The finite set Σ contains cell states. Each cell position maps to exactly one element from Σ. The tuple $N = (\vec{n}_1, \vec{n}_2, \ldots, \vec{n}_m)$ is the *neighborhood*. Its components $\vec{n}_i \in \mathbb{Z}^d$ are added to a cell position to get the m neighbors of a cell. Neighborhoods often contain the zero vector (so a cell's neighborhood includes itself) and are often symmetric on each axis, but these are not required conditions. The *local update rule* is denoted by $f : \Sigma^m \to \Sigma$, and maps all states of the neighbors of one cell to the next state for that cell.

A *configuration* is a mapping $c : \mathbb{Z}^d \to \Sigma$ that assigns a state to each cell. Given a configuration $c \in \Sigma^{\mathbb{Z}^d}$, we simultaneously apply the local update rule to each cell and its neighborhood to produce the next configuration c'. That is, $c'(\vec{n}) = f[c(\vec{n}+\vec{n}_1), c(\vec{n}+\vec{n}_2), \ldots, c(\vec{n}+\vec{n}_m)]$, for all $\vec{n} \in \mathbb{Z}^d$. This transformation is called the *global transition function*, denoted by $\mathcal{C} : \Sigma^{\mathbb{Z}^d} \to \Sigma^{\mathbb{Z}^d}$. We use $\mathcal{C}$ to denote both the cellular automaton tuple and its global transition function when it is clear from the context which meaning is intended. For a configuration c and a global transition function $\mathcal{C}$, the image of c under $\mathcal{C}$ is denoted by $\mathcal{C}(c)$.

Patterns and Periodic Configurations. An $\vec{r}$-*periodic* configuration c satisfies $c(\vec{n} + \vec{r}) = c(\vec{n})$, $\forall \vec{n} \in \mathbb{Z}^d$, for some $\vec{r} \in \mathbb{Z}^d$. A configuration is *totally periodic* if it is $\vec{r}_i$-periodic for each vector in a set of d linearly independent vectors $R = \{\vec{r}_1, \ldots, \vec{r}_d\}$.

The components of a vector $\vec{r} \in \mathbb{Z}^d$ are given with respect to the standard basis $\{\vec{e}_1, \ldots, \vec{e}_d\}$ and are denoted by $\vec{r} = (r_1, \ldots, r_d)$. Therefore, the components of a vector $\vec{r}_i \in R$ over the standard basis are denoted by $\vec{r}_i = ((\vec{r}_i)_1, \ldots, (\vec{r}_i)_d)$. Alternatively, we can refer to the jth component of vector $\vec{r}_i$ as $(\vec{r}_i)_j = \vec{r}_i \cdot \vec{e}_j$.

A *pattern* is a mapping $p : D \to \Sigma$ over some finite or infinite domain $D \subseteq \mathbb{Z}^d$. In this work, we primarily consider finite patterns $p : \mathbb{Z}_m^d \to \Sigma$ for some $m \in \mathbb{Z}^+$, having a domain that can clearly tile $\mathbb{Z}^d$. A *subpattern* of a pattern $p : D \to \Sigma$ is a pattern $p' : D' \to \Sigma$ with $D' \subseteq D$, and $p'(\vec{n}) = p(\vec{n})$, $\forall \vec{n} \in D'$. Since $\mathbb{Z}^d$ is a valid pattern domain, we can also refer to subpatterns of configurations.

Given a vector $\vec{r} \in \mathbb{Z}^d$, the *translation determined by* $\vec{r}$, denoted by $\tau_{\vec{r}}$, is defined as the map $\tau_{\vec{r}} : \Sigma^{\mathbb{Z}^d} \to \Sigma^{\mathbb{Z}^d}$, which takes a configuration c into another configuration $c' = \tau_{\vec{r}}(c)$ with $c'(\vec{n}) = c(\vec{n} + \vec{r})$, $\forall \vec{n} \in \mathbb{Z}^d$. We can extend this to patterns, where for some pattern $p : D \to \Sigma$, the translated pattern $p' = \tau_{\vec{r}}(p)$ has a new domain $D' = \{\vec{n} - \vec{r} \mid \vec{n} \in D\}$, and $p'(\vec{n}) = p(\vec{n} + \vec{r})$, $\forall \vec{n} \in D'$.

Constructing Configurations from Patterns. Given a pattern p with domain $\mathbb{Z}_m^d$ for some $m \in \mathbb{Z}^+$, denote by $c(p)$ the configuration with $c(p)(\vec{n}) = p(\vec{\ell})$, where $\ell_i \equiv n_i \bmod m$ for all $\vec{n} \in \mathbb{Z}^d$, $\vec{\ell} \in \mathbb{Z}_m^d$ and $i \in \{1, \ldots, d\}$. This results in a configuration where any repeated translation of the pattern domain by $\pm m$ along any axis has the same state assignment under $c(p)$. Note that $c(p)$ is totally periodic on vectors $m \cdot \vec{e}_i$, $\forall i \in \{1, \ldots, d\}$, where the $\vec{e}_i$ are the standard basis vectors of $\mathbb{Z}^d$. Due to the ease of tiling and concise notation, we refer to patterns with domain $D = \mathbb{Z}_m^d$, $m \in \mathbb{Z}^+$ as *regular*, and otherwise *irregular*. The ability to construct a configuration from a regular pattern in this way is stated below.

364   K. Ambrose et al.

Proposition 1. *Given $\mathbb{Z}^d$ with standard basis vectors $\{\vec{e}_1, \ldots, \vec{e}_d\}$ and a regular pattern $p : D \to \Sigma$ where $D = \mathbb{Z}_m^d$, for some $m \in \mathbb{Z}^+$, we can construct a totally periodic configuration $c(p) : \mathbb{Z}^d \to \Sigma$ matching p on D and being $(m \cdot \vec{e}_i)$-periodic $\forall\, i \in \{1, \ldots, d\}$.*

Note that any finite pattern with domain D can be translated to have domain D' so that $D' \subseteq \mathbb{Z}_m^d$, for some $m \in \mathbb{Z}^+$. For an irregular pattern $p : D \to \Sigma$ with $D \subset \mathbb{Z}_m^d$, we can create a new pattern $p' : \mathbb{Z}_m^d \to \Sigma$ with $p'(\vec{n}) = p(\vec{n})$, $\forall\, \vec{n} \in D$, and with cells in $\mathbb{Z}_m^d \setminus D$ assigned arbitrary states, so that $c(p')$ constructs a valid configuration. This is demonstrated in Fig. 1(b).

Example 1. To construct a configuration from any pattern, we can find a translated pattern $p : D \to \Sigma$ with $D \subseteq \mathbb{Z}_m^d$, $m \in \mathbb{Z}^+$, and assign states to cells in $\mathbb{Z}_m^d \setminus D$ as in Fig. 1(b). In this example, the irregular pattern p is translated so the domain is within $\mathbb{Z}_6^2$. The cells outside of p are each assigned a state, and dimmed to visually differentiate them.

Alternatively, we may start with an irregular pattern p that can tile $\mathbb{Z}^d$ by translation, i.e., translating p along each of d linearly independent vectors in a set R, to produce a totally periodic configuration $c : \mathbb{Z}^d \to \Sigma$. Assume that p has been translated to have no negative cell coordinates. Instead of setting m so that $\mathbb{Z}_m^d$ minimally bounds the irregular pattern p, and assigning arbitrary states to missing cells as in Fig. 1(b), we can take another approach.

Consider translating p along integer multiples of the vectors in R to assign missing states. Figure 1(c) demonstrates finding a regular pattern q with domain $\mathbb{Z}_m^d$ and with $c(q) = c$, for a specific $m \in \mathbb{Z}^+$, determined by the method given in Theorem 1. Here, $R = \{\left(\begin{smallmatrix} 0 \\ 4 \end{smallmatrix}\right), \left(\begin{smallmatrix} 3 \\ 1 \end{smallmatrix}\right)\}$ and $m = 12 = |0 \cdot 1 - 3 \cdot 4|$. $\Diamond$

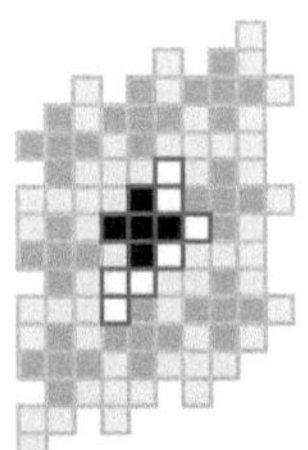

(a) A pattern p that can tile $\mathbb{Z}^2$ by translation.

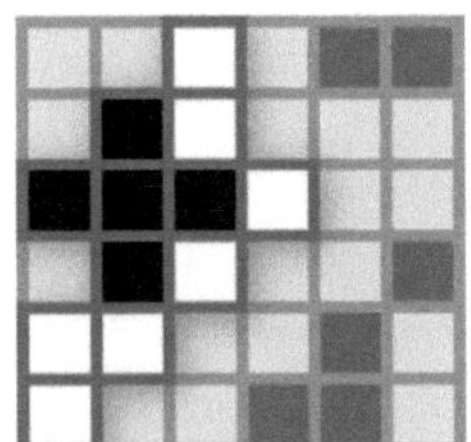

(b) A pattern over $\mathbb{Z}_6^2$ with subpattern p.

(c) A pattern over $\mathbb{Z}_{12}^2$ preserving neighbors.

Fig. 1. Creating regular patterns (b) and (c), from irregular pattern (a).

Extracting Patterns from Configurations. For an $\vec{r}$-periodic configuration that is totally periodic, $\vec{r}$ does not have to be axis-aligned. It is less clear in such cases that we can extract a pattern p with domain $\mathbb{Z}_m^d$ from which we can reconstruct c as $c(p)$.

In addition to constructing a regular pattern from an irregular one, Fig. 1(c) represents extraction of a regular pattern from a totally periodic configuration. Theorem 1 states that it is possible to find this kind of representative regular pattern p with $c = c(p)$, for any totally periodic configuration c, and gives a construction method.

Theorem 1. *From any totally periodic configuration c over $\mathbb{Z}^d$, we can extract a regular pattern p such that $c = c(p)$.*

Proof. Given a totally periodic d-dimensional configuration c, we have that c is $\vec{r}_i$-periodic for each vector $\vec{r}_i$, where $1 \leq i \leq d$, in a set of d linearly independent vectors $R = \{\vec{r}_1, \ldots, \vec{r}_d\}$. Let M be the invertible integer matrix with columns $\vec{r}_1, \ldots, \vec{r}_d$, and let $m = |\det(M)| \in \mathbb{Z}^+$. Then since $M^{-1} = \det(M)^{-1}\mathrm{adj}(M)$ and $\mathrm{adj}(M)$ is an integer matrix, we have that mM^{-1} is an integer matrix.

The ith column of $M(mM^{-1})$ is the linear combination of vectors $\vec{r}_1, \ldots, \vec{r}_d$ with the ith column of mM^{-1} as coefficients. Since $M(mM^{-1}) = mI_d$, it follows that the columns of mI_d are integer linear combinations of the columns of the vectors in R. This means that c is $(m \cdot \vec{e}_i)$-periodic for all $1 \leq i \leq d$, and so $c = c(p)$ for p having domain $D = \mathbb{Z}_m^d$ and matching c on D. $\qquad\square$

The next example illustrates the construction used in the proof.

Example 2. If $\vec{r}_1 = \left(\begin{smallmatrix}1\\2\end{smallmatrix}\right)$ and $\vec{r}_2 = \left(\begin{smallmatrix}3\\4\end{smallmatrix}\right)$, then $M = \left(\begin{smallmatrix}1&3\\2&4\end{smallmatrix}\right)$ and $m = 2$. The inverse $M^{-1} = \frac{1}{-2}\left(\begin{smallmatrix}4&-3\\-2&1\end{smallmatrix}\right)$, so $mM^{-1} = \left(\begin{smallmatrix}-4&3\\2&-1\end{smallmatrix}\right)$. Then $M(2M^{-1}) = \left(\begin{smallmatrix}1&3\\2&4\end{smallmatrix}\right)\left(\begin{smallmatrix}-4&3\\2&-1\end{smallmatrix}\right) = 2I_2$. From this, we have $-4\vec{r}_1 + 2\vec{r}_2 = 2\vec{e}_1$ and $3\vec{r}_1 - 1\vec{r}_2 = 2\vec{e}_2$. $\qquad\Diamond$

The next corollary directly follows from Theorem 1, but the construction can be simplified for the 1-dimensional case.

Corollary 1. *From any 1-dimensional periodic configuration c, we can find a finite pattern p with domain $D = \mathbb{Z}_m^1$ for some $m \in \mathbb{Z}^+$, such that $c(p) = c$.*

Given that c is $\vec{r}$-periodic where $\vec{r}$ is a single component vector, we can choose m to be the value of that component.

For example, let c be a $\vec{4}$-periodic configuration where $c(\vec{0}) = 1$, $c(\vec{1}) = 1$, $c(\vec{2}) = 0$, and $c(\vec{3}) = 1$. Then we have $c(p) = c$ for the pattern p with domain $D = \mathbb{Z}_4^1 = \{\vec{0}, \vec{1}, \vec{2}, \vec{3}\}$. For conciseness, when dealing with two-state patterns with domain $\mathbb{Z}_m^1$ for some m, we may refer to the pattern as a binary string such as 1101, or a tuple (p_1, p_2, p_3, p_4). Note that p_1 here refers to the state at Cell $\vec{0}$.

Images of Periodic Configurations. As noted in [15], application of a CA function preserves total periodicity. This is given by the following proposition.

Proposition 2. *If a configuration $c \in \Sigma^{\mathbb{Z}^d}$ is $\vec{r}$-periodic for some vector $\vec{r} \in \mathbb{Z}^d$, then for any d-dimensional cellular automaton $\mathcal{C}$ with state set Σ, the configuration $\mathcal{C}(c)$ is also $\vec{r}$-periodic.*

Proof. The global transition function $\mathcal{C}$ is defined as a local update rule applied to the relative neighborhoods for each cell. If c is $\vec{r}$-periodic for some $\vec{r} \in \mathbb{Z}^d$, then for any $\vec{n}, \vec{t} \in \mathbb{Z}^d$, we have $c(\vec{n} + \vec{t}) = c(\vec{n} + \vec{r} + \vec{t})$. Using $\vec{t}$ in the neighborhood N, we see that the local update rule has the same input for Cell $\vec{n}$ as for Cell $(\vec{n} + \vec{r})$. Then $C(c)(\vec{n}) = C(c)(\vec{n} + \vec{r})$, establishing the $\vec{r}$-periodicity of $C(c)$.   □

It follows that every cellular automaton maps a totally periodic configuration to another totally periodic configuration, that is periodic on the same vectors. For successive configurations, this set of vectors is not guaranteed to stay minimal, in the sense that the total vector length is minimal while describing all periodicities. For example, a $\vec{4}$-periodic 1-dimensional configuration repeating the state sequence 1000 may be succeeded by a $\vec{2}$-periodic configuration repeating 01.

3   Pattern Graphs of a Cellular Automaton

For a general d-dimensional cellular automaton with k states, a pattern with domain $D \subseteq \mathbb{Z}_m^d$ for some $m \in \mathbb{Z}^+$ can be represented by a pattern in d dimensions. For two and three dimensions this is quite a straightforward visual. In two dimensions our pattern becomes a grid of size $m_1 \times m_2$, and in three dimensions our pattern becomes a rectangular prism of size $m_1 \times m_2 \times m_3$, where $m_i \leq m$ represents the domain of the ith dimension. Extending this to d dimensions gives us a d-dimensional pattern of size $\prod_{i=1}^d m_i$. When we take into account the fact that our automaton has k states, we get a total of $k^{\prod_{i=1}^d m_i}$ possible patterns. Notice, when $m_i = m \forall i$, then we get k^{m^d} as the total number of possible patterns over d-dimensions. We will primarily deal with regular patterns to simplify notation. Thus, k^{m^d} is the total number of possible patterns over $\mathbb{Z}_m^d$.

For a k-state d-dimensional cellular automaton, a pattern over domain $\mathbb{Z}_m^d$ can be represented as rank d tensors with states in $\{1, 2, \ldots, k\}$. The set of all patterns with domain $\mathbb{Z}_m^d$ is denoted by P_m^d (or P when understood).

Mapping Patterns into Patterns. Given a d-dimensional cellular automaton $\mathcal{C}$ with states $\Sigma = \{1, 2, \ldots, k\}$ and a pattern p over $\mathbb{Z}_m^d$, i.e., $p \in P_m^d$, we can construct a totally periodic d-dimensional configuration $c(p)$ as stated in Proposition 1. By Proposition 2, applying the global transition function $\mathcal{C}$ to our totally periodic configuration $c(p)$, denoted by $\mathcal{C}(c(p))$, results in another totally periodic configuration $c' = c(q)$, for some pattern q having the same domain as p, and potentially different state assignments. Thus, given a pattern p over $\mathbb{Z}_m^d$ and a cellular automaton $\mathcal{C}$, we can obtain a pattern q over $\mathbb{Z}_m^d$ from the next configuration. This process also extends to subsequent configurations. Hence, for regular patterns p, q, we use the notation $\mathcal{C}(p) = q$ as a shorthand with $p, q \in P_m^d$. All positions are assigned a state, so $|P_m^d| = |\Sigma|^{|\mathbb{Z}_m^d|}$. Thus, P_m^d is the set of all patterns over $\mathbb{Z}_m^d$ with state set Σ.

Pattern Graphs. In the above notation, given $m \in \mathbb{Z}^+$ and a cellular automaton $\mathcal{C} = (d, \Sigma, N, f)$, one can say that the cellular automaton is a function

$\mathcal{C} : P_m^d \to P_m^d$, or simply $\mathcal{C} : P \to P$ when understood, that maps patterns into patterns. We can construct a state transition diagram similar to [18] with these patterns as vertices. To emphasize that we are dealing with patterns, we refer to this as a *pattern graph* and define it as follows.

Definition 1. Given $n \in \mathbb{Z}^+$ and a cellular automaton $\mathcal{C} = (d, \Sigma, N, f)$, let $P = P_n^d$ be the set of all patterns over $\mathbb{Z}_n^d$ with state set Σ. Define the *pattern graph* $G_P^{\mathcal{C}} = (V, E)$, where $V = P$ and $E = \{(p, q) \mid p, q \in P \text{ and } \mathcal{C}(p) = q\}$.

Thus, we can create a graph representing the possible patterns and their subsequent patterns for any cellular automaton. For a given $n \in \mathbb{Z}^+$ and a cellular automaton $\mathcal{C}$, the pattern graph $G_P^{\mathcal{C}}$ is a one-out directed graph, since the global transition function $\mathcal{C}$ maps every pattern $p \in P$ into a unique image, pattern q. It has $|\Sigma|^{n^d}$ vertices, and therefore the same number of edges.

An example subgraph of a pattern graph is given in Fig. 2. It depicts two converging paths in a 2-dimensional pattern graph with over 33 million vertices.

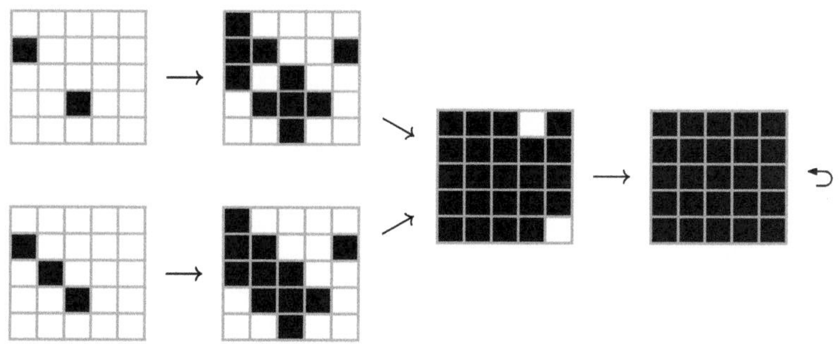

Fig. 2. A subgraph of a pattern graph with $n = 5$, $d = 2$, and $\Sigma = \{0, 1\}$.

Since pattern graphs can be constructed for any $n \in \mathbb{Z}^+$ and any cellular automaton $\mathcal{C}$, they provide a framework for studying cellular automata, as well as for connecting cellular automata to other structures.

Elementary Cellular Automata. One-dimensional binary cellular automata can be represented by cells arranged in a line, where the cells can be either in State 0 or in State 1. This makes their configurations bi-infinite binary sequences. If in addition, the neighborhood for the local update rule consists of a cell and its left and right neighbors, $\mathcal{C}$ is called an *elementary cellular automaton*. Defined by S. Wolfram, elementary CA have been studied extensively, see [4,22].

More precisely, an elementary cellular automaton $\mathcal{C} = (d, S, N, f)$, where $d = 1$, cells are located at each vector $\vec{n} \in \mathbb{Z}$, the set of cell states is $\Sigma = \{0, 1\}$, the neighborhood is $N = (\vec{n}_1, \vec{n}_2, \vec{n}_3) = (-\vec{1}, \vec{0}, \vec{1})$, and each cell position maps to exactly one element from Σ. Since there are 2^3 possible neighborhoods, there are $2^{2^3} = 256$ possible elementary CA. They are identified by a number from 0 to 255, where the binary representation of the number is the sequence that would be produced when applying the rule to neighborhoods that are the three digit binary representations of $7, 6, \ldots, 1, 0$, as shown in Fig. 3. We follow the standard representation of black and white cells to mean that they have been mapped to State 1 or State 0, respectively.

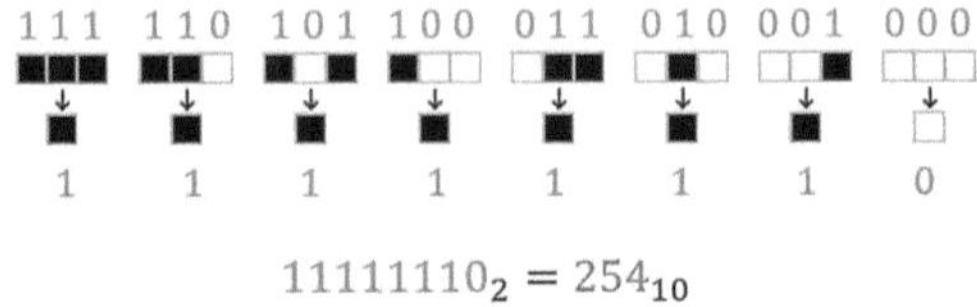

$$11111110_2 = 254_{10}$$

Fig. 3. Creating a binary sequence from the local update rule for Rule 254.

For a binary one-dimensional cellular automaton such as an elementary CA, a contiguous pattern of length n can be represented with a binary string of length n. Similarly, we will often represent it as n contiguous black or white cells. A one-dimensional configuration that is n-periodic can be characterized by a pattern of n contiguous cells. There are 2^n such patterns that represent each n-periodic one-dimensional configuration with binary cell states, counting configurations that are shifts of each other as different configurations.

Each element $p \in P_n^1$ (or simply P_n) corresponds to a pattern of n consecutive cells. For $p = (p_1, \ldots, p_n)$, $p_i \in \{0, 1\}$ is the state of the cell at position $\vec{i}$. As in the process for a d-dimensional cellular automaton and a pattern over $\mathbb{Z}_m^d$, given a pattern p of length $n \geq 1$, we can get another pattern q by looking at the configuration after applying the global transition function. Here, $p \in P_n$, where P_n is the set of all n-letter binary words, thus $|P_n| = 2^n$.

Pattern graphs on patterns of length n can have as little as one component and as many as 2^n components. One can use [23] to generate a pattern graph for a specific n and a given elementary CA. Two pattern graphs for patterns of length four, one with a single component and one with $16 = 2^4$ components are shown on Fig. 4.

4   Reaction Systems and Zero-Context Graphs

Reaction systems are a mathematical model that abstracts the dynamics of biochemical reactions and their interaction with the environment. They were introduced by Grzegorz Rozenberg and Andrzej Ehrenfeucht in [8] and are based on the observation that biochemical reactions require the presence of reactants, and absence of inhibitors in order for the reaction to take place. We recall commonly used reaction systems notation described in [8, 11, 12].

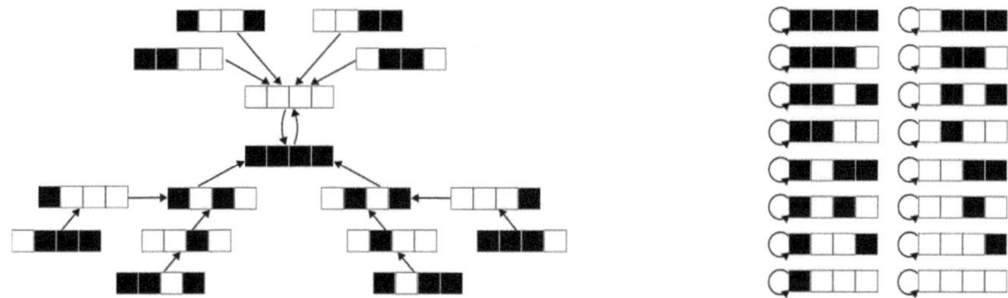

Fig. 4. Pattern graphs of Rules 37 and 204 for $n = 4$.

A *reaction system* is a pair $\mathcal{A} = (S, A)$ where S is a finite nonempty set called the *background set*, and $A \subseteq \mathcal{P}(S) \times \mathcal{P}(S) \times \mathcal{P}(S)$ is a set of *reactions* in S with $\mathcal{P}(S)$ denoting the set of all subsets of S. A *reaction* is a triple $a = (R_a, I_a, P_a)$ of subsets of S called the *reactant set of a, the inhibitor set of a*, and *the product set of a*, respectively. In the *generalized reaction systems* (RS) model, these subsets are allowed to be empty. A model of RS that requires R_a and I_a to be nonempty has also been widely studied, since it represents a more realistic reaction system, see for example [2,9,11,16]. A characteristic difference in their graphs can be found in Proposition 5.4. in [11], which does not hold for generalized RS.

For a subset of the entities $X \subseteq S$ and a reaction $a = (R_a, I_a, P_a)$, a is *enabled in X* if and only if $R_a \subseteq X$ and $I_a \cap X = \varnothing$. The *result of a on X* is denoted by $\mathrm{res}_a(X)$ and equals P_a if a is enabled in X and $\varnothing$ otherwise. The result of a reaction system $\mathcal{A}$ on X is $\mathrm{res}_\mathcal{A}(X) = \bigcup_{a \in A} \mathrm{res}_a(X)$. We note that for a reaction system $\mathcal{A}$, $\mathrm{res}_\mathcal{A}$ is a function on the power set of the background set of entities, $\mathrm{res}_\mathcal{A} : \mathcal{P}(S) \to \mathcal{P}(S)$, called *the result function of $\mathcal{A}$*.

Example 3. We recall a popular reaction systems example, first shown in [2]. Define the reaction system $\mathcal{A} = (S, A)$ to have background set $S = \{1, 2, 3, 4\}$ and the following reactions in set A:

$a_1 = (\{1\}, \{3\}, \{2\})$, $a_2 = (\{2\}, \{1\}, \{1\})$, $a_3 = (\{2\}, \{3\}, \{3\})$,

$a_4 = (\{3\}, \{1, 2\}, \{1, 2, 4\})$, $a_5 = (\{4\}, \{3\}, \{1, 2\})$, $a_6 = (\{1, 3\}, \{2, 4\}, \{2, 3\})$.

In the subset $X_1 = \{2, 3, 4\}$, only reaction a_2 is enabled, so $\mathrm{res}_\mathcal{A}(\{2, 3, 4\}) = \{1\}$. No reactions are enabled in $X_2 = \{1, 2, 3\}$, hence $\mathrm{res}_\mathcal{A}(\{1, 2, 3\}) = \varnothing$. And in $X_3 = \{1, 2, 4\}$, reactions $a_1, a_3,$ and a_5 are all enabled, producing $\mathrm{res}_\mathcal{A}(\{1, 2, 4\}) = \{2\} \cup \{3\} \cup \{1, 2\} = \{1, 2, 3\}$. $\diamond$

Zero-Context Graphs. The dynamical behavior of a reaction system can be modeled by a one-out graph with vertices, the subsets of the background set, and edges determined by the result function.

Starting from a subset of the entities $X \subseteq S$ and applying the result function $\mathrm{res}_\mathcal{A}$ to it, the reaction system produces products $Y \subseteq S$, after which the result function acts on Y and so on. The zero-context graphs defined in [11] provide a useful tool to study these dynamics.

Definition 2. For a reaction system $\mathcal{A} = (S, A)$, the *zero-context graph of* $\mathcal{A}$ is defined as the one-out graph $G_{\mathcal{A}}^0 = (\mathscr{P}(S), E)$, where $E = \{(X, \text{res}_{\mathcal{A}}(X)) \mid X \in \mathscr{P}(S)\}$ is the set of directed edges.

Zero-context graphs describe the behavior of a reaction system's reactions and products in a step-wise manner. They are also a useful tool to study the global transition graphs of RS which take into account adding context at each step, see [11]. In this work, we use them for providing a connection with cellular automata and also refer to them as *RS graphs* for short.

Example 4. The zero-context graph for the reaction system $\mathcal{A}$ from Example 3 is shown in Fig. 5. The vertices are all the subsets of S, while the edges are determined by the result function $\text{res}_{\mathcal{A}}$. For instance, $(X_1, \text{res}_{\mathcal{A}}(X_1)) = (\{2, 3, 4\}, \{1\})$ is an edge in $E(G_{\mathcal{A}}^0)$. Additionally, $(X_2, \text{res}_{\mathcal{A}}(X_2)) = (\{1, 2, 3\}, \varnothing)$ and $(X_3, \text{res}_{\mathcal{A}}(X_3)) = (\{1, 2, 4\}, \{1, 2, 3\})$ are also edges in the graph. $\quad\Diamond$

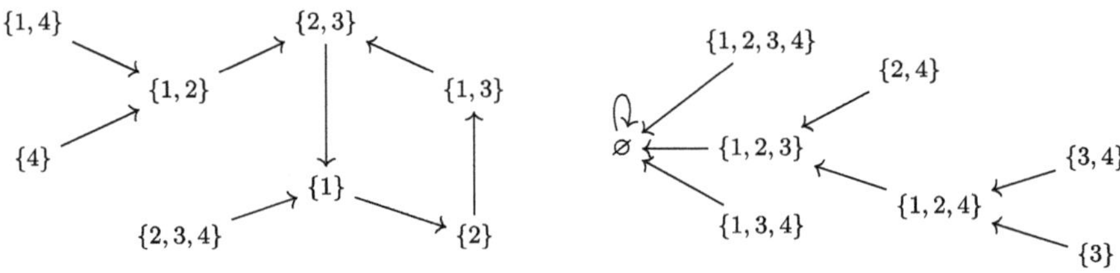

Fig. 5. The zero-context graph generated by reaction system $\mathcal{A}$ in Example 3.

Functions on Power Sets as Reaction Systems. Using the concept of maximally inhibited reactions, first defined by Salomaa in [20], it was shown in [12] that any function defined on a power set of a finite set S can be transformed into a reaction system, $\mathcal{A}$, with S as its background set. More specifically, if S is a finite set and $g : \mathscr{P}(S) \to \mathscr{P}(S)$, consider the graph of g, defined as $G_g = (V, E)$, where $V = \mathscr{P}(S)$ and $E = \{(X, Y) \mid g(X) = Y\}$. Then the graph G_g can be transformed into a reaction system $\mathcal{A} = (S, A)$ using *maximally inhibited reactions* $(X, S \setminus X, Y)$ over S, and then, the $2^{|S|}$ maximally inhibited reactions can be minimized through a logic minimization tool.

Example 5. Suppose a function $g : \mathscr{P}(S) \to \mathscr{P}(S)$ is given, such that G_g is the graph depicted in Fig. 5. For example, since $g(\{3, 4\}) = \{1, 2, 4\}$, the graph G_g has the edge $(\{3, 4\}, \{1, 2, 4\})$. This edge then transforms into the *maximally inhibited reaction* $(\{3, 4\}, \{1, 2\}, \{1, 2, 4\}) \in A$. This produces 2^4 maximally inhibited reactions which were later minimized to the set of six reactions A, listed in Example 3, using the Espresso logic minimization tool (see [12]). $\quad\Diamond$

For the rest of the paper, we will refer to such functions g as *power set functions*. That is, if g is a function on $\mathscr{P}(S)$ for some nonempty finite set S, g will be called a *power set function (over S)*.

3Binary Representation of RS Graphs. It is well known that there is a one-to-one correspondence between the subsets of a finite set S and the binary strings of length $|S|$. Given a finite non-empty set S with $|S| = n$, order the elements in S. For simplicity, assume that S is the set $\{1, \ldots, n\}$. Consider the set of binary strings of length n, denoted by P_n, and define $\phi : \mathcal{P}(S) \to P_n$ as follows: $\phi(X) = p_1 p_2 \ldots p_n$ such that $p_i = 1$ if $i \in X$ and $p_i = 0$ otherwise. It is well-known that ϕ is a bijection. Using ϕ we can represent the graph from Fig. 5 as the graph in Fig. 6. Note that any zero-context graph $G_\mathcal{A}^0$, as well as any graph of a power set function G_g, can be represented this way.

5   Reaction Systems Of CA on Periodic Configurations

In this section, we establish a connection between cellular automata and reaction systems. More specifically, we use a transformation from pattern graphs to zero-context graphs through graphs of power set functions to transform an elementary cellular automaton acting on periodic configurations to a reaction system.

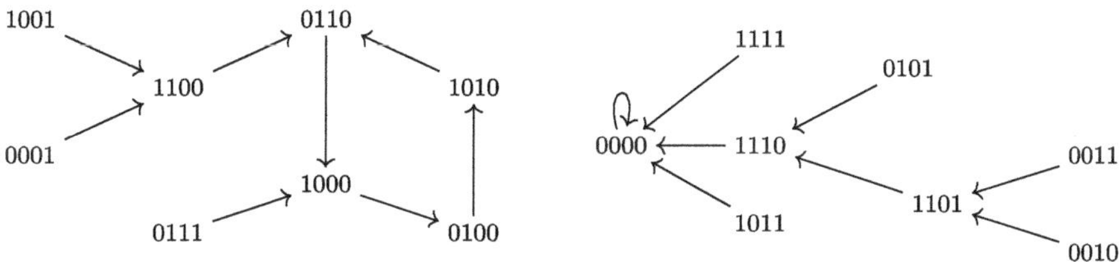

Fig. 6. The binary representation of the zero-context graph from Fig. 5.

Given an elementary cellular automaton $\mathcal{C}$ and $n \in \mathbb{Z}^+$, construct the pattern graph $G_P^\mathcal{C}$, where $P = P_n$. Recall that it is a one-out graph with 2^n vertices, P_n, that are all the binary strings of length n. Let $S = \{1, 2, \ldots, n\}$ and map the vertices P_n by ϕ^{-1} to convert them to subsets of S. Let ϕ' be the extension of ϕ^{-1} to $G_P^\mathcal{C}$, which preserves the edges. That is, ϕ' is an isomorphism on graphs that maps $G_P^\mathcal{C}$ to the graph G_g, a graph representing a power set function $g : \mathcal{P}(S) \to \mathcal{P}(S)$. Then, as detailed in the previous section and in Example 5, use maximally inhibited reactions to define a reaction system $\mathcal{A}' = (S, A')$. Note that $G_P^\mathcal{C}$ is precisely the zero-context graph $G_{\mathcal{A}'}^0$ of the reaction system $\mathcal{A}'$. Finally, one can minimize the maximally-inhibited reactions of A' through the Espresso logic minimization tool to obtain a minimized reaction system $\mathcal{A} - (S, A)$ of the RS $\mathcal{A}' = (S, A')$ (see [12]). This establishes the following result.

Theorem 2. *For every elementary cellular automaton $\mathcal{C}$ and every $n \in \mathbb{Z}^+$, there exists a reaction system $\mathcal{A}$ whose zero-context graph $G_\mathcal{A}^0$ is isomorphic to the pattern graph $G_P^\mathcal{C}$, where $P = P_n$.*

The above theorem states that every elementary cellular automaton that acts on all n-periodic configurations can be transformed into a reaction system that exhibits the same dynamics.

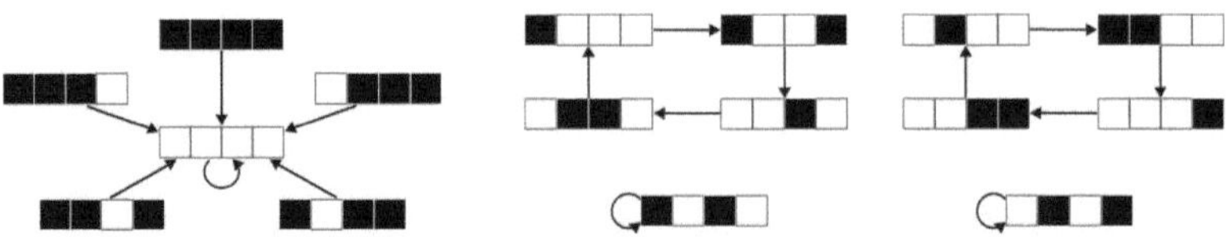

Fig. 7. The pattern graph for Rule 6, with $n = 4$.

Example 6. In this example, we find a minimized reaction system that produces the same graph structure as Rule 6 with $n = 4$, shown in Fig. 7. Since the patterns have 4 cells, $n = 4$ and $|P_4| = 16$. Hence, we will use $S = \{1, 2, 3, 4\}$ as the background set. Converting the binary sequences of P_4 to subsets with ϕ^{-1} and applying ϕ' produces the graph G_g of a power set function with edges

$$
\begin{array}{llll}
\varnothing \mapsto \varnothing & \{4\} \mapsto \{3, 4\} & \{2, 3\} \mapsto \{1\} & \{1, 2, 4\} \mapsto \varnothing \\
\{1\} \mapsto \{1, 4\} & \{1, 2\} \mapsto \{4\} & \{2, 4\} \mapsto \{2, 4\} & \{1, 3, 4\} \mapsto \varnothing \\
\{2\} \mapsto \{1, 2\} & \{1, 3\} \mapsto \{1, 3\} & \{3, 4\} \mapsto \{2\} & \{2, 3, 4\} \mapsto \varnothing \\
\{3\} \mapsto \{2, 3\} & \{1, 4\} \mapsto \{3\} & \{1, 2, 3\} \mapsto \varnothing & \{1, 2, 3, 4\} \mapsto \varnothing
\end{array}
$$

Representing presence of element i as v_i, absence as $\overline{v}_i$, and production of element i as φ_i, we construct the following logical formulas that correspond to the set of maximally inhibited reactions determined by the edges of G_g (see [12]).

$$\varphi_1 = v_1\overline{v}_2\overline{v}_3\overline{v}_4 + \overline{v}_1 v_2\overline{v}_3\overline{v}_4 + v_1\overline{v}_2 v_3\overline{v}_4 + \overline{v}_1 v_2 v_3\overline{v}_4$$

$$\varphi_2 = \overline{v}_1 v_2\overline{v}_3\overline{v}_4 + \overline{v}_1\overline{v}_2 v_3\overline{v}_4 + \overline{v}_1 v_2\overline{v}_3 v_4 + \overline{v}_1\overline{v}_2 v_3 v_4$$

$$\varphi_3 = \overline{v}_1\overline{v}_2 v_3\overline{v}_4 + \overline{v}_1\overline{v}_2\overline{v}_3 v_4 + v_1\overline{v}_2 v_3\overline{v}_4 + v_1\overline{v}_2\overline{v}_3 v_4$$

$$\varphi_4 = v_1\overline{v}_2\overline{v}_3\overline{v}_4 + \overline{v}_1\overline{v}_2\overline{v}_3 v_4 + v_1 v_2\overline{v}_3\overline{v}_4 + \overline{v}_1 v_2\overline{v}_3 v_4$$

These are then minimized with Espresso using Logic Friday [19], as in [12], and then translated to reactions. The minimized logical formulas are as follows.

$$\varphi'_1 = \overline{v}_1 v_2\overline{v}_4 + v_1\overline{v}_2\overline{v}_4 \qquad \varphi'_2 = \overline{v}_1\overline{v}_2 v_3 + \overline{v}_1 v_2\overline{v}_3$$

$$\varphi'_3 = \overline{v}_2\overline{v}_3 v_4 + \overline{v}_2 v_3\overline{v}_4 \qquad \varphi'_4 = \overline{v}_1\overline{v}_3 v_4 + v_1\overline{v}_3\overline{v}_4$$

Here each term, for example $\overline{v}_1 v_2\overline{v}_4$, is used in only one φ'_i. Thus, each of the product sets is a singleton. The corresponding RS is $\mathcal{A} = (S, A)$ with background set $S = \{1, 2, 3, 4\}$ and a set of reactions $A = \{a_1, a_2, \ldots, a_8\}$ as stated below.

$$
\begin{array}{llll}
a_1 = (\{2\}, \{1, 4\}, \{1\}) & a_3 = (\{3\}, \{1, 2\}, \{2\}) & a_5 = (\{4\}, \{2, 3\}, \{3\}) & a_7 = (\{4\}, \{1, 3\}, \{4\}) \\
a_2 = (\{1\}, \{2, 4\}, \{1\}) & a_4 = (\{2\}, \{1, 3\}, \{2\}) & a_6 = (\{3\}, \{2, 4\}, \{3\}) & a_8 = (\{1\}, \{3, 4\}, \{4\})
\end{array}
$$

This is represented as a zero-context graph in Fig. 8. $\Diamond$

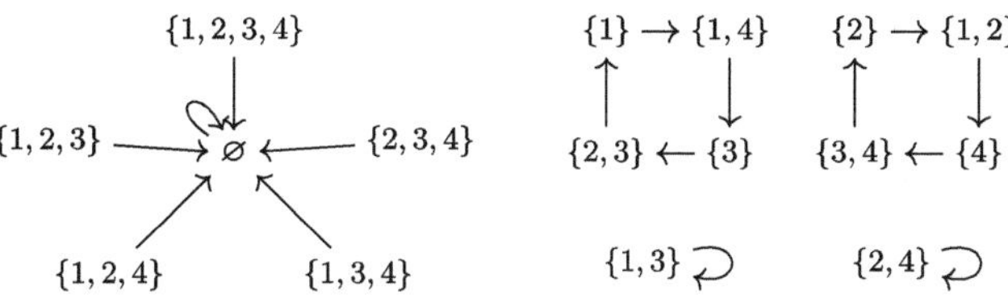

Fig. 8. The zero-context graph generated by reaction system $\mathcal{A}$ in Example 6.

The converse of Theorem 2, however, does not hold. As a counterexample, consider the reaction system from Example 3 and its zero-context graph shown in Fig. 5. Convert it with ϕ to the graph in Fig. 6. To see that there is no elementary CA that has this graph as its $G_P^{\mathcal{C}}$, where $P = P_4$, consider the edges $(1010, 0110)$ and $(1011, 0000)$. An elementary CA will have to send 101 to 1 for the first edge and send 101 to 0 for the second edge. Hence, no such $\mathcal{C}$ exists. In this graph, we also have the edges $(1000, 0100)$ and $(0100, 1010)$. Since CA functions commute with translation, then if a CA function $\mathcal{C}$ maps the periodic configuration represented by 1000 to 0100, it must also map 0100 to 0010. Hence, no CA function $\mathcal{C}$ exists, even without restriction to elementary CA.

6   Cellular Automata with More States

A cellular automaton must have at least two states to represent any nontrivial dynamics. There is a natural mapping from a pattern of n binary cells to subsets of a size n set (the map ϕ^{-1} as described above) but we may also want to consider cellular automata with three or more states.

For a k-state cellular automaton, there are k^n patterns of n cells. We can create a surjective function from this set of patterns to subsets of a reaction system background set S, provided $2^{|S|}$ is at most k^n. One approach is to take an ordering of all patterns and all subsets of S, and map pattern i to subset i. If $k^n > 2^{|S|}$, we can use a modular alignment of the orderings to handle the remaining patterns. Then pattern i (with $0 \le i < k^n$) maps to subset $j = i$ mod $2^{|S|}$, where subsets are ordered $0 \le j < 2^{|S|}$.

If k is a power of 2, a maximally sized S will have exactly k^n subsets, and $|S| = \log_2 k^n$ elements ($2^{|S|} = k^n$). We can define the orderings of patterns and subsets such that the concatenation of the uniform length binary string representations of cell states across all n cells maps to the subsets of S by ϕ^{-1}.

Example 7. Consider patterns of 3 cells with state set $\{0, 1, 2, 3\}$. There are 4^3 such patterns, which can be mapped to the 2^6 subsets of $S = \{1, 2, 3, 4, 5, 6\}$.

The pattern $p = (3, 0, 2)$ first has its states converted to binary strings, producing $(11, 00, 10)$. They are then concatenated as 110010, and through ϕ^{-1}, bijectively mapped to the subset $\{1, 2, 5\}$. $\Diamond$

Representing Arbitrary Reaction Systems. So far, all of our considered translations between reaction systems and cellular automata have been unable to represent every reaction system. This is because spatial translation of a cell configuration must commute with CA function application. With a direct mapping between subset inclusion and a pattern of binary cell states, a local update rule is not always able to identify which element it is computing the presence of.

A simple solution is to let the cellular automaton state set be the power set of the reaction system background set S, or equivalently, let it be the set of binary strings of length $|S|$. Then with a size 1 neighborhood, the reaction system dynamics can be represented using single cells. The neighborhood size may be expanded to model spatially local interaction between reaction systems with differing initial states. A local update rule may be designed such that each cell generates an interactive process, where elements diffused from its neighbors are used as context.

Another option is to include a sequence delimiter subpattern that is identifiable in the local update rule input for any cell, and for any valid configuration. Unlike the single-cell option, this requires restricting the domain of allowable configurations and CA functions, but still allows for a bijection between these restricted domains and all reaction systems.

With 3 cell states, we can have the sequence delimiter subpattern be a single cell in State 2, and map presence of elements to binary cell states as previously described. The neighborhood must include $|S| + 1$ cells to ensure that each application of the local update rule includes the delimiter cell.

Theorem 3. *For every reaction system $\mathcal{A} = (S, A)$, there exists a cellular automaton $\mathcal{C} = (d, \Sigma, N, f)$ with $d = 1$, $|\Sigma| = 3$, and $|N| = |S| + 1$, and a pair of functions δ_1, δ_2, such that for all $X \subseteq S$, we have $\delta_2(\mathcal{C}(\delta_1(X))) = res_A(X)$.*

Proof. Let $\mathcal{A} = (S, A)$ be a reaction system, with an ordered background set $S = \{s_1, s_2, \ldots, s_n\}$. Define a cellular automaton $\mathcal{C} = (d, \Sigma, N, f)$ with $d = 1$, $\Sigma = \{0, 1, 2\}$, where 2 is the delimiter state, and the neighborhood $N = (\vec{0}, \vec{1}, \ldots, \vec{n})$. Let the local update rule be $f(\sigma_0, \sigma_1, \ldots, \sigma_n)$, defined as follows, where for Cell $\vec{i}$, σ_0 is the state of Cell $\vec{i}$, σ_1 is the state of Cell $\vec{i} + 1$, and so on.

Position relative to the delimiter is determined by which of the $\sigma_j = 2$, with the function δ_1 and the construction of f ensuring that there is always exactly one cell in any neighborhood in State 2. Let $\sigma_i = 2$, for some i. Then for all $1 \le j \le n$, and with $k \equiv i + j \mod (n)$, σ_k corresponds to element s_j. If $\sigma_0 = 2$, then the result of f is 2 to preserve the delimiter. If $\sigma_0 \ne 2$, then the result of f is 1 if the element corresponding to σ_0 is in $res_A(X)$, and 0 otherwise.

Finally, let δ_1 map a set $X \subseteq S$ to the periodic configuration $c(p)$, as defined in Proposition 1, where p is the pattern with 2 prepended to $\phi(X)$. Let δ_2 be the function mapping an $(n \overset{\rightarrow}{+} 1)$-periodic configurations of this restricted form back to a subset of S. Clearly, $\delta_2(\mathcal{C}(\delta_1(X))) = res_A(X)$. $\square$

This construction generalizes to any size background set S and any power set function over $\mathcal{P}(S)$. An example of this construction is given below.

Example 8. Consider representing a binary XOR function. First, as a reaction system, let the background set be $S = \{x, y, z\}$. Then, we can construct a set of reactions A, such that $\mathrm{res}_A(\varnothing) = \mathrm{res}_A(\{x, y\}) = \varnothing$ and $\mathrm{res}_A(\{x\}) = \mathrm{res}_A(\{y\}) = \{z\}$. The results are the same if z is included in an input set.

Let $(\sigma_0, \sigma_1, \sigma_2, \sigma_3)$ be the local rule input for some Cell i in a configuration c. If this is equal to $(0, 2, 0, 1)$, then σ_2 corresponds to presence of x, σ_3 to y, and σ_0 to z. Then the result should be 1, corresponding to $\mathrm{res}_A(\{y\}) = \{z\}$, since the result of f in this case becomes the state of the cell representing presence of z. Cell $(i - 1)$ would have $(\sigma_0, \sigma_1, \sigma_2, \sigma_3) = (1, 0, 2, 0)$, which determines that σ_0 corresponds to presence of y, and will map to 0. For this configuration, we have $\delta_1(X) = c$ with $X = \{y\}$, and $\delta_2(\mathcal{C}(c)) = \{z\}$. $\Diamond$

It is also possible to use this subpattern delimiter method with only 2 cell states. Instead of a single cell in State 2, we can use the subpattern 100. To ensure that it is always identifiable, add a cell that is always in State 1 between every two cells that represent element presence. Then for the background set $S = \{a, b, c\}$, presence of elements a, c is represented with the pattern 10011011. The required pattern size may be reduced with a different pattern-avoiding encoding. If higher-dimensional patterns are used, the encoding must be designed such that the delimiting subpattern is uniquely identifiable in every cell's local rule input, and for every encoded subset.

Although the delimiter method may seem more complicated, encoding n element presences in this way is essentially equivalent to encoding them in a single cell. With the single-cell approach, the task of maintaining groups of n distinguishable bits is deferred to implementation.

7   Conclusion

In this paper, we have introduced a framework connecting cellular automata and reaction systems. More specifically, we define pattern graphs to serve as a tool that captures the dynamics of cellular automata and relates it to the behavior of reaction systems through their zero-context graphs.

Pattern graphs can be defined for any cellular automaton with a set of possible patterns of given size serving as vertices and the CA determining the edges. The patterns can be viewed as extracted from totally periodic configurations or used to create such. In the one-dimensional binary case, we demonstrated how a bijection between patterns and subsets enables a direct transformation from pattern graphs of cellular automata to zero-context graphs of reaction systems.

We also showed that the converse transformation does not always result in a pattern graph of a CA and a future direction of investigation is to find combinatorial and structural properties of RS that can be transformed into pattern graphs. We presented techniques to remedy this and showed that by increasing the number of states or increasing the length of the binary sequences corresponding to a cell, a reaction system's zero-context graph can be transformed to a subgraph of a pattern graph, enabling transformation from a reaction system to a cellular automaton.

Acknowledgments. The authors gratefully acknowledge grant support from the UNF College of Arts and Sciences through the Joan Van Vleck Exceptional Service Award.

Disclosure of Interests. The authors have no competing interests to declare that are relevant to the content of this article.

References

1. Bottoni, P., Labella, A., Rozenberg, G.: Reaction systems with influence on environment. J. Membr. Comput. **1**(1), 3–19 (2019). https://doi.org/10.1007/s41965-018-00005-8
2. Brijder, R., Ehrenfeucht, A., Main, M.G., Rozenberg, G.: A tour of reaction systems. Int. J. Found. Comput. Sci. **22**(7), 1499–1517 (2011)
3. Brodo, L., et al.: Causal analysis of positive reaction systems. Int. J. Softw. Tools Technol. Transfer **26**, 509–526 (2024)
4. Cook, M.: Universality in elementary cellular automata. Complex Syst. **15**(1), 1–40 (2004)
5. Corolli, L., Maj, C., Marini, F., Besozzi, D., Mauri, G.: An excursion in reaction systems: from computer science to biology. Theoret. Comput. Sci. **454**, 95–108 (2012)
6. Csuhaj-Varjú, E., Vaszil, G.: Variants of distributed reaction systems. Nat. Comput. **23**, 269–284 (2024)
7. Dennunzio, A., Formenti, E., Margara, L.: On the dynamical behavior of cellular automata on finite groups. IEEE Access **12**, 122061–122077 (2024)
8. Ehrenfeucht, A., Rozenberg, G.: Reaction systems. Fund. Inform. **75**(1–4), 263–280 (2007)
9. Ehrenfeucht, A., Kleijn, J., Koutny, M., Rozenberg, G.: Minimal reaction systems. In: Priami, C., Petre, I., de Vink, E. (eds.) Transactions on Computational Systems Biology XIV. LNCS, vol. 7625, pp. 102–122. Springer, Heidelberg (2012). https://doi.org/10.1007/978-3-642-35524-0_5
10. Farrell, R., Genova, D., Strickley, D. : Minimizing cycles in reaction systems. In: Cho, DJ., Kim, J. (eds.) UCNC 2024, LNCS, vol. 14776, 237–252 (2024)
11. Genova, D., Hoogeboom, H.J., Jonoska, N.: A graph isomorphism condition and equivalence of reaction systems. Theoret. Comput. Sci. **701**, 109–119 (2017)
12. Genova, D., Hoogeboom, H.J., Prodanoff, Z.: Extracting reaction systems from function behavior. J. Membr. Comput. **2**(3), 194–206 (2020). https://doi.org/10.1007/s41965-020-00045-z
13. Genova, D., Hoogeboom, H.J., Kleijn, J.: Functional equivalence and a cover relation for reactions. Theoret. Comput. Sci. **1004**, 114633 (2024)
14. Ivanov, S., Petre, I.: Controllability of reaction systems. J. Membr. Comput. **2**, 290–302 (2020)
15. Kari, J.: Theory of cellular automata: a survey. Theoret. Comput. Sci. **334**(1–3), 3–33 (2005)
16. Kleijn, J., Koutny, M., Mikulski, Ł, Rozenberg, G.: Reaction systems, transition systems, and equivalences. In: Böckenhauer, H.-J., Komm, D., Unger, W. (eds.) Adventures Between Lower Bounds and Higher Altitudes. LNCS, vol. 11011, pp. 63–84. Springer, Cham (2018). https://doi.org/10.1007/978-3-319-98355-4_5
17. Manzoni, L., Porreca, A.E., Rozenberg, G.: Facilitation in reaction systems. J. Membr. Comput. **2**, 149–161 (2020)

18. Martin, O.C., Odlyzko, A.M., Wolfram, S.: Algebraic properties of cellular automata. Commun. Math. Phys. **93**, 219–258 (1984)
19. Rickman, S.: Logic Friday (Version 1.1.4) [Computer software]. Sontrak MicroDesigns (2012)
20. Salomaa, A.: On state sequences defined by reaction systems. In: Constable, R.L., Silva, A. (eds.) Logic and Program Semantics. LNCS, vol. 7230, pp. 271–282. Springer, Heidelberg (2012). https://doi.org/10.1007/978-3-642-29485-3_17
21. Von Neumann, J., Burks, A.W.: Theory of Self-Reproducing Automata. University of Illinois Press, Urbana (1966)
22. Wolfram, S.: Statistical mechanics of cellular automata. Rev. Mod. Phys. **55**, 601–644 (1983)
23. Wolfram, S.: "Cellular Automaton State Transition Diagrams" Wolfram Demonstrations Project (2007)

Polynomial Simulations of CRN Models with Trimolecular Void Step-Cycle CRNs

Austin Luchsinger[(✉)], Aiden Massie, Robert Schweller, Evan Tomai, and Tim Wylie

University of Texas Rio Grande Valley, Edinburg, USA
`austin.luchsinger@utrgv.edu`

Abstract. We investigate the computational power of Step-Cycle Chemical Reaction Networks (CRNs) when restricted to void reactions of size at most $(3, 1)$. Step-Cycle CRNs extend the previously introduced step CRN model by repeatedly cycling through a fixed sequence of species additions and reaction phases. We show that even under the severe constraint of trimolecular void rules—which can only delete or preserve species—the model retains full computational power. In particular, we prove that (3,1) void Step-Cycle CRNs can polynomially simulate (1) any general CRN, (2) any general Step CRN, and (3) any general Step-Cycle CRN. Ultimately, these results demonstrate that the Step-Cycle model retains its complete expressive power even when restricted to $(3, 1)$-size void rules.

Keywords: Chemical Reaction Networks · Simulations · Petri-nets · Vector Addition Systems

1 Introduction

Chemical Reaction Networks (CRNs) [5,6] are a well-established model of abstract molecular computing. CRNs transform complex real-world chemical dynamics into a simpler system consisting of molecular *species*, which change through applications of *reactions*. CRNs have been shown to be computationally equivalent to other important models of distributed systems, including Petri-nets [20] and Vector Addition Systems [17], as shown in [11,15].

There is a long history of proposed extensions to the models of distributed systems mentioned above. These extensions include inhibiting transitions by the presence of certain species [1,9,13,16], establishing a priority of firing some transitions over others [16,21,22], and enabling parallel firings of transitions [8,23]. It is known that each of these extensions allows the systems to detect when a species is absent (not possible in the traditional CRNs), which immediately unlocks Turing-universal computation [7,19].

This research was supported in part by National Science Foundation Grant CCF-2329918.

© The Author(s), under exclusive license to Springer Nature Switzerland AG 2026
E. Formenti and L. Manzoni (Eds.): UCNC 2025, LNCS 16364, pp. 378–393, 2026.
https://doi.org/10.1007/978-3-032-15641-9_25

This paper focuses on one such extension, the Step-Cycle CRN model, which reflects cyclic protocols often seen in molecular systems and laboratory practices. In particular, we explore the expressive power of such systems when restricted to simple *void* rules that only ever remove species. The following paragraphs provide background on void rules, step-cycles, polynomial simulation, and our main contributions.

Void Reactions. Perhaps the simplest form of reactions in CRNs are *void* reactions, which do not generate new species. Void reactions can be categorized into two types: *true void* reactions that only delete reactant species, and *catalyst void* reactions that may preserve some reactant species. While conceptually simple and relevant to practical implementation, restricting systems to only use void rules can drastically limit computational power. For example, the reachability problem (which asks if a target configuration $\vec{B}$ can be reached from some initial configuration $\vec{A}$) is known to be Ackermann-complete for general CRNs [12,18], but becomes polynomial-time solvable when restricted to true void rules of size $(2,0)$ [2].

Step-Cycle CRNs. Of particular interest to this paper is a recently introduced extension for CRNs known as the *Step* CRN model [3]. Motivated by real-world laboratory practices, the Step CRN augments a traditional CRN by introducing new copies of species into a configuration once no more reactions can be applied to it. The authors of [3,4] demonstrated that Step CRNs, even when restricted to true void reactions of small sizes such as $(2,0)$ or $(3,0)$, can compute threshold circuits. In [14], it was then proven that the addition of only one step strengthens the CRN reachability problem with only $(2,0)$ rules to NP-complete.

The *Step-Cycle* CRN is a simple modification to the Step CRN model in which the system continually loops through its steps. The cyclic nature of execution makes this model particularly relevant for automating recurring operations in wet lab experiments and other time-structured settings. All of the results of this paper involve Step-Cycle CRNs that are restricted to using void rules of at most size $(3,1)$.

Polynomial Simulation of CRNs. Although the aforementioned CRN extensions have been shown to be Turing universal, universality alone does not capture how efficiently these models can simulate one another. For this reason, recent work [7] has introduced *polynomial simulation* as a framework for comparing models not just in power, but in practical encoding complexity.

Under this framework, one CRN system T' simulates another T if configurations of T can be represented as configurations of T', and the dynamics of T are faithfully reproduced by bounded sequences of transitions in T'. A simulation is *polynomially efficient* if the number of species, rule size, volume growth, and transition steps remain polynomially bounded with respect to the simulated system. This approach provides a more nuanced comparison between CRN models. It allows us to reason about simulation overhead and efficiency, rather than treating all universal models as equally expressive. In this work, we use

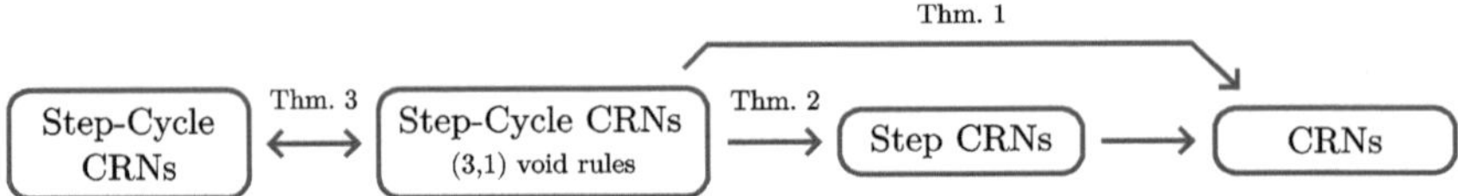

Fig. 1. Our simulation results (and corresponding theorems) with $(3,1)$ void Step-Cycle CRNs. An arrow pointing from a system T' to another system T represents T' simulating T under polynomial simulation. Note that other than the $(3,1)$ void rule system, every system in the hierarchy is a generalization from left to right.

this framework to analyze the power of Step-Cycle CRNs that are restricted to using void rules of at most size (3,1).

Our Contributions. In this paper, we show that $(3,1)$ void rule step-cycle CRNs can perform polynomial simulation of:

1. general CRNs,
2. general Step CRNs, and
3. general Step-Cycle CRNs.

Thus, even when restricting this experimentally motivated step-cycle model to some of the simplest void-rule reactions, full expressive power is maintained. These results and the corresponding theorems are visualized in Fig. 1.

2   Preliminaries

We first describe the CRN model, as well as its extensions discussed in this paper and void reactions, then describe the framework of polynomial simulation.

2.1   Chemical Reaction Network Models

We formally define the CRN models discussed in this paper as follows. Example systems for each CRN model are shown in Fig. 2.

CRNs. A *chemical reaction network* (CRN) $\mathcal{C} = (\Lambda, \Gamma)$ is defined by a finite set of species Λ , and a finite set of reactions Γ where each reaction is a pair $(\vec{R}, \vec{P}) \in \mathbb{N}^\Lambda \times \mathbb{N}^\Lambda$, sometimes written $\vec{R} \longrightarrow \vec{P}$, that denotes the *reactant* species consumed by the reaction and the *product* species generated by the reaction. For example, given $\Lambda = \{a, b, c\}$, the reaction $((2,0,0), (0,1,1))$ represents $2a \longrightarrow b + c$.

A *configuration* $\vec{C} \in \mathbb{N}^\Lambda$ of a CRN assigns integer counts to every species $\lambda \in \Lambda$, and we use notation $\vec{C}[\lambda]$ to denote that count. For a species $\lambda \in \Lambda$, we denote the configuration consisting of a single copy of λ and no other species as $\vec{\lambda}$. It is often useful to reference the set of species whose counts are not zero in a given configuration. In such cases, the notation $\{\vec{C}\}$ is used. Formally,

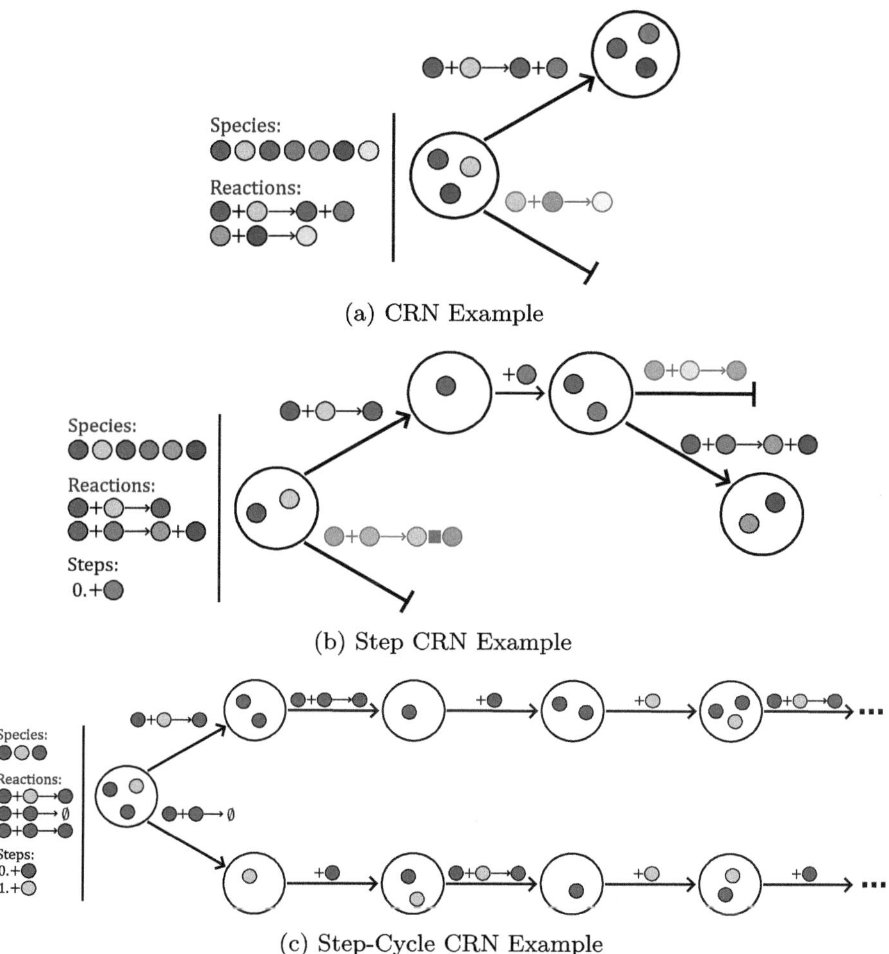

(a) CRN Example

(b) Step CRN Example

(c) Step-Cycle CRN Example

Fig. 2. Example systems for the CRN models discussed in this paper. The blurred portions of the figure represent invalid reaction sequences. We also use colors for the alphabet of the species set of a system. (a) CRN model. (b) Step CRN model with 1 step. Note that the step species are only added once the system reaches a terminal state. (c) Step-Cycle CRN model with 2 steps. Note how the system adds the species of Step 0 again even after the addition of the species of Step 1.

$\{\overrightarrow{C}\} = \{\lambda \in \Lambda \mid \overrightarrow{C}[\lambda] > 0\}$, and when convenient and clear from the context, we further use $\{\overrightarrow{C}\}$ to denote the configuration (vector) representation in which each element has a single copy. Finally, let $|\overrightarrow{C}| = \sum_{\lambda \in \Lambda} C[\lambda]$ denote the total number of copies of all species in a configuration, sometimes referred to as the *volume* of $\overrightarrow{C}$.

A reaction $(\overrightarrow{R}, \overrightarrow{P})$ is said to be *applicable* in configuration $\overrightarrow{C}$ if $\overrightarrow{R} \leq \overrightarrow{C}$; in other words, a reaction is applicable if $\overrightarrow{C}$ has at least as many copies of each species as $\vec{R}$. If the reaction $(\vec{R}, \vec{P})$ is applicable, it results in configuration $\vec{C'} = \vec{C} - \vec{R} + \vec{P}$ if it occurs, and we write $\vec{C} \to \vec{C'}$. If there exists a finite sequence of configurations such that $\vec{C} \to \vec{C}_1 \to \ldots \to \vec{C}_n \to \vec{D}$, then we say that $\vec{D}$ is *reachable* from $\vec{C}$ and we write $\vec{C} \rightsquigarrow \vec{D}$. A configuration is said to be *terminal* if no reactions are applicable. We denote the set of reachable configurations of a CRN as $REACH_C$. We define the subset of reachable configurations that are terminal as $TERM_C$. When considering computation in CRNs, we use the notion of output-stable computation from [10].

Definition 1 (Discrete Chemical Reaction Network). *A discrete chemical reaction network (CRN) is an ordered pair (Λ, Γ) where Λ is an ordered alphabet of species, and Γ is a set of rules over Λ.*

Definition 2 (CRN Dynamics). *For a CRN (Λ, Γ) and configurations $\vec{A}$ and $\vec{B}$, we say that $\vec{A} \rightarrow_{crn}^{(\Lambda, \Gamma)} \vec{B}$ if there exists a rule $(\vec{R}, \vec{P}) \in \Gamma$ such that $\vec{R} \leq \vec{A}$, and $\vec{A} - \vec{R} + \vec{P} = \vec{B}$.*

Step CRNs. A step CRN is an augmentation of a basic CRN in which additional copies of species may be added to the system once it reaches a terminal configuration within a *step*. Once a step has passed (i.e. all copies of the new species are added), the system transitions to the next step if it exists. Formally, a step CRN of k steps is defined as $((\Lambda, \Gamma), (\vec{S}_0, \vec{S}_1, \ldots, \vec{S}_{k-1}))$, where the first element is a normal CRN (Λ, Γ) and the second is a sequence of length-$|\Lambda|$ vectors of non-negative integers denoting how many copies of each species type to add after each step. We define a *step-configuration* $\vec{C}_i$ for a step CRN as a valid configuration $\vec{C}$ over (Λ, Γ) along with an integer $i \in \{0, \ldots, k-1\}$ that denotes the configuration's step.

Definition 3 (Step Dynamics). *For a step CRN $((\Lambda, \Gamma), (\vec{S}_0, \vec{S}_1, \ldots, \vec{S}_{k-1}))$ and step-configurations $\vec{A}_i$ and $\vec{B}_j$, we say that $\vec{A}_i \rightarrow_s^{(\Lambda, \Gamma)} \vec{B}_j$ if either*

1. *$i = j$, there exists a rule $(\vec{R}, \vec{P}) \in \Gamma$ s.t. $\vec{R} \leq \vec{A}_i$, and $\vec{A}_i - \vec{R} + \vec{P} = \vec{B}_j$, or*
2. *$i + 1 = j$, $\vec{A}_i$ is terminal, and $\vec{A}_i + \vec{S}_i = \vec{B}_j$.*

Step-Cycle CRNs. A *step-cycle CRN* is a step CRN that infinitely repeats steps 0 through $k-1$: that is, once $\vec{S}_{k-1}$ is added to the terminal configuration in the $(k-1)^{th}$ step, the resulting configuration is treated as the new initial configuration for the step CRN. A *step-cycle CRN* of k steps is again formally defined as an ordered pair $((\Lambda, \Gamma), (\vec{S}_0, \vec{S}_1, \ldots, \vec{S}_{k-1}))$. We also define a *step-configuration* $\vec{C}_i$ for a step-cycle CRN as a valid configuration $\vec{C}$ over (Λ, Γ) where $i \in \{0, \ldots, k-1\}$.

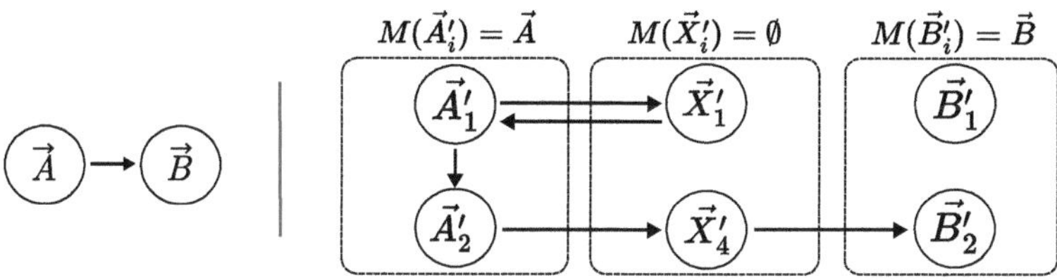

Fig. 3. (Left) A system T with states $\vec{A}$ and $\vec{B}$, and and transition $\vec{A} \to_T \vec{B}$. (Right) A state-space diagram for system T' that simulates T. Here, each arrow represents some transition $\vec{X} \to_{T'} \vec{Y}$ under the dynamics of T'. Observe how T follows T' and T' models T.

Definition 4 (Step-Cycle Dynamics). *For a step-cycle CRN $((\Lambda, \Gamma), (\vec{S}_0, \vec{S}_1, \ldots, \vec{S}_{k-1}))$ and step-configurations $\vec{A}_i$ and $\vec{B}_j$, we say that $\vec{A}_i \to_{sc}^{(\Lambda, \Gamma)} \vec{B}_j$ if either*

1. *$i = j$, there exists a rule $(\vec{R}, \vec{P}) \in \Gamma$ s.t. $\vec{R} \le \vec{A}_i$, and $\vec{A}_i - \vec{R} + \vec{P} = \vec{B}_j$, or*
2. *$(i + 1) \mod k = j$, $\vec{A}_i$ is terminal, and $\vec{A}_i + \vec{S}_i = \vec{B}_j$.*

Void Reactions. A *void* reaction is any rule that does not create any new copies of any species type, and can only delete or preserve existing species copies. Thus, a void reaction either has no products or has products that are a subset of its reactants (in which case these products are termed *catalysts*). The formal definitions of void reactions and rule size are as follows.

Definition 5 (Void Reactions). *A reaction $(\vec{R}, \vec{P})$ is a void reaction if $\vec{P} - \vec{R}$ has no positive entries and at least one negative entry. There are two classes of void reactions, catalytic and true void. In catalytic void reactions, one or more reactants remain and one or more reactant is deleted after the rule is applied. In true void reactions, there are no products remaining.*

Definition 6 (Size-(i, j) Reactions). *A reaction $(\vec{R}, \vec{P})$ is said to be a size-(i, j) reaction if $(i, j) = (|\vec{R}|, |\vec{P}|)$. A reaction is trimolecular if $i = 3$, bimolecular if $i = 2$, and unimolecular if $i = 1$.*

2.2 Simulation

To define the concept of one system $T' = (\Lambda', \Gamma')$ simulating another system $T = (\Lambda, \Gamma)$ we introduce *configuration maps* and *representative configurations*. A configuration map is a polynomial-time computable function $M : \text{configs}_{T'} \to \text{configs}_T \bigcup \emptyset$ mapping at least one element in $\text{configs}_{T'}$ to each element in configs_T, and the *representative* configurations for a configuration

$\vec{C} \in \text{configs}_T$ are $[\![\vec{C}]\!] = \{\vec{C'} \mid \vec{C} = M(\vec{C'})\}$, also computable in polynomial time.[1] We say a configuration map M is *valid* if for all $\vec{C'}, \vec{D'} \in \text{configs}_{T'}$ where $M(C') \neq \emptyset$ and $M(\vec{D'}) = \emptyset$, if $\vec{C'} \rightsquigarrow \vec{D'}$ then $\vec{D'} \notin TERM_{T'}$. Finally, we define the concept of *macro transitionable* for representative configurations $\vec{A'}$ and $\vec{B'}$, denoted $\vec{A'} \Rightarrow_{T'} \vec{B'}$, to mean that $\vec{B'}$ is reachable from $\vec{A'}$ in system T' through a sequence of k intermediate configurations $\langle \vec{A'}, \vec{X'_1}, \ldots, \vec{X'_k}, \vec{B'} \rangle$ such that $M(\vec{A'}) \neq \emptyset$, $M(\vec{B'}) \neq \emptyset$, and each $M(\vec{X'_i}) \in \{M(A'), \emptyset\}$. Note that k can be zero, resulting in the sequence $\langle \vec{A'}, \vec{B'} \rangle$.

Intuition. Each representative configuration set $[\![\vec{C}]\!]$ is a particular collection (of at least 1 configuration) that adheres to the strictest modeling of system T within system T'. That is, any $\vec{C'} \in [\![\vec{C}]\!]$ must be able to grow into anything that $\vec{C} = M(\vec{C'})$ can grow into, and no $\vec{C'} \in [\![\vec{C}]\!]$ may grow into something that $\vec{C} = M(\vec{C'})$ cannot grow into (Fig. 3).

For a given configuration map and collection of representative configurations, we define the concepts of *following* and *modeling*, followed by our definition of *simulation*.

Definition 7 (It Follows). *We say system T follows system T' if whenever $\vec{A'} \Rightarrow_{T'} \vec{B'}$ and $M(\vec{A'}) \neq M(\vec{B'})$, then $M(\vec{A'}) \rightarrow_T M(\vec{B'})$.*

Definition 8 (Models). *We say system T' models system T if $\vec{A} \rightarrow_T \vec{B}$ implies that $\forall \vec{A'} \in [\![\vec{A}]\!]$, $\exists \vec{B'} \in [\![\vec{B}]\!]$ such that $\vec{A'} \Rightarrow_{T'} \vec{B'}$.*

Definition 9 (Simulation). *A system T' simulates a system T if there exists a valid, polynomial time computable function $M : \text{configs}_{T'} \rightarrow \text{configs}_T$ and polynomial time computable set $[\![\vec{C}]\!] = \{\vec{C'} \mid M(\vec{C'}) = \vec{C}\}$ for each $\vec{C} \in \text{configs}_T$, such that:*

1. *T follows T'.*
2. *T' models T.*

Polynomial Simulation. We say a simulation is *polynomial efficient* if the simulating system adheres to the following:

polynomial species and rules. The number of species and rules is at most polynomial in the number of species and rules of the simulated system.

polynomial rule size. The maximum rule size (number of products plus number of reactants) is polynomial in the maximum rule size of the simulated system.

---

[1] We write $M(\vec{C'}) = \emptyset$ to mean that the mapping is undefined, which is not the same as $M(\vec{C'}) = \vec{0}$.

polynomial transition sequences. For all B such that $A \to_T B$, the expected number of transitions taken to perform a macro transition from $M(A) \Rightarrow_{T'} M(B)$, conditioned that $M(A)$ does macro transition to $M(B)$, has expected number of transitions polynomial in the number of rules and species of the simulated system based on a uniform sampling of applicable rules.

polynomial volume. Each $\overrightarrow{C'} \in [\![\overrightarrow{C}]\!]$ has a volume that is polynomially bounded by the volume of C, and for any macro transition $A' \Rightarrow_{T'} B'$, any intermediate configuration within this macrotransition has volume polynomially bounded in the volume of $M(A')$ and $M(B')$.

3 Simulation Results

Here, we show our simulation results achieved by void Step-Cycle CRNs with rule size at most $(3, 1)$. Due to space constraints, detailed explanations of the constructions are presented in this section, while the complete formal simulation proofs are left for the full version of this paper.

3.1 Table Notation

Here, we provide an overview of notation used for the added species and reactions illustrated in our tables in this section.

In detailing the added species created for the simulating system, we may describe that we add a species using the universal quantifier, such as for each reaction γ_i in Γ or each step $\overrightarrow{S}_l \in \mathcal{C}_S$. If the added species shares the same index i/l, then this indicates that we create a unique species for each reaction, each of which we add one copy of. If the species does not have an index, this means that we add $|\Gamma|$ copies of that species instead. We may also further indicate that we add species for each individual species in the reactants and products (indicated by $\mathcal{R}_i$ and $\mathcal{P}_i$ respectively) for each reaction, or for each step species for each step-configuration. A similar case is made for the reactions created for the simulating system.

For example, if the added species of a step is described as: $\forall \gamma_i \in \Gamma : \overrightarrow{x}, \overrightarrow{G}_i$, then the step introduces $|\Gamma|$ copies of $\overrightarrow{x}$, but a single copy for each of $\overrightarrow{G}_1, \cdots, \overrightarrow{G}_{|\Gamma|}$. Similarly, if the reaction is described as $\forall \gamma_i \in \Gamma : \overrightarrow{G} + \overrightarrow{G}_i + \overrightarrow{g}_i \to \emptyset$, then we create a reaction set consisting of $\overrightarrow{G} + \overrightarrow{G}_1 + \overrightarrow{g}_1 \to \emptyset, \cdots, \overrightarrow{G} + \overrightarrow{G}_{|\Gamma|} + \overrightarrow{g}_{|\Gamma|} \to \emptyset$.

Theorem 1. *Step-Cycle CRNs can simulate any given CRN under polynomial simulation, even when restricted to at most size-$(3, 1)$ void rules.*

Proof. Given a CRN $\mathcal{C} = (\Lambda, \Gamma)$ and a configuration $\overrightarrow{C}$ of $\mathcal{C}$, we construct a Step-Cycle CRN $\mathcal{C}'_{SC} = (\Lambda', \Gamma', (\overrightarrow{S}'_0, \ldots, \overrightarrow{S}'_8))$ and a configuration $\overrightarrow{C'}$ of $\mathcal{C}'_{SC}$ that simulates $\mathcal{C}$ over $\overrightarrow{C}$ as follows.

Species and Configuration. We first add the species from $\mathcal{C}$ with no changes. For the new set of species, we first create $G, G_1, \cdots, G_{|\Gamma|}$, and $g_1, \cdots, g_{|\Gamma|}$. G functions as a global species that randomly selects a rule $\gamma_i \in \Gamma$ to attempt to apply a rule in the set of species for all $\lambda_i \in \Lambda$. However, a single copy of both G_i and g_i must be present for G to select γ_i. $G'_1, \cdots, G'_{|\Gamma|}$ are also made; these species are used to ensure that, if γ_i was found to not be applicable, it will not be selected again until another separate reaction is successfully applied. We also construct the species $E_1^1, \ldots, E_{|\Gamma|}^{|\mathcal{R}_{|\Gamma|}|}$ for use in determining if the count of any reactant species $r_j \in \mathcal{R}_i$ is zero. a_y and a_n are created to verify if applying γ_i is either valid or invalid, respectively. We also create the species c and w for assistance in the above processes. $R_1^1, \ldots, R_{|\Gamma|}^{|\mathcal{R}_{|\Gamma|}|}$ and $P_1^1, \ldots, P_{|\Gamma|}^{|\mathcal{P}_{|\Gamma|}|}$ are made for use in the actual application of γ_i. The pair of species y_1 and y_2 are used to perform "clean-up" operations for excessive species. Finally, we create x and z. The former determines if all reactions have been selected without any successful rule application; if this is true, then z ensures that G can no longer select a reaction, effectively indicating a terminal configuration for $\mathcal{C}'_{\mathcal{SC}}$. The final species set of $\mathcal{C}'_{\mathcal{SC}}$ is then $\Lambda' = \Lambda \cup \{G, G_1, \cdots, G_{|\Gamma|}, g_1, \cdots, g_{|\Gamma|}, G'_1, \cdots, G'_{|\Gamma|}, E_1^1, \ldots, E_{|\Gamma|}^{|\mathcal{R}_{|\Gamma|}|},$ $R_1^1, \ldots, R_{|\Gamma|}^{|\mathcal{R}_{|\Gamma|}|}, P_1^1, \ldots, P_{|\Gamma|}^{|\mathcal{P}_{|\Gamma|}|}, a_y, a_n, c, w, x, y_1, y_2, z\}$.

Given a configuration of $\mathcal{C}$ $\vec{C}$, let the configuration of $\mathcal{C}'_{\mathcal{SC}}$ be $\vec{C}' = \vec{C} \cup \{\vec{G}_1, \cdots, \vec{G}_{|\Gamma|}, \vec{g}_1, \cdots, \vec{g}_{|\Gamma|}\}$.

Steps and Reactions. We first note that we construct the steps and reactions of $\mathcal{C}'_{\mathcal{SC}}$ such that a specific subset of the reactions will only be applicable within specific steps. Therefore, our descriptions here will mainly describe the steps in order, along with any *relevant rules* that are applicable only in those steps. The full set of added species and reactions for each step is illustrated in Table 1.

The first steps (0–4) involve selecting a random reaction and determining the validity of applying it within the set of species for all $\lambda_i \in \Lambda$. In Step 0, we only add one copy of $\vec{G}$. Then the only initially applicable reaction is an instance of Reaction 1, which represents $\vec{G}$ selecting a random reaction $\gamma_i \in \Gamma$ to try simulating. We then move to Step 1, where we add a single copy of $\vec{E}_j^i$ for each individual reactant r_j for each reaction in Γ. If a $\vec{E}_j^i$ species represents a reactant that is *not* in γ_i, it will be filtered out by Reaction 2. In Step 2, we add a single copy of $\vec{w}$. $\vec{w}$ performs two operations in this step: it first deletes all $\vec{g}_i$ copies with Reaction 3, and then checks if each reactant species in γ_i is present in the configuration by Reaction 4. If the check is completely successful, then no $\vec{E}_i^j$ copy will be present after this step, indicating that γ_i can be applied, and at least one otherwise. We then progress to Step 3 and add a single copy of $\vec{a}_y$. If any $\vec{E}_i^j$ copy is still present, then $\vec{a}_y$ will be deleted (Reaction 5), but is preserved otherwise. Finally, in Step 4, we add a copy of $\vec{a}_n$. If $\vec{a}_y$ is present, then $\vec{a}_n$ is removed through Reaction 6. Else, $\vec{a}_n$ remains in the configuration.

Step 5 will apply γ_i if allowed. Here, we add one copy of species representing each individual reactant and product for each reaction in Γ, both in their "nor-

Table 1. Steps and reactions for a $(3, 1)$ void Step-Cycle CRN $\mathcal{C}'_{SC}$ simulating a CRN $\mathcal{C}$.

Step	Additions		Relevant Rules	
0		$\vec{G}$	$\forall \gamma_i \in \Gamma :$	1. $\vec{G} + \vec{G}_i + \vec{g}_i \to \emptyset$ 24. $\vec{z} + \vec{G} \to \vec{z}$
1	$\forall \gamma_i \in \Gamma, \forall r_j \in \mathcal{R}_i :$	$\vec{E}_i^j$	$\forall \gamma_i \in \Gamma, \forall r_j \in \mathcal{R}_i :$	2. $\vec{G}_i + \vec{E}_i^j \to \vec{G}_i$
2		$\vec{w}$	$\forall \gamma_i \in \Gamma, \forall r_j \in \mathcal{R}_i :$	3. $\vec{w} + \vec{g}_i \to \vec{w}$ 4. $\vec{w} + \vec{r}_j + \vec{E}_i^j \to \vec{w}$
3		$\vec{a}_y$	$\forall \gamma_i \in \Gamma, \forall r_j \in \mathcal{R}_i :$	5. $\vec{E}_i^j + \vec{a}_y + \vec{w} \to \vec{E}_i^j$ 25. $\vec{a}_y + \vec{w} + \vec{z} \to \emptyset$
4		$\vec{a}_n$		6. $\vec{a}_y + \vec{a}_n + \vec{w} \to \vec{a}_y$
5	$\forall \gamma_i \in \Gamma :$ $\forall r_j \in \mathcal{R}_i : \vec{r}_j, \vec{R}_i^j$ $\forall p_j \in \mathcal{P}_i : \vec{p}_j, \vec{P}_i^j$		$\forall \gamma_i \in \Gamma :$	7. $\vec{G}_i + \vec{r}_j + \vec{R}_i^j \to \vec{G}_i$ $\forall r_j \in \mathcal{R}_i :$ 8. $\vec{E}_i^j + \vec{r}_j + \vec{R}_i^j \to \emptyset$ 9. $\vec{a}_y + \vec{r}_j + \vec{R}_i^j \to \vec{a}_y$ $\forall p_j \in \mathcal{P}_i :$ 10. $\vec{G}_i + \vec{p}_j + \vec{P}_i^j \to \vec{G}_i$ 11. $\vec{a}_n + \vec{p}_j + \vec{P}_i^j \to \vec{a}_n$
6	$\forall \gamma_i \in \Gamma :$	$\vec{x}, \vec{G}_i'$	$\forall \gamma_i \in \Gamma :$	12. $\vec{x} + \vec{G}_i' + \vec{G}_i \to \emptyset$
7	$\forall \gamma_i \in \Gamma :$	$\vec{c}$	$\forall \gamma_i \in \Gamma :$	13. $\vec{a}_y + \vec{c} + \vec{G}_i' \to \vec{a}_y$
8	$\forall \gamma_i \in \Gamma :$	$\vec{g}_i$ $\vec{y}_1$	$\forall \gamma_i \in \Gamma :$	$\forall r_j \in \mathcal{R}_i :$ 14. $\vec{y}_1 + \vec{E}_i^j \to \vec{y}_1$ 15. $\vec{y}_1 + \vec{R}_i^j \to \vec{y}_1$ $\forall p_j \in \mathcal{P}_i :$ 16. $\vec{y}_1 + \vec{P}_i^j \to \vec{y}_1$ 17. $\vec{G}_i' + \vec{g}_i \to \vec{G}_i'$ 18. $\vec{y}_1 + \vec{x} \to \vec{y}_1$ 19. $\vec{y}_1 + \vec{a}_y \to \vec{y}_1$ 20. $\vec{y}_1 + \vec{a}_n \to \vec{y}_1$ 21. $\vec{y}_1 + \vec{c} \to \vec{y}_1$
9	$\forall \gamma_i \in \Gamma :$	$\vec{y}_2, \vec{z}$ $\vec{G}_i$	$\forall \gamma_i \in \Gamma :$	22. $\vec{y}_1 + \vec{y}_2 \to \emptyset$ 23. $\vec{z} + \vec{g}_i \to \vec{g}_i$
			Return to Step 0.	

mal" form and in their $\vec{R}_i^j$ or $\vec{P}_i^j$ form. Regarding reactions, we first perform a similar procedure to Step 1, in which each reactant r_j and product p_j not from γ_i is filtered out through Reactions 7 and 10. Then, if γ_i is found to *not* be applicable by the presence of $\vec{a}_n$, then we delete all product species p_i and $\vec{P}_i^j$ with Reaction 11. Additionally, we use Reaction 8 to delete all reactants whose corresponding E_i^j is still present (Reaction 9). The effect is that any reactants of γ_i that *were* present in $\vec{C'}$ in Step 0 are restored back; those that aren't are removed. Alternatively, if γ_i can be applied by the presence of $\vec{a}_y$, then the added reactant species r_i and $\vec{R}_i^j$ will be deleted once again (Reaction 9) and the product species are left unchanged.

The final steps (6–9) constitute of "clean-up" operations, as well as ensuring that our reaction, if not applied, cannot be chosen again until another reaction is successfully applied. In step 6, we add a copy of $\vec{G}'_i$ for each reaction, as well as $|\Gamma|$ copies of $\vec{x}$. Here, $\vec{x}$ and G'_i will delete all present copies of $\vec{G}_i$ through Reaction 12. Note that if $\mathcal{C}'_{SC}$ selects a reaction γ_i from Step 0, then at least one copy of $\vec{x}$, and thus a copy of $\vec{G}'_i$, is guaranteed to remain. Then in Step 7, $|\Gamma|$ copies of $\vec{c}$ are introduced. Only if $\vec{a}_y$ is present will it delete a copy of $\vec{c}$ and the $\vec{G}'_i$ copy if γ_i is selected. Moving to Step 8, we add a copy of $\vec{g}_i$ for each reaction and one copy of $\vec{y}_1$. $\vec{y}_1$ will remove any remaining excessive species with Reactions 14–16 and 18–21. In Reaction 17, $\vec{g}_i$ will be deleted if the corresponding $\vec{G}'_i$ species is still present; this ensures that γ_i cannot be selected again in Step 0 until another reaction is applied due to the missing $\vec{g}_i$ species required for Reaction 1. Finally, in Step 9, we add a copy of $\vec{y}_2$, one copy of $\vec{z}$, and a copy of $\vec{G}_i$ for each reaction. $\vec{y}_1$ and $\vec{y}_2$ will delete themselves with Reaction 22. Additionally, if at least one copy of $\vec{g}_i$ exists, then $\vec{z}$ will be deleted (Reaction 23). If $\vec{z}$ remains present after this step, it indicates that we have exhausted all of our reaction choices and none of them can be applied.

We define the step-configuration mapping as follows. If a single copy of $\vec{G}$ is present in the step-configuration of $\mathcal{C}_{SC}$ $\vec{C}'_i$, then we map $\vec{C}'_i$ to a configuration of $\mathcal{C}$ in which the number of all species $\lambda_i \in \Lambda$ in $\vec{C}$ is equivalent to the count of all species $\lambda_i \in \Lambda$ in $\vec{C}'_i$. We note that, by the construction of $\mathcal{C}_{SC}$, $\vec{G}$ can only be present in the first step configuration $\vec{C}'_0$, so any configurations with a defined mapping can be simply referred to as $\vec{C}'$.

$$M(\vec{C}'_i) = \begin{cases} \sum_{\lambda' \in \Lambda} \vec{C}'_i[\lambda'] \cdot \vec{\lambda}' & \text{if } \vec{G} \in \{\vec{C}'_i\} \\ \emptyset & \text{otherwise.} \end{cases}$$

Polynomial Simulation. Let n and m be the maximum number of reactants and products of any reaction in Γ, respectively. $\mathcal{C}'_{SC}$ has $\mathcal{O}(|\Lambda| + |\Gamma| \cdot (n + m))$ species and $\mathcal{O}(|\Gamma|(n + m))$ reactions. $\mathcal{C}'_{SC}$ has a maximum constant rule size of $(3, 1)$. The sequence of transitions is at most $\mathcal{O}(|\Gamma|(n + m))$. The volume of any representative configuration differs from the original volume by only by a polynomial amount $\mathcal{O}(|\Gamma|)$. The volume of any configuration along a macrotransition will differ from the original volume by at most $\mathcal{O}(|\Gamma|(n + m))$.

Theorem 2. *Step-Cycle CRNs can simulate any given Step CRN under polynomial simulation, even when restricted to at most $(3, 1)$ void rules.*

Proof. Given a Step CRN $\mathcal{C}_{\mathcal{S}} = (\Lambda, \Gamma, (\vec{S}_0, \ldots, \vec{S}_k))$ and a configuration of $\mathcal{C}_{\mathcal{S}}$ $\vec{C}_S$, we construct a Step-Cycle CRN $\mathcal{C}'_{SC} = (\Lambda, \Gamma, (\vec{S}'_0, \ldots, \vec{S}'_{12}))$ and a configuration of $\mathcal{C}'_{SC}$ $\vec{C}'$ that simulates $\mathcal{C}_{\mathcal{S}}$ over $\vec{C}_S$ as follows.

First, we create the same species as described Theorem 1. We now construct new species exclusively for the step functionality of $\mathcal{C}_{\mathcal{S}}$. First, for each step $\vec{S}_l \in \mathcal{C}_{\mathcal{S}}$, we construct the species S_l, d_l, and E_{l+1}. S_l is used to record the

Table 2. Steps and reactions for a (3,1) rule size void Step-Cycle CRN $\mathcal{C}'_{\mathcal{SC}}$ simulating a Step-CRN $\mathcal{C}_{\mathcal{S}}$.

Step	Additions		Relevant Rules	
	Steps 0-6 follow from Table 1.			
7	$\forall \gamma_i \in \Gamma:$ $\qquad\qquad \vec{c}$ $\forall \vec{S}_l \in \mathcal{C}_{\mathcal{S}} : \forall s_j \in \vec{S}_l : \vec{s}_j, \vec{S}_l^j, \vec{E}_{l+1}, \vec{b}$		$\forall \gamma_i \in \Gamma:$ $\forall \vec{S}_l \in \mathcal{C}_{\mathcal{S}}:$	13. $\vec{a}_y + \vec{c} + \vec{G}'_i \to \vec{a}_y$ 14. $\vec{S}_l + \vec{s}_j + \vec{S}_l^j \to \vec{S}_l$ 15. $\vec{S}_l + \vec{E}_{l+1} + \vec{b} \to \vec{S}_l$ 16. $\vec{x} + \vec{s}_j + \vec{S}_l^j \to \vec{x}$ 17. $\vec{x} + \vec{E}_l + \vec{b} \to \vec{x}$
8	$\forall \gamma_i \in \Gamma:$ $\qquad \begin{matrix} \vec{y} \\ \vec{g}_i \end{matrix}$		$\forall \gamma_i \in \Gamma:$ $\quad \forall r_j \in \mathcal{R}_i:$ $\forall p_j \in \mathcal{P}_i:$ $\forall \vec{S}_l \in \mathcal{C}_{\mathcal{S}}:$	18. $\vec{y}_1 + \vec{E}_i^j \to \vec{y}$ 19. $\vec{y}_1 + \vec{R}_i^j \to \vec{y}_1$ 20. $\vec{y}_1 + \vec{P}_i^j \to \vec{y}_1$ 21. $\vec{G}'_i + \vec{g}_i \to \vec{G}'_i$ 22. $\vec{y}_1 + \vec{a}_y \to \vec{y}_1$ 23. $\vec{y}_1 + \vec{a}_n \to \vec{y}_1$ 36. $\vec{y}_1 + \vec{b} \to \vec{y}_1$ 37. $\vec{y}_1 + \vec{c} \to \vec{y}_1$ 26. $\vec{y}_1 + \vec{S}_l^j \to \vec{y}_1$
9	$\forall \vec{S}_l \in \mathcal{C}_{\mathcal{S}}:$ $\qquad \vec{S}_l$		$\forall \vec{S}_l \in \mathcal{C}_{\mathcal{S}}:$	27. $\vec{S}_l + \vec{S}_l + \vec{E}_l \to \emptyset$
10	$\forall \vec{S}_l \in \mathcal{C}_{\mathcal{S}}:$ $\qquad \vec{d}_l$		$\forall \vec{S}_l \in \mathcal{C}_{\mathcal{S}}$	28. $\vec{S}_l + \vec{S}_l + \vec{d}_l \to \vec{S}_l$ 29. $\vec{S}_l + \vec{d}_l + \vec{x} \to \vec{x}$
11	$\qquad\qquad \vec{y}_2$ $\forall \gamma_i \in \Gamma:$ $\qquad \vec{g}_i$		$\forall \vec{S}_l \in \mathcal{C}_{\mathcal{S}}:$ $\forall \gamma_i \in \Gamma:$	30. $\vec{y}_2 + \vec{d}_l \to \vec{y}_2$ 31. $\vec{y}_2 + \vec{x} \to \vec{y}_2$ 32. $\vec{y}_2 + \vec{y}_1 \to \vec{y}_2$ 33. $\vec{G}'_i + \vec{g}_i \to \vec{G}'_i$
12	$\qquad\qquad \vec{y}_2, \vec{z}$ $\forall \gamma_i \in \Gamma:$ $\qquad \vec{G}_i$			34. $\vec{y}_2 + \vec{y}_2 \to \emptyset$ 35. $\vec{z} + \vec{g}_i \to \vec{g}_i$
	Return to Step 0.			

current step-configuration the system is simulating, which is represented by the *missing* S_l copy. Additionally, E_{l+1} will help perform the transition from step $\vec{S}_l$ to $\vec{S}_{l+1}$ when needed while d_l removes any excessive species involved simulating the steps. Furthermore, for each species $s_j \in \vec{S}_l$, we create the species S_l^j to ensure only the step species will be affected by reactions related to the simulated steps. Finally, the species b is made to assist in smaller tasks for $\mathcal{C}'_{\mathcal{SC}}$.

The final species set of $\mathcal{C}'_{\mathcal{SC}}$ is then $\Lambda' = \Lambda \cup \{G, G_1, \cdots, G_{|\Gamma|}, g_1, \cdots, g_{|\Gamma|}, G'_1, \cdots, G'_{|\Gamma|}, E_1^1, \cdots, E_{|\Gamma|}^{|\mathcal{R}_{|\Gamma|}|}, R_1^1, \cdots, R_{|\Gamma|}^{|\mathcal{R}_{|\Gamma|}|}, P_1^1, \cdots, P_{|\Gamma|}^{|\mathcal{P}_{|\Gamma|}|}, \vec{S}_0, \cdots, \vec{S}_k, \vec{S}_0^1, \cdots, \vec{S}_k^{|\vec{S}_k|}, \vec{E}_1, \cdots, \vec{E}_{k+1}, a_y, a_n, b, c, w, x, y_1, y_2, z\}$.

Constructing the configuration is the same as from Theorem 1, with the only new addition of the species $S_1, \cdots, S_k$; S_0 is intentionally excluded to represent $\mathcal{C}'_{\mathcal{SC}}$ starting at Step 0. Then, let the configuration of $\mathcal{C}'_{\mathcal{SC}}$ be $\vec{C} \cup \{\vec{G}_1, \cdots, \vec{G}_{|\Gamma|}, \vec{g}_1, \cdots, \vec{g}_{|\Gamma|}, \vec{S}_1, \cdots, \vec{S}_k\}$.

Steps and Reactions. Steps 0–6 and its respective relevant reactions are constructed as described from Table 1, with the exceptions of modifying Reaction 24 to $\vec{z} + \vec{G} \to \emptyset$ and removing Reaction 25 to ensure that we can "reset" our reaction choices back to normal once a new step has been reached. The full set of added species and reactions for steps 7–12 are presented in Table 2.

Step 7 adds the step species from $\vec{S}_l$ if allowed. This is determined by the configuration of $\mathcal{C}'_{SC}$ following Step 6. Recall that, if no copy of $\vec{x}$ remains present following Step 6, then this indicates no reaction γ_i was selected, and the simulated system $\mathcal{C}_S$ must be at a terminal step-configuration.

We add $|\Gamma|$ copies of $\vec{c}$ and a copy of $\vec{S}_l^{\,j}, \vec{E}_{l+1}$ and $\vec{b}$ for each step-configuration $\vec{S}_l \in \mathcal{C}_S$ and a copy of $\vec{s}_j$ for all step species. If a copy of $\vec{x}$ remains present in Step 7, all copies of $\vec{s}_j, \vec{S}_l^{\,j}$ and $\vec{E}_{l+1}$ copies are removed by Reactions 16 and 17, preventing any step species from being added. Else, each present $\vec{S}_l$ copy removes the E_{l+1} copy and their respective step species in its "normal" ($\vec{s}_j$) and $\vec{S}_l^{\,j}$ form. This leaves one copy for all step species in $\vec{S}_l$ and $\vec{E}_{l+1}$. Additionally, $\vec{a}_y$ must be present, and thus Reaction 13 removes all $\vec{G}'_i$ copies, allowing $\vec{G}$ to select reactions when introduced again.

Steps 8–9 performs some initial clean-up operations and the transition from $\vec{S}_l$ to $\vec{S}_{l+1}$ if required. First, Step 8 introduces $\vec{y}_1$ and $|\Gamma|$ copies of $\vec{g}_i$ and largely functions similar to Step 7 from the construction of 1. The only changes are that $\vec{y}_1$ does not remove $\vec{x}$ and now removes any remaining copies of $\vec{b}, \vec{c}$, and $\vec{S}_l^{\,j}$ by Reactions 36, 37, and 26 respectively. In Step 9, we add a copy of $\vec{S}_l$ for each step. In the case that the step-configuration $\vec{S}_l$ has become terminal in $\mathcal{C}'_{SC}$, a copy of each of $\vec{S}_l$ and E_{l+1} must be present. It then removes $\vec{S}_{l+1}$ by Reaction 27, representing preparation for the next step.

The final steps (10–12) perform the final set of "clean-up" operations. Step 10 adds one copy of $\vec{d}_l$ for each step-configuration. If $\mathcal{C}'_{SC}$ has performed a transition to a new configuration, then $\vec{d}_l$ will remove the excessive $\vec{S}_l$ copies added in Step 9 by Reaction 28. Else, then a remaining copy of $\vec{x}$ will remove the $\vec{d}_l$ copy and extra copy of $\vec{S}_l$ with Reaction 29. Step 11 and 12 essentially follow the same function as Steps 8 and 9 of the construction in Theorem 1; they perform the last "clean-up" reactions.

We define the step-configuration mapping as follows. Because we are simulating a CRN system with step-configurations, we must map both the count of all species $\lambda_i \in \Lambda$ and the step-configuration $\vec{S}_i$. If a single copy of $\vec{G}$ is present in the step-configuration of $\mathcal{C}_{SC}$ $\vec{C}'_i$, then we map $\vec{C}'_i$ to a step-configuration of $\mathcal{C}_S$ in which the number of all species $\lambda_i \in \Lambda$ in $\vec{C}$ is equivalent to the count of all species $\lambda_i \in \Lambda$ in $\vec{C}'_i$. Additionally, the specific step of $\mathcal{C}_S$ is represented by the one missing copy from all of $\vec{S}_0, \cdots, \vec{S}_k$. A special configuration mapping is present for the final step configuration $\vec{S}_{k-1}$. If a copy for all of $\vec{S}_0, \cdots, \vec{S}_k$ is present in $\vec{C}'_i$, then $\vec{C}'_i$ only maps to the step-configuration $\vec{S}_{k-1}$.

$$M(\overrightarrow{C_i}') = \begin{cases} (\sum_{\lambda' \in \Lambda} \overrightarrow{C_i}'[\lambda'] \cdot \overrightarrow{\lambda}', i) & \text{if } \overrightarrow{G} \in \{\overrightarrow{C_i}\}' \wedge \overrightarrow{S}_i \notin \{\overrightarrow{C'}\} \\ (\sum_{\lambda' \in \Lambda} \overrightarrow{C}'_{k-1}[\lambda'] \cdot \overrightarrow{\lambda}', k-1) & \text{if } \overrightarrow{G} \in \{\overrightarrow{C_i}\}' \wedge \overrightarrow{S}_0, \cdots, \overrightarrow{S}_{k-1} \in \{\overrightarrow{C'}\} \\ \emptyset & \text{otherwise.} \end{cases}$$

Polynomial Simulation. Let a be the number of steps in the simulated CRN, n be the maximum number of reactants of a reaction in Γ, m be the maximum number of products of a reaction in Γ, and p be the maximum number of introduced copies in a step in $\mathcal{C}'_{SC}$. $\mathcal{C}'_{SC}$ uses $\mathcal{O}(|\Lambda| + |\Gamma|(n+m))$ species and $\mathcal{O}(a \cdot p + |\Gamma|(n+m))$ reactions with maximum constant rule size of $(3,1)$. The worst case transition sequence has $\mathcal{O}(|\Gamma|)$ length. Any configuration of $\mathcal{C}_{SC}'$ has at most an $\mathcal{O}(|\Gamma| + a)$ blowup in volume.

Theorem 3. *Step-Cycle CRNs can simulate any given Step-Cycle CRN under polynomial simulation, even when restricted to at most $(3,1)$ void rules.*

Proof. Given a Step-Cycle CRN $\mathcal{C}_{SC} = (\Lambda, \Gamma, (\overrightarrow{S}_0, \ldots, \overrightarrow{S}_k))$ and a configuration of $\mathcal{C}_{SC}$ $\overrightarrow{C}_{SC}$, we construct a Step-Cycle CRN $\mathcal{C}'_{SC} = (\Lambda, \Gamma, (\overrightarrow{S}'_0, \ldots, \overrightarrow{S}'_{12}))$ and a configuration of $\mathcal{C}'_{SC}$ $\overrightarrow{C}'$ that simulates $\mathcal{C}_{SC}$ over $\overrightarrow{C}_{SC}$ as follows.

Species and Configuration. First, we create the same species made in Theorem 2. We then include E_0 to allow $\mathcal{C}'_{SC}$ to simulate returning to $\overrightarrow{S}_0$ following the additions of all species in $\overrightarrow{S}_k$.

The configuration is constructed the same as in Theorem 2, which is $\overrightarrow{C} \cup \{\overrightarrow{G}_1, \cdots, \overrightarrow{G}_{|\Gamma|}, \overrightarrow{g}_1, \cdots, \overrightarrow{g}_{|\Gamma|}, \overrightarrow{S}_1, \cdots, \overrightarrow{S}_k\}$.

Steps and Reactions. The steps and reactions are again constructed nearly the same as in Theorem 2. The only new modification is the specific variant of Reaction 15 $\overrightarrow{S}_k + \overrightarrow{E}_{k+1} + b \rightarrow \emptyset$ from Step 7 is changed to $\overrightarrow{S}_k + \overrightarrow{E}_0 + b \rightarrow \emptyset$. The effect of this new change is that, upon $\mathcal{C}'_{SC}$ adding the step species of $\overrightarrow{S}_k$, the $\overrightarrow{E}_0$ copy is not removed. Therefore, in Step 9, when an additional copy of $\overrightarrow{S}_0$ is added, Reaction 27 is executed, leaving no $\overrightarrow{S}_0$ copies remaining. This indicates $\overrightarrow{S}_0$ as the next step-configuration to simulate; $\overrightarrow{E}_1$ then becomes the next deletion species to survive, eventually triggering Reaction 27 on $\overrightarrow{S}_1$.

The configuration mapping is structured the same as in Theorem 2, although we remove the special mapping case for the final step:

$$M(\overrightarrow{C_i}') = \begin{cases} (\sum_{\lambda' \in \Lambda} \overrightarrow{C_i}'[\lambda'] \cdot \overrightarrow{\lambda}', i) & \text{if } \overrightarrow{G} \in \{\overrightarrow{C_i}\}' \wedge \overrightarrow{S}_i \notin \{\overrightarrow{C'}\} \\ \emptyset & \text{otherwise.} \end{cases}$$

Corollary 1. *Step-Cycle CRNs are Turing universal, even when restricted to at most $(3,1)$ void rules.*

Proof. Follows from Theorem 3 and the fact that general Step-Cycle CRNs are Turing universal [7].

4   Conclusion

We have demonstrated that the Step-Cycle CRN model retains a vast computational power, even when restricted to using only trimolecular void reactions, through simulating CRNs, step CRNs, and Step-Cycle CRNs under polynomial simulation. Nevertheless, there are still several open questions to pursue.

For example, can other extended CRN models—such as those with inhibitors, priorities, or synchrony—also achieve polynomial simulation using only small void reactions? Furthermore, can a Step-Cycle CRN, under the same rule restriction, simulate other models such as inhibitory CRNs [9], coarse-rate CRNs, or instruction-parallel CRNS? How do these instruction-parallel CRNs compare to the ability to add a large number of species in a single "step"?

Also, our simulation definition allows for nondeterministic simulation of deterministic systems. While *polynomial efficient* simulation requires that all macro transitions are expected to complete in a polynomial number of steps, this does permit cyclic reaction sequences that may never complete the macro transition. The question of whether (3,1) void rule Step-Cycle CRNs can deterministically simulate general Step-Cycle CRNs remains open.

References

1. Agerwala, T.: Complete model for representing the coordination of asynchronous processes. Technical report, Johns Hopkins Univ., Baltimore, Md. (USA) (1974)
2. Alaniz, R.M., et al.: Reachability in restricted chemical reaction networks (2022). arXiv:2211.12603
3. Anderson, R., et al.: Computing threshold circuits with void reactions in step chemical reaction networks. In: 10th Conference on Machines, Computations and Universality, MCU'24 (2024)
4. Anderson, R., et al.: Computing threshold circuits with bimolecular void reactions in step chemical reaction networks. In: International Conference on Unconventional Computation and Natural Computation, UCNC'24, pp. 253–268. Springer (2024)
5. Aris, R.: Prolegomena to the rational analysis of systems of chemical reactions. Arch. Ration. Mech. Anal. **19**(2), 81–99 (1965). https://doi.org/10.1007/BF00282276
6. Aris, R.: Prolegomena to the rational analysis of systems of chemical reactions ii. some addenda. Arch. Rational Mech. Anal. **27**(5), 356–364 (1968). https://doi.org/10.1007/BF00251438
7. Bajaj, D., et al.: Polynomial equivalence of extended chemical reaction models. Under Submission (2025)
8. Berthomieu, B., Zaitsev, D.A.: Sleptsov nets are turing-complete. Theor. Comput. Sci. **986**, 114346 (2024). https://www.sciencedirect.com/science/article/pii/S030439752300659X, https://doi.org/10.1016/j.tcs.2023.114346
9. Calabrese, K., Doty, D.: Rate-independent continuous inhibitory chemical reaction networks are turing-universal. In: International Conference on Unconventional Computation and Natural Computation, pp. 104–118. Springer (2024)
10. Chen, H.-L., Doty, D., Soloveichik, D.: Deterministic function computation with chemical reaction networks. Nat. Comput. **13**, 517–534 (2014)

11. Cook, M., Soloveichik, D., Winfree, E., Bruck, J.: Programmability of chemical reaction networks. In: Algorithmic Bioprocesses, pp. 543–584. Springer (2009)
12. Czerwiński, W., Orlikowski, Ł.: Reachability in vector addition systems is ackermann-complete. In: 2021 IEEE 62nd Annual Symposium on Foundations of Computer Science (FOCS), pp. 1229–1240. IEEE(2022)
13. Esparza, J., Nielsen, M.: Decidability issues for petri nets–a survey. arXiv preprint: arXiv:2411.01592 (2024)
14. Bin, F., et al.: Brief announcement: reachability in deletion-only chemical reaction networks. In Proc. of the Symposium on Algorithmic Foundations of Dynamic Networks, SAND (2025)
15. Hack, M.H.T.: Decidability questions for Petri Nets. PhD thesis, Massachusetts Institute of Technology (1976)
16. Hack, M.H.T.: Petri net language. Technical report, Massachusetts Institute of Technology (1976)
17. Karp, R.M., Miller, R.E.: Parallel program schemata: a mathematical model for parallel computation. In: 8th Annual Symposium on Switching and Automata Theory (SWAT 1967), pp. 55–61. IEEE (1967)
18. Leroux, J.: The reachability problem for Petri nets is not primitive recursive. In: 2021 IEEE 62nd Annual Symposium on Foundations of Computer Science (FOCS), pp. 1241–1252. IEEE (2022)
19. Peterson, J.L.: Petri net theory and the modeling of systems (1981)
20. Petri, C.A.: Communication with automata (1966)
21. Senum, P., Riedel, M.: Rate-independent constructs for chemical computation. PLoS ONE **6**(6), e21414 (2011)
22. Shea, A., Fett, B., Riedel, M.D., Parhi, K.: Writing and compiling code into biochemistry. In: Biocomputing 2010, pp. 456–464. World Scientific (2010)
23. Thomas, P.B.: Petri net: a modeling tool for the coordination of asynchronous processes. Technical report, Tennessee Univ., Knoxville (USA); Oak Ridge Gaseous Diffusion Plant, Tenn.(USA) (1976)

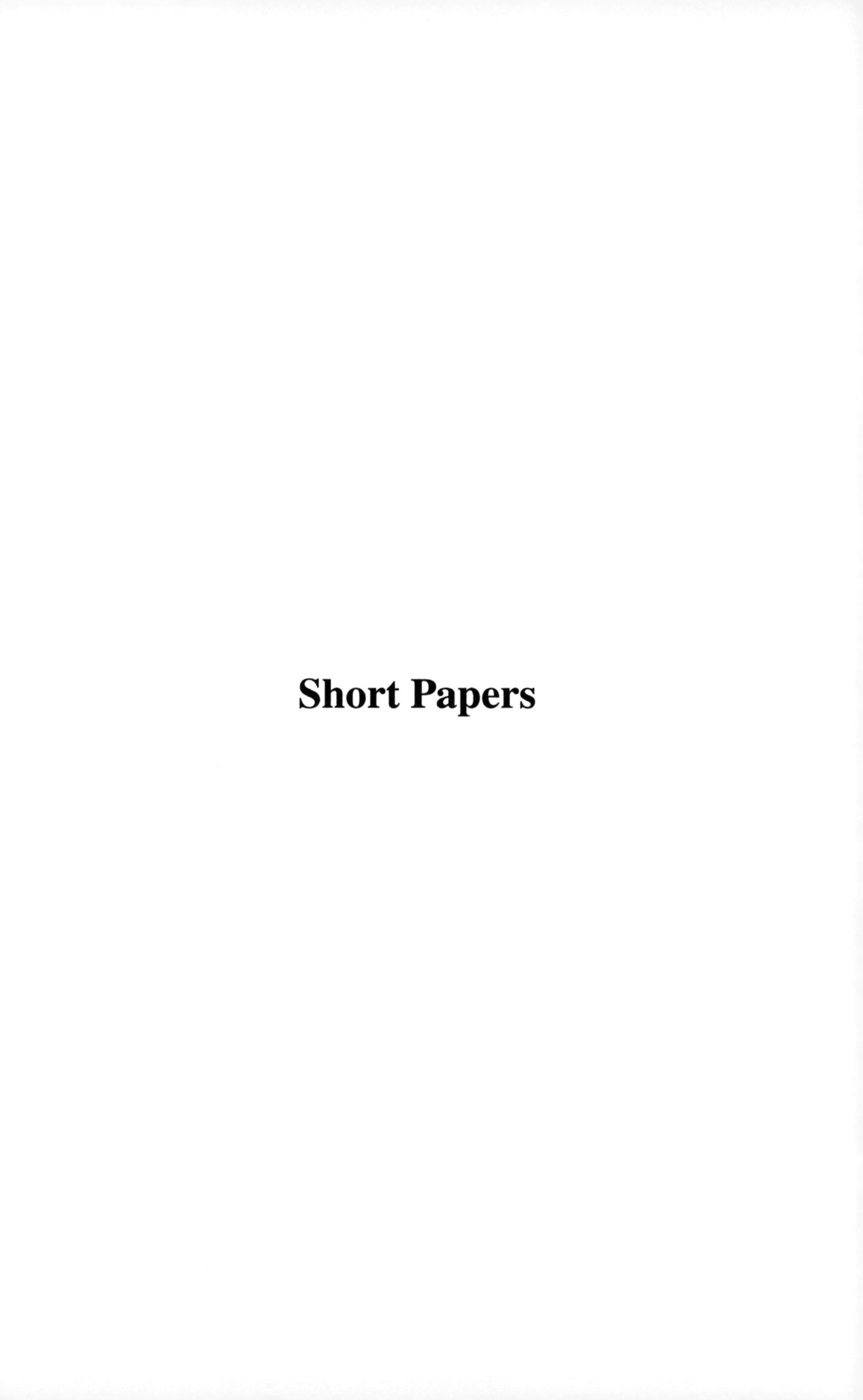

Short Papers

A Selective Dual-Railing Technique
for General-Purpose Analog Computers

Nicholas Haisler[2], Xiang Huang[1], Andrei Migunov[2(✉)],
Khalid Mohammed[2], and Garrett Provence[2]

[1] Department of Computer Science, University of Illinois Springfield,
Springfield, IL, USA
`xhuan5@uis.edu`
[2] Department of Mathematics and Computer Science, Drake University,
Des Moines, IA, USA
`{nicholas.haisler,andrei.migunov,khalid.mohammed,`
`garrett.provence}@drake.edu`

Abstract. This paper describes an efficient variation of a well-known technique for converting a *general-purpose analog computer* (GPAC) to a *chemical reaction network* (CRN). If an input system has n variables, the existing *dual-railing* technique requires $2n$ variables in the resulting system. As the resulting CRNs are often fed into downstream processes which may themselves be exponential in the number of variables [7], it is frequently worthwhile to try to minimize the number of variables generated in this conversion. This paper presents a technique that results in fewer than n new variables, by rewriting only those whose positive terms are 'infected' by ill-formed variables.

1 Introduction

Chemical Reaction Networks (CRNs) are a popular programming language for nanotechnological applications, allowing one to describe computation as a process of interactions between abstract molecular species. Moreover, CRNs have a natural *hardware language* in which they can be physically realized at the nano-scale: DNA Strand Displacement (DSD) cascades [11].

The ease of use of a 'high-level' language may depend on how comfortably one can manipulate and modify a program in order to change its behavior in an intended way. While the prospect of one day compiling high-level digital languages to CRNs or DSDs is also interesting [15], it is common and sometimes more convenient to convert other *analog* languages to CRNs.

With this in mind, ordinary differential equations (ODEs) are a well-studied and widely used model (see any standard text, e.g. [13]). In the world of analog computation, ODEs are the basis of the *general purpose analog computer* (GPAC) model, originally devised by Shannon [3,10]. Both GPACs and CRNs are Turing-complete [1,4] but the GPAC comes with the many conveniences of ODEs: tricks, substitution techniques, and many centuries of examples, in particular of the computation of numbers and functions.

© The Author(s), under exclusive license to Springer Nature Switzerland AG 2026
E. Formenti and L. Manzoni (Eds.): UCNC 2025, LNCS 16364, pp. 397–402, 2026.
https://doi.org/10.1007/978-3-032-15641-9_26

In this sense, we find it useful to regard CRNs as GPACs: one can express any chemical system in terms of its *mass action* kinetics, in which the concentration of each molecular species is tracked by the solution of a system of ODEs. If we want to manipulate CRNs, we might begin with a GPAC, program it to do what we want (for example, to compute $\sqrt{2}$ as the limit of one of its component functions), and then convert it to a CRN while preserving this property.

Not every GPAC is by default a CRN, as GPACs can have negative values and there cannot be negative quantities of chemical species. Moreover, in this paper we are concerned specifically with *bounded* systems whose variable values always lie within some interval $(-\beta, \beta)$ for $\beta > 0$. This is a realistic restriction that ensures that quantities do not 'overflow' to infinity, and that the time complexity of terminating computations can be meaningfully measured [2].

Section 2 reviews some formal definitions. In Sect. 3, we explore the *dual-railing* method, which transforms a GPAC into a CRN by encoding each variable as a difference of non-negative variables. In Sect. 4, we discuss a more efficient version of this method and demonstrate the improvement with an example.

2   Preliminaries

Definition 1. *A* General Purpose Analog Computer (GPAC) *is a system of ordinary differential equations of the form* $\mathbf{y}'(t) = \mathbf{p}(\mathbf{y}(t))$, *together with a set of initial values* $\mathbf{y}(\mathbf{0})$, *where* $\mathbf{p}$ *is a vector of polynomials. In this sense, a GPAC is a* polynomial initial value problem. *Such systems are always polynomial, autonomous, and first-order.*

Definition 2. *A* Chemical Reaction Network (CRN) *is a pair* (S, R) *where* S *is a set of (usually, capitalized and single-letter) names of molecular species, and* R *is a set of reactions in which those species participate. Each reaction* ρ *has the following form, where* $k_\rho \in \mathbb{N}$ *is its* rate constant*:*

$$\rho : a_1 X_1 + \ldots + a_n X_n \xrightarrow{k_\rho} b_1 Y_1 + \ldots + b_m Y_m.$$

Theorem 1. (CRN-implementable Functions [5], Theorem 3.1 and 3.2). *A function is CRN-implementable if and only if it is GPAC-implementable with the further restriction that the derivative of each component* x_i *in its variable vector* $\mathbf{x} = (x_1, x_2, \cdots, x_n)$ *is of the form*

$$x_i' = p - q x_i, \quad \text{where } p, \ q \text{ are positive polynomials.} \tag{1}$$

3   Naive Dual-Railing

Here we describe the well-known dual-railing method for converting a GPAC into a CRN. To *dual-rail a variable* x is to split it into two variables, x^+ and x^-, ensuring that $x(t) = x^+(t) - x^-(t)$ for all $t \in [0, \infty)$, and that both x^+ and x^- are CRN-implementable. The dual-railing method is really a collection of related approaches that use this idea, and in examining a dual-railing approach we consider two factors: how many variables of a system are modified, and in what way are these modified. In our sense, a *naive* approach modifies every single variable even if, as we will show in the next section, this is not always necessary.

The other factor concerns the specific way of defining the new variables. One folkloric and common [9,14] option is to use *fast annihilation reactions*: as dual species X^+ and X^- accumulate, their excess can be shed via $X^+ + X^- \xrightarrow{k} W$, where W is a waste species. This introduces a term $-kx^+x^-$ in the new variables' ODEs. Another approach scales the $-x^+x^-$ term by the entire ODE x' [8], in which case it is known that the resulting system is bounded [6]. It is often assumed that the fast annihilation reaction method results in a bounded system for sufficiently large k, but this has not yet been shown:

Problem 1. Given $\mathbf{x}$, is there a constant k so that the dual-railed system based on annihilation reactions ($Z = k$ in Eq. (2)) is bounded?

Construction 1. *(Naive Dual-Railing). If* $\mathbf{x} = (x_1, x_2, ..., x_n)$ *is a GPAC consisting of n variables with ODEs*

$$
\begin{aligned}
x_1' &= p_1(x_1, ..., x_n), \\
x_2' &= p_2(x_1, ..., x_n), \\
&... \\
x_n' &= p_n(x_1, ..., x_n),
\end{aligned}
$$

then we define a new system $\hat{\mathbf{x}} = (x_1^+, x_1^-, x_2^+, x_2^-, ..., x_n^+, x_n^-)$ *as follows:*

1. *For each i, define a new polynomial $\hat{p}_i(\hat{\mathbf{x}}) = p_i(x_1^+ - x_1^-, x_2^+ - x_2^-, ..., x_n^+ - x_n^-)$ by substituting the differences $x_i^+ - x_i^-$ for each x_i everywhere in p_i.*
2. *Next, split this polynomial up into $\hat{p}_i = \hat{P}_i^+ - \hat{P}_i^-$, where $\hat{P}_i^+$ consists of all the positive terms in $\hat{p}_i$, and $\hat{P}_i^-$ consists of all the positive parts of the negative terms in $\hat{p}_i$.*
3. *Finally, define the CRN-implementable ODEs for the new variables by:*

$$
\begin{aligned}
x_i^{+'} &= \hat{P}_i^+ - x_i^+ \cdot x_i^- \cdot Z, \\
x_i^{-'} &= \hat{P}_i^- - x_i^+ \cdot x_i^- \cdot Z,
\end{aligned}
\tag{2}
$$

where either $Z = \hat{P}_i^+ + \hat{P}_i^-$ or $Z = k$ (where k is a sufficiently large constant), based on the chosen approach.

As a result, for every original variable, $x_i' = p_i = \hat{P}_i^+ - \hat{P}_i^- = x_i^{+'} - x_i^{-'}$.

4   Selective Dual-Railing

Next, we demonstrate an improvement based on the observation that not every variable necessarily needs to be dual-railed. As before, let $\mathbf{x} = (x_1, x_2, ..., x_n)$ be a GPAC with n variables. We identify a vertex in a graph with its variable in the ODE system.

Definition 3. *A variable x_i is* ill *if its equation x_i' is ill-formed, i.e., fails to have the form of Eq. (1).*

Definition 4. *If x_a, x_b are variables of $\mathbf{x}$, we say that x_a* affects *x_b if x_a appears anywhere in the positive terms of x_b', i.e. if there is a positive monomial $T = x_a S$ in x_b' where S is some monomial or a constant.*

Definition 5. *x_a can* infect *x_b if x_a affects x_b and there is a positive term in x_b' where x_a appears but x_b does not.*

Note that if x_b' has a positive term where x_b appears, then splitting up the other variables in that term can only produce negative terms that *also* contain x_b, thus it does not harm CRN-implementability.

Definition 6. *The* infection graph *is the directed graph $I(\mathbf{x}) = (V, E)$ where V is the set of variables of $\mathbf{x}$ and $E = \{(x_a, x_b) \mid x_a \text{ can infect } x_b\}$.*

As an example, consider the following system, noting that only one variable, a, has an ill-formed ODE expression (the term $-d$ does not include a):

$$\mathbf{x} = \begin{cases} a' = b - ab - d \\ b' = c + a - ab^2 \\ c' = bc - bc^2 + 0.1d \\ d' = c - cd \end{cases} \tag{3}$$

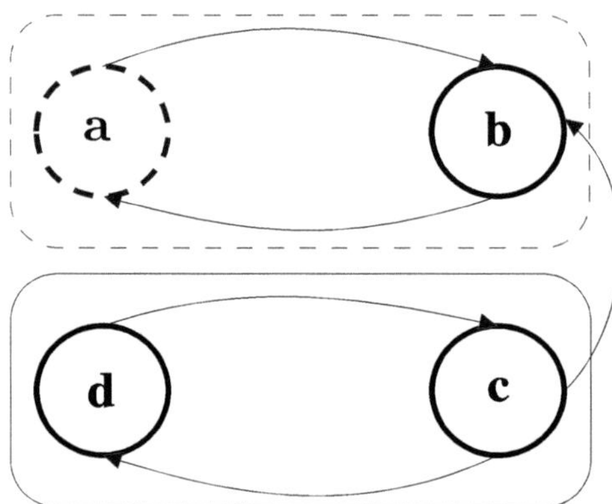

Fig. 1. $I(\mathbf{x})$, of Eq. 3, with one infected (dashed) and one uninfected (solid) SCC.

Construction 1 results in 8 variables. However, if we look at the strongly connected components of $I(\mathbf{x})$ in Fig. 1, we see that only two variables require rewriting: a and b, requiring 6 variables total for the system.

Below is an algorithm that selectively dual-rails $\mathbf{x}$:

Algorithm 1: Selective Dual-Railing

Input: A GPAC $x = (x_1, x_2, \ldots, x_n)$.

Output: A CRN-implementable system.

1   Construct the infection graph $I(\mathbf{x}) = (V, E)$.

2   Compute the set $S(I)$ of strongly connected components (SCCs) of I via Tarjan's algorithm.

3   Build the induced acyclic graph of SCCs, with edges inherited from I.

4   Identify the set ILL $\subseteq \{x_1, \ldots, x_n\}$ of ill-formed variables.

5   **foreach** $S \in S(I)$ **do**

6   **if** S *is reachable from an SCC* S' *with* $S' \cap \text{ILL} \neq \emptyset$ **then**

7   Label S infected.

8   Let $R \subseteq V$ be the set of variables belonging to infected SCCs.

9   **foreach** $x_i \in R$ **do**

10   Define the polynomial $\hat{p}_i$ and the sets $\hat{P}_i^+$ and $\hat{P}_i^-$ by substituting $(y_j, y_j^+ - y_j^-)$ for all $y_j \in R$ as in Construction 1.

11   Apply dual-railing to x_i' as in Eq. (2).

12   **foreach** $x_i \notin R$ **do**

13   **foreach** $y_i \in R$ **do**

14   Substitute $(y_i^+ - y_i^-)$ for all instances of y_i in x_i'.

15   **return** the modified ODE system.

The values of original system variables are preserved via the relationships $x_i = (x_i^+ - x_i^-)$, as is boundedness if one uses $Z = \hat{P}_i^+ - \hat{P}_i^-$. All ill-formed variables, and those that depend on them, are dual-railed - thus, become CRN-implementable. Variables x_i that are not dual-railed remain CRN-implementable: although they may undergo substitutions, by Definition 5 these cannot result in negative terms that fail to include x_i.

One may use, for example, Tarjan's algorithm [12] to identify strongly connected components in $I(\mathbf{x})$. Algorithm 1 introduces a number of new variables equal to the size of the union of infected SCCs of $I(\mathbf{x})$, typically fewer than n. The algorithm parses the input system in linear time with respect to its string representation. It performs symbolic transformations that may introduce new variables or expand terms. The total runtime is output-sensitive, and in the worst case can be exponential in the input size due to polynomial expansion. In practice, the output is typically much smaller, and the algorithm runs in time linear in the input and output sizes combined.

5   Conclusion

While the standard dual-railing method for bounded systems (Construction 1) results in n new variables, the method of the previous section often results in fewer. It can be worthwhile to use it when one is concerned about the number of variables and about preserving boundedness, and especially if one suspects that the initial system has some well-formed variables that are decoupled from ill-formed ones.

References

1. Bournez, O., Campagnolo, M.L., Graça, D.S., Hainry, E.: The general purpose analog computer and computable analysis are two equivalent paradigms of analog computation. In: Proceedings of the 3rd International Conference on Theory and Applications of Models of Computation, pp. 631–643. Springer Berlin Heidelberg (2006)
2. Bournez, O., Graça, D.S., Pouly, A.: Polynomial time corresponds to solutions of polynomial ordinary differential equations of polynomial length. J. ACM **64**(6), 38 (2017)
3. Bush, V.: The differential analyzer. A new machine for solving differential equations. J. Franklin Inst. **212**(4), 447–488 (1931)
4. Fages, F., Le Guludec, G., Bournez, O., Pouly, A.: Strong turing completeness of continuous chemical reaction networks and compilation of mixed analog-digital programs. In: Feret, J., Koeppl, H. (eds.) CMSB 2017. LNCS, vol. 10545, pp. 108–127. Springer, Cham (2017). https://doi.org/10.1007/978-3-319-67471-1_7
5. Hárs, V., Tóth, J.: On the inverse problem of reaction kinetics. In: Colloquia Mathematica Societatis János Bolyai, 30: Qualitative Theory of Differential Equations, pp. 363–379 (1981)
6. Huang, X.: Chemical reaction networks: computability, complexity, and randomness. PhD thesis, Iowa State University (2020)
7. Huang, X., Huls, R.N.: Computing real numbers with large-population protocols having a continuum of equilibria. In: 28th International Conference on DNA Computing and Molecular Programming (2022)
8. Huang, X., Klinge, T.H., Lathrop, J.I.: Real-time equivalence of chemical reaction networks and analog computers. In: International Conference on DNA Computing and Molecular Programming, pp. 37–53. Springer (2019)
9. Klinge, T.H., Lathrop, J.I., Osera, P.M., Rogers, A.: Reactamole: functional reactive molecular programming. Nat. Comput. **23**(3), 477–495 (2024)
10. Shannon, C.E.: Mathematical theory of the differential analyzer. Stud. Appl. Math. **20**(1–4), 337–354 (1941)
11. Soloveichik, D., Seelig, G., Winfree, E.: DNA as a universal substrate for chemical kinetics. In: Goel, A., Simmel, F.C., Sosík, P. (eds.) DNA 2008. LNCS, vol. 5347, pp. 57–69. Springer, Heidelberg (2009). https://doi.org/10.1007/978-3-642-03076-5_6
12. Tarjan, R.: Depth-first search and linear graph algorithms. SIAM J. Comput. **1**(2), 146–160 (1972)
13. Teschl, G.: Ordinary Differential Equations and Dynamical Systems. American Mathematical Society (2012)
14. Vasic, M., Chalk, C., Khurshid, S., Soloveichik, D.: Deep molecular programming: a natural implementation of binary-weight ReLU neural networks. In: International Conference on Machine Learning, pp. 9701–9711. PMLR (2020)
15. Vasić, M., Soloveichik, D., Khurshid, S.: CRN++: molecular programming language. Nat. Comput. **19**(2), 391–407 (2020)

Partial Information Decomposition
of the Radical Pair Spin Dynamics
in Avian Magnetoreception Mechanism

Dragana Laketić[(✉)]

Norwegian University of Science and Technology, Trondheim, Norway
`dragana.laketic@ntnu.no`

Abstract. This short paper presents the initial investigation of the biologically inspired coherent spin dynamics for use in an unconventional in-materio computing scenario. The radical pair mechanism is analyzed from the information-theoretic perspective based on the previously developed framework and further investigated regarding the structure of multivariate information, an approach recently adopted for describing emergent phenomena. Some initial simulation results indicate that the entropic relations within the investigated system hold for the addressed framework. This knowledge provides directions for manipulation of physical substrates with radical pairs toward a computational goal.

Keywords: quantum biology · information theory · emergence · hierarchical levels

1   Introduction

Recently, a number of fine processes of quantum nature have been recognized to enable important biological functions [9]. Such is one of the mechanisms which aid avian navigation in the geomagnetic field. Proven in vitro, it is based on the quantum dynamics of the radical pair (RP) spins of the cryptochrome molecule in the retina of the bird's eye. Physical substrates which may generate RPs have already been used for sensing. However, the challenge we address is whether such substrates could be used for computations. In particular, we address the challenge of the possible use within a Computation-in-Materio paradigm. In order to find out the most suitable ways of the material manipulation, we start with the investigation of the fine information structure of the coherent RP spin dynamics.

This short, work-in-progress paper provides initial insights and directions for further analysis of the radical pair (RP) mechanism of avian magnetoreception. Based on the previously established information-theoretic model of the system [6], summarized in Sect. 2, its information structure is further analyzed in the context of multivariate information, Sect. 3, and supported by some initial results from a simple simulation of processes at the quantum level. Section 4 presents a brief discussion and some directions for future work.

© The Author(s), under exclusive license to Springer Nature Switzerland AG 2026
E. Formenti and L. Manzoni (Eds.): UCNC 2025, LNCS 16364, pp. 403–408, 2026.
https://doi.org/10.1007/978-3-032-15641-9_27

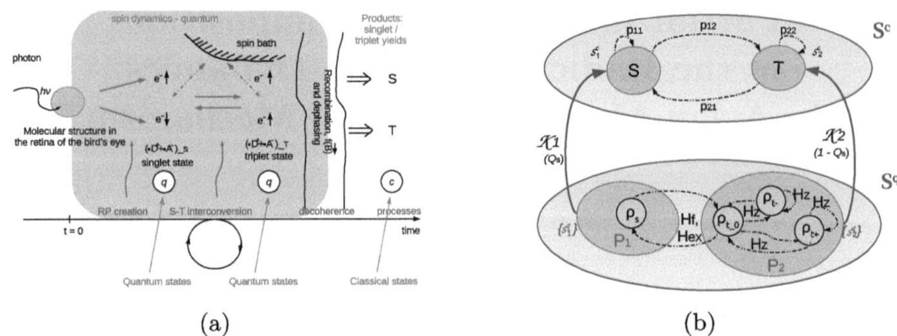

Fig. 1. Schematic view of the timeline of RP creation (a) and state space descriptions of the spin dynamics at quantum and classical level (b), adopted from [6].

2 Information-Theoretic Approach to RP Mechanism of Avian Magnetoreception

The RP mechanism in the context of avian magnetoreception [14] is schematically shown in Fig. 1. RPs, denoted as D^+ and A^-, are formed when photons of a particular wavelength hit cryptochrome molecules in the bird's retina causing the displacement of one of the electrons from the electron pair. The electron pair transitions from the singlet, s, to the triplet, t, state, a known Zeeman effect [4]. Their further dynamics is guided by the interactions with the surrounding spin bath and the total magnetic field including a geomagnetic component. The s or t state of the electron pair at the time the cryptochrome molecule participates in a chemical reaction determines the type of the chemical yield, S or T. The ratio of S and T encodes the information about the geomagnetic field carried via optical nerve to be processed by higher neural structures in the bird's brain.

The dynamics of such open quantum systems can be described by the master equation [1]. One of its forms provided in [2,5] considers the interaction of the three spins: two spins of the RP electron pair and the nearest nuclear spin:

$$\frac{\partial}{\partial t}\rho = -i[H,\rho] - \frac{(k_S + k_T)(Q_S\rho + \rho Q_S - 2Q_S\rho Q_S)}{2} \tag{1}$$

The state of the three spins is given by ρ; H is the Hamiltonian with components: H_m, the (geo)magnetic field, H_Z Zeeman interactions responsible for the transitions between s and t states; k_S and k_T the recombination rates; Q_S and Q_T projection operators for which it holds $Q^n_{S/T} = Q_{S/T}, \quad n > 1, Q_S + Q_T = 1$.

In [6] the RP dynamics is set within the information-theoretic framework of dynamical hierarchies [10] as schematically shown in Fig. 1. The two levels are distinguished: the quantum level where the three spins are quantum entities given by mathematical description as $\mathscr{S}^q = \{s_i^q(t) = \rho_i(t) : \rho_i(t) \xrightarrow{H,B} \rho_j(t+1)\}_{i,j=1}^{N_q}$, and the classical level at which the chemical yields S and T are observed, $\mathscr{S}^c = \{s_i^c(t) : s_i^c(t) \xrightarrow{f_T} s_j^c(t+1)\}_{i,j=1}^{N_c}$.

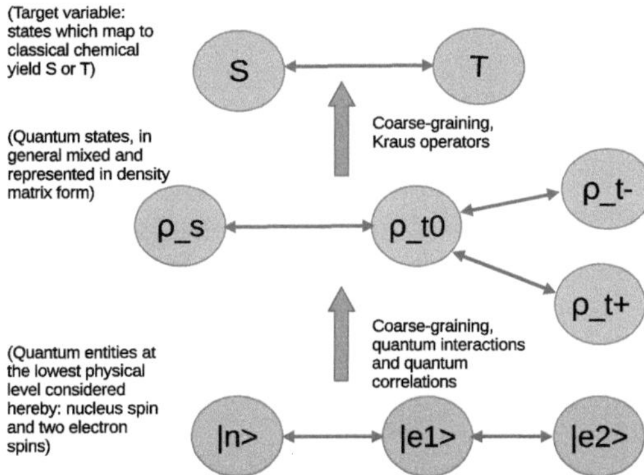

Fig. 2. Quantum and classical levels at which the states of the three spin dynamics are considered and the corresponding coarse-grainings.

For the description at the quantum level, S^q, the considered quantum states are the tensor product of the nuclear spin, $|n\rangle$, and the two electron spins, $|e_1\rangle$ and $|e_2\rangle$, which can be ρ_s or ρ_{t*} dependent on the state of the electron pair: $\{|s\rangle = \frac{1}{\sqrt{2}}(|\uparrow\downarrow\rangle - |\downarrow\uparrow\rangle), |t_0\rangle = \frac{1}{\sqrt{2}}(|\uparrow\downarrow\rangle + |\downarrow\uparrow\rangle), |t_-\rangle = |\downarrow\downarrow\rangle, |t_+\rangle = |\uparrow\uparrow\rangle\}$. The transition function f_T is given as a set of probabilities: $f_T = \{p_{ij} : p_{ij} = p(s^c(t) = s_i^c, s^c(t+1) = s_j^c)\}_{i,j=1}^{N_c}$.

The coarse-graining [3] from quantum to classical level is given by the description function $d_{qc} = \{\mathcal{K}_1, \mathcal{K}_2\}$, where Kraus operators $\mathcal{K}_1, \mathcal{K}_2$ act on the entities which assume the collective state of the RP electron spins. An even finer level of the information-theoretic description considers individual spins. For the sake of clarity, the coarse-grainings are schematically shown in Fig. 2.

3 Partial Information Decomposition of the RP Dynamics

The coherent spin dynamics of RP described in [6] and illustrated in Fig. 1 can be interpreted as the system whose dynamics is based on the three source variables which represent the state of the three spins, i.e., $|n\rangle$, $|e_1\rangle$, $|e_2\rangle$, and which results in a target variable which is the ratio of the classical chemical yields S and T. A deeper analysis of statistical correlations among its parts and how various interactions can be manipulated by the identified Hamiltonian and environmental components is expected to lead to the directions as of how the physical properties of the substrate can be used within the Computation-in-Materio computing scenario. The first step in such an endeavor is to learn about the structure of information within the system. For this purpose, we find the partial information decomposition (PID) [13] to be a suitable approach, in particular for systems with emergent properties [11]. The three types of information are recognized with respect to how much of the information is provided by individual source

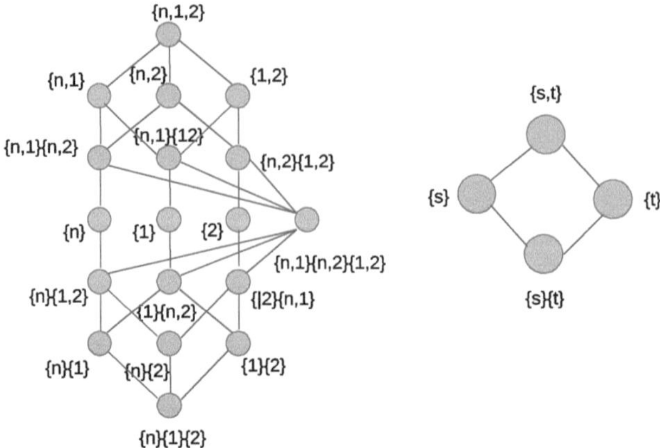

Fig. 3. Lattice of partial information atoms: on the left, a lattice of poset $(\{n, 1, 2\}, \preccurlyeq)$ and, on the right, a lattice of poset $(\{s, t\}, \preccurlyeq)$. The latter describes the structure of information for the RP mechanism analyzed in [6] and corresponds to the coarse-graining in the upper part of Fig. 2. The former refers to the information structure of three entities at the quantum level and corresponds to the coarse-graining in the lower part of Fig. 2. Due to the quantum nature of the entities at this physical level, a special kind of quantum correlations must be considered when addressing quantum information content of the system.

variables and all possible partitions of the partitions of the set of source variables: redundant, unique and synergistic information. Moreover, a measure of redundancy was introduced [13] for providing how much of redundant information each subset of the partitions of the set of sources contains. This relation enables the establishment of the partial ordering $\preccurlyeq$ for the set of all partitions of the partition of the set of sources. In mathematical terms, for such poset it is possible to define a lattice such as shown in Fig. 3. The meaning of the structure of information is that as we climb up the lattice, the content of the synergistic information increases: it is 0 at the bottom for individual sources and highest at the top node, which stands for the synergy of all sources. In addition, moving in the opposite direction, a partial information (PI) function can be defined recursively based on the measure of redundancy. In this way, the PI of each node, a unique information contributed by the collection of sources related to the node, can be calculated recursively due to the partial ordering $\preccurlyeq$. This information-theoretic framework has been used for the analysis of information structure in a variety of systems which exhibit emergent properties [11, 12].

The information-theoretic model considered so far assumes the states of two electron spins in their collective form, which reflects their energy spectrum and the level of entanglement. Therefore, to get the full picture of the structure of information within the considered system, further investigation is needed into the realm of quantum correlations. For this purpose, the quantum dynamics of the RP given by Eq. 1 was simulated in QuTip [7]. The time evolution of the

Table 1. Simulation results: numerical values which reflect the rate of transitions between S and T states.

Simulation set	$|\mathbf{B}|$, $[\mu s]$	First time $Q_S = Q_T$ $[\mu s]$
Abuja	33.65	1.3
Rome	46.5	1.7
Kirkenes	51.7	1.8

system was tested for the varying strengths of the geomagnetic field. In particular, the three different values were considered for the corresponding components of the system Hamiltonian. The collapse operators were based on the projection operators Q_S, Q_T according to $Q_S = 1/4 - S_1 * S_2$, $Q_T = 3/4 + S_1 * S_2$, where S_1, S_2 represent tensor products of the three spin operators. The values of the recombination rates k_S and k_T and the remaining parameters were as in [5].

The results show the rate of transitions between the s and t states (the plots not provided due to the lack of space). Some numerical results are shown in Table 1 where it can be seen that the strength of the geomagnetic field affects the rate of spin transitions between s and t states. This supports the fact that the Hamiltonian component which causes these transitions depends on the strength of the geomagnetic field, as indicated in Fig. 1.

In addition, von Neumann entropies of various collections of sources, as illustrated in the left lattice in Fig. 3, were monitored at each time evolution step alongside the mutual and relative entropies. The results indicate highly correlated states, which is as expected due to the spin entanglement. However, at this stage of investigation, the results are inconclusive.

4   Discussion and Future Work

Investigation of the RP mechanism, one of the mechanisms of avian magnetoreception, was revisited within the information-theoretic framework. The fine information structure of the multivariate variable which represents the ratio of the chemical yields and which carries information about the bird's geo-position is set within the framework of PID. Simulations have provided some initial insights into the dynamics at the quantum level and confirmed some expected behaviors. Since spin states s and t directly influence the rate of production of chemical yields S and T and, therefore, their ratio, a classical observable, this insight provides the knowledge to tune the transition probabilities p_{ij} at the higher classical level.

However, additional simulations are needed to provide more accurate insights into quantum dynamics regarding all aspects of the magnetic field and the particular molecular environment in which the RP is immersed. The available quantum computing resources are suitable platforms for this kind of simulations and, indeed, some results have already been achieved in this direction [8] although for smaller molecules.

The RP mechanism addressed in this short article was based on the in vitro confirmed mechanism of avian magnetoreception. However, the principles underlying this mechanism affect a wide range of biological processes [15] and certainly deserve to be investigated in greater detail. One possible direction is indicated in this short paper.

Acknowledgments. The author would like to thank CrossLabs for the online available talk The Many Faces of Emergence, with Dr. Fernando Rosas.

References

1. Breuer, H.P., Petruccione, F.: The Theory of Open Quantum Systems. Oxford University Press (2002)
2. Dellis, A., Kominis, I.: The quantum Zeno effect immunizes the avian compass against the deleterious effects of exchange and dipolar interactions. Biosystems **107**, 153–157 (2012)
3. Flack, J.C.: Coarse-graining as a downward causation mechanism. Philos. Trans. Roy. Soc. A (2017)
4. Griffiths, D.J., Schroeter, D.F.: The Zeeman Effect, Chapter 7.4, Introduction to Quantum Mechanics, Third Edition. Cambridge University Press (2018). https://doi.org/10.1017/9781316995433
5. Kominis, I.: Quantum Zeno effect explains magnetic-sensitive radical-ion-pair reactions. Phys. Rev. E **80**, 056115 (2009)
6. Laketić, D.: Hyperdescriptions of quantum dynamics - a case study for avian magnetoreception. In: Artificial Life Conference Proceedings (2021)
7. Lambert, N., et al.: QuTiP 5: the quantum toolbox in Python (2024). https://doi.org/10.48550/arXiv.2412.04705
8. Liepuoniute, I., Doney, K.D., Robledo-Moreno, J., Job, J.A., Friend, W.S., Jones, G.O.: Quantum-centric study of methylene singlet and triplet states (2024). https://arxiv.org/abs/2411.04827
9. McFadden, J., Al-Khalili, J.: The origins of quantum biology. Proc. Roy. Soc. A **474**, 20180674 (2018)
10. McGregor, S., Fernando, C.: Levels of description: a novel approach to dynamical hierarchies. Artif. Life **11**(4), 459–472 (2005)
11. Rosas, et al.: Reconciling emergences: an information-theoretic approach to identify causal emergence in multivariate data. PLoS Comput. Biol. (2020)
12. Varley, T.F., Hoel, E.: Emergence as the conversion of information: a unifying theory. Phylosophical Trans. Roy. Soc. A (2022). https://doi.org/10.1098/rsta.2021.0150
13. Williams, P.L., Beer, R.D.: Nonnegative decomposition of multivariate information (2010). https://doi.org/10.48550/arXiv.1004.2515
14. Wiltschko, R., Wiltschko, T.: Magnetoreception in birds. J. R. Soc. Interface **16**, 20190295 (2019)
15. Zadeh-Haghighi, H., Simon, C.: Magnetic field effects in biology from the perspective of the radical pair mechanism. J. Roy. Soc., Interface (2022). https://doi.org/10.1098/rsif.2022.0325

On the Composition of Cellular Automata

Firas Ben Ramdhane[1,2] and Giuliamaria Menara[1(✉)]

[1] Department of Informatics, Systems and Communications, University of
Milano-Bicocca, Milano, Italy
`{firas.benramdhane,giuliamaria.menara}@unimib.it`
[2] University of Sfax, Faculty of Sciences of Sfax, Sfax, Tunisia

Abstract. This short paper is about the behavior of the cellular
automaton (CA) obtained by the composition of two or more cellular
automata. The general question we aim to face is: what is the relationship between a certain property of the CA obtained by the composition
and the same (or other) property of each single CA appearing in the composition? We show that if P is one of the main basic and set-theoretic
properties, the CA resulting from the composition has a certain property P iff both CA components have P, while this equivalence generally
no longer holds when P is one of the main dynamical properties. Some
important questions naturally arise from these results.

Keywords: Cellular Automata · Composition · Discrete Dynamical
Systems

1 Introduction

Cellular Automata (CA) are bio-inspired formal models for complex systems
(for an introduction to CA theory see [17,18], while for other natural computing
models see [11,13,23], for instance). Although they are defined by a finite local
rule over a finite alphabet, CA can exhibit rich and complex global behaviors. For
that reason, they are used in many fields for different purposes [2,6,14,22,24].

The composition of any two CA over the same alphabet gives rise to a new
CA whose properties are often related in a non trivial way to those of its two
CA components. Understanding how the set-theoretic and dynamical properties of the two CA components influence those of their composition (and vice
versa) represents a new and significant research direction that is important from
several points of view. From a theoretical perspective and regarding the emergence of complex behaviors, such an understanding provides information on the
algebraic structure of the CA monoid under composition, helps to classify CA
behaviors (with respect to their stability/instability under the composition), and,
hence, sheds light on the emergence of complexity from simple (composed) local
interactions. Indeed, when composed, two CA can exhibit a greater or weaker
complexity, making the composition a formal tool for studying complex systems.

From a point of view of applications, the CA compositions naturally arise
in the *modular* design of systems based on CA and that are used, for instance,

© The Author(s), under exclusive license to Springer Nature Switzerland AG 2026
E. Formenti and L. Manzoni (Eds.): UCNC 2025, LNCS 16364, pp. 409–415, 2026.
https://doi.org/10.1007/978-3-032-15641-9_28

in cryptography, simulations of real phenomena, parallel computation, *etc.*. In particular, composing simple CA to attain complex or tailored global behavior is a crucial design strategy in applications concerning for instance secure communication protocols and neural CA architectures.

In this work, we study how both basic and set-theoretic CA properties –such as injectivity (and, hence, reversibility), surjectivity, permutativity, closingness, and openness– and dynamical properties –such as sensitivity to the initial conditions, topological transitivity and mixing, chaos, equicontinuity and almost equicontinuity, and positive expansivity– behave under composition. In particular, we show that if P is one of the above mentioned basic and set-theoretic properties, the CA resulting from the composition has a certain property P iff both CA components have P, while this equivalence generally no longer holds when P is one of the main dynamical properties. Some important questions naturally arise from these results. Essentially, they concern the identification of the conditions on one or both the CA components under which the dynamical properties are preserved, amplified, weakened, or even destroyed by the CA composition.

2   Basic Notions and Background

Let A be a finite set (also called *alphabet*). A *configuration* is any element of $A^{\mathbb{Z}}$. For any configuration $c \in A^{\mathbb{Z}}$ and any integer $i \in \mathbb{Z}$, the value of c in position i is denoted by c_i. The set $A^{\mathbb{Z}}$ is as usual equipped with the standard Tychonoff distance d. A *one-dimensional CA* (or, briefly, a *CA*) over A is a pair $(A^{\mathbb{Z}}, F)$, where $F\colon A^{\mathbb{Z}} \to A^{\mathbb{Z}}$ is the uniformly continuous transformation (called *global rule*) defined as $\forall c \in A^{\mathbb{Z}}, \forall i \in \mathbb{Z}, F(c)_i = f(c_{i-r}, \ldots, c_{i+r})$, for some fixed natural number $r \in \mathbb{N}$ (called *radius*) and some fixed function $f\colon A^{2r+1} \to A$ (called *local rule* of radius r). Equivalently, a pair $(A^{\mathbb{Z}}, F)$ is a CA iff F is continuous and $F \circ \sigma = \sigma \circ F$, where $\sigma\colon A^{\mathbb{Z}} \to A^{\mathbb{Z}}$, called *shift* map, is defined for any $c \in A^{\mathbb{Z}}$ and any $i \in \mathbb{Z}$ as $\sigma(c)_i = c_{i+1}$. The shift map is the global rule of a CA. A function $f\colon A^{2r+1} \to A$ is said to be rightmost (leftmost) permutative if $r > 0$ and for any $u \in A^{2r}$ it holds that for any $b \in A$ there exists a unique $a \in A$ such that $f(ua) = b$ (resp., $f(au) = b$). In the sequel, we will sometimes identify any CA with its global rule.

A CA $(A^{\mathbb{Z}}, F)$ is surjective (resp., injective, resp., open) if F is surjective (resp., injective, resp., open). Let $q \in A$. A configuration c is said to be q–finite, if the number of positions i in which $c_i \neq q$ is finite. Recall that injective CA are surjective and any CA is surjective iff it is *injective on the finite configurations*, i.e., $F(c) \neq F(c')$ for any pair c, c' of distinct q-finite configurations [20,21]. Two configurations c and c' are said to be left (resp., right) asymptotic, briefly, l-a (resp., r-a) if there exists $k \in \mathbb{Z}$ such that $c_i = c'_i$ for all $i < k$ (resp., $i > k$). A CA is right (resp., left) closing if $F(c) \neq F(c')$ for any pair c, c' of distinct l-a (resp., r-a) configurations. Left (or right) closing CA are surjective and have DPO (see later).

A CA $(A^{\mathbb{Z}}, F)$ is *topologically transitive*, or, simply *transitive*, if for any pair of nonempty open subsets $U, V \subseteq A^{\mathbb{Z}}$ there exists a natural $h > 0$ such that

$F^h(U) \cap V \neq \emptyset$, while it is said to be *topologically mixing*, or, simply *mixing*, if the latter intersection condition holds ultimately. A CA $(A^{\mathbb{Z}}, F)$ has *dense periodic orbits* (DPO) if the set of its periodic points is dense in $A^{\mathbb{Z}}$ [8], where a periodic point is any $c \in A^{\mathbb{Z}}$ such that $F^h(c) = c$ for some natural $h > 0$. A CA $(A^{\mathbb{Z}}, F)$ is *sensitive to the initial conditions* or, simply *sensitive*, if there exists $\epsilon > 0$ such that for any $\delta > 0$ and $c \in A^{\mathbb{Z}}$ there is $c' \in A^{\mathbb{Z}}$ with $0 < d(c', c) < \delta$ such that $d(F^h(c'), F^h(c')) \geq \epsilon$ for some natural h. A CA $(A^{\mathbb{Z}}, F)$ is said to be *equicontinuous* if for any $\epsilon > 0$ there exists $\delta > 0$ such that for all $c, c' \in A^{\mathbb{Z}}$, $d(c', c) < \delta$ implies that $\forall k \in \mathbb{N}$, $d(F^k(c'), F^k(c')) < \epsilon$. A weaker form is *almost equicontinuity* (see [17,18] for the exact definition) and it holds that a CA is almost equicontinous iff it is not sensitive. Sensitivity and equicontinuity represent the main features of strongly unstable and stable discrete dynamical systems, respectively. The former is also the well-known basic component and essence of chaotic behavior. Indeed, sensitivity, transitivity and DPO are the features that together define the popular notion of *chaos* due to Devaney [12]. CA with a leftmost or rightmost permutative local rule are chaotic [4,7]. We recall that a CA $(A^{\mathbb{Z}}, F)$ is *positively expansive* if for some constant $\varepsilon > 0$ it holds that for any pair of distinct $c, c' \in A^{\mathbb{Z}}$ there exists a natural number ℓ such that $d(F^\ell(c), F^\ell(c')) \geq \varepsilon$. CA positive expansivity is a condition of strong chaos. Indeed, positive expansivity is stronger than sensitivity and any positively expansive CA is also transitive (even mixing) and, at the same time, it has DPO.

We now recall what *linear* CA are. The linear background foresees that the alphabet A is $\mathbb{K}^n$, where $\mathbb{K} = \mathbb{Z}/m\mathbb{Z}$ for some natural $m > 1$. Clearly, in that case both $\mathbb{K}^n$ and $(\mathbb{K}^n)^{\mathbb{Z}}$ become $\mathbb{K}$-modules in the componentwise way. We will denote by $\mathbb{L}_m$ the set of Laurent polynomials with coefficients in $\mathbb{Z}/m\mathbb{Z}$. A local rule $f: (\mathbb{K}^n)^{2r+1} \rightarrow \mathbb{K}^n$ of radius r is said to be *linear* if it is defined by $2r + 1$ matrices $A_{-r}, \ldots, A_r \in \mathbb{K}^{n \times n}$ as follows: $\forall (x_{-r}, \ldots, x_r) \in (\mathbb{K}^n)^{2r+1}$, $f(x_{-r}, \ldots, x_r) = \sum_{i=-r}^{r} A_i \cdot x_i$.

A one-dimensional *linear CA (LCA)* over $\mathbb{K}^n$ is a CA F based on a linear local rule. The *matrix associated with F* is $A = \sum_{i=-r}^{r} A_i X^{-i} \in (\mathbb{L}_m)^{n \times n}$. Easy-to-check characterizations of the above mentioned set-theoretic and dynamical properties have been provided for both LCA over $\mathbb{Z}/m\mathbb{Z}$ [5,16,19], and more recently characterizations have been provided for LCA over $(\mathbb{Z}/m\mathbb{Z})^n$ [9].

3   Composing CA

We start by stating that, if P is basic or set-theoretic CA property the composition of two CA has a property P iff both the component CA have P.

Theorem 1. *Let $(A^{\mathbb{Z}}, F)$ and $(A^{\mathbb{Z}}, G)$ be any two CA and let f and g their corresponding local rules. Let h be the local rule of $F \circ G$. The following hold:*

(i) $F \circ G$ is injective iff both F and G are injective;
(ii) $F \circ G$ is surjective iff both F and G are surjective;
(iii) $F \circ G$ is right-closing (or left-closing) iff both F and G are, too;
(iv) $F \circ G$ is open iff both F and G are open;
(v) h is rightmost (or leftmost) permutative iff both f and g are, too.

Proof. In each item of the statement, the implication "$\Leftarrow$" is trivial. (i) Assume that $F \circ G$ is injective. It immediately follows that G is injective but, since G is a CA, it is also surjective. We are going to show that F is injective. Let $d, d' \in A^{\mathbb{Z}}$ with $d \neq d'$. Since G is bijective, there exist $c, c' \in A^{\mathbb{Z}}$ with $c \neq c'$ and such that $G(c) = d$ and $G(c') = d'$. Now, by the assumption on $F \circ G$, we get $F(G(c)) \neq F(G(c'))$, i.e., $F(d) \neq F(d')$. Therefore, both F and G are injective. (ii) Assume now that $F \circ G$ is surjective. It immediately follows that F is surjective. We are going to show that G is surjective, i.e., G is injective on the finite configurations. For a sake of argument, suppose that there exist two distict finite configurations $c, c' \in A^{\mathbb{Z}}$ such that $G(c) = G(c')$. Hence, we get that $F(G(c)) = F(G(c'))$, the latter contradicting the fact that $F \circ G$ is injective on the finite configurations. So, both F and G are surjective. (iii) Assume now that $F \circ G$ is right-closing (the proof regarding left closingness is symmetric). First of all, we prove that G is right-closing. For a sake of argument, suppose that there exist two distinct l-a configurations $c, c' \in A^{\mathbb{Z}}$ such that such that $G(c) = G(c')$. Hence, we get that $F(G(c)) = F(G(c'))$, the latter contradicting that $F \circ G$ is right-closing. Therefore, G is right-closing. We now show that F is right-closing, too. For a sake of argument, suppose that there exist two distinct l-a configurations $d, d' \in A^{\mathbb{Z}}$ such that $F(d) = F(d')$. Since G is right-closing, by [18, Prop. 5.44], one can build two distinct l-a configurations $c, c' \in A^{\mathbb{Z}}$ such that $G(c) = d$ and $G(c') = d'$. So, there exist two distinct l-a configurations $c, c' \in A^{\mathbb{Z}}$ such that $G(c) = d \neq d' = G(c')$. Hence, we get that $F(G(c)) = F(G(c'))$, the latter contradicting that $F \circ G$ is right-closing. (iv) It follows from iii). v) It follows from the definitions. $\square$

Unlike the basic and set-theoretic CA properties dealt with in Theorem 1, the dynamical ones are more complicated to manage as illustrated in the following.

Proposition 1. *Let P be any dynamical property among the following ones: sensitivity, transitivity, mixing, chaos, positive expansivity, equicontinuity, and almost equicontinuity. There exist two CA on the same alphabet such that the following equivalence does not hold: $F \circ G$ has property P iff both F and G also have P. In particular, both the directions of the equivalence certainly do not hold for transitivity, mixing, and chaos. Moreover, if $P \neq$ sensitivity certainly there exist two CA such that their composition has the property P, but none of them has P, while if $P \in \{transitivity, mixing, chaos, sensitivity\}$, there exist two CA with property P, but their composition does not have P.*

Proof. Let $\mathcal{H}$ be a positively expansive CA of radius r. As far as the CA $F = \sigma^r \circ \mathcal{H}$ and $G = \sigma^{-r} \circ \mathcal{H}$ are concerned, by [1], it holds that neither F nor G are positively expansive but $F \circ G = \mathcal{H}^2$ is. Consider now any sujective and almost equicontinuous CA $(A^{\mathbb{Z}}, \mathcal{E})$. We know that $\mathcal{E}$ has DPO. Regarding the CA $F = \sigma \circ \mathcal{E}$ and $G = \sigma^{-1} \circ \mathcal{E}$, it holds that they are both mixing (and so transitive and sensitive) CA having DPO (and so they are both chaotic, too), but $F \circ G = \mathcal{E}^2$ is an equicontinuous CA, and, hence, $F \circ G$ is not mixing (neither transitive, neither sensitive, and, hence, neither chaotic). The previous scenario also shows the existence of two CA that are not almost equicontinuous (or, not equicontinuous, when $\mathcal{E}$ is a surjective and equicontinuous CA) but their composition is. On the other hand, if $F = \mathcal{E} \times \sigma$ and $G = \sigma \times \mathcal{E}$, both F and G have DPO, but neither F nor G are mixing (thus, transitive) CA, and, hence, neither F nor G are chaotic, while $F \circ G = (\mathcal{E} \circ \sigma) \times (\sigma \circ \mathcal{E})$ is a mixing (and then transitive) CA with DPO and, thus, $F \circ G$ is also chaotic. $\qquad\square$

Some questions naturally arise regarding dynamical properties.

Question 1. What are the conditions on two CA F and G in order that $F \circ G$ has property P iff both F and G also have P?

Question 2. Suppose to know that a CA F has a certain property P. What properties must G satisfy in order that $F \circ G$ (or $G \circ F$) has a certain property Q (possibly $P = Q$)?

Linear CA certainly represent a good context to begin answering the above questions. Indeed, easy-to-check algebraic characterizations have been provided for all the above mentioned asymptotic properties in terms (of the characteristic polynomial) of the matrix defining them. Therefore, Questions 1 and 2 can be reformulated on the basis of the characterization of P (and, possibly, Q) in an algebraic equivalent form involving the matrices defining F, G, and $F \circ G$.

4   Conclusions

We have dealt with the composition of CA. In particular, we have shown that, when composed, two CA can exhibit a greater or weaker complexity, making the composition a formal tool for studying complex systems. Some important questions naturally arise from these results. Essentially, they concern the identification of the conditions on one or both the component CA under which the dynamical properties are preserved, amplified, weakened, or even destroyed by the CA composition. These are important questions that deserve to be answered. We believe that a first step to achieve this result is to consider the class of linear CA where all the properties are characterized by easy-to-check algebraic conditions on their local rule and, hence, questions become easier to manage. Future investigations can concern the above questions as far as non-uniform CA [3] or asynchronous CA [10, 15] are considered and also over computational properties.

Acknowledgments. This work was supported by the HORIZON-MSCA-2022-SE-01 project 101131549 "Application-driven Challenges for Automata Networks and Complex Systems (ACANCOS)" and by the PRIN 2022 PNRR project "Cellular Automata Synthesis for Cryptography Applications (CASCA)" (P2022MPFRT) funded by the European Union - Next Generation EU.

Disclosure of Interests. The authors have no competing interests to declare that are relevant to the content of this article.

References

1. Blanchard, F., Maass, A.: Dynamical properties of expansive one-sided cellular automata. Israel J. Math. **99**, 149–174 (1997)
2. Cattaneo, G., Dennunzio, A., Farina, F.: A full cellular automaton to simulate predator-prey systems. In: El Yacoubi, S., Chopard, B., Bandini, S. (eds.) ACRI 2006. LNCS, vol. 4173, pp. 446–451. Springer, Heidelberg (2006). https://doi.org/10.1007/11861201_52
3. Cattaneo, G., Dennunzio, A., Formenti, E., Provillard, J.: Non-uniform cellular automata. In: Dediu, A.H., Ionescu, A.M., Martín-Vide, C. (eds.) LATA 2009. LNCS, vol. 5457, pp. 302–313. Springer, Heidelberg (2009). https://doi.org/10.1007/978-3-642-00982-2_26
4. Cattaneo, G., Dennunzio, A., Margara, L.: Chaotic subshifts and related languages applications to one-dimensional cellular automata. Fund. Inform. **52**(1–3), 39–80 (2002)
5. Cattaneo, G., Dennunzio, A., Margara, L.: Solution of some conjectures about topological properties of linear cellular automata. Theoret. Comput. Sci. **325**(2), 249–271 (2004)
6. Chopard, B., Dupuis, A., Masselot, A., Luthi, P.O.: Cellular automata and lattice boltzmann techniques: an approach to model and simulate complex systems. Adv. Complex Syst. **5**(2–3), 103–246 (2002)
7. Dennunzio, A.: From one-dimensional to two-dimensional cellular automata. Fund. Inform. **115**(1), 87–105 (2012)
8. Dennunzio, A., Di Lena, P., Formenti, E., Margara, L.: Periodic orbits and dynamical complexity in cellular automata. Fund. Inform. **126**(2–3), 183–199 (2013)
9. Dennunzio, A., Formenti, E., Grinberg, D., Margara, L.: Chaos and ergodicity are decidable for linear cellular automata over $(\mathbb{Z}/m\mathbb{Z})^n$. Inf. Sci. **539**, 136–144 (2020)
10. Dennunzio, A., Formenti, E., Manzoni, L.: Computing issues of asynchronous CA. Fund. Inform. **120**(2), 165–180 (2012)
11. Dennunzio, A., Formenti, E., Manzoni, L., Porreca, A.E.: Ancestors, descendants, and gardens of eden in reaction systems. Theor. Comput. Sci. **608**, 16–26 (2015)
12. Devaney, R.L.: An Introduction to Chaotic Dynamical Systems. Addison-Wesley, Addison-Wesley advanced book program (1989)
13. Ehrenfeucht, A., Rozenberg, G.: Reaction systems. Fundam. Informaticae **75**(1–4), 263–280 (2007)
14. Farina, F., Dennunzio, A.: A predator-prey cellular automaton with parasitic interactions and environmental effects. Fund. Inform. **83**(4), 337–353 (2008)
15. Fatès, N.: A guided tour of asynchronous cellular automata. J. Cell. Autom. **9**(5–6), 387–416 (2014)
16. Ito, M., Osato, N., Nasu, M.: Linear cellular automata over $_m$. J. Comput. Syst. Sci. **27**, 125–140 (1983)
17. Kari, J.: Theory of cellular automata: a survey. Theoret. Comput. Sci. **334**(1–3), 3–33 (2005)
18. Kůrka, P.: Topological and symbolic dynamics. Volume 11 of Cours Spécialisés, Société Mathématique de France (2004)
19. Manzini, G., Margara, L.: A complete and efficiently computable topological classification of d-dimensional linear cellular automata over $_m$. Theoret. Comput. Sci. **221**(1–2), 157–177 (1999)
20. Moore, E.F.: Machine models of self-reproduction. Proc. Symp. Appl. Math. **14**, 13–33 (1962)

21. Myhill, J.: The converse to Moore's garden-of-eden theorem. Proc. Am. Math. Soc. **14**, 685–686 (1963)
22. Nandi, S., Kar, B.K., Chaudhuri, P.P.: Theory and applications of cellular automata in cryptography. IEEE Trans. Comput. **43**(12), 1346–1357 (1994)
23. Paun, G.: Computing with membranes. J. Comput. Syst. Sci. **61**(1), 108–143 (2000)
24. del Rey, Á.M., Mateus, J.P., Sánchez, G.R.: A secret sharing scheme based on cellular automata. Appl. Math. Comput. **170**(2), 1356–1364 (2005)

A Comparison of Polynomial-Based Tree Clustering Methods

Pengyu Liu[1]([envelope]) [iD], Mariel Vázquez[2] [iD], and Nataša Jonoska[3]([envelope]) [iD]

[1] University of Rhode Island, Kingston, RI 02881, USA
`pengyu.liu@uri.edu`
[2] University of California, Davis, Davis, CA 95616, USA
[3] University of South Florida, Tampa, FL 33620, USA
`jonoska@usf.edu`

Abstract. Tree structures appear in many fields of the life sciences, including phylogenetics, developmental biology and nucleic acid structures. Trees can be used to represent RNA secondary structures, which directly relate to the function of non-coding RNAs. Recent developments in sequencing technology and artificial intelligence have yielded numerous biological data that can be represented with tree structures. This requires novel methods for tree structure data analytics. Tree polynomials provide a computationally efficient, interpretable and comprehensive way to encode tree structures as matrices, which are compatible with most data analytics tools. Machine learning methods based on the Canberra distance between tree polynomials have been introduced to analyze phylogenies and nucleic acid structures. In this paper, we compare the performance of different distances in tree clustering methods based on a tree distinguishing polynomial. We also implement two basic autoencoder models for clustering trees using the polynomial. We find that the distance based methods with entry-level normalized distances have the highest clustering accuracy among the compared methods.

Keywords: Tree polynomial · Tree clustering · Structure analytics

1 Introduction

Tree structures are important in many areas of life sciences. Phylogenetic trees record evolutionary information and patterns [1]. Tree representations capture the essence of RNA secondary structures [2]. Cell lineage trees record the developmental history of an organism from a stem cell [3]. The advancements of sequencing technology and artificial intelligence applications in structure folding have produced a myriad of structural data [4,5]. The abundance of structural data emphasizes the need for structural data analytics methods.

In [6], the author defined a complete polynomial invariant of trees. This polynomial provides computationally efficient, interpretable and comprehensive methods for representing tree structures in a way compatible with data analytics tools. We call this polynomial the *tree polynomial* or *polynomial encoding* of

© The Author(s), under exclusive license to Springer Nature Switzerland AG 2026
E. Formenti and L. Manzoni (Eds.): UCNC 2025, LNCS 16364, pp. 416–421, 2026.
https://doi.org/10.1007/978-3-032-15641-9_29

trees. Polynomial encoding together with distance-based machine learning methods have been introduced to infer evolutionary information [7] and to analyze RNA structures and predict the formation of non-canonical nucleic acid structures [8]. In these applications, the Canberra distance [9] was implemented for measuring the similarity between trees. Here, we compare the performance of different distances in tree clustering with polynomial encoding. In addition, we implement two basic autoencoder models for clustering trees using tree polynomials and test their accuracy.

2   Methods

2.1   Random Tree Generation

We cluster random rooted binary trees generated by the beta-splitting model [10]. The model is parametrized with one parameter β. We use three values $\beta = -1.5$, $\beta = -1$ and $\beta = 0$ to generate random trees. These three values of β correspond to the proportional to distinguishable arrangements (PDA) model, the Aldous branching model and the Yule model, respectively [11], and they are known to generate trees with distinct topologies [12]. We choose these rooted binary trees to control the experiment. We generate 100 sets of random trees. Each set contains 300 random trees with 100 leaf vertices. In each set, there are 100 trees generated by parameter $\beta = -1.5$, $\beta = -1$, and $\beta = 0$, respectively, forming three generated clusters of trees.

2.2   Tree Polynomial

We use the tree distinguishing polynomial [6] to encode rooted trees. The definition of the polynomial is recursive from the leaf vertices to the root of a tree. Each leaf vertex v has a polynomial $P(v, x, y) = x$. An internal vertex v with k child vertices $v_1, v_2, \ldots, v_k$ has the polynomial defined by formula (1).

$$P(v, x, y) = y + \prod_{i=1}^{k} P(v_i, x, y) \tag{1}$$

The polynomial at the root is the polynomial that encodes the entire tree.

2.3   Polynomial-Based Tree Clustering Methods

We can represent a tree polynomial $P(T, x, y)$ by its coefficient matrix C_T. The entry $c^{(i,j)}$ at the $(i-1)$'th row and $(j-1)$'th column of C_T is the coefficient in the term $c^{(i,j)} x^i y^j$ of $P(T, x, y)$ [7].

Tree Polynomial Distances with K-Medoids Clustering Here, we implement 6 distances of polynomial encoding and apply the k-medoids algorithm [13] for tree clustering. Let P_1, P_2 be two tree polynomials and C_1, C_2 be the corresponding coefficient matrices. We denote the entries in C_1 and C_2 by $c_1^{(i,j)}$ and $c_2^{(i,j)}$, respectively.

Euclidean Distances. The Euclidean distance and the normalized Euclidean distance between C_1 and C_2 are defined by formula (2) and (3), respectively.

$$d_E(C_1, C_2) = \sqrt{\sum_{0 \leq i,j \leq n} (c_1^{(i,j)} - c_2^{(i,j)})^2} \tag{2}$$

$$d_{\bar{E}}(C_1, C_2) = \sqrt{\sum_{0 \leq i,j \leq n} \frac{(c_1^{(i,j)} - c_2^{(i,j)})^2}{\max(c_1^{(i,j)}, c_2^{(i,j)})^2}} \tag{3}$$

Manhattan Distances. The Manhattan distance and the normalized Manhattan distance between C_1 and C_2 are defined by formula (4) and (5), respectively.

$$d_M(C_1, C_2) = \sum_{0 \leq i,j \leq n} |c_1^{(i,j)} - c_2^{(i,j)}| \tag{4}$$

$$d_{\bar{M}}(C_1, C_2) = \sum_{0 \leq i,j \leq n} \frac{|c_1^{(i,j)} - c_2^{(i,j)}|}{\max(c_1^{(i,j)}, c_2^{(i,j)})} \tag{5}$$

Canberra and Bray-Curtis Distances. The Canberra distance [9] and the Bray-Curtis distance [14] between C_1 and C_2 are defined by formula (6) and (7), respectively.

$$d_C(C_1, C_2) = \sum_{0 \leq i,j \leq n} \frac{|c_1^{(i,j)} - c_2^{(i,j)}|}{(c_1^{(i,j)} + c_2^{(i,j)})} \tag{6}$$

$$d_{BC}(C_1, C_2) = \frac{\sum_{0 \leq i,j \leq n} |c_1^{(i,j)} - c_2^{(i,j)}|}{\sum_{0 \leq i,j \leq n} (c_1^{(i,j)} + c_2^{(i,j)})} \tag{7}$$

The distances defined in (3), (5), (6) are entry-level normalized. If the denominator in a summand equals 0, then the summand is defined to be 0. For the distance defined in (7), the distance is 0 if the denominator is 0.

Clustering Experiment and Accuracy. For each of the 100 sets of random trees, we compute the pairwise polynomial distance between the 100 random trees in the set using the 6 distances. This results in a 100-by-100 distance matrix for each of the distances. Then we apply the k-medoids algorithm on the distance matrix to perform tree clustering. We compute the clustering accuracy by comparing the predicted clusters to the generated clusters using the majority rule. We repeat the k-medoids clustering for 10 times to each distance matrix and compute the mean clustering accuracy.

Tree Polynomial Autoencoders with K-Means Clustering. We construct two autoencoder neural network models [15] to cluster tree polynomials.

Linear Autoencoder Model. The encoder consists of 4 sequential linear layers with decreasing dimensions: 1024, 256, 64, and a latent layer of size 16. Each of the first 3 layers is followed by a ReLU activation function. The decoder has a mirrored architecture, with a final sigmoid activation function. The model was trained using the Adam optimizer with learning rate 0.001 and mean squared error (MSE) loss over 500 epochs.

Convolutional Autoencoder Model. The encoder consists of 3 convolutional layers with 16, 32, and 64 filters, respectively, each with a 3×3 kernel, stride of 2, and padding of 1. Each convolutional layer is followed by a ReLU activation function. The resulting feature maps are flattened and pass through a linear layer to produce a latent vector of size 16. The decoder mirrors this structure with a final sigmoid activation function. The model was trained for 500 epochs using the Adam optimizer with a learning rate of 0.001 and MSE loss.

Clustering Experiment and Accuracy. For each of the 100 sets of random trees, we use the autoencoder models to cluster tree polynomials of the 100 random trees in the set. The coefficient matrix of each tree polynomial is normalized by dividing each entry with the maximum value of the matrix. Note that for a tree with 100 leaf vertices, the coefficient matrix is of size 101×101. For the linear autoencoder, each normalized coefficient matrix is flattened to a vector of size 10201. For the convolutional autoencoder, the coefficient matrices stay unchanged.

For each autoencoder model, we extract the latent vector of size 16 for each of the random trees, and apply the k-means algorithm [16] on the 100 latent vectors to perform tree clustering. We compute the clustering accuracy by comparing the predicted clusters to the generated clusters using the majority rule. We repeat the k-means clustering for 10 times to each set of 100 latent vectors and compute the mean clustering accuracy.

3 Results

We display the clustering accuracy of the polynomial-based tree clustering methods in Fig. 1. The mean accuracy over the 100 sets of random trees is 0.62 for using the Euclidean distance in the polynomial-based tree clustering method, 0.89 for the normalized Euclidean distance, 0.62 for the Manhattan distance, 0.90 for the normalized Manhattan distance, 0.90 for the Canberra distance, and 0.85 for the Bray-Curtis distance. For clustering methods based on the linear and convolutional autoencoders, the mean accuracy is 0.79 and 0.76, respectively.

We observe that the distance-based methods with entry-level normalized distances have the highest accuracy among the compared tree clustering methods. These include the normalized Euclidean distance, the normalized Manhattan distance and the Canberra distance. The first two normalize the difference between each pair of corresponding entries in a coefficient matrix by the maximum of the pair. The Canberra distance normalizes the difference between each pair of corresponding entries by the sum of the pair. Methods based on the Euclidean and

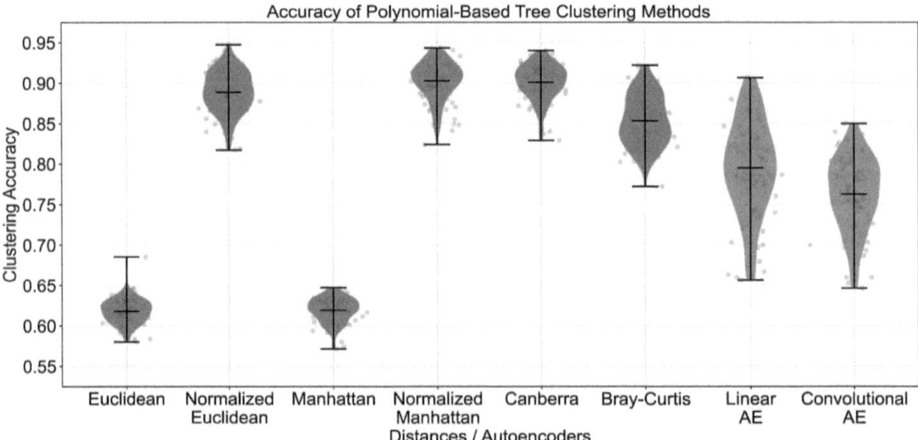

Fig. 1. This figure displays the distribution of the accuracy of 8 polynomial-based tree clustering methods. The distance-based methods are colored in blue, and the autoencoder-based methods are in red. Each dot in a violin plot represents the accuracy for one set of random trees. The bars in each violin plot shows the mean, maximum and minimum observed accuracy over the 100 sets of random trees. (Color figure online)

the Manhattan distance have low accuracy because the unnormalized distance between two coefficient matrices can be dominated by the large differences of a few entries.

4   Conclusion and Future Work

We benchmarked the clustering accuracy of 8 polynomial-based tree clustering methods. We found that distance-based methods with entry-level normalized distances have the highest accuracy.

We used a small dataset of random rooted binary trees generated by the beta-splitting model with three different parameters, as they are known to have distinct tree topologies. Future work will include investigating stochastic processes that generate non-binary trees with distinct topologies and benchmarking the accuracy for clustering non-binary trees. We will also test these methods for clustering large datasets of non-coding RNA secondary structures.

We introduced two basic autoencoder models for clustering trees using tree polynomials and tested their accuracy with the k-means clustering algorithm. These autoencoder-based methods did not perform as well as the distance-based methods with entry-level normalized distances. This indicates a need for more structure data and novel neural network models more tailored for tree polynomials. The definition of tree polynomials depends on polynomial multiplication, so the coefficients of a tree polynomial are related. For example, the coefficients $a + b$ and ab in $(x + ay)(x + by) = x^2 + (a + b)xy + aby^2$ are related. This suggests that transformers or other attention-based models can be well-suited for performing structural data analytics using tree polynomials. We will explore these models and examine their performance in future work.

Acknowledgments. P.L. was supported by the startup funds of the University of Rhode Island. M.V. was supported by the NSF grant DMS/NIGMS#2054347. N.J. was supported by NFS grant DMS/NIGMS#2054321.

Disclosure of Interests. The authors declare no competing interests.

References

1. Semple, C., Steel, M.: Phylogenetics. Oxford University Press, New York (2003)
2. Schlick, T.: Adventures with RNA graphs. Methods **143**(1), 16–33 (2018)
3. Santella, A., Kovacevic, I., Herndon, L.A., Hall, D.H., et al.: Digital development: a database of cell lineage differentiation in C. elegans with lineage phenotypes, cell-specific gene functions and a multiscale model. Nucleic Acids Res. **44**(D1), D781–D785 (2016)
4. Kozlov, A., Alves, J.M., Stamatakis, A., et al.: Cell Phy: accurate and fast probabilistic inference of single-cell phylogenies from scDNA-seq data. Genome Biol. **23**, 37 (2022)
5. Abramson, J., Adler, J., Dunger, J., et al.: Accurate structure prediction of biomolecular interactions with AlphaFold 3. Nature **630**, 493–500 (2024)
6. Liu, P.: A tree distinguishing polynomial. Discret. Appl. Math. **288**, 1–8 (2021)
7. Liu, P., Biller, P., Gould, M., Colijn, C.: Analyzing phylogenetic trees with a tree lattice coordinate system and a graph polynomial. Syst. Biol. **71**(6), 1378–1390 (2022)
8. Liu, P., Lusk, J., Jonoska, N., Vázquez, M.: Tree polynomials identify a link between co-transcriptional R-loops and nascent RNA folding. PLoS Comput. Biol. **20**(12), e1012669 (2024)
9. Lance, G.N., Williams, W.T.: A general theory of classificatory sorting strategies: II. Clustering systems. Comput. J. **10**(3), 271–277 (1967)
10. Aldous, D.: Probability distributions on cladograms. In: Aldous, D., Pemantle, R. (eds.) Random Discrete Structures, The IMA Volumes in Mathematics and its Applications, vol. 76, pp. 1–18. Springer, New York, NY (1996)
11. Blum, M.G.B., François, O.: Which random processes describe the tree of life? a large-scale study of phylogenetic tree imbalance. Syst. Biol. **55**(4), 685–691 (2006)
12. Colijn, C., Plazzotta, G.: A metric on phylogenetic tree shapes. Syst. Biol. **67**(1), 113–126 (2018)
13. Kaufman, L., Rousseeuw, P.J.: Finding groups in data: an introduction to cluster analysis. Wiley, Hoboken, NJ (2005)
14. Bray, J.R., Curtis, J.T.: An ordination of the upland forest communities of southern Wisconsin. Ecol. Monogr. **27**(4), 325–349 (1957)
15. Baldi, P., Hornik, K.: Neural networks and principal component analysis: learning from examples without local minima. Neural Netw. **2**(1), 53–58 (1989)
16. Bishop, C.M.: Pattern recognition and machine learning. Springer, New York (2006)

Analog-Hybrid Implementation for Reconfigurable CPGs

Shrish Roy[✉] and Lucas Wetzel

anabrid GmbH, Am Stadtpark 3, 12167 Berlin, Germany
`roy@anabrid.com`

Abstract. Central Pattern Generators (CPGs) are neural circuits capable of autonomously producing rhythmic output patterns without requiring rhythmic input. They maintain stable phase relationships among constituent oscillators, adapt to sensory feedback and descending modulation, and exhibit resilience to perturbations. This work presents a novel hybrid Analog-Digital CPG architecture that leverages the continuous, low-latency dynamics of analog oscillators alongside the flexibility, reconfigurability, and precision of digital control. The proposed system features tunable parameters and a modular design, enabling real-time adaptation to sensory inputs and environmental conditions. This approach provides a versatile framework for generating robust, adaptable gait patterns, with applications spanning autonomous robotics, soft robotics, and human-assistive technologies.

1　Introduction

Central Pattern Generators (CPG) are neural circuits found in vertebrate and invertebrate nervous systems that produce rhythmic outputs without requiring rhythmic sensory or central input. These neural networks generate coordinated patterns of rhythmic activity that underlie various motor behaviors such as locomotion, respiration, mastication, and rhythmic movements.

The essential biological understanding is that central pattern generators comprise neural structures endowed with inherent oscillatory properties that are capable of autonomously generating rhythmic patterns in the absence of rhythmic input, maintaining phase relationships between constituent oscillators, adapting to sensory feedback and descending commands, and exhibiting robust behavior despite perturbations. Auke Ijspeert's research at the École Polytechnique Fédérale de Lausanne (EPFL) has significantly advanced bio-inspired robotics through the development of CPG models for locomotion control [1]. His work includes the implementation of CPGs in robotic platforms such as the Pleurobot, a salamander-inspired robot utilizing a microcontroller programmed with mathematical models of a salamander's spinal neural network, effectively emulating CPG functionality.

© The Author(s), under exclusive license to Springer Nature Switzerland AG 2026
E. Formenti and L. Manzoni (Eds.): UCNC 2025, LNCS 16364, pp. 422–427, 2026.
https://doi.org/10.1007/978-3-032-15641-9_30

1.1 Network of Coupled Hopf Oscillators for CPG Networks

Instead of constructing models based on individual neuronal behavior, the dynamics are conceptualized through coupled oscillators, i.e.*Phase Oscillators*:

$$\dot{\theta}_i = \omega_i + \sum_j K_{ij} \sin(\theta_j - \theta_i - \phi_{ij}), \tag{1}$$

where, θ_i represents the phase of oscillator i, ω_i denotes its intrinsic frequency, K_{ij} is the coupling strength, and ϕ_{ij} indicates the phase bias between oscillators. Or the *Amplitude-Phase Oscillators*: This framework encompasses both phase and amplitude dynamics, rendering them particularly relevant for robotics applications where precise timing and activation strength are critical.

Hopf Oscillators: A notably sophisticated representation is the Hopf oscillator, which can be articulated in Cartesian coordinates:

$$\dot{x} = (\mu - r^2)x - \omega y, \tag{2}$$

$$\dot{y} = (\mu - r^2)y + \omega x, \tag{3}$$

wherein $r^2 = x^2 + y^2$, μ governs the oscillation amplitude, and ω determines the frequency. The Hopf oscillator's elegance is manifested in its bifurcation characteristics: when $\mu < 0$, the system attains a stable fixed point at the origin; conversely, when $\mu > 0$, this fixed point becomes unstable, precipitating a stable limit cycle with radius $\sqrt{\mu}$.

In developing models for CPG networks, we integrate several Hopf oscillators:

$$\dot{x}_i = (\mu - r_i^2)x_i - \omega_i y_i + \sum_j K_{ij}(x_j - x_i), \tag{4}$$

$$\dot{y}_i = (\mu - r_i^2)y_i + \omega_i x_i + \sum_j K_{ij}(y_j - y_i), \tag{5}$$

where the terms $K_{ij}(x_j - x_i)$ and $K_{ij}(y_j - y_i)$ denote diffusive coupling. This mathematical framework embodies crucial features for dynamical systems: structural stability preserves the limit cycle against perturbations, smooth convergence leads initial conditions toward the limit cycle consistently, and controllable oscillation frequency allows amplitude-independent adjustments. Additionally, entrainment facilitates oscillator synchronization with external stimuli, enhancing sensory feedback, while phase coupling enables coordinated movements through defined phase relationships [4].

2 Analog-Hybrid Implementation for CPGs

In the 1990s and early 2000s, analog circuits were explored to mimic neuronal oscillations. Nakada et al. [2] created analog CMOS neurons with integrate-and-fire behavior, forming compact CPG networks with low power usage by using transistors in the subthreshold regime. However, issues like component mismatch

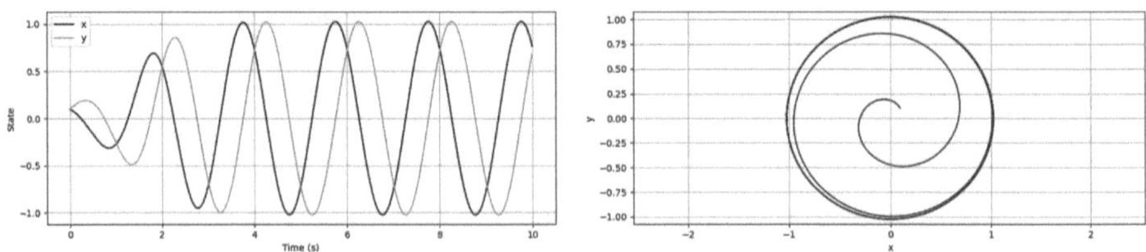

Fig. 1. The Hopf oscillator generates a stable oscillation pattern. Note how it converges to a circle in the phase portrait, with radius $\sqrt{\mu}$.

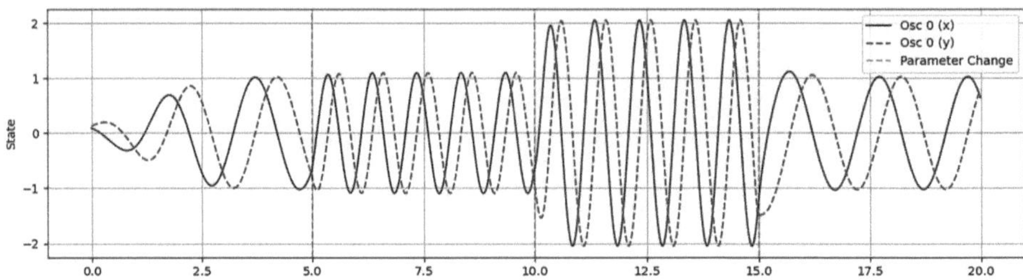

Fig. 2. Example for transitions between different gait parameters. Starting with 0.5 Hz, changing to 1 Hz at time at t=5.0 s. At t=10.0 s: Changing amplitude to 2.0. At t=15.0 s: Changing frequency to 0.5 Hz and amplitude to 1.0. Notice how the transitions for frequency- and amplitude changes are continuous and smooth. the frequency changes at t=5 s and t=15 s. The amplitude changes at t=10 s and t=15 s. This demostrates the smooth transitions in gaits when parameters are changed in real-time.

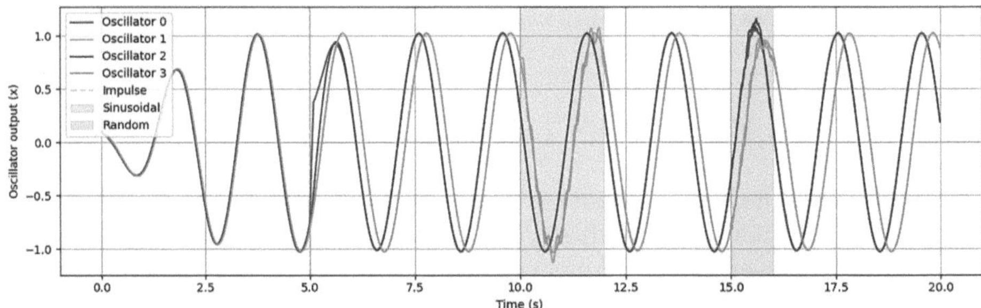

Fig. 3. At t=5.0 s: Applying strong impulse to oscillator 0. At t=10.0 s: Applying sinusoidal disturbance to oscillator 1. At t=15.0 s: Applying random noise to all oscillators. This demonstrates the self correcting and robustness of coupled Hopf oscillators against small perturbations.

and drift hindered precise frequency tuning, limiting scalability. Recent advancements with novel materials and devices address these issues. Dutta et al. [3] used vanadium dioxide (VO_2), a phase-change material with an insulator-to-metal transition when electrically stimulated, to make compact relaxation oscillators.

3   Hybrid Analog-Digital Architectures and Implementation

Hybrid analog-digital architectures utilize analog circuits for efficient and fast rhythmic generation and adaptability, while digital microcontrollers manage sensing and reconfiguration. An analog CPG platform can output joint angle trajectories, with a digital supervisor adjusting frequency and patterns via converters. This combines the low power and real-time processing of analog systems with the flexibility of digital systems. Analog CPGs are advantageous over digital ones due to their lower power consumption, compactness for small robots, fast signal processing, and natural signal representation for biomimetic controllers.

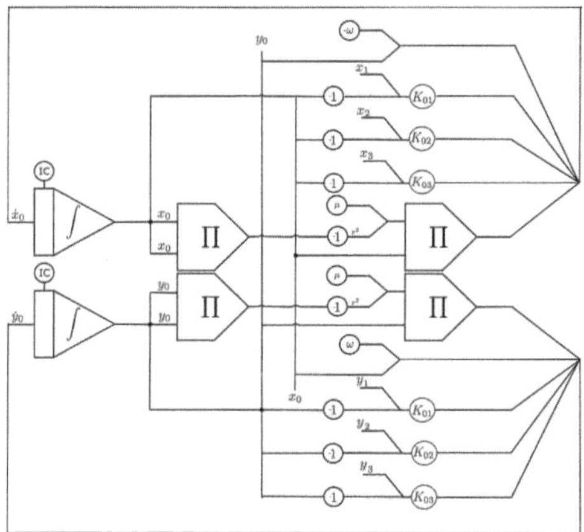

Fig. 4. Analog Circuit block diagram for system of 4 reconfigurable coupled Hopf oscillators. Shown here is one of the four oscillators. The diagram consists of Analog Integrators and Multipliers and scaling coefficients, this is an analog circuit representation of equations (4) and (5).

3.1   Applications, Further Research and Preliminary Results

Sensor signals, including muscle-sensors, EMGs, terrain data, and perturbation inputs, inform movement templates and real-time frequency and amplitude cues. These inputs integrate in a hybrid CPG core, combining analog oscillations with digital control for adaptable rhythmic patterns. The resulting gait signals can guide swarm-robot locomotion, bio-mimetic animal gaits, adaptive neuro-prostheses, orthoses, or other applications. This architecture offers a flexible framework for robust, tunable gait behaviors in robotics and human-assistive technology (see Fig. 5) [4–6].

As part of ongoing research, further experiments, testing, implementations, and improvements of Fig (4) are currently underway and being tested on anabrid

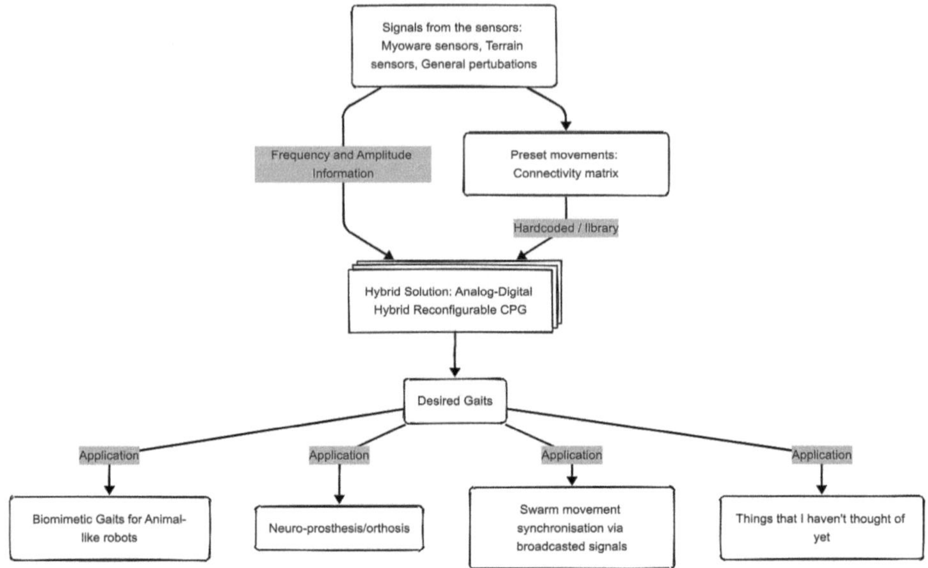

Fig. 5. A flow chart of the control process architecture involved, beginning from receiving signals from the sensors to the CPG producing the desired gaits.

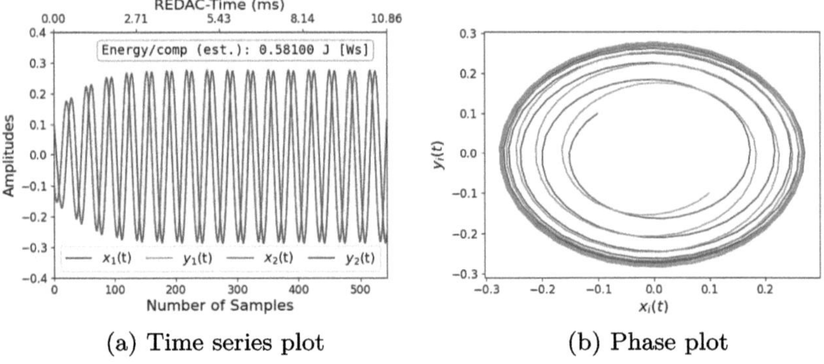

(a) Time series plot

(b) Phase plot

Fig. 6. Test of two **in-phase** synchronized hopf oscillators implemented on the REDAC, on the left is the time series plot showing the x and y values as seen in eq (4) and eq (5).

GmbH's Reconfigurable Discrete Analog Computer (REDAC). We implemented two mutually coupled Hopf oscillators with coupling strength $K_{ij} = 0.01$, $\mu=$ -0.02 and $\omega = 0.95$ Hz (angular frequency) and similar time series and phase plots as shown in fig (1) have been generated in Fig. (6) and Fig. (7).

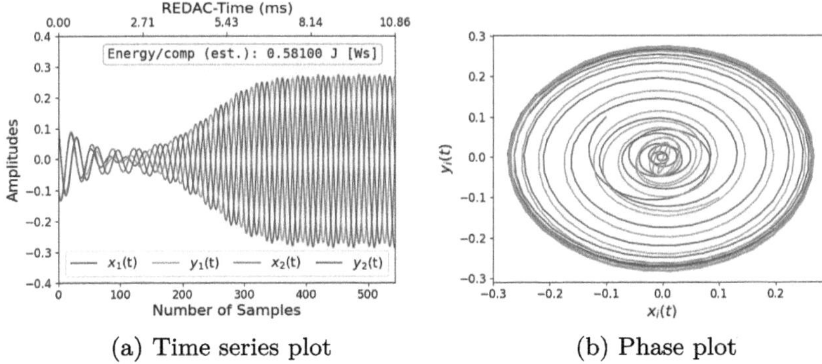

(a) Time series plot   (b) Phase plot

Fig. 7. Test of two **anti-phase** synchronized hopf oscillators implemented on the REDAC, on the left is the time series plot showing the x and y values as seen in eq (4) and eq (5).

References

1. Ijspeert, A.: Central pattern generators for locomotion control in animals and robots: a review. Neural Netw., **21**(4), 642–653 (2008)
2. Nakada, K.: An analog CMOS central pattern generator for interlimb coordination in quadruped locomotion. IEEE Trans. Neural Netw. **14**(5), 1356–1365 (2003)
3. Dutta, S., et al.: Programmable coupled oscillators for synchronized locomotion. Nat. Commun. **10**(1), 3299 (2019)
4. Righetti, L., Buchli, J., Ijspeert, A.J.: Dynamic hebbian learning in adaptive frequency oscillators. Physica D **216**(2), 269–281 (2006)
5. Righetti, L., Buchli, J., Ijspeert, A.J.: Adaptive frequency oscillators and applications. Open Cybern. Systemics J. **3**, 64–69 (2009)
6. Righetti, L., Buchli, J., Ijspeert, A.J.: From dynamic hebbian learning for oscillators to adaptive central pattern generators. In Proceedings of 3rd International Symposium on Adaptive Motion in Animals and Machines–AMAM 2005. Verlag ISLE, Ilmenau (2005)

Author Index

© The Editor(s) (if applicable) and The Author(s), under exclusive license
to Springer Nature Switzerland AG 2026
E. Formenti and L. Manzoni (Eds.): UCNC 2025, LNCS 16364, pp. 429–430, 2026.
https://doi.org/10.1007/978-3-032-15641-9

MIX
Papier aus verantwortungsvollen Quellen
Paper from responsible sources
FSC® C105338

If you have any concerns about our products,
you can contact us on
ProductSafety@springernature.com

In case Publisher is established outside the EU,
the EU authorized representative is:
Springer Nature Customer Service Center GmbH
Europaplatz 3, 69115 Heidelberg, Germany

Printed by Libri Plureos GmbH
in Hamburg, Germany